MATLAB® for Engineering Applications

William J. Palm III
University of Rhode Island

MATLAB® FOR ENGINEERING APPLICATIONS

Published by McGraw-Hill Education, 2 Penn Plaza, New York, NY 10121. Copyright © 2019 by McGraw-Hill Education. All rights reserved. Printed in the United States of America. No part of this publication may be reproduced or distributed in any form or by any means, or stored in a database or retrieval system, without the prior written consent of McGraw-Hill Education, including, but not limited to, in any network or other electronic storage or transmission, or broadcast for distance learning.

Some ancillaries, including electronic and print components, may not be available to customers outside the United States.

This book is printed on acid-free paper.

1 2 3 4 5 6 7 8 9 LCR 21 20 19 18

ISBN 978-1-260-08471-9
MHID 1-260-08471-X

Cover Image: ©Getty Images/iStockphoto

mheducation.com/highered

To my sisters, Linda and Chris, and to my parents, Lillian and William

ABOUT THE AUTHOR

William J. Palm III is Emeritus Professor of Mechanical Engineering at the University of Rhode Island. In 1966 he received a B.S. from Loyola College in Baltimore, and in 1971 a Ph.D. in Mechanical Engineering and Astronautical Sciences from Northwestern University in Evanston, Illinois.

During his 44 years as a faculty member, he taught 19 courses. One of these is a freshman MATLAB course, which he helped develop. He has authored eight textbooks dealing with modeling and simulation, system dynamics, control systems, and MATLAB. These include *System Dynamics,* 3rd ed. (McGraw-Hill, 2014). He wrote a chapter on control systems in the *Mechanical Engineers' Handbook,* 3rd ed. (M. Kutz, ed., Wiley, 2016), and was a special contributor to the fifth editions of *Statics* and *Dynamics,* both by J. L. Meriam and L. G. Kraige (Wiley, 2002).

Professor Palm's research and industrial experience are in control systems, robotics, vibrations, and system modeling. He was the Director of the Robotics Research Center at the University of Rhode Island from 1985 to 1993, and is the coholder of a patent for a robot hand. He served as Acting Department Chair from 2002 to 2003. His industrial experience is in automated manufacturing; modeling and simulation of naval systems, including underwater vehicles and tracking systems; and design of control systems for underwater-vehicle engine-test facilities.

CONTENTS

Numbered Examples

PREFACE

Formerly used mainly by specialists in signal processing and numerical analysis, MATLAB® has achieved widespread and enthusiastic acceptance throughout the engineering community. Many engineering schools require a course based entirely or in part on MATLAB early in the curriculum. MATLAB is programmable and has the same logical, relational, conditional, and loop structures as other programming languages. Thus it can be used to teach programming principles. In most schools, MATLAB is the principal computational tool used throughout the curriculum. In some technical specialties, such as signal processing and control systems, it is the standard software package for analysis and design.

The popularity of MATLAB is partly due to its long history, and thus it is well developed and well tested. People trust its answers. Its popularity is also due to its user interface, which provides an easy-to-use interactive environment that includes extensive numerical computation and visualization capabilities. Its compactness is a big advantage. For example, you can solve a set of many linear algebraic equations with just three lines of code, a feat that is impossible with traditional programming languages. MATLAB is also extensible; currently more than 30 "toolboxes" in various application areas can be used with MATLAB to add new commands and capabilities.

MATLAB is available for MS Windows and Macintosh personal computers and for other operating systems. It is compatible across all these platforms, which enables users to share their programs, insights, and ideas. This text is based on release R2017b of the software. This includes MATLAB version 9.3. Some of the material in Chapter 9 is based on the Control System toolbox, Version 10.3. Chapter 10 is based on Version 9.0 of Simulink®, and Chapter 11 is based on Version 8.0 of the Symbolic Math toolbox.

TEXT OBJECTIVES AND PREREQUISITES

This text is intended as a stand-alone introduction to MATLAB. It can be used in an introductory course, as a self-study text, or as a supplementary text. The text's material is based on the author's experience in teaching a required two-credit semester course devoted to MATLAB for engineering freshmen. In addition, the text can serve as a reference for later use. The text's many tables and its referencing system in an appendix have been designed with this purpose in mind. A secondary objective is to introduce and reinforce the use of problem-solving methodology as practiced by the engineering profession in general and as applied to the use of computers to solve problems in particular. This methodology is introduced in Chapter 1.

®MATLAB and Simulink are registered trademarks of The MathWorks, Inc.

The reader is assumed to have some knowledge of algebra and trigonometry; knowledge of calculus is not required for the first seven chapters. Some knowledge of high school chemistry and physics, primarily simple electric circuits, and basic statics and dynamics, is required to understand some of the examples.

TEXT ORGANIZATION

In addition to updating material from the previous edition to include new features, new functions, and changes in syntax and function names, the text incorporates the many suggestions made by reviewers and other users. More examples and homework problems have been added.

The text consists of 11 chapters. The first five chapters constitute a basic course in MATLAB. The remaining six chapters are independent of each other and cover more advanced applications of MATLAB, the Control Systems toolbox, Simulink, and the Symbolic Math toolbox.

Chapter 1 gives an overview of MATLAB features, including its windows and menu structures. It also introduces the problem-solving methodology.

Chapter 2 introduces the concept of an array, which is the fundamental data element in MATLAB, and describes how to use numeric arrays, cell arrays, and structure arrays for basic mathematical operations.

Chapter 3 discusses the use of functions and files. MATLAB has an extensive number of built-in math functions, and users can define their own functions and save them as a file for reuse.

Chapter 4 discusses programming with MATLAB and covers relational and logical operators, conditional statements, `for` and `while` loops, and the `switch` structure. A major application of the chapter's material is in simulation, to which a section is devoted.

Chapter 5 deals with two- and three-dimensional plotting. It first establishes standards for professional-looking, useful plots. In the author's experience, beginning students are not aware of these standards, so they are emphasized. The chapter then covers MATLAB commands for producing different types of plots and for controlling their appearance. The Live Editor, which is a major addition to MATLAB, is covered in Section 5.1.

Chapter 6 covers function discovery, which uses data plots to discover a mathematical description of the data and is a useful tool for model building. It is a common application of plotting, and a separate section is devoted to this topic. The chapter also treats polynomial and multiple linear regression as part of its modeling coverage.

Chapter 7 reviews basic statistics and probability and shows how to use MATLAB to generate histograms, perform calculations with the normal distribution, and create random number simulations. The chapter concludes with linear and cubic spline interpolation.

Chapter 8 covers the solution of linear algebraic equations, which arise in applications in all fields of engineering. This coverage establishes the terminology and some important concepts required to use the computer methods properly.

The chapter then shows how to use MATLAB to solve systems of linear equations that have a unique solution. Underdetermined and overdetermined systems are also covered.

Chapter 9 covers numerical methods for calculus and differential equations. Numerical integration and differentiation methods are treated. Ordinary differential equation solvers in the core MATLAB program are covered, as well as the linear system solvers in the Control System toolbox. For those readers not familiar with differential equations, this chapter provides some background for Chapter 10.

Chapter 10 introduces Simulink, which is a graphical interface for building simulations of dynamic systems. Simulink has increased in popularity and has seen increased use in industry. The MathWorks provides Simulink support packages for computer hardware such as LEGO© MINDSTORMS©, Arduino©, and Raspberry Pi©, which are popular with researchers and hobbyists for controlling drones and robots. These packages let you develop and simulate algorithms that run standalone on the supported hardware. They include a library of Simulink blocks for configuring and accessing the hardware's sensors, actuators, and communication interfaces. You can also tune parameters live from your Simulink model while your algorithm runs on the hardware. The MathWorks supports an active user community online where you can see applications and download files. Chapter 10 discusses some of the robotic vehicle applications.

Chapter 11 covers symbolic methods for manipulating algebraic expressions and for solving algebraic and transcendental equations, calculus, differential equations, and matrix algebra problems. The calculus applications include integration and differentiation, optimization, Taylor series, series evaluation, and limits. Laplace transform methods for solving differential equations are also introduced. This chapter requires the use of the Symbolic Math toolbox.

Appendix A contains a guide to the commands and functions introduced in the text. Appendix B is an introduction to producing animation and sound with MATLAB. While not essential to learning MATLAB, these features are helpful for generating student interest. Appendix C summarizes functions for creating formatted output. Appendix D is a list of references. Appendix E, which is available on the text's website, contains some suggestions for course projects and is based on the author's experience in teaching a freshman MATLAB course. Answers to selected problems and an index appear at the end of the text.

All figures, tables, equations, and exercises have been numbered according to their chapter and section. For example, Figure 3.4–2 is the second figure in Chapter 3, Section 4. This system is designed to help the reader locate these items. The end-of-chapter problems are the exception to this numbering system. They are numbered 1, 2, 3, and so on to avoid confusion with the in-chapter exercises.

SPECIAL REFERENCE FEATURES

The text has the following special features, which have been designed to enhance its usefulness as a reference.

- Throughout each of the chapters, numerous tables summarize the commands and functions as they are introduced.
- Appendix A is a complete summary of all the commands and functions described in the text, grouped by category, along with the number of the page on which they are introduced.
- At the end of each chapter is a list of the key terms introduced in the chapter, with a reference to where they are introduced.
- The index has four sections: a listing of MATLAB symbols, an alphabetical list of MATLAB commands and functions, a list of Simulink block names, and an alphabetical list of topics.

PEDAGOGICAL AIDS

The following pedagogical aids have been included:

- Each chapter begins with an overview.
- Test Your Understanding exercises appear throughout the chapters near the relevant text. These relatively straightforward exercises allow readers to assess their grasp of the material as soon as it is covered. In most cases the answer to the exercise is given with the exercise. Students should work these exercises as they are encountered.
- Each chapter ends with numerous problems, grouped according to the relevant section.
- Each chapter contains numerous practical examples. The major examples are numbered.
- Each chapter has a summary section that reviews the chapter's objectives.
- Answers to many end-of-chapter problems appear at the end of the text. These problems are denoted by an asterisk next to their number (for example, **15**[*]).

Two features have been included to motivate the student toward MATLAB and the engineering profession:

- Most of the examples and the problems deal with engineering applications. These are drawn from a variety of engineering fields and show realistic applications of MATLAB. A guide to these examples appears on page vii.
- The facing page of each chapter contains a photograph of a recent engineering achievement that illustrates the challenging and interesting opportunities that await engineers in the 21st century. A description of the achievement, its related engineering disciplines, and a discussion of how MATLAB can be applied in those disciplines accompanies each photo.

ONLINE RESOURCES

An Instructor's Manual is available online for instructors who have adopted this text. This manual contains the complete solutions to all of the Test Your Understanding exercises and to all of the chapter problems. The text website also has downloadable files containing PowerPoint slides keyed to the text and suggestions for projects.

ELECTRONIC TEXTBOOK OPTIONS

McGraw-Hill offers this text as an ebook. To talk about the ebook options, please contact your McGraw-Hill sales rep.

MATLAB INFORMATION

For MATLAB and Simulink product information, please contact:

The MathWorks, Inc.
3 Apple Hill Drive
Natick, MA, 01760-2098 USA
Tel: 508-647-7000
Fax: 508-647-7001
E-mail: info@mathworks.com
Web: www.mathworks.com
How to buy: www.mathworks.com/store

ACKNOWLEDGMENTS

Many individuals are due credit for this text. Working with faculty at the University of Rhode Island in developing and teaching a freshman course based on MATLAB has greatly influenced this text. Email from many users contained useful suggestions. The author greatly appreciates their contributions.

The MathWorks, Inc., has always been very supportive of educational publishing. I especially want to thank Naomi Fernandes of The MathWorks, Inc., for her help. Thomas Scaife, Jolynn Kilburg, Laura Bies, and Lora Neyens of McGraw-Hill Education and Kate Scheinman efficiently handled the manuscript reviews and guided the text through production.

My sisters, Linda and Chris, and my mother, Lillian, have always been there, cheering my efforts. My father was always there for support before he passed away. Finally, I want to thank my wife, Mary Louise, and my children, Aileene, Bill, and Andy, for their understanding and support of this project.

William J. Palm III
Kingston, Rhode Island
September 2017

Source: NASA

Engineering in the 21st Century. . .

Remote Exploration

It will be many years before humans can travel to other planets. In the meantime, unmanned probes have been rapidly increasing our knowledge of the universe. Their use will increase in the future as our technology develops to make them more reliable and more versatile. Better sensors are expected for imaging and other data collection. Improved robotic devices will make these probes more autonomous, and more capable of interacting with their environment, instead of just observing it.

NASA's planetary rover *Sojourner* landed on Mars on July 4, 1997, and excited people on Earth while they watched it successfully explore the Martian surface to determine wheel-soil interactions, to analyze rocks and soil, and to return images of the lander for damage assessment.

Then in early 2004, two improved rovers, *Spirit* and *Opportunity,* landed on opposite sides of the planet. In one of the major discoveries of the 21st century, they obtained strong evidence that water once existed on Mars in significant amounts. Although planned to operate for only 90 Martian days, *Spirit* operated for over five years, became stuck in 2009, and stopped communicating in 2010. The rover likely lost power due to excessively cold internal temperatures. *Opportunity* was still active as of 2016, having already exceeded its planned operational life by over twelve Earth years, almost 50 times its designed lifespan.

The rover *Curiosity* landed on Mars in 2012 less than 2.4 km (1.5 mi) from its intended target after a 563,000,000 km (350,000,000 mi) journey. It was designed to investigate the Martian climate and geology; to assess whether the Gale crater ever had an environment suitable for microbial life, and to determine the habitability of the site for future human exploration. *Curiosity* has a mass of 899 kg (1,982 lb) including 80 kg (180 lb) of instruments. The rover is 2.9 m (9.5 ft) long by 2.7 m (8.9 ft) wide by 2.2 m (7.2 ft) in height.

In addition to its scientific instruments, *Curiosity*'s major systems include a radioisotope thermoelectric generator for power, a temperature management system using electric heaters with a pumped fluid system, two computers, a navigation system with cameras, and several communication systems. Equipped with six 50 cm (20 in) diameter wheels in a rocker-bogie suspension that gives a ground clearance of 60 cm (24 in), *Curiosity* can navigate obstacles nearly 65 cm (26 in) high. Depending on conditions, the maximum average speed is estimated to be 200 m (660 ft) per day when using the automatic navigation system.

All engineering disciplines were involved with the rover projects. From the design of the rocket propulsion of the launch vehicles and the calculation of the interplanetary trajectories, to the design of the rovers' systems, MATLAB was used in many of these applications, and it is well suited to assist designers of future probes and autonomous vehicles like the Mars rovers. ■

CHAPTER 1

An Overview of MATLAB®*

This is the most important chapter in the book. By the time you have finished this chapter, you will be able to use MATLAB to solve many kinds of problems. Section 1.1 provides an introduction to MATLAB as an interactive calculator. Section 1.2 covers the main menus and the Toolstrip. Section 1.3 introduces built-in functions, arrays, and plots. Section 1.4 discusses how to create, edit, and save MATLAB programs. Section 1.5 introduces the extensive MATLAB Help System and Section 1.6 introduces the methodology of engineering problem solving.

How to Use This Book

The book's chapter organization is flexible enough to accommodate a variety of users. However, it is important to cover at least the first four chapters, in that order. Chapter 2 covers *arrays,* which are the basic building blocks in MATLAB.

---

*MATLAB is a registered trademark of The MathWorks, Inc.

Chapter 3 covers file usage, functions built into MATLAB, and user-defined functions. Chapter 4 covers programming using relational and logical operators, conditional statements, and loops.

Chapters 5 through 11 are independent chapters that can be covered in any order. They contain in-depth discussions of how to use MATLAB to solve several common types of problems. Chapter 5 covers two- and three-dimensional plots in greater detail. Chapter 6 shows how to use plots to build mathematical models from data. Chapter 7 covers probability, statistics, and interpolation applications. Chapter 8 treats linear algebraic equations in more depth by developing methods for the overdetermined and underdetermined cases. Chapter 9 introduces numerical methods for calculus and ordinary differential equations. Simulink®*, the topic of Chapter 10, is a graphical user interface for solving differential equation models. Chapter 11 covers symbolic processing with the MATLAB Symbolic Math toolbox, with applications to algebra, calculus, differential equations, transforms, and special functions.

Reference and Learning Aids

The book has been designed as a reference as well as a learning tool. The special features useful for these purposes are as follows.

- Throughout each chapter margin notes identify where new terms are introduced.
- Throughout each chapter short Test Your Understanding exercises appear. Where appropriate, answers immediately follow the exercise so you can measure your mastery of the material.
- Homework exercises conclude each chapter. These usually require greater effort than the Test Your Understanding exercises.
- Each chapter contains tables summarizing the MATLAB commands introduced in that chapter.
- At the end of each chapter is
  - A summary of what you should be able to do after completing that chapter
  - A list of key terms you should know
- Appendix A contains tables of MATLAB commands, grouped by category, with the appropriate page references.
- The index has four parts: MATLAB symbols, MATLAB commands, Simulink blocks, and topics.

1.1 MATLAB Interactive Sessions

We now show how to start MATLAB, how to make some basic calculations, and how to exit MATLAB.

---

*Simulink and MuPAD are registered trademarks of The MathWorks, Inc.

Conventions

In this text we use `typewriter` font to represent MATLAB commands, any text that you type in the computer, and any MATLAB responses that appear on the screen, for example, `y = 6*x`. Variables in normal mathematics text appear in italics, for example, $y = 6x$. We use boldface type for three purposes: to represent vectors and matrices in normal mathematics text (for example, $\mathbf{Ax = b}$), to represent an action on the keyboard (for example, press **Enter**), and to represent the name of a screen menu or an item a menu when it is the object of an action (for example, click on **File**). It is assumed that you press the **Enter** key after you type a command. We do not show this action with a separate symbol.

Starting MATLAB

To start MATLAB on a Windows system, double-click on the MATLAB icon. You will then see the MATLAB *Desktop*. The Desktop manages the Command window and a Help Browser as well as other tools. The Desktop may appear differently in different versions of MATLAB, but the basic features should be similar to those discussed here. The default appearance of the Desktop in MATLAB version R2017b is shown in Figure 1.1–1. Four windows appear. These are the Command window in the center, the Workspace window in the right, the Details window in the lower left, and the Current Folder window in the upper left. Across the top of the Desktop are a row of menu names and a row of icons called the *Toolstrip*. The default Desktop shows three tabs: HOME, PLOT, and APPS. Use of these tabs is discussed in Section 1.2. To the right of the tabs is a box showing the Shortcut button that enables you to create easy access to commonly used procedures. The remaining items in the box are used for more advanced features and are initially inactive. We will describe the various menus later in this chapter.

DESKTOP

You use the *Command window* to communicate with the MATLAB program, by typing instructions of various types called *commands, functions,* and

**COMMAND
WINDOW**

Figure 1.1–1 The default MATLAB Desktop for version R2017b.

statements. Later we will discuss the differences between these types, but for now, to simplify the discussion, we will call the instructions by the generic name *commands*. MATLAB displays the prompt (>>) to indicate that it is ready to receive instructions. Before you give MATLAB instructions, make sure the cursor is located just after the prompt. If it is not, use the mouse to move the cursor. The prompt in the Student Edition looks like EDU >>. We will use the normal prompt symbol >> to illustrate commands in this text. The Command window in Figure 1.1–1 shows some commands and the results of the calculations. We will cover these commands later in this chapter.

Three other windows appear in the default Desktop. The Current Folder window is much like a file manager window; you can use it to access files. Double-clicking on a file name with the extension .m will open that file in the MATLAB Editor. The Editor is discussed in Section 1.4. Figure 1.1–1 shows the files in the author's folder Examples.

Underneath the Current Folder window is the Details window. It displays the first comment (if any) in the file. Note that four file types are shown in the Current Folder. These in order are a MATLAB script file, a JPEG figure file, a MATLAB user-defined file, and a Simulink model file. These have the extensions .m, .jpg, m, and .mdl, respectively. Each file type has its own icon. We will cover m files in this chapter. The other file types will be covered in later chapters. You can have other file types in the folder.

VARIABLE

The Workspace window appears to the right. The Workspace window displays the variables created in the Command window. Double-click on a *variable* name to open the Variables Editor, which is discussed in Chapter 2.

You can alter the appearance of the Desktop if you wish. For example, to eliminate a window, just click on its Close-window button (✕) in its upper right-hand corner. To undock, or separate the window from the Desktop, click on the button containing a curved arrow. An undocked window can be moved around on the screen. You can manipulate other windows in the same way. To restore the default configuration, click on **Layout** in the toolbar, and select **Default.**

Test Your Understanding

T1.1–1   Experiment with your Desktop. Type ver at the prompt to see what MATLAB version you are using and to see details about your computer. If you are not using version R2017a, find the windows discussed in this section. Examine the toolbar to locate items similar to those shown in Figure 1.1–1.

Entering Commands and Expressions

To see how simple it is to use MATLAB, try some practice. Make sure the cursor is at the prompt in the Command window. To divide 8 by 10, type 8/10 and press **Enter** (the symbol / is the MATLAB symbol for division). Your entry and the

MATLAB response look like the following on the screen (we call this interaction between you and MATLAB an *interactive session*, or simply a *session*). Remember, the symbol >> automatically appears on the screen; you do not type it.

```
>> 8/10
ans =
      0.8000
```

MATLAB indents the numerical result. MATLAB uses high precision for its computations, but by default it usually displays its results using four decimal places except when the result is an integer.

If you make a mistake, for now just press **Enter** and retype the line correctly. Ignore for now any error messages you may see.

Using Variables   MATLAB assigns the most recent answer to a variable called ans, which is an abbreviation for *answer*. A variable in MATLAB is a symbol used to contain a value. You can use the variable ans for further calculations; for example, using the MATLAB symbol for multiplication (*), we obtain

```
>> 5*ans
ans =
      4
```

Note that the variable ans now has the value 4.

You can use variables to write mathematical expressions. Instead of using the default variable ans, you can assign the result to a variable of your own choosing, say, r, as follows:

```
>> r = 8/10
r =
      0.8000
```

This is called an *assignment statement*. The variable, and only the variable, is always on the left of the = symbol. This symbol is called the *assignment* or *replacement* operator, and it cannot be used the same way as the equals sign is used in mathematics. The previous entry means "assign the value of 8/10 to the variable r".

If you now type r at the prompt and press **Enter,** you will see

```
>> r
r =
      0.8000
```

thus verifying that the variable r has the value 0.8. You can use this variable in further calculations. For example,

```
>> s=20*r
s =
      16
```

A common mistake is to forget the multiplication symbol * and type the expression as you would in algebra, as $s = 20r$. If you do this in MATLAB, you will get an error message.

Spaces in the line improve its readability; for example, you can put a space before and after the = symbol and the multiplication symbol * if you want. So you could type

```
>> s = 20 * r
```

MATLAB ignores spaces when making its calculations, with one exception that we will discuss in Chapter 2.

Order of Precedence

SCALAR

A *scalar* is a single number. A *scalar variable* is a variable that contains a single number. MATLAB uses the symbols + - * / ^ for addition, subtraction, multiplication, division, and exponentiation (power) of scalars. These are listed in Table 1.1–1. For example, typing x = 8 + 3*5 returns the answer x = 23. Typing 2^3 - 10 returns the answer ans = -2. The *forward slash* (/) represents *right division*, which is the normal division operator familiar to you. Typing 15/3 returns the result ans = 5.

MATLAB has another division operator, called *left division,* which is denoted by the *backslash* (\). The left division operator is useful for solving sets of linear algebraic equations, as we will see. A good way to remember the difference between the right and left division operators is to note that the slash slants toward the denominator. For example, 7/2 = 2\7 = 3.5.

PRECEDENCE

The mathematical operations represented by the symbols + - * / \ and ^ follow a set of rules called *precedence*. Mathematical expressions are evaluated starting from the left, with the exponentiation operation having the highest order of precedence, followed by multiplication and division with equal precedence, followed by addition and subtraction with equal precedence.

Parentheses can be used to alter this order. Evaluation begins with the innermost pair of parentheses and proceeds outward. Table 1.1–2 summarizes these rules. For example, note the effect of precedence on the following session.

Table 1.1–1 Scalar arithmetic operations

Symbol	Operation	MATLAB form
^	exponentiation: a^b	a^b
*	multiplication: a^b	a*b
/	right division: $a/b = \frac{a}{b}$	a/b
\	left division: $a\backslash b = \frac{b}{a}$	a\b
+	addition: $a + b$	a+b
−	subtraction: $a - b$	a-b

```
>>8 + 3*5
ans =
        23
>>(8 + 3)*5
ans =
        55
>>4^2 - 12 - 8/4*2
ans =
        0
>>4^2 - 12 - 8/(4*2)
ans =
        3
>>3*4^2 + 5
ans =
        53
>>(3*4)^2 + 5
ans =
        149
>>27^(1/3) + 32^(0.2)
ans =
        5
>>27^(1/3) + 32^0.2
ans =
        5
>>27^1/3 + 32^0.2
ans =
        11
>>4^(1/2)
ans =
        2
>>4^(-1/2)
ans =
        0.5
```

To avoid mistakes, feel free to insert parentheses wherever you are unsure of the effect precedence will have on the calculation. Use of parentheses also improves

Table 1.1–2 Order of precedence

Precedence	Operation
First	Parentheses, evaluated starting with the innermost pair.
Second	Exponentiation, evaluated from left to right.
Third	Multiplication and division with equal precedence, evaluated from left to right.
Fourth	Addition and subtraction with equal precedence, evaluated from left to right.

the readability of your MATLAB expressions. For example, parentheses are not needed in the expression 8 + (3*5), but they make clear our intention to multiply 3 by 5 before adding 8 to the result.

Parentheses must be *balanced*, which means that there must be an equal number of left-facing and right-facing parentheses. However, just because they are balanced does not mean the expression is correct. For example, to evaluate the expression

$$y = (x - 3)(x - 2)^2$$

the following code gives the correct answer.

```
y = (x - 3)*(x - 2)^2
```

However, if you type by mistake

```
y = (x - 3*(x - 2))^2
```

the parentheses are balanced and MATLAB will not give an error message but the answer will be incorrect. For example, if $x = 8$, the correct answer is 180, but the previous code gives 100.

Test Your Understanding

T1.1–2   Use MATLAB to compute the following expressions.

a. $6\left(\frac{10}{13}\right) + \frac{18}{5(7)} + 5\left(9^2\right)$

b. $6(35^{1/4}) + 14^{0.35}$

(Answers: a. 410.1297 b. 17.1123.)

T1.1–3   What answer is produced by the following MATLAB expressions?

a. 25^-1

b. 25^-1/2

c. 25^(-1/2)

d. 4^3/2

(Answers: a. 0.04 b. 0.02 c. 0.2 d. 32.)

Proper Use of the Assignment Operator

It is important to understand that the = symbol in MATLAB works differently than the equals sign you know from mathematics. When you type x = 3, you tell MATLAB to assign the value 3 to the variable x. This usage is no different than in mathematics. However, in MATLAB we can also type something like this: x = x + 2. This tells MATLAB to add 2 to the current value of x, and to replace the current value of x with this new value. If x originally had the value 3, its new value would be 5. This use of the = operator is different from its use in mathematics. For example, the mathematics equation $x = x + 2$ is invalid because it implies that $0 = 2$.

In MATLAB the variable on the *left-hand* side of the = operator is replaced by the value generated by the *right-hand* side. Therefore, one variable, and only one variable, must be on the left-hand side of the = operator. Thus in MATLAB you cannot type 6 = x. Another consequence of this restriction is that you cannot write in MATLAB expressions like the following:

```
>>x+2=20
```

The corresponding equation $x + 2 = 20$ is acceptable in algebra and has the solution $x = 18$, but MATLAB cannot solve such an equation without additional commands (these commands are available in the Symbolic Math toolbox, which is described in Chapter 11).

Another restriction is that the right-hand side of the = operator must have a computable value. For example, if the variable y has not been assigned a value, then the following will generate an error message in MATLAB.

```
>>x = 5 + y
```

In addition to assigning known values to variables, the assignment operator is very useful for assigning values that are not known ahead of time, or for changing the value of a variable by using a prescribed procedure. The following example shows how this is done.

Volume of a Circular Cylinder

EXAMPLE 1.1–1

The volume of a circular cylinder of height h and radius r is given by $V = \pi r^2 h$. A particular cylindrical tank is 15 m tall and has a radius of 8 m. We want to construct another cylindrical tank with a volume 20 percent greater but having the same height. How large must its radius be?

■ Solution
First solve the cylinder equation for the radius r. This gives

$$r = \sqrt{\frac{V}{\pi h}} = \left(\frac{V}{\pi h}\right)^{1/2}$$

The session is shown below. First we assign values to the variables r and h representing the radius and height. Then we compute the volume of the original cylinder and increase the volume by 20 percent. Finally we solve for the required radius. For this problem we can use the MATLAB built-in constant pi .

```
>>r = 8;
>>h = 15;
>>V = pi*r^2*h;
>>V = V + 0.2*V;
>>r = (V/(pi*h))^(1/2)
r =
    8.7636
```

Thus the new cylinder must have a radius of 8.7636 m. Note that the original values of the variables r and V are replaced with the new values. This is acceptable as long as we do not wish to use the original values again. Note how precedence applies to the line $V = pi*r^2*h;$. It is equivalent to $V = pi*(r^2)*h;$.

The expression $r = (V/(pi*h))^(1/2)$ is an example of the use of *nested parentheses* where the inner pair makes clear our intention to multiply pi by h before dividing their product into V. The outer pair of parentheses is required to indicate the target of the square root operation. You may always use nested parentheses to indicate your intentions. Make sure they are used in balanced pairs; otherwise you will get an "unbalanced parentheses warning."

Variable Names

WORKSPACE

The term *workspace* refers to the names and values of any variables in use in the current work session. Variable names must begin with a letter; the rest of the name can contain letters, digits, and underscore characters, but no spaces. MATLAB is case-sensitive. Thus the following names represent five different variables: speed, Speed, SPEED, Speed_1, and Speed_2. There is a large, but finite limit to the number of characters in a name. This can depend on the particular MATLAB version. Type namelengthmax to determine this limit. MATLAB ignores any extra characters.

Managing the Work Session

Table 1.1–3 summarizes some commands and special symbols for managing the work session. A semicolon at the end of a line suppresses printing the results to the screen. If a semicolon is not put at the end of a line, MATLAB displays the

Table 1.1–3 Commands for managing the work session

Command	Description
clc	Clears the Command window.
clear	Removes all variables from memory.
clear var1 var2	Removes the variables var1 and var2 from memory.
exist('name')	Determines if a file or variable exists having the name 'name'.
quit	Stops MATLAB.
who	Lists the variables currently in memory.
whos	Lists the current variables and sizes and indicate if they have imaginary parts.
:	Colon; generates an array having regularly spaced elements.
,	Comma; separates elements of an array.
;	Semicolon; suppresses screen printing; also denotes a new row in an array.
...	Ellipsis; continues a line.

results of the line on the screen. Even if you suppress the display with the semi-colon, MATLAB still retains the variable's value.

You can put several commands on the same line if you separate them with a comma if you want to see the results of the previous command or semicolon if you want to suppress the display. For example,

```
>>x=2;y=6+x,x=y+7
y =
     8
x =
    15
```

Note that the first value of x was not displayed. Note also that the value of x changed from 2 to 15.

If you need to type a long line, you can use an *ellipsis,* by typing three periods, to delay execution. For example,

```
>>NumberOfApples = 10; NumberOfOranges = 25;
>>NumberOfPears = 12;
>>FruitPurchased = NumberOfApples + NumberOfOranges …
+NumberOfPears
FruitPurchased =
     47
```

Tab Completion

MATLAB suggests corrections for *syntax errors*, which are incorrect expressions in the MATLAB language. Suppose you mistakenly typed the line

```
>>x = 1 + 2(6 + 5)
```

If you pressed **Enter**, MATLAB responds with an error message and asks if you meant to type x = 1 + 2*(6+5). But if you did not yet press **Enter,** instead of retyping the entire line, press the left-arrow key (←) several times to move the cursor and add the missing t, then press **Enter.**

The left-arrow (←) and right-arrow (→) keys move left and right through a line one *character* at a time. To move through one *word* at a time, press **Ctrl** and → simultaneously to move to the *right;* press **Ctrl** and ← simultaneously to move to the *left*. Press **Home** to move to the beginning of a line; press **End** to move to the end of a line.

You can use the *tab completion* feature to reduce the amount of typing. MATLAB automatically completes the name of a function, variable, or file if you type the first few letters of the name and press the **Tab** key. If the name is unique, it is automatically completed. For example, in the session listed earlier, if you type Fruit and press **Tab,** MATLAB completes the name and displays FruitPurchased. Press **Enter** to display the value of the variable, or continue editing to create a new executable line that uses the variable FruitPurchased.

The tab completion feature also corrects for misspelling. If you type `fruit` and press **Tab,** MATLAB correctly displays `FruitPurchased`.

If there is more than one name that starts with the letters you typed, MATLAB displays these names when you press the **Tab** key. Use the mouse to select the desired name from the pop-up list by double-clicking on its name.

Command History

The Pop-up Command History displays commands recently used in the Command Window. By default it displays in response to the up-arrow (↑) in the Command Window. You can use it for recalling, viewing, filtering, and searching recently used commands in the Command Window To retrieve a command in the list, use the up arrow key to highlight the desired command and then press **Enter**, or use the mouse to select it. To retrieve a command using a partial match, type any part of the command at the prompt, and then press the up-arrow key. Marks the same color as error messages appear on the left side of the Command History to indicate commands that generate errors.

Deleting and Clearing

Press **Del** to delete the character at the cursor; press **Backspace** to delete the character before the cursor. Press **Esc** to clear the entire line; press **Ctrl** and **k** simultaneously to delete (*kill*) to the end of the line.

MATLAB retains the last value of a variable until you quit MATLAB or clear its value. Overlooking this fact commonly causes errors in MATLAB. For example, you might prefer to use the variable x in a number of different calculations. If you forget to enter the correct value for x, MATLAB uses the last value, and you get an incorrect result. You can use the `clear` function to remove the values of *all* variables from memory, or you can use the form `clear var1 var2` to clear the variables named `var1` and `var2`. The effect of the `clc` command is different; it clears the Command window of everything in the window display, but the values of the variables remain.

You can type the name of a variable and press **Enter** to see its current value. If the variable does not have a value (i.e., if it does not exist), you see an error message. You can also use the `exist` function. Type `exist('x')` to see if the variable x is in use. If a 1 is returned, the variable exists; a 0 indicates that it does not exist. The `who` function lists the names of all the variables in memory, but does not give their values. The form `who var1 var2` restricts the display to the variables specified. The wildcard character * can be used to display variables that match a pattern. For instance, `who A*` finds all variables in the current workspace that start with A. The `whos` function lists the variable names and their sizes and indicates whether they have nonzero imaginary parts.

The difference between a function and a command or a statement is that functions have their arguments enclosed in parentheses. Commands, such as `clear`, need not have arguments; but if they do, they are not enclosed in parentheses, for example, `clear x`. Statements cannot have arguments; for example, `clc` and `quit` are statements.

Press **Ctrl-C** to cancel a long computation without terminating the session. You can quit MATLAB by typing `quit`. You can also click on the **File** menu, and then click on **Exit MATLAB.**

Predefined Constants

MATLAB has several predefined special constants, such as the built-in constant `pi` we used in Example 1.1–1. Table 1.1–4 lists them. The symbol `Inf` stands for ∞, which in practice means a number so large that MATLAB cannot represent it. For example, typing `5/0` generates the answer `Inf`. The symbol `NaN` stands for "not a number." It indicates an undefined numerical result such as that obtained by typing `0/0`. The symbol `eps` is the smallest number which, when added to 1 by the computer, creates a number greater than 1. We use it as an indicator of the accuracy of computations.

The symbols `i` and `j` denote the imaginary unit, $i = j = \sqrt{-1}$. where we use them to create and represent complex numbers, such as `x = 5 + 8i`.

Try not to use the names of special constants as variable names. Although MATLAB allows you to assign a different value to these constants, it is not good practice to do so.

Complex Number Operations

MATLAB handles complex number algebra automatically. For example, the number $c_1 = 1 - 2i$ is entered as follows: `c1 = 1-2i`. You can also type `c1 = Complex(1, -2)`.

> **Caution:** Note that an asterisk is not needed between `i` or `j` and a number, although it is required with a variable, such as `c2 = 5 - i*c1`. This convention can cause errors if you are not careful. For example, the expressions `y = 7/2*i` and `x = 7/2i` give two different results: $y = (7/2)i = 3.5i$ and $x = 7/(2i) = -3.5i$.

Addition, subtraction, multiplication, and division of complex numbers are easily done. For example,

```
>>s = 3+7i;w = 5-9i;
>>w+s
```

Table 1.1–4 Special variables and constants

Command	Description
ans	Temporary variable containing the most recent answer.
eps	Specifies the accuracy of floating point precision.
i,j	The imaginary unit $\sqrt{-1}$.
Inf	Infinity.
NaN	Indicates an undefined numerical result.
pi	The number π.

```
ans =
     8.0000 - 2.0000i
>>w*s
ans =
     78.0000 + 8.0000i
>>w/s
ans =
     -0.8276 - 1.0690i
```

Test Your Understanding

T1.1–4   Given $x = -5 + 9i$ and $y = 6 - 2i$, use MATLAB to show that $x + y = 1 + 7i$, $xy = -12 + 64i$, and $x/y = -1.2 + 1.1i$.

Formatting Commands

The `format` command controls how numbers appear on the screen. Table 1.1–5 gives the variants of this command. MATLAB uses many significant figures in its calculations, but we rarely need to see all of them. The default MATLAB display format is the `short` format, which uses four decimal digits. You can display more by typing `format long`, which gives 16 digits. To return to the default mode, type `format short`.

   You can force the output to be in scientific notation by typing `format short e`, or `format long e`, where e stands for the number 10. Thus the output `6.3792e+03` stands for the number 6.3792×10^3. The output `6.3792e-03` stands for the number 6.3792×10^{-3}. Note that in this context e does *not* represent the number e, which is the base of the natural logarithm. Here e stands for "exponent." It is a poor choice of notation, but MATLAB follows conventional computer programming standards that were established many years ago.

   Use `format bank` only for monetary calculations; it does not recognize imaginary parts.

Table 1.1–5 Numeric display formats

Command	Description and example
format short	Four decimal digits (the default); 13.6745.
format long	16 digits; 17.27484029463547.
format short e	Five digits (four decimals) plus exponent; 6.3792e+03.
format long e	16 digits (15 decimals) plus exponent; 6.379243784781294e−04.
format bank	Two decimal digits; 126.73.
format +	Positive, negative, or zero; +.
format rat	Rational approximation; 43/7.
format compact	Suppresses some blank lines.
format loose	Resets to less compact display mode.

The Live Editor

With the MATLAB Live Editor, which was added in R2016a, you can create and run *live scripts*. Live scripts combine code, output, and formatted content together in a single interactive environment. Formatted content includes formatted text, plots, images, hyperlinks, and equations. You can create an interactive narrative to be shared.

The Live Editor enables you to work more efficiently because you can write, execute, and test code without leaving the environment, and you can run blocks of code individually or the whole file. You can see the results and graphics next to the code that produced them, and you can see errors at the file location where they occur.

The best way to learn more is to type Live Editor in the documentation search box in the top right of the Desktop.

1.2 The Toolstrip

The *Desktop* manages the Command window and other MATLAB tools. The default appearance of the R2017b Desktop is shown in Figure 1.1–1. Across the top of the Desktop is the *Toolstrip* which contains a row of three tabs labeled HOME, PLOTS, and APPS. To the right of the tabs is the *Quick Access toolbar*, which contains frequently used options such as cut, copy, and paste. This toolbar is customizable. To the right of this toolbar is the Search Documentation box.

The Toolstrip looks like Figure 1.1–1 when the HOME tab is clicked. Below the tabs are various menu names and a row of icons called the *toolbar*. See Figure 1.2–1.

If you click another tab, the Toolstrip will change. Also, other tabs may appear. For example, if you open a file, the EDITOR, PUBLISH, and VIEW tabs will appear. The PLOTS tab will open a plotting toolbar, which will be discussed in Chapter 5. The APPS tab opens a gallery of applications from the MATLAB family of products, such as any installed MATLAB toolboxes.

The HOME Tab Menus

Most of your interaction will be in the Command window with the HOME tab active. This toolbar is shown in Figure 1.2–1. It deals with the following general categories of operations:

    Files: Enables you to create, open, find, and compare files. To create a new script file, click the **New Script** icon. This opens the Editor, and displays the EDITOR, PUBLISH, and VIEW tabs. The Editor enables

Figure 1.2–1 The MATLAB Toolstrip after selecting the HOME tab.

you to create a new program file, called a *script file*. This is one type of file, called an *M-file*, that we will examine in Section 1.4. The **New** icon opens other types of files, such as figure files, that we will discuss later. The **Compare** icon lets you compare the contents of two files.

Variables: Enables you to create variables by importing data or by using the *Variables Editor*. Clicking the **New Variable** icon opens the VARI-ABLES and VIEW tabs, and displays a grid in which you type the variable's values. You can also open and clear variables, and save the contents of your workspace.

Code: Enables you to analyze, run, time, and clear commands in your pro-grams.

Simulink: Starts the Simulink program. Simulink is an optional add-on to MATLAB, and is discussed in Chapter 10. You will not see this icon if Simulink is not installed on your system.

Environment: The **Layout** icon enables you to configure the layout of the Desktop, as discussed in Section 1.1. You can set preferences for how MATLAB displays information, and manage add-on programs.

Resouces: The **Help** icon accesses the help system, which is discussed in Section 1.5. The remaining icons let you request help from the Math-Works and the MATLAB community, and engage in self-learning with the MATLAB Academy.

1.3 Built-In Functions, Arrays, and Plots

This section discusses functions that are built in to MATLAB, and it introduces arrays, which are the basic building blocks in MATLAB. The section also shows how to handle files and to generate plots from arrays.

Built-In Functions

ARGUMENT

MATLAB has hundreds of built-in functions. One of these is the *square root* function, sqrt. A pair of parentheses is used after the function's name to enclose the value—called the function's *argument*—that is operated on by the function. For example, to compute the square root of 9 and assign its value to the variable r, you type r = sqrt(9). Note that this expression is equivalent to r = (9)^(1/2) but is more compact.

Table 1.3–1 lists some of the commonly used built-in functions. Chapter 3 gives extensive coverage of the built-in functions. MATLAB users can create their own functions for their special needs. Creation of user-defined functions is covered in Chapter 3.

For example, to compute sin x, where x has a value in radians, you type sin(x). To compute cos x, type cos(x). The exponential function e^x is computed from exp(x). The natural logarithm, ln x, is computed by typing log(x). (Note the spelling difference between mathematics text, ln, and MATLAB syntax, log.) You compute the base-10 logarithm by typing log10(x).

Table 1.3–1 Some commonly used mathematical functions

Function	MATLAB syntax*
e^x	`exp(x)`
$\sqrt{x}$	`sqrt(x)`
$\ln x$	`log(x)`
$\log 10^x$	`log10(x)`
$\cos x$	`cos(x)`
$\sin x$	`sin(x)`
$\tan x$	`tan(x)`
$\cos^{-1}x$	`acos(x)`
$\sin^{-1}x$	`asin(x)`
$\tan^{-1}x$	`atan(x)`

*The MATLAB trigonometric functions listed here use radian measure. Trigonometric functions ending in `d`, such as `sind(x)` and `cosd(x)`, take the argument `x` in degrees. Inverse functions such as `atand(x)` return values in degrees. MATLAB also has the four-quadrant inverse tangent functions `atan2(y, x)` and `atan2d(y, x)`.

The inverse sine, or arcsine, is obtained by typing `asin(x)`. It returns an answer in radians, not degrees. The function `asind(x)` returns degrees.

The inverse tangent, or arctangent, is obtained by typing `atan(x)`. It returns an answer in radians, not degrees. The function `atand(x)` returns degrees. You must be careful when using either inverse tangent function. For example, typing `atand(1)` returns 45 degrees but the tangent of −135 degrees is also 1. So you must know the correct quadrant in order to interpret the answer correctly.

MATLAB has the four quadrant inverse tangent function, `atan2(y,x)`, that automatically computes the radian angle in the correct quadrant of a line from the origin (0,0) to a point whose coordinates are *(x, y)*. The function `atan2d(y,x)` returns the answer in degrees. So typing `atan2d(-1,-1)` returns −135 degrees.

Arrays

One of the strengths of MATLAB is its ability to handle a collection of numbers, called an *array*, as if it were a single variable. We will also use arrays for generating plots.

ARRAY

A numerical array is an ordered collection of numbers (a set of numbers arranged in a specific order). An example of an array variable is one that contains the numbers 0, 4, 3, and 6, in that order. With one exception noted later, you must use *square brackets* to define the variable `x` to contain this collection. You must separate the elements by either commas or spaces or both. For example, type `x = [0, 4, 3, 6]`. Commas are preferred to improve readability and avoid mistakes.

Note that the variable `y` defined as `y = [6, 3, 4, 0]` is *not* the same as `x` because the order is different. The reason for using the brackets is as follows. If you were to type `x = 0, 4, 3, 6`, MATLAB would treat this as four separate inputs and would assign the value 0 to `x` and ignore the inputs `4, 3, 6`. Using parentheses instead of square brackets will generate an error message.

The array [0, 4, 3, 6] can be considered to have one row and four columns, and it is a subcase of a *matrix,* which has multiple rows and columns. As we will see, matrices are also denoted by square brackets, unlike in some mathematics texts that use parentheses.

We can add the two arrays x and y to produce another array z by typing the single line z = x + y. To compute z, MATLAB adds all the corresponding numbers in x and y to produce z. The resulting array z contains the numbers 6, 7, 7, 6. You can subtract arrays in a similar way, but array multiplication and division requires more detailed treatment, which we will see in Chapter 2.

You need not type all the numbers in the array if they are regularly spaced. Instead, you type the first number and the last number, with the spacing in the middle, separated by colons. For example, the numbers 0, 0.1, 0.2,..., 10 can be assigned to the variable u by typing u = 0:0.1:10. In this application of the *colon operator,* the brackets can be used for improved readability but are not required.

Some of the power of MATLAB comes for its ability to use simple code to perform operations on arrays containing many values. For example, to compute $w = 5 \sin u$ for $u = 0, 0.1, 0.2,..., 10$, the session is

```
>>u = 0:0.1:10;
>>w = 5*sin(u);
```

The single line w = 5*sin(u) computed the formula $w = 5 \sin u$ 101 times, once for each value in the array u, to produce an array z that has 101 values.

You can see all the u values by typing u after the prompt; or, for example, you can see the seventh value by typing u(7). The number 7 is called an *array index,* because it points to a particular element in the array. An example session follows.

ARRAY INDEX

```
>>u(7)
ans =
       0.6000
>>w(7)
ans =
       2.8232
```

Arrays such as the ones created thus far that display on the screen as a single row of numbers with more than one column are called *row arrays.* You can create *column arrays,* which have more than one row, by using a semicolon to separate the rows. For example, typing r = [0; 4; 3; 6] creates a column array with four rows and one column.

You can use the length function to determine how many values are in an array. For example, continue the previous session as follows:

```
>>m = length(w)
m =
     101
```

Arrays and Polynomial Roots

Many applications require us to solve for the roots of a polynomial. The familiar quadratic formula gives the solution for the roots of a quadratic (second-degree) polynomial. Formulas exist for the roots of third- and fourth-degree polynomials but they are complicated. MATLAB contains sophisticated algorithms for finding roots of high-degree polynomials.

We can describe a polynomial in MATLAB with an array whose elements are the polynomial's coefficients, *starting with the coefficient of the highest power of the variable*. For example, the polynomial $4x^3 - 8x^2 + 7x - 5$ would be represented by the array $[4,-8,7,-5]$. The *roots* of the polynomial $f(x)$ are the values of x such that $f(x) = 0$. Polynomial roots can be found with the roots(a) function, where a is the polynomial's coefficient array. The result is a *column* array that contains the polynomial roots. For example, to find the roots of $x^3 - 7x^2 + 40x - 34 = 0$, the session is

```
>>a = [1,-7,40,-34];
>>roots(a)
ans =
      3.0000 + 5.000i
      3.0000 - 5.000i
      1.0000
```

The roots are $x = 1$ and $x = 3 \pm 5i$. The two commands could have been combined into the single command roots([1,-7,40,-34]).

Test Your Understanding

T1.3–1 Use MATLAB to determine how many elements are in the array cos(0):0.02:log10(100). Use MATLAB to determine the 25th element. (Answer: 51 elements and 1.48.)

T1.3–2 Use MATLAB to find the roots of the polynomial $290 - 11x + 6x^2 + x^3$. (Answer: $x = -10, 2 \pm 5i$.)

Plotting with MATLAB

Arrays are used to create plots in MATLAB. MATLAB contains many powerful functions for easily creating plots of several different types, such as rectilinear, logarithmic, surface, and contour plots. As a simple example, let us plot the function $y = 5\cos(2x)$ for $0 \leqslant x \leqslant 7$. We choose to use an increment of 0.01 to generate a large number of x values in order to produce a smooth curve. The function plot(x,y) generates a plot with the x values on the horizontal axis (the abscissa) and the y values on the vertical axis (the ordinate). The session is

```
>>x = 0:0.01:7;
>>y = 3*cos(2*x);
>>plot(x,y),xlabel('x'),ylabel('y')
```

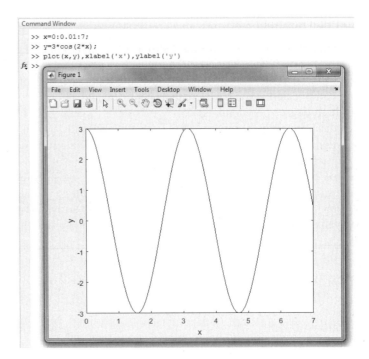

Figure 1.3–1 A graphics window showing a plot.

GRAPHICS
WINDOW

The plot appears on the screen in a *graphics window*, named **Figure 1,** as shown in Figure 1.3–1. The `xlabel` function places the text in single quotes as a label on the horizontal axis. The `ylabel` function performs a similar function for the vertical axis. When the plot command is successfully executed, a graphics window automatically appears. If a hard copy of the plot is desired, the plot can be printed by selecting **Print** from the **File** menu on the graphics window. The window can be closed by selecting **Close** on the **File** menu in the graphics window. You will then be returned to the prompt in the Command window.

Other useful plotting functions are `title` and `gtext`. These functions place text on the plot. Both accept text within parentheses and single quotes, as with the `xlabel` function. The `title` function places the text at the top of the plot; the `gtext` function places the text at the point on the plot where the cursor is located when you click the left mouse button.

OVERLAY PLOTS

You can create multiple plots, called *overlay plots*, by including another set or sets of values in the `plot` function. For example, to plot the functions $y = 2\sqrt{x}$ and $z = 4\sin(3x)$ for $0 \le x \le 5$ on the same plot, the session is

```
>>x = 0:0.01:5;
>>y = 2*sqrt(x);
>>z = 4*sin(3*x);
>>plot(x,y,x,z),xlabel('x'),gtext('y'),gtext('z')
```

After the plot appears on the screen, the program waits for you to position the cursor and click the mouse button, once for each `gtext` function used. Use the `gtext` function to place the labels y and z next to the appropriate curves.

You can also distinguish curves from one another by using different line types for each curve. For example, to plot the z curve using a dashed line, replace the `plot(x,y,x,z)` function in the above session with `plot(x,y,x,z, '- -')`. Other line types can be used. These are discussed in Chapter 5.

Sometimes it is useful or necessary to obtain the coordinates of a point on a plotted curve. The function `ginput` can be used for this purpose. Place it at the end of all the plot and plot formatting statements, so that the plot will be in its final form. The command `[x,y] = ginput(n)` gets n points and returns the x and y coordinates in the vectors x and y, which have a length n. Position the cursor using a mouse, and press the mouse button. The returned coordinates have the same scale as the coordinates on the plot.

In cases where you are plotting *data,* as opposed to functions, you should use a *data marker* to plot each data point (unless there are very many data points). To mark each point with a plus sign `+`, the required syntax for the `plot` function is `plot(x,y,'+')`. You can connect the data points with lines if you wish. In that case, you must plot the data twice, once with a data marker and once without a marker.

DATA MARKER

For example, suppose the data for the independent variable is `x = [15:2:23]` and the dependent variable values are `y = [20, 50, 60, 90, 70]`. To plot the data with plus signs, use the following session:

```
>>x = 15:2:23;
>>y = [20, 50, 60, 90, 70];
>>plot(x,y,'+',x,y),xlabel('x'),ylabel('y'), grid
```

The `grid` command puts grid lines on the plot. Other data markers are available. These are discussed in Chapter 5.

Table 1.3–2 summarizes these plotting commands. We will discuss other plotting functions, and the Plot Editor, in Chapter 5.

Table 1.3–2 Some MATLAB plotting commands

Command	Description
`[x,y] = ginput(n)`	Enables the mouse to get n points from a plot, and returns the x and y coordinates in the vectors x and y, which have a length n.
`grid`	Puts grid lines on the plot.
`gtext('text')`	Enables placement of text with the mouse.
`plot(x,y)`	Generates a plot of the array y versus the array x on rectilinear axes.
`title('text')`	Puts text in a title at the top of the plot.
`xlabel('text')`	Adds a text label to the horizontal axis (the abscissa).
`ylabel('text')`	Adds a text label to the vertical axis (the ordinate).

Test Your Understanding

T1.3–3 Plot the function $y = 3x^2 + 2$ over the interval $0 \leq x \leq 10$.

T1.3–4 Use MATLAB to plot the function $s = 2 \sin(3t + 2) + \sqrt{5t + 1}$ over the interval $0 \leq t \leq 5$. Put a title on the plot, and properly label the axes. The variable s represents speed in feet per second; the variable t represents time in seconds.

T1.3–5 Use MATLAB to plot the functions $y = 4\sqrt{6x + 1}$ and $z = 5e^{0.3x} - 2x$ over the interval $0 \leq x \leq 1.5$. Properly label the plot and each curve. The variables y and z represent force in newtons; the variable x represents distance in meters.

1.4 Working with Files

Thus far we have shown how to use MATLAB in an interactive session. However, for more detailed applications, we eventually will want to save our work and perhaps our code to be reused. We can do this by using files, of which MATLAB has several types.

File Types

MAT-FILE

ASCII-FILE

DATA FILE

MATLAB uses several types of files that enable you to save session results, data, and programs. Program files that you create are saved with the extension .m, and thus are called M-files. *MAT-files* have the extension .mat and are used to save the names and values of variables created during a MATLAB session.

Because they are *ASCII files*, M-files can be created using just about any word processor. MAT-files are *binary* files that are generally readable only by the software that created them. MAT-files contain a machine signature that allows them to be transferred between machine types such as MS Windows and Macintosh machines.

The third type of file we will be using is a *data file*, specifically an ASCII data file, that is, one created according to the ASCII format. You may need to use MATLAB to analyze data stored in such a file created by a spreadsheet program, a word processor, or a laboratory data acquisition system, or in a data file you share with someone else.

Saving and Retrieving Your Workspace Variables

If you want to continue a MATLAB session at a later time, you can either click on the Save Workspace icon on the Toolstrip or use the `save` command. If you use the icon you will be asked to enter a filename; the default name is matlab. Typing `save(myfile)` causes MATLAB to save the workspace variables, that is, the variable names, their sizes, and their values, in a binary file called `myfile. mat`, which MATLAB can read. To retrieve your workspace variables, either click on the Import Data icon or type `load(myfile)`. You can then continue your session as before. If the saved file contains the variables A, B, and C for example, then loading the file places these variables back into the workspace and overwrites any existing variables having the same name. To load just some of your variables, say, var1 and var2, type load(myfile, var1, var2).

To save just some of your variables, say, var1 and var2, type save(myfile, var1, var2). You need not type the variable names to retrieve them; just type load(myfile).

Directories and Path　It is important to know the location of the files you use with MATLAB. File location frequently causes problems for beginners. Suppose you use MATLAB on your home computer and save a file to removable medium such as flash drive. If you bring that drive to use with MATLAB on another computer, say, in a public computer lab, you must make sure that MATLAB knows how to find your files. When you save your files you must know where they are saved, especially in a public lab. This procedure depends on the specific lab, so you need to get that information from the lab manager.

Files are stored in *folders,* also called *directories*. Folders can have subfolders below them. For example, you may wish to store your files in the folder c:\ matlab\mywork. Then the \mywork folder is a subfolder under the folder c:\ matlab. The *path* tells us and MATLAB how to find a particular file.

The path is shown in the window above the Current Folder window in the default Desktop (see Figure 1.1–1). The path can be changed by clicking on the path shown until the desired subfolder appears (assuming it already exists). You can also type pwd to see the path. this shows the *top* folder in the *search path*, which is the complete list of folders that MATLAB searches when trying to find a file.

Before we show how to create and save programs in M-files, we need to discuss how MATLAB looks for variables, commands, and files. Suppose you have saved the file problem1.m in the folder c:\matlab\homework. When you type problem1,

1. MATLAB first checks to see if problem1 is a variable and if so, displays its value.
2. If not, MATLAB then checks to see if problem1 is one of its own commands, and executes it if it is.
3. If not, MATLAB then looks in the current folder for a file named problem1.m and executes problem1 if it finds it.
4. If not, MATLAB then searches the folders in its *search path*, in order, for problem1.m and then executes it if found. When files with the same name appear in multiple folders on the search path, MATLAB uses the file named problem1 found in the folder nearest to the top of the search path. So the *order* of folders on the search path is important.

You can display the MATLAB search path by typing path. You need to make sure that problem1.m is in a folder that is in the search path, otherwise MATLAB will not find the file and will generate an error message.

PATH

SEARCH PATH

If you have saved your file on a removable medium and you bring it to a public computer lab, if you cannot change the search path, an alternative procedure is to copy your file to a folder in the search path. However, there are several pitfalls with this approach: (1) if you change the file during your session, you might forget to copy the revised file back to your medium, and (2) someone else can access your work!

CURRENT DIRECTORY

The `what` command displays a list of the MATLAB-specific files in the *current directory*. The `what dirname` command does the same for the directory `dirname`. Type `which item` to display the full path name of the function `item` or the file `item` (include the file extension). If `item` is a variable, then MATLAB identifies it as such.

You can add a directory to the search path by using the `addpath` command. To remove a directory from the search path, use the `rmpath` command. The Set Path tool is a graphical interface for working with files and directories. Type `pathtool` to start the browser. To save the path settings, click on **Save** in the tool. To restore the default search path, click on **Default** in the browser.

These commands are summarized in Table 1.4–1.

Creating Script Files

You can perform operations in MATLAB in two ways:

1. In the interactive mode, in which all commands are entered directly in the Command window.
2. By running a MATLAB program stored in a *script* file. This type of file contains MATLAB commands, so running it is equivalent to typing all the commands, one at a time, at the Command window prompt. You can run the file by typing its name at the Command window prompt.

Table 1.4–1 System, directory, and file commands

Command	Description
`addpath dirname`	Adds the directory `dirname` to the search path.
`cd dirname`	Changes the current directory to `dirname`.
`dir`	Lists all files in the current directory.
`dir dirname`	Lists all the files in the directory `dirname`.
`path`	Displays the MATLAB search path.
`pathtool`	Starts the Set Path tool.
`pwd`	Displays the current directory.
`rmpath dirname`	Removes the directory `dirname` from the search path.
`what`	Lists the MATLAB-specific files found in the current working directory. Most data files and other non-MATLAB files are not listed. Use `dir` to get a list of all files.
`what dirname`	Lists the MATLAB-specific files in directory `dirname`.
`which item`	Displays the path name of `item` if `item` is a function or file. Identifies `item` as a variable if so.

When the problem to be solved requires many commands or a repeated set of commands, or has arrays with many elements, the interactive mode is inconvenient. Fortunately, MATLAB allows you to write your own programs to avoid this difficulty. You write and save MATLAB programs in M-files, which have the extension .m; for example, program1.m.

MATLAB uses two types of M-files: *script files* and *function files*. You can use the Editor built into MATLAB to create M-files. Because they contain commands, script files are sometimes called *command files*. Function files are discussed in Chapter 3.

SCRIPT FILES

The symbol % designates a *comment*, which is not executed by MATLAB. Comments are used mainly in script files for the purpose of documenting the file. The comment symbol may be put anywhere in the line. MATLAB ignores everything to the right of the % symbol. For example, consider the following session.

COMMENT

```
>>% This is a comment.
>>x = 2+3 % So is this.
x =
    5
```

Note that the portion of the line before the % sign is executed to compute x.

Here is a simple example that illustrates how to create, save, and run a script file, using the Editor built into MATLAB. However, you may use another text editor to create the file. The sample file is shown below. It computes the cosine of the square root of several numbers and displays the results on the screen.

```
% Program Example_1.m
% This program computes the cosine of
% the square root and displays the result.
x = sqrt(13:3:25);
y = cos(x)
```

To create this new M-file when in the Command window, select **New Script** from the **HOME** tab. You will then see a new edit window and the EDITOR tab will appear, as shown in Figure 1.4–1. Type in the file shown previously, using the keyboard and the EDITOR menu. When finished, select **Save** from the EDITOR menu. In the dialog box that appears, replace the default name provided (usually named Untitled) with the name Example_1, and click on **Save.** The Editor will automatically provide the extension .m and save the file in the MATLAB current directory.

Once the file has been saved, in the MATLAB Command window type the script file's name Example_1 to execute the program. You should see the result displayed in the Command window. Figure 1.4–1 shows a screen containing the resulting Command window display and the Editor/Debugger opened to display the script file.

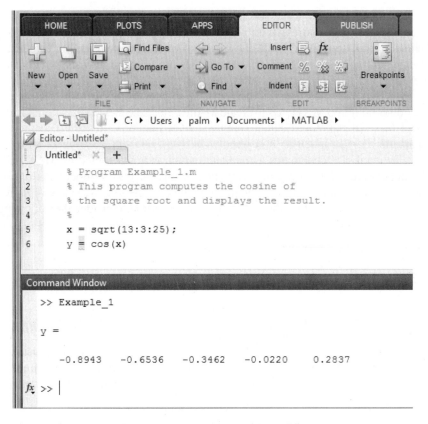

Figure 1.4–1 The MATLAB Command window with the Editor open.

Effective Use of Script Files

Create script files to avoid the need to retype lengthy and commonly used procedures. Here are some other things to keep in mind when using script files:

1.   The name of a script file must follow the MATLAB convention for naming variables.

2.   Recall that typing a variable's name at the Command window prompt causes MATLAB to display the value of that variable. Thus, do not give a script file the same name as a variable it computes because MATLAB will not be able to execute that script file more than once, unless you clear the variable.

3.   Do not give a script file the same name as a MATLAB command or function. You can check to see if a command, function, or file name already exists by using the `exist` command as discussed in Section 1.1.

Note that not all functions supplied with MATLAB are built-in functions. Some are M-files depending on the MATLAB version. For example, in earlier releases of MATLAB, the `plot` function was an M-file, but is now a built-in function. The function `mean.m` is supplied but is not a built-in function. The command `exist('mean')` will return a 2. The `sqrt` function is built-in, so typing `exist('sqrt')` will return a 5. You may think of built-in functions as primitives that form the basis for other MATLAB functions. You cannot view the entire file of a built-in function in a text editor, only the comments.

Programming Style

Comments may be put anywhere in the script file. However, because the first comment line before any executable statement is the line searched by the `lookfor` command, discussed later in this chapter, consider putting keywords that describe the script file in this first line (called the H1 line). A suggested structure for a script file is the following.

1. *Comments section* In this section put comment statements to give
   a. The name of the program and any keywords in the first line.
   b. The date created and the creators' names in the second line.
   c. The definitions of the variable names for every input and output variable. Divide this section into at least two subsections, one for input data and one for output data. A third, optional section may include definitions of variables used in the calculations. *Be sure to include the units of measurement for all input and all output variables!*
   d. The name of every user-defined function called by the program.
2. *Input section* In this section put the input data and/or the input functions that enable data to be entered. Include comments where appropriate for documentation.
3. *Calculation section* Put the calculations in this section. Include comments where appropriate for documentation.
4. *Output section* In this section put the functions necessary to deliver the output in whatever form required. For example, this section might contain functions for displaying the output on the screen. Include comments where appropriate for documentation.

The programs in this text often omit some of these elements to save space. In those cases the in-text discussion associated with the program provides the required documentation.

Controlling Input and Output

MATLAB provides several useful commands for obtaining input from the user and for formatting the output (the results obtained by executing the MATLAB commands). Table 1.4–2 summarizes these commands.

Table 1.4–2 Input/output commands

Command	Description
`disp(A)`	Displays the contents, but not the name, of the array **A**.
`disp('text')`	Displays the text string enclosed within single quotes.
`format`	Controls the screen's output display format (see Table 1.1–5).
`x = input('text')`	Displays the text in quotes, waits for user input from the keyboard, and stores the value in **x**.
`x = input('text','s')`	Displays the text in quotes, waits for user input from the keyboard, and stores the input as a string in **x**.

The `disp` function (short for "display") can be used to display the value of a variable but not its name. Its syntax is `disp(A)`, where **A** represents a MATLAB variable name. The `disp` function can also display text such as a message to the user. You enclose the text within single quotes. For example, the command `disp('The predicted speed is:')` causes the message to appear on the screen. This command can be used with the first form of the `disp` function in a script file as follows (assuming the value of Speed is 63):

```
disp('The predicted speed is:')
disp(Speed)
```

When the file is run, these lines produce the following on the screen:

```
The predicted speed is:
   63
```

The `input` function displays text on the screen, waits for the user to enter something from the keyboard, and then stores the input in the specified variable. For example, the command `x = input('Please enter the value of x:')` causes the message to appear on the screen. If you type 5 and press **Enter,** the variable x will have the value 5.

STRING VARIABLE

A *string variable* is composed of text (alphanumeric characters). If you want to store a text input as a string variable, use the other form of the input command. For example, the command `Calendar = input('Enter the day of the week:','s')` prompts you to enter the day of the week. If you type Wednesday, this text will be stored in the string variable `Calendar`.

Example of a Script File

The following is a simple example of a script file that shows the preferred program style. The speed of a falling object dropped with no initial velocity is given as a function of time t by $v = gt$, where g is the acceleration due to gravity. In SI units, $g = 9.81$ m/s². We want to compute and plot v as a function of t for

$0 \le t \le t_{final}$, where t_{final} is the final time entered by the user. The script file is the following.

```
% Program Falling_Speed.m: plots speed of a falling object.
% Created on March 1, 2016 by W. Palm III
%
% Input Variable:
% tfinal = final time (in seconds)
%
% Output Variables:
% t = array of times at which speed is computed (seconds)
% v = array of speeds (meters/second)
%
% Parameter Value:
g = 9.81; % Acceleration in SI units
%
% Input section:
tfinal = input('Enter the final time in seconds:');
%
% Calculation section:
dt = tfinal/500;
t = 0:dt:tfinal; % Creates an array of 501 time values.
v = g*t;
%
% Output section:
plot(t,v),xlabel('Time (seconds)'), ylabel('Speed (meters/second)')
```

After creating this file, you save it with the name `Falling_Speed.m`. To run it, you type `Falling_Speed` (without the `.m`) in the Command window at the prompt. You will then be asked to enter a value for t_{final}. After you enter a value and press **Enter,** you will see the plot on the screen.

Test Your Understanding

T1.4–1 The surface area A of a sphere depends on its radius r as follows: $A = 4\pi r^2$. Write a script file that prompts the user to enter a radius, computes the surface area, and displays the result.

T1.4-2 The length c of the hypotenuse of a right triangle, whose side lengths are a and b, is given by

$$c^2 = a^2 + b^2$$

Write a script file that prompts the user to enter the side lengths a and b, computes the hypotenuse length, and displays the result.

Debugging Script Files

Debugging a program is the process of finding and removing the "bugs," or errors, in a program. Such errors usually fall into one of the following categories.

1. Syntax errors such as omitting a parenthesis or comma, or spelling a command name incorrectly. MATLAB usually detects the more obvious errors and displays a message describing the error and its location.
2. Errors due to an incorrect mathematical procedure, called *runtime errors.* They do not necessarily occur every time the program is executed; their occurrence often depends on the particular input data. A common example of a runtime error is division by zero.

To locate an error, try the following:

1. Always test your program with a simple version of the problem, whose answers can be checked by hand calculations.
2. Display any intermediate calculations by removing semicolons at the end of statements.
3. Use the debugging features of the Editor, which are introduced in Chapter 4. However, one advantage of MATLAB is that it requires relatively simple programs to accomplish many types of tasks. Thus you probably will not need to use the debugging features of the Editor for the problems encountered in this text.

1.5 The MATLAB Help System

To explore the more advanced features of MATLAB not covered in this book, you will need to know how to use effectively the MATLAB Help System. MATLAB has these options to get help for using MathWorks products.

1. *Function Browser* This provides quick access to the documentation for MATLAB functions.
2. *Help Icon* Click on the **Help** icon under the HOME tab to view documentation, examples, and a support web site.
3. *Help Functions* The functions help, lookfor, and doc can be used to display syntax information for a specified function.
4. *Other Resources* For additional help, you can run demos, contact technical support, search documentation for other MathWorks products, view a list of other books, and participate in a newsgroup.

The Function Browser

To activate the Function Browser, click on the fx icon to the left of the prompt. Figure 1.5–1 shows the resulting menu after selecting plot two levels down under the **Graphics** category. Scroll down to see the entire documentation of the plot function.

Figure 1.5–1 The Function Browser after **plot** has been selected.

Help Functions

Three MATLAB functions can be used for accessing online information about MATLAB functions.

The help Function The help function is the most basic way to determine the syntax and behavior of a particular function. For example, typing help log10 in the Command window produces the following display:

```
LOG10 Common (base 10) logarithm.
  LOG10(X) is the base 10 logarithm of the elements of X.
  Complex results are produced if X is not positive.

  See also LOG, LOG2, EXP, LOGM.
```

Note that the display describes what the function does, warns about any unexpected results if nonstandard argument values are used, and directs the user to other related functions.

   All the MATLAB functions are organized into logical groups, upon which the MATLAB directory structure is based. For instance, all elementary mathematical functions such as log10 reside in the elfun directory, and the polynomial functions reside in the polyfun directory. To list the names of all the functions in that directory, with a brief description of each, type help polyfun. If you are unsure of what directory to search, type help to obtain an extensive list of all the directories, with a description of the function category each represents.

The lookfor Function The lookfor function allows you to search for functions on the basis of a keyword. It searches through the first line of Help text, known as the H1 line, for each MATLAB function, and returns all the H1 lines containing a specified keyword. For example, MATLAB does not have a function named sine. So the response from typing help sine is the same as from typing help sin. (In earlier releases the response was "sine.m not found", which was perhaps more useful).

However, typing `lookfor sine` produces over a dozen matches, depending on which toolboxes you have installed. For example, you will see, among others,

```
ACOS      Inverse cosine, result in radians
ACOSD     Inverse cosine, result in degrees
ACOSH     Inverse hyperbolic cosine
ASIN      Inverse sine, result in radians
...
SIN       Sine of argument in radians
...
```

From this list you can find the correct name for the sine function. Note that all words containing sine are returned, such as cosine. Adding `−all` to the `lookfor` function searches the entire Help entry, not just the H1 line.

The `doc` Function   Typing `doc function_name` displays the documentation for the specified function `function_name`. For example, typing `doc sqrt` displays the documentation page for the function `sqrt`.

The MathWorks Website

If your computer is connected to the Internet, you can access The MathWorks, Inc., the home of MATLAB. You can use electronic mail to ask questions, make suggestions, and report possible bugs. You can also use a solution search engine at The MathWorks website to query an up-to-date database of technical support information. The website address is http://www.mathworks.com.

The Help system is very powerful and detailed, so we have only described its basics. You can, and should, use the Help system to learn how to use its features in greater detail.

Test Your Understanding

T1.5–1   Use the Help system to learn about the built-in function `nthroot`. Use it to calculate the cube root of 64.

T1.5–2   Find out how many hyperbolic functions are supported by MATLAB.

T1.5–3   Type why at the command prompt. Is it a built-in function? What does it do?

1.6 Problem-Solving Methodologies

Designing new engineering devices and systems requires a variety of problem-solving skills. (This variety is what keeps engineering from becoming boring!) When you are solving a problem, it is important to plan your actions ahead of time. You can waste many hours by plunging into the problem without a plan of attack. Here we present a plan of attack, or *methodology*, for solving engineering problems in general. Because solving engineering problems often requires a computer solution and because the examples and exercises in this text require you to develop a computer solution (using MATLAB), we also discuss a methodology for solving computer problems in particular.

Steps in Engineering Problem Solving

Table 1.6–1 summarizes the methodology that has been tried and tested by the engineering profession for many years. These steps describe a general problem-solving procedure. Simplifying the problem sufficiently and applying the appropriate fundamental principles is called *modeling,* and the resulting mathematical description is called a *mathematical model,* or just a *model.* When the modeling is finished, we need to solve the mathematical model to obtain the required answer. If the model is highly detailed, we might need to solve it with a computer program. Most of the examples and exercises in this text require you to develop a computer solution (using MATLAB) to problems for which the model has already been developed. Thus we will not always need to use all the steps shown in Table 1.6–1. Greater discussion of engineering problem solving can be found in [Eide, 2008].[*]

MODEL

Table 1.6–1 Steps in engineering problem solving

1. Understand the purpose of the problem.
2. Collect the known information. Realize that some of it might later be found unnecessary.
3. Determine what information you must find.
4. Simplify the problem only enough to obtain the required information. State any assumptions you make.
5. Draw a sketch and label any necessary variables.
6. Determine which fundamental principles are applicable.
7. Think generally about your proposed solution approach and consider other approaches before proceeding with the details.
8. Label each step in the solution process.
9. If you solve the problem with a program, hand check the results using a simple version of the problem. Checking the dimensions and units and printing the results of intermediate steps in the calculation sequence can uncover mistakes.
10. Perform a "reality check" on your answer. Does it make sense? Estimate the range of the expected result and compare it with your answer. Do not state the answer with greater precision than is justified by any of the following:

    (a) The precision of the given information.

    (b) The simplifying assumptions.

    (c) The requirements of the problem.

    Interpret the mathematics. If the mathematics produces multiple answers, do not discard some of them without considering what they mean. The mathematics might be trying to tell you something, and you might miss an opportunity to discover more about the problem.

*References appear in Appendix D.

Example of Problem Solving

Consider the following simple example of the steps involved in problem solving. Suppose you work for a company that produces packaging. You are told that a new packaging material can protect a package when dropped, provided that the package hits the ground at less than 25 ft/sec. The package's total weight is 20 lb, and it is rectangular with dimensions of 12 by 12 by 8 in. You must determine whether the packaging material provides enough protection when the package is carried by delivery persons.

The steps in the solution are as follows:

1. *Understand the purpose of the problem.* The implication here is that the packaging is intended to protect against being dropped while the delivery person is carrying it. It is not intended to protect against the package falling off a moving delivery truck. In practice, you should make sure that the person giving you this assignment is making the same assumption. Poor communication is the cause of many errors!

2. *Collect the known information.* The known information is the package's weight, dimensions, and maximum allowable impact speed.

3. *Determine what information you must find.* Although it is not explicitly stated, you need to determine the maximum height from which the package can be dropped without damage. You need to find a relationship between the speed of impact and the height at which the package is dropped.

4. *Simplify the problem only enough to obtain the required information. State any assumptions you make.* The following assumptions will simplify the problem and are consistent with the problem statement as we understand it:

   a. The package is dropped from rest with no vertical or horizontal velocity.

   b. The package does not tumble (as it might when dropped from a moving truck). The given dimensions indicate that the package is not thin and thus will not "flutter" as it falls.

   c. The effect of air drag is negligible.

   d. The greatest height from which the delivery person could drop the package is 6 ft (and thus we ignore the existence of a delivery person 8 ft tall!).

   e. The acceleration g due to gravity is constant (because the distance dropped is only 6 ft).

5. *Draw a sketch and label any necessary variables.* Figure 1.6–1 is a sketch of the situation, showing the height h of the package, its mass m, its speed v, and the acceleration due to gravity g.

6. *Determine which fundamental principles are applicable.* Because this problem involves a mass in motion, we can apply Newton's laws. From physics we know that the following relations result from Newton's laws and the basic kinematics of an object falling a short distance under the influence of gravity, with no air drag or initial velocity:

   a. Height versus time to impact t_i: $h = \frac{1}{2} g t_i^2$.

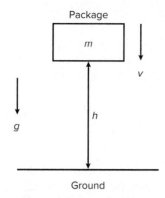

Figure 1.6–1 Sketch of the
dropped-package problem.

    b. Impact speed v_i versus time to impact: $v_i = g\,t_i$.

    c. Conservation of mechanical energy: $mgh = \frac{1}{2}mv_i^2$.

7. *Think generally about your proposed solution approach and consider other approaches before proceeding with the details.* We could solve the second equation for t_i and substitute the result into the first equation to obtain the relation between h and v_i. This approach would also allow us to find the time to drop t_i. However, this method involves more work than necessary because we need not find the value of t_i. The most efficient approach is to solve the third relation for h.

$$h = \frac{1}{2}\frac{v_i^2}{g} \tag{1.6–1}$$

Notice that the mass m cancels out of the equation. The mathematics just told us something! It told us that the mass does not affect the relation between the impact speed and the height dropped. Thus we do not need the weight of the package to solve the problem.

8. *Label each step in the solution process.* This problem is so simple that there are only a few steps to label:

    a. Basic principle: conservation of mechanical energy

$$h = \frac{1}{2}\frac{v_i^2}{g}$$

    b. Determine the value of the constant g: $g = 32.2$ ft/sec^2.

    c. Use the given information to perform the calculation and round off the result consistent with the precision of the given information:

$$h = \frac{1}{2}\frac{25^2}{32.2} = 9.7 \text{ ft}$$

    Because this text is about MATLAB, we might as well use it to do this simple calculation. The session looks like this:

```
>>g=32.2;
>>vi=25;
```

```
>>h=vi^2/(2*g)
h =
    9.7050
```

9. *Check the dimensions and units.* This check proceeds as follows, using Equation (1.6–1),

$$[\text{ft}] = \left[\frac{1}{2}\right] \frac{[\text{ft/sec}]^2}{[\text{ft/sec}^2]} = \frac{[\text{ft}]^2}{[\text{sec}]^2} \frac{[\text{sec}]^2}{[\text{ft}]} = [\text{ft}]$$

which is correct.

10. *Perform a reality check and precision check on the answer.* If the computed height were negative, we would know that we did something wrong. If it were very large, we might be suspicious. However, the computed height of 9.7 ft does not seem unreasonable.

If we had used a more accurate value for g, say $g = 32.17$, then we would be justified in rounding the result to $h = 9.71$. However, given the need to be conservative here, we probably should round the answer *down* to the nearest foot. So we probably should report that the package will not be damaged if it is dropped from a height of less than 9 ft.

The mathematics told us that the package mass does not affect the answer. The mathematics did not produce multiple answers here. However, many problems involve the solution of polynomials with more than one root; in such cases we must carefully examine the significance of each.

Steps for Obtaining a Computer Solution

If you use a program such as MATLAB to solve a problem, follow the steps shown in Table 1.6–2. Greater discussion of modeling and computer solutions can be found in [Starfield, 1990] and [Jayaraman, 1991].

MATLAB is useful for doing numerous complicated calculations and then automatically generating a plot of the results. The following example illustrates the procedure for developing and testing such a program.

Table 1.6–2 Steps for developing a computer solution

1. State the problem concisely.
2. Specify the data to be used by the program. This is the *input*.
3. Specify the information to be generated by the program. This is the *output*.
4. Work through the solution steps by hand or with a calculator; use a simpler set of data if necessary.
5. Write and run the program.
6. Check the output of the program with your hand solution.
7. Run the program with your input data and perform a "reality check" on the output. Does it make sense? Estimate the range of the expected result and compare it with your answer.
8. If you will use the program as a general tool in the future, test it by running it for a range of reasonable data values; perform a reality check on the results.

Piston Motion

EXAMPLE 1.6–1

Figure 1.6–2a shows a piston, connecting rod, and crank for an internal combustion engine. When combustion occurs, it pushes the piston down. This motion causes the connecting rod to turn the crank, which causes the crankshaft to rotate. We want to develop a MATLAB program to compute and plot the distance d traveled by the piston as a function of the angle A, for given values of lengths L_1 and L_2. Such a plot would help the engineers designing the engine to select appropriate values for lengths L_1 and L_2.

We are told that typical values for these lengths are $L_1 = 1$ ft and $L_2 = 0.5$ ft. Because the mechanism's motion is symmetrical about $A = 0$, we need consider only angles in the range $0 \leq A \leq 180°$. Figure 1.6–2b shows the geometry of the motion. From this figure we can use trigonometry to write the following expression for d:

$$d = L_1 \cos B + L_2 \cos A \tag{1.6-2}$$

Thus to compute d given the lengths L_1 and L_2 and the angle A, we must first determine the angle B. We can do so using the law of sines, as follows:

$$\frac{\sin A}{L_1} = \frac{\sin B}{L_2}$$

Solve this for B:

$$\sin B = \frac{L_2 \sin A}{L_1}$$

$$B = \sin^{-1}\left(\frac{L_2 \sin A}{L_1}\right) \tag{1.6-3}$$

(a)   (b)

Figure 1.6–2 A piston, connecting rod, and crank for an internal combustion engine.

Equations (1.6–2) and (1.6–3) form the basis of our calculations. Develop and test a MATLAB program to plot d versus A.

■ Solution

Here are the steps in the solution, following those listed in Table 1.6–2.

1. *State the problem concisely.* Use Equations (1.6–2) and (1.6–3) to compute d; use enough values of A in the range $0 \le A \le 180°$ to generate an adequate (smooth) plot.

2. *Specify the input data to be used by the program.* The lengths L_1 and L_2 and the angle A are given.

3. *Specify the output to be generated by the program.* A plot of d versus A is the required output.

4. *Work through the solution steps by hand or with a calculator.* You could have made an error in deriving the trigonometric formulas, so you should check them for several cases. You can check for these errors by using a ruler and protractor to make a scale drawing of the triangle for several values of the angle A; measure the length d; and compare it to the calculated values. Then you can use these results to check the output of the program.

    Which values of A should you use for the checks? Because the triangle "collapses" when $A = 0°$ and $A = 180°$, you should check these cases. The results are $d = L_1 - L_2$ for $A = 0°$ and $d = L_1 + L_2$ for $A = 180°$. The case $A = 90°$ is also easily checked by hand, using the Pythagorean theorem; for this case $d = \sqrt{L_1^2 - L_2^2}$. You should also check one angle in the quadrant $0° < A < 90°$ and one in the quadrant $90° < A < 180°$. The following table shows the results of these calculations using the given typical values: $L_1 = 1$, $L_2 = 0.5$ ft.

A (degrees)	d (ft)
0	1.5
60	1.15
90	0.87
120	0.65
180	0.5

5. *Write and run the program.* The following MATLAB session uses the values $L_1 = 1$, $L_2 = 0.5$ ft.

```
>>L_1 = 1;
>>L_2 = 0.5;
>>R = L_2/L_1;
>>A_d = 0:0.5:180;
>>A_r = A_d*(pi/180);
>>B = asin(R*sin(A_r));
>>d = L_1*cos(B)+L_2*cos(A_r);
>>plot(A_d,d),xlabel('A (degrees)'), ...
        ylabel('d (feet)'),grid
```

Note the use of the underscore (_) in variable names to make the names more meaningful. The variable A_d represents the angle *A* in degrees. Line 4 creates an array of numbers 0, 0.5, 1, 1.5,...,180. Line 5 converts these degree values to radians and assigns the values to the variable A_r. This conversion is necessary because MATLAB trigonometric functions use radians, not degrees. (A common oversight is to use degrees.) MATLAB provides the built-in constant pi to use for π. Line 6 uses the inverse sine function asin.

The plot command requires the label and grid commands to be on the same line, separated by commas. The line-continuation operator, called an *ellipsis,* consists of three periods. This operator enables you to continue typing the line after you press **Enter.** Otherwise, if you continued typing without using the ellipsis, you would not see the entire line on the screen. Note that the prompt is not visible when you press **Enter** after the ellipsis.

The grid command puts grid lines on the plot so that you can read values from the plot more easily. The resulting plot appears in Figure 1.6–3.

6. *Check the output of the program with your hand solution.* Read the values from the plot corresponding to the values of A given in the preceding table. You can use the ginput function to read values from the plot. The values should agree with one another, and they do.

7. *Run the program and perform a reality check on the output.* You might suspect an error if the plot showed abrupt changes or discontinuities. However, the plot

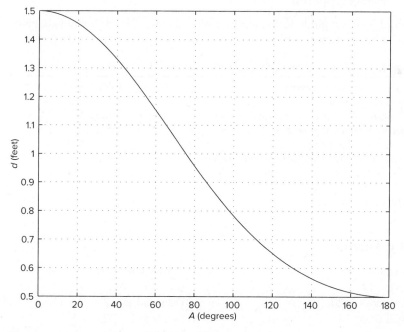

Figure 1.6–3 Plot of the piston motion versus crank angle.

is smooth and shows that d behaves as expected. It decreases smoothly from its maximum at $A = 0°$ to its minimum at $A = 180°$.

8. *Test the program for a range of reasonable input values.* Test the program using various values for L_1 and L_2, and examine the resulting plots to see whether they are reasonable. Something you might try on your own is to see what happens if $L_1 \leq L_2$. Should the mechanism work the same way it does when $L_1 > L_2$? What does your intuition tell you to expect from the mechanism? What does the program predict?

1.7 Summary

You should now be familiar with basic operations in MATLAB. These include

■ Starting and exiting MATLAB
■ Computing simple mathematical expressions
■ Managing variables

You should also be familiar with the MATLAB menu and toolbar system.

The chapter gives an overview of the various types of problems MATLAB can solve. These include

■ Using arrays and polynomials
■ Creating plots
■ Creating script files

Table 1.7–1 is a guide to the tables of this chapter. The following chapters give more details on these topics.

Table 1.7–1 Guide to commands and features introduced in this chapter

Scalar arithmetic operations	Table 1.1–1
Order of precedence	Table 1.1–2
Commands for managing the work session	Table 1.1–3
Special variables and constants	Table 1.1–4
Numeric display formats	Table 1.1–5
Some commonly used mathematical functions	Table 1.3–1
Some MATLAB plotting commands	Table 1.3–2
System, directory, and file commands	Table 1.4–1
Input/output commands	Table 1.4–2

Key Terms

Argument,   18

Array,   19

Array index,   20

ASCII file,   24

Command window,   5

Comment,   27

Current directory,   26

Data file,   24

Data marker,   23

Debugging,   32

Desktop,   5

Graphics window,   22

Problems

Answers to problems marked with an asterisk are given at the end of the text.

Section 1.1

1.   Make sure you know how to start and quit a MATLAB session. Use MATLAB to make the following calculations, using the values $x = 10$, $y = 3$. Check the results by using a calculator.

 a. $u = x + y$ b. $v = xy$ c. $w = x/y$

 d. $z = \sin x$ e. $r = 8 \sin y$ f. $s = 5 \sin 2y$

2.*   Suppose that $x = 2$ and $y = 5$. Use MATLAB to compute the following.

 a. $\dfrac{yx^3}{x - y}$ b. $\dfrac{3x}{2y}$ c. $\dfrac{3}{2}xy$ d. $\dfrac{x^5}{x^5 - 1}$

3.   Suppose that $x = 3$ and $y = 4$. Use MATLAB to compute the following, and check the results with a calculator.

 a. $\left(1 - \dfrac{1}{x^5}\right)^{-1}$ b. $3\pi x^2$ c. $\dfrac{3y}{4x - 8}$ d. $\dfrac{4(y - 5)}{3x - 6}$

4.   Evaluate the following expressions in MATLAB for the given value of x. Check your answers by hand.

 a. $y = 6x^3 + \dfrac{4}{x}$, $x = 3$ b. $y = \dfrac{x}{4}3$, $x = 7$

 c. $y = \dfrac{(4x)^2}{25}$, $x = 9$ d. $y = 2\dfrac{\sin x}{5}$, $x = 4$

 e. $y = 7(x^{1/3}) + 4x^{0.58}$, $x = 30$

5.   Assuming that the variables a, b, c, d, and f are scalars, write MATLAB statements to compute and display the following expressions. Test your statements for the values $a = 1.12$, $b = 2.34$, $c = 0.72$, $d = 0.81$, and $f = 19.83$.

$$x = 1 + \frac{a}{b} + \frac{c}{f^2} \qquad s = \frac{b - a}{d - c}$$

$$r = \frac{1}{\dfrac{1}{a} + \dfrac{1}{b} + \dfrac{1}{c} + \dfrac{1}{d}} \qquad y = ab\frac{1}{c}\frac{f^2}{2}$$

6.   Use MATLAB to calculate

a. $\dfrac{3}{4}(6)(7^2) + \dfrac{4^5}{7^3 - 145}$

b. $\dfrac{48.2(55) - 9^3}{53 + 14^2}$

c. $\dfrac{27^2}{4} + \dfrac{319^{4/5}}{5} + 60(14)^{-3}$

Check your answers with a calculator.

7.   Use MATLAB to compute the following expressions.

a. 16^{-1} b. $16^{-1/2}$ c. $16^{(-1/2)}$ d. $64^{3/2}$

8.   What answer is produced by the following MATLAB expressions?

a. `100^-1`   b. `100^-1/2`   c. `100^(-1/2)`   d. `100^3/2`

9.   The functions `realmax` and `realmin` give the largest and smallest possible numbers that can be handled by MATLAB. Calculations generating numbers that are too large or too small result in *overflow* and *underflow*. Usually this does not present a problem if you arrange the calculation sequence properly. Type `realmax` and `realmin` in MATLAB to determine the upper and lower limits for your system. For example, suppose you have the variables $a = 3 \times 10^{150}$, $b = 5 \times 10^{200}$.

a. Use MATLAB to calculate $c = ab$.

b. Suppose $d = 5 \times 10^{-200}$, use MATLAB to calculate $f = d/a$.

c. Use MATLAB to calculate the product $x = abd$ two ways, i) by calculating the product directly as x = `a*b*d` and then ii) by splitting up the calculation as y = `b*d` and then x = `a*y`. Compare the results.

10.   The volume of a circular cylinder of height h and radius r is given by $V = \pi r^2 h$. A particular cylindrical tank is 10 m tall and has a radius of 6 m. We want to construct another cylindrical tank with a volume 30 percent greater but having the same radius. How high must the tank be?

11.   The volume of a sphere is given by $V = 4\pi r^3/3$, where r is the radius. Use MATLAB to compute the radius of a sphere having a volume 40 percent greater than that of a sphere of radius 4 ft.

12.* Suppose that $x = -7 - 5i$ and $y = 4 + 3i$. Use MATLAB to compute

a. $x + y$ b. xy c. x/y

13.   Use MATLAB to compute the following. Check your answers by hand.

a. $(3 + 6i)(-7 - 9i)$ b. $\dfrac{5 + 4i}{5 - 4i}$ c. $\dfrac{3}{2}i$ d. $\dfrac{3}{2i}$

14.   Evaluate the following expressions in MATLAB, for the values $x = 5 + 8i$, $y = -6 + 7i$. Check your answers by hand.

a. $u = x + y$ b. $v = xy$ c. $w = x/y$

d. $z = e^x$ e. $r = \sqrt{y}$ f. $s = xy^2$

15. The *ideal gas law* provides one way to estimate the pressure exerted by a gas in a container. The law is

$$P = \frac{nRT}{V}$$

More accurate estimates can be made with the *van der Waals* equation

$$P = \frac{nRT}{V - nb} - \frac{an^2}{V^2}$$

where the term nb is a correction for the volume of the molecules and the term an^2/V^2 is a correction for molecular attractions. The values of a and b depend on the type of gas. The gas constant is R, the *absolute* temperature is T, the gas volume is V, and the number of gas molecules is indicated by n. If $n = 1$ mol of an ideal gas were confined to a volume of $V = 22.41$ L at $0°C$ (273.2 K), it would exert a pressure of 1 atm. In these units, $R = 0.08206$.

   For chlorine (Cl_2), $a = 6.49$ and $b = 0.0562$. Compare the pressure estimates given by the ideal gas law and the van der Waals equation for 1 mol of Cl_2 in 22.41 L at 273.2 K. What is the main cause of the difference in the two pressure estimates, the molecular volume or the molecular attractions?

16. The *ideal gas law* relates the pressure P, volume V, absolute temperature T, and amount of gas n. The law is

$$P = \frac{nRT}{V}$$

where R is the gas constant.

   An engineer must design a large natural gas storage tank to be expandable to maintain the pressure constant at 2.2 atm. In December when the temperature is $4°F$ ($-15°C$), the volume of gas in the tank is 28,500 ft^3. What will the volume of the same quantity of gas be in July when the temperature is $88°F$ ($31°C$)? (*Hint:* Use the fact that n, R, and P are constant in this problem. Note also that $K = °C + 273.2$.)

Section 1.3

17. Use MATLAB to calculate:
    a. e^2 b. $\log 2$ c. $\ln 2$ d. $\sqrt[4]{600}$

18. Use MATLAB to calculate:
    a. $\cos(\pi/2)$ b. $\cos 80°$
    c. $\cos^{-1} 0.7$ in radians   d. $\cos^{-1} 0.6$ in degrees

19. Use MATLAB to calculate:
    a. $\tan^{-1} 2$
    b. $\tan^{-1} 100$

    c. The angle corresponding to $x = 2, y = 3$

    d. The angle corresponding to $x = -2, y = 3$

    e. The angle corresponding to $x = 2, y = -3$

20. Suppose x takes on the values $x = 1, 1.2, 1.4, \ldots, 5$. Use MATLAB to compute the array y that results from the function $y = 7 \sin(4x)$. Use MATLAB to determine how many elements are in the array y and the value of the third element in the array y.

21. Use MATLAB to determine how many elements are in the array `sin(-pi/2):0.05:cos(0)`. Use MATLAB to determine the 10th element.

22. Use MATLAB to calculate

    a. $e^{(-2.1)^3} + 3.47 \, \log(14) + \sqrt[4]{287}$

    b. $(3.4)^7 \log(14) + \sqrt[4]{287}$

    c. $\cos^2\left(\dfrac{4.12\pi}{6}\right)$

    d. $\cos\left(\dfrac{4.12\pi}{6}\right)^2$

    Check your answers with a calculator.

23. Use MATLAB to calculate

    a. $6\pi \tan^{-1}(12.5) + 4$ *b.* $5 \tan[3 \sin^{-1}(13/5)]$

    c. $5 \ln(7)$ *d.* $5 \log(7)$

    Check your answers with a calculator.

24. The Richter scale is a measure of the intensity of an earthquake. The energy E (in joules) released by the quake is related to the magnitude M on the Richter scale as follows.

$$E = 10^{4.4} 10^{1.5M}$$

How much more energy is released by a magnitude 7.6 quake than a 5.6 quake?

25.* Use MATLAB to find the roots of $13x^3 + 182x^2 - 184x + 2503 = 0$.

26. Use MATLAB to find the roots of the polynomial $70x^3 + 24x^2 - 10x + 20$.

27. Use MATLAB to plot the function $T = 6 \ln t - 7e^{0.2t}$ over the interval $1 \le t \le 3$. Put a title on the plot and properly label the axes. The variable T represents temperature in degrees Celsius; the variable t represents time in minutes.

28. Use MATLAB to plot the functions $u = 2 \log_{10}(60x + 1)$ and $v = 3 \cos(6x)$ over the interval $0 \le x \le 2$. Properly label the plot and each curve. The variables u and v represent speed in miles per hour; the variable x represents distance in miles.

29. The Fourier series is a series representation of a periodic function in terms of sines and cosines. The Fourier series representation of the function

$$f(x) = \begin{cases} 1 & 0 < x < \pi \\ -1 & -\pi < x < 0 \end{cases}$$

is

$$\frac{4}{\pi}\left(\frac{\sin x}{1} + \frac{\sin 3x}{3} + \frac{\sin 5x}{5} + \frac{\sin 7x}{7} + \cdots\right)$$

Plot on the same graph the function $f(x)$ and its series representation, using the four terms shown.

30. A *cycloid* is the curve described by a point P on the circumference of a circular wheel of radius r rolling along the x axis. The curve is described in parametric form by the equations

$$x = r(\phi - \sin \phi)$$
$$y = r(1 - \cos \phi)$$

Use these equations to plot the cycloid for $r = 10$ in. and $0 \le \phi \le 4\pi$.

31. A boat moves at 20 km/hr along a straight path described by $y = 11x/15 + 43/3$, starting at $x = -10$, $y = 7$. Plot the angle (in degrees) of the line of sight from an observer at the coordinate origin to the boat as a function of time for 3 hours.

Section 1.4

32. Determine which search path MATLAB uses on your computer. If you use a lab computer as well as a home computer, compare the two search paths. Where will MATLAB look for a user-created M-file on each computer?

33. A fence around a field is shaped as shown in Figure P33. It consists of a rectangle of length L and width W and a right triangle that is symmetric about the central horizontal axis of the rectangle. Suppose the width W is known (in meters) and the enclosed area A is known (in square meters). Write a MATLAB script file in terms of the given variables W and A to determine the length L required so that the enclosed area is A. Also determine the total length of fence required. Test your script for the values $W = 6$ m and $A = 80$ m^2.

Figure P33

34. The four-sided figure shown in Figure P34 consists of two triangles having a common side a. The law of cosines for the top triangle states that

$$a^2 = b_1^2 + c_1^2 - 2b_1c_1\cos A_1$$

and a similar equation can be written for the bottom triangle. Develop a procedure for computing the length of side c_2 if you are given the lengths of sides b_1, b_2, and c_1 and the angles A_1 and A_2 in degrees. Write a script file to implement this procedure. Test your script, using the following values: $b_1 = 180$ m, $b_2 = 165$ m, $c_1 = 115$ m, $A_1 = 120°$, and $A_2 = 100°$.

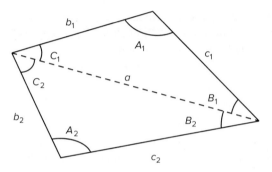

Figure P34

35. Write a script file to compute the three roots of the cubic equation

$$x^3 + ax^2 + bx + c = 0$$

Use the `input` function to let the user enter values for a, b, and c.

Section 1.5

36. Use the MATLAB Help facilities to find information about the following topics and symbols: plot, label, cos, cosine, :, and *.

37. Use the MATLAB Help facilities to determine what happens if you use the `sqrt` function with a negative argument.

38. Use the MATLAB Help facilities to determine what happens if you use the `exp` function with an imaginary argument.

Section 1.6

39. *a.* With what initial speed must you throw a ball vertically for it to reach a height of 20 ft? The ball weighs 1 lb. How does your answer change if the ball weighs 2 lb?

b. Suppose you want to throw a steel bar vertically to a height of 20 ft. The bar weighs 2 lb. How much initial speed must the bar have to reach this height? Discuss how the length of the bar affects your answer.

40. Consider the motion of the piston discussed in Example 1.6–1. The piston *stroke* is the total distance moved by the piston as the crank angle varies from 0° to 180°.

a. How does the piston stroke depend on L_1 and L_2?

b. Suppose $L_2 = 0.5$ ft. Use MATLAB to plot the piston motion versus crank angle for two cases: $L_1 = 0.6$ ft and $L_1 = 1.4$ ft. Compare each plot with the plot shown in Figure 1.6–3. Discuss how the shape of the plot depends on the value of L_1.

Engineering in the 21st Century. . .

Innovative Construction

We tend to remember the great civilizations of the past in part by their public works, such as the Egyptian pyramids and the medieval cathedrals of Europe, which were technically challenging to create. Perhaps it is in our nature to "push the limits," and we admire others who do so. The challenge of innovative construction continues today. As space in our cities becomes scarce, many urban planners prefer to build vertically rather than horizontally. The newest tall buildings push the limits of our abilities, not only in structural design but also in areas that we might not think of, such as elevator design and operation, aerodynamics, and construction techniques. The photo above shows the 1149-ft-high Las Vegas Stratosphere Tower, the tallest observation tower in the United States. It required many innovative techniques in its assembly.

Designers of buildings, bridges, and other structures will use new technologies and new materials, some based on nature's designs. Pound for pound, spider silk is stronger than steel, and structural engineers hope to use cables of synthetic spider silk fibers to build earthquake-resistant suspension bridges. *Smart* structures, which can detect impending failure from cracks and fatigue, are now close to reality, as are *active* structures that incorporate powered devices to counteract wind and other forces. Various MATLAB toolboxes are useful for such projects. These include the toolboxes in the following families: Partial Differential Equations (for structural design), Signal Processing (for smart structures), Control Systems (for active structures), and Computational Finance (for cost analysis of large projects). ■

CHAPTER 2

Numeric, Cell, and Structure Arrays

One of the strengths of MATLAB is the capability to handle collections of items, called *arrays*, as if they were a single entity. The array-handling feature means that MATLAB programs can be very short.

The array is the basic building block in MATLAB. The following classes of arrays are available in MATLAB 7:

Array
numeric | character | logical | cell | structure | function handle | Java

So far we have used only numeric arrays, which are arrays containing only numeric values. Within the numeric class are the subclasses *single* (single precision), *double* (double precision), *int8, int16,* and *int32* (signed 8-bit, 16-bit, and 32-bit integers), and *uint8, uint16,* and *uint32* (unsigned 8-bit, 16-bit, and 32-bit integers). A character array is an array containing strings. The elements of logical arrays are "true" or "false," which, although represented by the symbols 1 and 0, are not numeric quantities. We will study the logical arrays in Chapter 4. Cell arrays and structure arrays are covered in Sections 2.6 and 2.7. Function handles are treated in Chapter 3. The Java class is not covered in this text.

The first four sections of this chapter treat concepts that are essential to understanding MATLAB and therefore must be covered. Section 2.5 treats polynomial applications. Sections 2.6 and 2.7 introduce two types of arrays that are useful for some specialized applications.

2.1 One- and Two-Dimensional Numeric Arrays

We can represent the location of a point in three-dimensional space by three Cartesian coordinates x, y, and z. These three coordinates specify a *vector* **p**. (In mathematical text we often use boldface type to indicate vectors.) The set of *unit vectors* **i, j, k**, whose lengths are 1 and whose directions coincide with the x, y, and z axes, respectively, can be used to express the vector mathematically as follows: $\mathbf{p} = x\mathbf{i} + y\mathbf{j} + z\mathbf{k}$. The unit vectors enable us to associate the vector components x, y, z with the proper coordinate axes; therefore, when we write $\mathbf{p} = 5\mathbf{i} + 7\mathbf{j} + 2\mathbf{k}$, we know that the x, y, and z coordinates of the vector are 5, 7, and 2, respectively. We can also write the components in a specific order, separate them with a space, and identify the group with brackets, as follows: [5 7 2]. As long as we agree that the vector components will be written in the order x, y, z, we can use this notation instead of the unit-vector notation. In fact, MATLAB uses this style for vector notation. MATLAB allows us to separate the components with commas for improved readability if we desire so that the equivalent way of writing the preceding vector is [5, 7, 2]. This expression is a *row vector*, which is a horizontal arrangement of the elements.

ROW VECTOR

COLUMN VECTOR

We can also express the vector as a *column vector*, which has a vertical arrangement. A vector can have only one column, or only one row. Thus, a vector is a one-dimensional array. In general, arrays can have more than one column and more than one row.

Creating Vectors in MATLAB

The concept of a vector can be generalized to any number of components. In MATLAB a vector is simply a list of scalars, whose order of appearance in the list might be significant, as it is when specifying *xyz* coordinates. As another example, suppose we measure the temperature of an object once every hour. We can represent the measurements as a vector, and the 10th element in the list is the temperature measured at the 10th hour.

To create a row vector in MATLAB, you simply type the elements inside a pair of square brackets, separating the elements with a space or a comma. Brackets are required for arrays unless you use the colon operator to create the array. In this case you should not use brackets, but you can optionally use parentheses. The choice between a space or comma is a matter of personal preference, although the chance of an error is less if you use a comma. (You can also use a comma followed by a space for maximum readability.)

To create a column vector, you can separate the elements by semicolons; alternatively, you can create a row vector and then use the *transpose* notation (′), which converts a row vector into a column vector, or vice versa. For example:

TRANSPOSE

```
>>g = [3; 7; 9]
g =
    3
    7
    9
9>>g = [3, 7, 9]'
g =
    3
    7
    9
```

The third way to create a column vector is to type a left bracket ([) and the first element, press **Enter**, type the second element, press **Enter**, and so on until you type the last element followed by a right bracket (]) and **Enter**. On the screen this sequence looks like

```
>>g = [3
7
9]
g =
    3
    7
    9
```

Note that MATLAB displays row vectors horizontally and column vectors vertically.

You can create vectors by "appending" one vector to another. For example, to create the row vector u whose first three columns contain the values of r = [2, 4, 20] and whose fourth, fifth, and sixth columns contain the values of w = [9, -6,3], you type u = [r, w]. The result is the vector u = [2, 4, 20, 9, -6, 3].

The colon operator (:) easily generates a large vector of regularly spaced elements. Typing

```
>>x = m:q:n
```

creates a vector x of values with a spacing q. The first value is m. The last value is n if m - n is an integer multiple of q. If not, the last value is less than n. For example, typing x = 0:2:8 creates the vector x = [0, 2, 4, 6, 8], whereas typing x = 0:2:7 creates the vector x = [0, 2, 4, 6]. To create a row vector z consisting of the values from 5 to 8 in steps of 0.1, you type z = 5:0.1:8. If the increment q is omitted, it is presumed to be 1. Thus y = -3:2 produces the vector y = [-3, -2, -1, 0, 1, 2].

The increment q can be negative. In this case m should be greater than n. For example, u = 10:-2:4 produces the vector [10, 8, 6, 4].

The linspace command also creates a linearly spaced row vector, but instead you specify the number of values rather than the increment. The syntax is linspace(x1,x2,n), where x1 and x2 are the lower and upper limits and n is the number of points. For example, linspace(5, 8, 31) is equivalent to 5:0.1:8. If n is omitted, the spacing is 1.

The logspace command creates an array of logarithmically spaced elements. Its syntax is logspace(a, b, n), where n is the number of points between 10^a and 10^b. For example, x = logspace(-1, 1, 4) produces the vector x = [0.1000, 0.4642, 2.1544, 10.000]. If n is omitted, the number of points defaults to 50.

Two-Dimensional Arrays

MATRIX

An array having rows and columns is a two-dimensional array that is sometimes called a *matrix*. In mathematical text, if possible, vectors are usually denoted by boldface lowercase letters and matrices by boldface uppercase letters. An example of a matrix having three rows and two columns is

$$\mathbf{M} = \begin{bmatrix} 2 & 5 \\ -3 & 4 \\ -7 & 1 \end{bmatrix}$$

We refer to the *size* of an array by the number of rows and the number of columns. For example, an array with 3 rows and 2 columns is said to be a 3 × 2 array. *The number of rows is always stated first!* We sometimes represent a matrix $\mathbf{A}$ as $[a_{ij}]$ to indicate its elements a_{ij}. The subscripts i and j, called *indices*, indicate the row and column location of the element a_{ij}. *The row number must always come first!* For example, the element a_{32} is in row 3, column 2. Two matrices $\mathbf{A}$ and $\mathbf{B}$ are equal if they have the same size and if all their corresponding elements are equal, that is, $a_{ij} = b_{ij}$ for every value of i and j.

Creating Matrices

The most direct way to create a matrix is to type the matrix row by row, separating the elements in a given row with spaces or commas and separating the rows with semicolons. Brackets are required. For example, typing

```
>>A = [2,4,10;16,3,7];
```

creates the following matrix:

$$\mathbf{A} = \begin{bmatrix} 2 & 4 & 10 \\ 16 & 3 & 7 \end{bmatrix}$$

If the matrix has many elements, you can press **Enter** and continue typing on the next line. MATLAB knows you are finished entering the matrix when you type the closing bracket (]).

You can append a row vector to another row vector to create either a third row vector or a matrix (if both vectors have the same number of columns). Note the difference between the results given by [a, b] and [a; b] in the following session:

```
>>a = [1,3,5];
>>b = [7,9,11];
>>c = [a,b]
c =
    1 3 5 7 9 11
>> D = [a;b]
D =
    1 3 5
    7 9 11
```

Matrices and the Transpose Operation

The transpose operation interchanges the rows and columns. In mathematics text we denote this operation by the superscript T. For an $m \times n$ matrix $\mathbf{A}$ with m rows and n columns, $\mathbf{A}^T$ (read "A transpose") is an $n \times m$ matrix.

$$\mathbf{A} = \begin{bmatrix} -2 & 6 \\ -3 & 5 \end{bmatrix} \qquad \mathbf{A}^T = \begin{bmatrix} -2 & -3 \\ 6 & 5 \end{bmatrix}$$

If $\mathbf{A}^T = \mathbf{A}$, the matrix $\mathbf{A}$ is *symmetric*. Note that the transpose operation converts a row vector into a column vector, and vice versa.

If the array contains complex elements, the transpose operator ($'$) produces the *complex conjugate transpose*; that is, the resulting elements are the complex conjugates of the original array's transposed elements. Alternatively, you can use the *dot transpose* operator ($.'$) to transpose the array without producing complex conjugate elements, for example, A.'. If all the elements are real, the operators $'$ and $.'$ give the same result.

Array Addressing

Array indices are the row and column numbers of an element in an array and are used to keep track of the array's elements. For example, the notation v(5) refers to the fifth element in the vector v, and A(2, 3) refers to the element in row 2, column 3 in the matrix A. *The row number is always listed first!* This notation enables you to correct entries in an array without retyping the entire array. For example, to change the element in row 1, column 3 of a matrix **D** to 6, you can type D(1, 3) = 6.

The colon operator selects individual elements, rows, columns, or "subarrays" of arrays. Here are some examples:

- v(:) represents all the row or column elements of the vector v.
- v(2:5) represents the second through fifth elements; that is v(2), v(3), v(4), v(5).

■ A(:, 3) denotes all the elements in the third *column* of the matrix A.

■ A(3,:) denotes all the elements in the third *row* of A.

■ A(:, 2:5) denotes all the elements in the second through fifth columns of A.

■ A(2:3, 1:3) denotes all the elements in the second and third rows that are also in the first through third columns.

■ v = A(:) creates a vector v consisting of all the columns of A stacked from first to last.

■ A(end, :) denotes the last row in A, and A(:, end) denotes the last column.

You can use array indices to extract a smaller array from another array. For example, if you create the array **B**

$$\mathbf{B} = \begin{bmatrix} 2 & 4 & 10 & 13 \\ 16 & 3 & 7 & 18 \\ 8 & 4 & 9 & 25 \\ 3 & 12 & 15 & 17 \end{bmatrix} \tag{2.1-1}$$

by typing

```
>>B = [2, 4, 10, 13;16, 3, 7, 18;8, 4, 9, 25;3, 12, 15, 17];
```

and then type

```
>>C = B(2:3, 1:3);
```

you can produce the following array:

$$\mathbf{C} = \begin{bmatrix} 16 & 3 & 7 \\ 8 & 4 & 9 \end{bmatrix}$$

EMPTY ARRAY

The *empty array* contains no elements and is expressed as []. Rows and columns can be deleted by setting the selected row or column equal to the empty array. This step causes the original matrix to collapse to a smaller one. For example, A(3, :) = [] deletes the third row in A, while A(:, 2:4) = [] deletes the second through fourth columns in A. Finally, A([1 4],:) = [] deletes the first and fourth rows of A.

Suppose we type A = [6, 9, 4; 1, 5, 7] to define the following matrix:

$$\mathbf{A} = \begin{bmatrix} 6 & 9 & 4 \\ 1 & 5 & 7 \end{bmatrix}$$

Typing A(1, 5) = 3 changes the matrix to

$$\mathbf{A} = \begin{bmatrix} 6 & 9 & 4 & 0 & 3 \\ 1 & 5 & 7 & 0 & 0 \end{bmatrix}$$

Because **A** did not have five columns, its size is automatically expanded to accept the new element in column 5. MATLAB adds zeros to fill out the remaining elements.

MATLAB does not accept negative or zero indices, but you can use negative increments with the colon operator. For example, typing B = A(:, 5:-1:1) reverses the order of the columns in **A** and produces

$$\mathbf{B} = \begin{bmatrix} 3 & 0 & 4 & 9 & 6 \\ 0 & 0 & 7 & 5 & 1 \end{bmatrix}$$

Suppose that C = [-4, 12, 3, 5, 8]. Then typing B(2, :) = C replaces row 2 of B with C. Thus **B** becomes

$$\mathbf{B} = \begin{bmatrix} 3 & 0 & 4 & 9 & 6 \\ -4 & 12 & 3 & 5 & 8 \end{bmatrix}$$

Suppose that D = [3, 8, 5; 4, -6, 9]. Then typing E = D([2, 2, 2], :) repeats row 2 of D three times to obtain

$$\mathbf{E} = \begin{bmatrix} 4 & -6 & 9 \\ 4 & -6 & 9 \\ 4 & -6 & 9 \end{bmatrix}$$

Using **clear** to Avoid Errors

You can use the clear command to protect yourself from accidentally reusing an array that has the wrong dimension. Even if you set new values for an array, some previous values might still remain. For example, suppose you had previously created the 2×2 array A = [2, 5; 6, 9], and then you create the 5×1 arrays x = (1:5)' and y = (2:6)'. Note that parentheses are needed here to use the transpose operator. Suppose you now redefine A so that its columns will be x and y. If you then type A(:, 1) = x to create the first column, MATLAB displays an error message telling you that the number of rows in A and x must be the same. MATLAB thinks A should be a 2×2 matrix because A was previously defined to have only two rows and its values remain in memory. The clear command wipes A and all other variables from memory and avoids this error. To clear A only, type clear A before typing A(:,1) = x.

Some Useful Array Functions

MATLAB has many functions for working with arrays (see Table 2.1–1). Here is a summary of some of the more commonly used functions.

The max(A) function returns the algebraically greatest element in **A** if **A** is a vector having all real elements. It returns a row vector containing the greatest elements in each *column* if **A** is a matrix containing all real elements. If *any* of the elements are complex, max(A) returns the element that has the largest magnitude. The syntax [x, k] = max(A) is similar to max(A), but it stores the maximum values in the row vector **x** and their indices in the row vector **k**.

If A and B have the same size, C = max(A, B) creates an array the same size, having the maximum value from each corresponding location in A and B. For example, the following **A** and **B** matrices give the **C** matrix shown.

Table 2.1–1 Basic syntax of array functions*

Command	Description
find(x)	Computes an array containing the indices of the nonzero elements of the array **x**.
[u,v,w] = find(A)	Computes the arrays **u** and **v**, containing the row and column indices of the nonzero elements of the matrix **A**, and the array **w**, containing the values of the nonzero elements. The array **w** may be omitted.
length(A)	Computes either the number of elements of **A** if **A** is a vector or the largest value of m or n if **A** is an $m \times n$ matrix.
linspace(a,b,n)	Creates a row vector of n regularly spaced values between a and b.
logspace(a,b,n)	Creates a row vector of n logarithmically spaced values between a and b.
max(A)	Returns the algebraically largest element in **A** if **A** is a vector. Returns a row vector containing the largest elements in each column if **A** is a matrix. If any of the elements are complex, max(A) returns the elements that have the largest magnitudes.
[x,k] = max(A)	Similar to max(A) but stores the maximum values in the row vector **x** and their indices in the row vector **k**.
min(A)	Same as max(A) but returns minimum values.
[x,k] = min(A)	Same as [x,k] = max(A) but returns minimum values.
norm(x)	Computes a vector's geometric length $\sqrt{x_1^2 + x_2^2 + \cdots + x_n^2}$.
numel(A)	Returns the total number of elements in the array A.
size(A)	Returns a row vector [m n] containing the sizes of the $m \times n$ array **A**.
sort(A)	Sorts each column of the array **A** in ascending order and returns an array the same size as **A**.
sum (A)	Sums the elements in each column of the array **A** and returns a row vector containing the sums.

*Many of these functions have extended syntax. See the text and MATLAB help for more discussion.

$$\mathbf{A} = \begin{bmatrix} 1 & 6 & 4 \\ 3 & 7 & 2 \end{bmatrix} \quad \mathbf{B} = \begin{bmatrix} 3 & 4 & 7 \\ 1 & 5 & 8 \end{bmatrix} \quad \mathbf{C} = \begin{bmatrix} 3 & 6 & 7 \\ 3 & 7 & 8 \end{bmatrix}$$

The functions min(A) and [x, k] = min(A) are the same as max(A) and [x, k] = max(A) except that they return minimum values.

The function size(A) returns a row vector [m n] containing the sizes of the $m \times n$ array **A**. The length(A) function computes either the number of elements of **A** if **A** is a vector or the largest value of m or n if **A** is an $m \times n$ matrix. The function numel(A) returns the total number of elements in the array **A**.

For example, if

$$\mathbf{A} = \begin{bmatrix} 6 & 2 \\ -10 & -5 \\ 3 & 0 \end{bmatrix}$$

then max(A) returns the vector [6, 2]; min(A) returns the vector [-10, -5]; size(A) returns [3, 2]; numel(A) returns 6, and length(A) returns 3.

The sum(A) function sums the elements in each *column* of the array **A** and returns a row vector containing the sums. The sort(A) function sorts each *column* of the array **A** in ascending order and returns an array the same size as **A**.

If A has one or more complex elements, the max, min, and sort functions act on the absolute values of the elements and return the element that has the largest magnitude.

For example, if

$$A = \begin{bmatrix} 6 & 2 \\ -10 & -5 \\ 3+4i & 0 \end{bmatrix}$$

then max(A) returns the vector [-10, -5] and min(A) returns the vector [3 + 4i, 0]. (The magnitude of $3 + 4i$ is 5.)

The sort will be done in descending order if the form sort(A, 'descend') is used. The min, max, and sort functions can be made to act on the rows instead of the columns by transposing the array.

The complete syntax of the sort function is sort(A, dim, mode), where dim selects a dimension along which to sort and mode selects the direction of the sort, 'ascend' for ascending order and 'descend' for descending order. So, for example, sort(A, 2, 'descend') would sort the elements in each row of A in descending order.

The find(x) command computes an array containing the indices of the *nonzero* elements of the vector **x**. The syntax [u, v, w] = find(A) computes the arrays **u** and **v**, containing the row and column indices of the nonzero elements of the matrix **A**, and the array **w**, containing the values of the nonzero elements. The array **w** may be omitted.

For example, if

$$A = \begin{bmatrix} 6 & 0 & 3 \\ 0 & 4 & 0 \\ 2 & 7 & 0 \end{bmatrix}$$

then the session

```
>>A = [6, 0, 3; 0, 4, 0; 2, 7, 0];
>>[u, v, w] = find(A)
```

returns the vectors

$$\mathbf{u} = \begin{bmatrix} 1 \\ 3 \\ 2 \\ 3 \\ 1 \end{bmatrix} \qquad \mathbf{v} = \begin{bmatrix} 1 \\ 1 \\ 2 \\ 2 \\ 3 \end{bmatrix} \qquad \mathbf{w} = \begin{bmatrix} 6 \\ 2 \\ 4 \\ 7 \\ 3 \end{bmatrix}$$

The vectors **u** and **v** give the (row, column) indices of the nonzero values, which are listed in **w**. For example, the second entries in **u** and **v** give the indices (3, 1), which specifies the element in row 3, column 1 of **A**, whose value is 2.

These functions are summarized in Table 2.1–1.

Magnitude, Length, and Absolute Value of a Vector

The terms *magnitude, length,* and *absolute value* are often loosely used in everyday language, but you must keep their precise meaning in mind when using

MATLAB. The MATLAB `length` command gives the number of elements in the vector. The *magnitude* of a vector **x** having real elements $x_1, x_2, \ldots, x_n$ is a scalar, given by $\sqrt{x_1^2 + x_2^2 + \cdots + x_n^2}$, and is the same as the vector's geometric length. The *absolute value* of a vector **x** is a vector whose elements are the absolute values of the elements of **x**. For example, if x = [2,-4,5], its length is 3; its magnitude is $\sqrt{2^2 + (-4)^2 + 5^2} = 6.7082$; and its absolute value is [2,4,5]. The length, magnitude, and absolute value of x are computed by `lenqth(x)`, `norm(x)`, and `abs(x)`, respectively.

Test Your Understanding

T2.1–1   For the matrix **B**, find the array that results from the operation [B; B']. Use MATLAB to determine what number is in row 5, column 3 of the result.

$$\mathbf{B} = \begin{bmatrix} 2 & 4 & 10 & 13 \\ 16 & 3 & 7 & 18 \\ 8 & 4 & 9 & 25 \\ 3 & 12 & 15 & 17 \end{bmatrix}$$

T2.1–2   For the same matrix **B**, use MATLAB to (a) find the largest and smallest elements in **B** and their indices and (b) sort each column in **B** to create a new matrix **C**.

The Variable Editor

The MATLAB Workspace Browser provides a graphical interface for managing the workspace. You can use it to view, save, and clear workspace variables. It includes the *Variable Editor*, a graphical interface for working with variables, including arrays. To open the Workspace Browser, type `workspace` at the Command window prompt.

Keep in mind that the Desktop menus are context-sensitive. Thus their contents will change depending on which features of the browser and Variable Editor you are currently using. The Workspace Browser shows the name of each variable, its value, *array size*, and class. The icon for each variable illustrates its class.

ARRAY SIZE

From the Workspace Browser you can open the Variable Editor to view and edit a visual representation of two-dimensional numeric arrays, with the rows and columns numbered. To open the Variable Editor from the Workspace Browser, double-click on the variable you want to open. The Variable Editor opens, displaying the values for the selected variable. The Variable Editor and the Variables tab appear as shown in Figure 2.1–1. This tab enables you to insert, delete, transpose, and sort rows and columns.

Repeat the steps to open additional variables into the Variable Editor. In the Variable Editor, access each variable via its tab at the top of the window. You can also open the Variable Editor directly from the Command window by typing `openvar('var')`, where `var` is the name of the variable to be edited. Once an

Figure 2.1–1 The Variable Editor.

array is displayed in the Variable Editor, you can change a value in the array by clicking on its location, typing in the new value, and pressing **Enter**.

You can clear a variable from the Workspace Browser by right-clicking on it in the Browser, then selecting **Delete** in the pop-up menu.

2.2 Multidimensional Numeric Arrays

MATLAB supports multidimensional arrays. For more information, type `help datatypes`.

A three-dimensional array has the dimension $m \times n \times q$. A four-dimensional array has the dimension $m \times n \times q \times r$, and so forth. The first two dimensions are the row and column, as with a matrix. The higher dimensions are called *pages*. You can think of a three-dimensional array as layers of matrices. The first layer is page 1; the second layer is page 2, and so on. If A is a $3 \times 3 \times 2$ array, you can access the element in row 3, column 2 of page 2 by typing `A(3, 2, 2)`. To access all of page 1, type `A(:, :, 1)`. To access all of page 2, type `A(:, :, 2)`. The `ndims` command returns the number of dimensions. For example, for the array A just described, `ndims(A)` returns the value 3.

You can create a multidimensional array by first creating a two-dimensional array and then extending it. For example, suppose you want to create a three-dimensional array whose first two pages are

$$\begin{bmatrix} 4 & 6 & 1 \\ 5 & 8 & 0 \\ 3 & 9 & 2 \end{bmatrix} \quad \begin{bmatrix} 6 & 2 & 9 \\ 0 & 3 & 1 \\ 4 & 7 & 5 \end{bmatrix}$$

To do so, first create page 1 as a 3×3 matrix and then add page 2, as follows:

```
>>A = [4, 6, 1; 5, 8, 0; 3, 9, 2];
>>A (:, :, 2) = [6, 2, 9; 0, 3, 1; 4, 7, 5];
```

Another way to produce such an array is with the `cat` command. Typing `cat(n, A, B, C,...)` creates a new array by concatenating the arrays `A`, `B`, `C`, and so on along the dimension `n`. Note that `cat(1, A, B)` is the same as `[A; B]` and that `cat(2, A, B)` is the same as `[A, B]`. For example, suppose we have the 2 × 2 arrays **A** and **B**:

$$\mathbf{A} = \begin{bmatrix} 8 & 2 \\ 9 & 5 \end{bmatrix} \quad \mathbf{B} = \begin{bmatrix} 4 & 6 \\ 7 & 3 \end{bmatrix}$$

Then `C = cat(3, A, B)` produces a three-dimensional array composed of two layers; the first layer is the matrix **A**, and the second layer is the matrix **B**. The element `C(m, n, p)` is located in row *m*, column *n*, and layer *p*. Thus the element `C(2, 1, 1)` is 9, and the element `C(2, 2, 2)` is 3.

Multidimensional arrays are useful for problems that involve several parameters. For example, if we have data on the temperature distribution in a rectangular object, we could represent the temperatures as an array `T` with three dimensions.

2.3 Element-by-Element Operations

To increase the magnitude of a vector, multiply it by a scalar. For example, to double the magnitude of the vector `r = [3, 5, 2]`, multiply each component by 2 to obtain `[6, 10, 4]`. In MATLAB you type `v = 2*r`.

Multiplying a matrix **A** by a scalar *w* produces a matrix whose elements are the elements of **A** multiplied by *w*. For example:

$$3\begin{bmatrix} 2 & 9 \\ 5 & -7 \end{bmatrix} = \begin{bmatrix} 6 & 27 \\ 15 & -21 \end{bmatrix}$$

This multiplication is performed in MATLAB as follows:

```
>>A = [2, 9; 5, -7];
>>3*A
```

ARRAY OPERATIONS

MATRIX OPERATIONS

Thus multiplication of an array by a scalar is easily defined and easily carried out. However, multiplication of two *arrays* is not so straightforward. In fact, MATLAB uses two definitions of multiplication: (1) array multiplication and (2) matrix multiplication. Division and exponentiation must also be carefully defined when you are dealing with operations between two arrays. MATLAB has two forms of arithmetic operations on arrays. In this section we introduce one form, called *array operations*, which are also called *element-by-element* operations. In the next section we introduce *matrix operations*. Each form has its own applications, which we illustrate by examples.

Array Addition and Subtraction

Array addition can be done by adding the corresponding components. To add the arrays `r = [3, 5, 2]` and `v = [2, -3, 1]` to create `w` in MATLAB, you type `w = r + v`. The result is `w = [5, 2, 3]`.

When two arrays have identical size, their sum or difference has the same size and is obtained by adding or subtracting their corresponding elements. Thus $\mathbf{C} = \mathbf{A} + \mathbf{B}$ implies that $c_{ij} = a_{ij} + b_{ij}$ if the arrays are matrices. The array $\mathbf{C}$ has the same size as $\mathbf{A}$ and $\mathbf{B}$. For example,

$$\begin{bmatrix} 6 & -2 \\ 10 & 3 \end{bmatrix} + \begin{bmatrix} 9 & 8 \\ -12 & 14 \end{bmatrix} = \begin{bmatrix} 15 & 6 \\ -2 & 17 \end{bmatrix} \tag{2.3-1}$$

Array subtraction is performed in a similar way.

The addition shown in Equation (2.3–1) is performed in MATLAB as follows:

```
>>A = [6, -2; 10, 3];
>>B = [9, 8; -12, 14]
>>A+B
ans =
     15  6
     -2 17
```

Array addition and subtraction are associative and commutative. For addition these properties mean that

$$(\mathbf{A} + \mathbf{B}) + \mathbf{C} = \mathbf{A} + (\mathbf{B} + \mathbf{C}) \tag{2.3-2}$$

$$\mathbf{A} + \mathbf{B} + \mathbf{C} = \mathbf{B} + \mathbf{C} + \mathbf{A} = \mathbf{A} + \mathbf{C} + \mathbf{B} \tag{2.3-3}$$

Array addition and subtraction require that both arrays be the same size. The only exception to this rule in MATLAB occurs when we add or subtract a *scalar* to or from an array. In this case the scalar is added or subtracted from each element in the array. Table 2.3–1 gives examples.

Element-by-Element Multiplication

MATLAB defines element-by-element multiplication only for arrays that are the same size. The definition of the product x.*y, where x and y each have *n* elements, is

```
x.*y = [x(1)y(1), x(2)y(2)..., x(n)y(n)]
```

Table 2.3–1 Element-by-element operations

Symbol	Operation	Form	Example
+	Scalar-array addition	A + b	[6,3] + 2 = [8,5]
−	Scalar-array subtraction	A - b	[8,3] - 5 = [3,-2]
+	Array addition	A + B	[6,5] + [4,8] = [10,13]
−	Array subtraction	A - B	[6,5] - [4,8] = [2,-3]
.*	Array multiplication	A.*B	[3,5].*[4,8] = [12,40]
./	Array right division	A./B	[2,5]./[4,8] = [2/4,5/8]
.\	Array left division	A.\B	[2,5].\[4,8] = [2\4,5\8]
.^	Array exponentiation	A.^B	[3,5].^2 = [3^2,5^2]
			2.^[3,5] = [2^3,2^5]
			[3,5].^[2,4] = [3^2,5^4]

if x and y are row vectors. For example, if

$$x = [2, 4, -5] \qquad y = [-7, 3, -8] \qquad (2.3\text{--}4)$$

then z = x.*y gives

$$z = [2(-7), 4(3), -5(-8)] = [-14, 12, 40]$$

This type of multiplication is sometimes called *array* multiplication.

If u and v are column vectors, the result of u.*v is a column vector.

Note that x' is a column vector with size 3×1 and thus does not have the same size as y, whose size is 1×3. Thus for the vectors x and y the operations x'.*y and y.*x' are not defined in MATLAB and will generate an error message. With element-by-element multiplication, it is important to remember that the dot (.) and the asterisk (*) form *one* symbol (.*). It might have been better to have defined a single symbol for this operation, but the developers of MATLAB were limited by the selection of symbols on the keyboard.

The generalization of array multiplication to arrays with more than one row or column is straightforward. Both arrays must be the same size. The array operations are performed between the elements in corresponding locations in the arrays. For example, the array multiplication operation A.*B results in a matrix C that has the same size as A and B and has the elements $c_{ij} = a_{ij}b_{ij}$. For example, if

$$A = \begin{bmatrix} 11 & 5 \\ -9 & 4 \end{bmatrix} \qquad B = \begin{bmatrix} -7 & 8 \\ 6 & 2 \end{bmatrix}$$

then C = A.*B gives this result:

$$C = \begin{bmatrix} 11(-7) & 5(8) \\ -9(6) & 4(2) \end{bmatrix} = \begin{bmatrix} -77 & 40 \\ -54 & 8 \end{bmatrix}$$

Test Your Understanding

T2.3–1 Given the vectors

$$x = [6 \quad 5 \quad 10] \qquad\qquad y = [3 \quad 9 \quad 8]$$

do the following by hand, then check your answer using MATLAB.

a. Find the sum of **x** and **y**.

b. Find the array product w = x.*y.

c. Find the array product z = y.*x. Is z = w?

(Answers: a. [9, 14, 18]   b. [18, 45, 80]   c. The same.)

T2.3–2 Given the matrices

$$A = \begin{bmatrix} 6 & 4 \\ 5 & 3 \end{bmatrix} \qquad\qquad B = \begin{bmatrix} 5 & 2 \\ 7 & 9 \end{bmatrix}$$

do the following by hand, then check your answer using MATLAB.

a. Find the sum of **A** and **B**.

b. Find the array product w = A.*B.

c. Find the array product z = B.*A. Is z = w?

(Answers: a. [11, 6; 12, 12]   b. [30, 8; 35, 27]   c. yes.)

Vectors and Displacement

Suppose two divers start at the surface and establish the following coordinate system: x is to the west, y is to the north, and z is down. Diver 1 swims 55 ft west, 36 ft north, and then dives 25 ft. Diver 2 dives 15 ft, swims east 20 ft and then north 59 ft. (a) Find the distance between diver 1 and the starting point. (b) How far in each direction must diver 1 swim to reach diver 2? How far in a straight line must diver 1 swim to reach diver 2?

■ **Solution**

(a) Using the xyz coordinates selected, the position of diver 1 is $\mathbf{r} = 55\mathbf{i} + 36\mathbf{j} + 25\mathbf{k}$, and the position of diver 2 is $\mathbf{r} = -20\mathbf{i} + 59\mathbf{j} + 15\mathbf{k}$. (Note that diver 2 swam east, which is in the negative x direction.) The distance from the origin of a point xyz is given by $\sqrt{x^2 + y^2 + z^2}$, that is, by the magnitude of the vector pointing from the origin to the point xyz. This distance is computed in the following session.

```
>>r = [55, 36, 25]; w = [-20, 59, 15];
>>dist1 = sqrt (sum(r.*r))
dist1 =
    70.3278
```

The distance is approximately 70 ft. The distance could also have been computed from `norm(r)`.

(b) The location of diver 2 relative to diver 1 is given by the vector **v** pointing from diver 1 to diver 2. We can find this vector using vector subtraction: $\mathbf{v} = \mathbf{w} - \mathbf{r}$. Continue the above MATLAB session as follows:

```
>>v = w-r
v =
    -75    23    -10
>>dist2 = sqrt(sum(v.*v))
dist2 =
    79.0822
```

Thus to reach diver 2 by swimming along the coordinate directions, diver 1 must swim 75 ft east, 23 ft north, and 10 ft up. The straight-line distance between them is approximately 79 ft.

Vectorized Functions

The built-in MATLAB functions such as `sqrt(x)` and `exp(x)` automatically operate on array arguments to produce an array result the same size as the array argument `x`. Thus these functions are said to be *vectorized* functions.

Thus, when multiplying or dividing these functions, or when raising them to a power, you must use element-by-element operations if the arguments are arrays. For example, to compute $z = (e^y \sin x)\cos^2 x$, you must type `z = exp(y).*sin(x).*(cos(x)).^2`. Obviously, you will get an error message if the size of `x` is not the same as the size of `y`. The result `z` will have the same size as `x` and `y`.

| EXAMPLE 2.3–2 | Aortic Pressure Model |

The following equation is a specific case of one model used to describe the blood pressure in the aorta during systole (the period following the closure of the heart's aortic valve). The variable t represents time in seconds, and the dimensionless variable y represents the pressure difference across the aortic valve, normalized by a constant reference pressure.

$$y(t) = e^{-8t} \sin\left(9.7t + \frac{\pi}{2}\right)$$

Plot this function for $t \geq 0$.

Figure 2.3–1 Aortic pressure response for Example 2.3–2.

■ **Solution**

Note that if t is a vector, the MATLAB functions `exp(-8*t)` and `sin(9.7*t+pi/2)` will also be vectors the same size as t. Thus we must use element-by-element multiplication to compute $y(t)$.

We must decide on the proper spacing to use for the vector t and its upper limit. The sine function $\sin(9.7t + \pi/2)$ oscillates with a frequency of 9.7 rad/sec, which is $9.7/(2\pi) = 1.5$ Hz. Thus its period is $1/1.5 = 2/3$ sec. The spacing of t should be a small fraction of the period in order to generate enough points to plot the curve. Thus we select a spacing of 0.003 to give approximately 200 points per period.

The amplitude of the sine wave decays with time because the sine is multiplied by the decaying exponential e^{-8t}. The exponential's initial value is $e^0 = 1$, and it will be 2 percent of its initial value at $t = 0.5$ (because $e^{-8(0.5)} = 0.02$). Thus we select the upper limit of t to be 0.5. The session is

```
>>t = 0:0.003:0.5;
>>y = exp(-8*t).*sin(9.7*t+pi/2);
>>plot(t,y),xlabel('t (sec)'), ...
    ylabel('Normalized Pressure Difference y(t)')
```

The plot is shown in Figure 2.3–1. Note that we do not see much of an oscillation despite the presence of a sine wave. This is so because the period of the sine wave is greater than the time it takes for the exponential e^{-8t} to become essentially zero.

Element-by-Element Division

The definition of element-by-element division, also called array division, is similar to the definition of array multiplication except, of course, that the elements of one array are divided by the elements of the other array. Both arrays must be the same size. The symbol for array right division is `./`. For example, if

$$\mathbf{x} = [8, \ 12, \ 15] \qquad \mathbf{y} = [-2, \ 6, \ 5]$$

then z = x./y gives

$$\mathbf{z} = [8/(-2), \ 12/6, \ 15/5] = [-4, \ 2, \ 3]$$

Also, if

$$\mathbf{A} = \begin{bmatrix} 24 & 20 \\ -9 & 4 \end{bmatrix} \qquad \mathbf{B} = \begin{bmatrix} -4 & 5 \\ 3 & 2 \end{bmatrix}$$

then C = A./B gives

$$\mathbf{C} = \begin{bmatrix} 24/(-4) & 20/5 \\ -9/3 & 4/2 \end{bmatrix} = \begin{bmatrix} -6 & 4 \\ -3 & 2 \end{bmatrix}$$

The array left division operator (`.\`) is defined to perform element-by-element division using left division. Refer to Table 2.3–1 for examples. Note that A.\B is not equivalent to A./B.

Test Your Understanding

T2.3–3 Given the vectors

$$\mathbf{x} = [6 \quad 5 \quad 10] \qquad \mathbf{y} = [3 \quad 9 \quad 8]$$

do the following by hand, then check your answer using MATLAB.

a. Find the array quotient w = x./y.

b. Find the array quotient z = y./x.

(Answers: a. [2, 0.5556, 1.25]   b. [0.5, 1.8, 0.8])

T2.3–4 Given the matrices

$$\mathbf{A} = \begin{bmatrix} 6 & 4 \\ 5 & 3 \end{bmatrix} \qquad \mathbf{B} = \begin{bmatrix} 5 & 2 \\ 7 & 9 \end{bmatrix}$$

do the following by hand, then check your answer using MATLAB.

a. Find the array quotient C = A./B.

b. Find the array quotient D = B./A.

c. Find the array quotient E = A.\B.

d. Find the array quotient F = B.\A.

e. Are any of C, D, E, or F equal?

(Answers: a. [1.2, 2; 0.7143, 0.3333] b. [0.8333, 0.5; 1.4, 3] c. [0.8333, 0.5; 1.4, 3] d. [1.2, 2; 0.7143, 0.3333] e. C and F are the same; D and E are the same.)

| EXAMPLE 2.3–3 | | Transportation Route Analysis |

The following table gives data for the distance traveled along five truck routes and the corresponding time required to traverse each route. Use the data to compute the average speed required to drive each route. Find the route that has the highest average speed.

	1	2	3	4	5
Distance (mi)	560	440	490	530	370
Time (hr)	10.3	8.2	9.1	10.1	7.5

■ **Solution**

For example, the average speed on the first route is $560/10.3 = 54.4$ mi/hr. First we define the row vectors d and t from the distance and time data. Then, to find the average speed on each route using MATLAB, we use array division. The session is

```
>>d = [560, 440, 490, 530, 370]
>>t = [10.3, 8.2, 9.1, 10.1, 7.5]
>>speed = d./t
speed =
    54.3689    53.6585    53.8462    52.4752    49.3333
```

The results are in miles per hour. Note that MATLAB displays more significant figures than is justified by the three-significant-figure accuracy of the given data, so we should round the results to three significant figures before using them.

To find the highest average speed and the corresponding route, continue the session as follows:

```
>>[highest_speed, route] = max(speed)
highest_speed =
    54.3689
route =
    1
```

The first route has the highest speed.

If we did not need the speeds for every route, we could have solved this problem by combining two lines as follows: `[highest_speed, route] = max(d./t)`.

Element-by-Element Exponentiation

MATLAB enables us not only to raise arrays to powers but also to raise scalars and arrays to *array* powers. To perform exponentiation on an element-by-element basis, we must use the `.^` symbol. For example, if `x = [3, 5, 8]`, then typing `x.^3` produces the array $[3^3, 5^3, 8^3] = [27, 125, 512]$. If `x = 0:2:6`, then typing `x.^2` returns the array $[0^2, 2^2, 4^2, 6^2] = [0, 4, 16, 36]$. If

$$\mathbf{A} = \begin{bmatrix} 4 & -5 \\ 2 & 3 \end{bmatrix}$$

then `B = A.^3` gives this result:

$$\mathbf{B} = \begin{bmatrix} 4^3 & (-5)^3 \\ 2^3 & 3^3 \end{bmatrix} = \begin{bmatrix} 64 & -125 \\ 8 & 27 \end{bmatrix}$$

We can raise a scalar to an array power. For example, if `p = [2, 4, 5]`, then typing `3.^p` produces the array $[3^2, 3^4, 3^5] = [9, 81, 243]$. This example illustrates a common situation in which it helps to remember that `.^` is a *single* symbol; the dot in `3.^p` is not a decimal point associated with the number 3. The following operations, with the value of `p` given here, are equivalent and give the correct answer:

```
3.^p
3.0.^p
3..^p
(3).^p
3.^[2, 4, 5]
```

With array exponentiation, the power may be an array if the base is a scalar or if the power's dimensions are the same as the base dimensions. For example, if `x = [1, 2, 3]` and `y = [2, 3, 4]`, then `y.^x` gives the array `[2 9 6 4]`. If `A = [1, 2; 3, 4]`, then `2.^A` gives the array `[2, 4; 8, 16]`.

T2.3–5 Given the matrices

$$\mathbf{A} = \begin{bmatrix} 21 & 27 \\ -18 & 8 \end{bmatrix} \qquad \mathbf{B} = \begin{bmatrix} -7 & -3 \\ 9 & 4 \end{bmatrix}$$

find (a) their array product, (b) their array right division (**A** divided by **B**), and (c) **B** raised to the third power element by element.

(Answers: (a) [-147, -81; -162, 32], (b) [-3, -9; -2, 2], and (c) [-343, -27; 729, 64].)

EXAMPLE 2.3–4	Current and Power Dissipation in Resistors

The current i passing through an electrical resistor having a voltage v across it is given by Ohm's law, $i = v/R$, where R is the resistance. The power dissipated in the resistor is given by v^2/R. The following table gives data for the resistance and voltage for five resistors. Use the data to compute (a) the current in each resistor and (b) the power dissipated in each resistor.

	1	2	3	4	5
$R(\Omega)$	10^4	2×10^4	3.5×10^4	10^5	2×10^5
$v(V)$	120	80	110	200	350

■ **Solution**

(a) First we define two row vectors, one containing the resistance values and one containing the voltage values. To find the current $i = v/R$ using MATLAB, we use array division. The session is

```
>>R = [10000, 20000, 35000, 100000, 200000];
>>v = [120, 80, 110, 200, 350];
>>current = v./R
current =
    0.0120 0.0040 0.0031 0.0020 0.0018
```

The results are in amperes and should be rounded to three significant figures because the voltage data contains only three significant figures.

(b) To find the power $P = v^2/R$, use array exponentiation and array division. The session continues as follows:

```
>>power = v.^2./R
power =
    1.4400    0.3200    0.3457    0.4000    0.6125
```

These numbers are the power dissipation in each resistor in watts. Note that the statement $\texttt{v.\^{}2./R}$ is equivalent to $\texttt{(v.\^{}2)./R}$. Although the rules of precedence are unambiguous here, we can always put parentheses around quantities if we are unsure how MATLAB will interpret our commands.

A Batch Distillation Process

EXAMPLE 2.3–5

Consider a system for heating a liquid benzene/toluene solution to distill a pure benzene vapor. A particular batch distillation unit is charged initially with 100 mol of a 60 percent mol benzene/40 percent mol toluene mixture. Let L (mol) be the amount of liquid remaining in the still, and let x (mol B/mol) be the benzene mole fraction in the remaining liquid. Conservation of mass for benzene and toluene can be applied to derive the following relation [Felder, 1986].

$$L = 100 \left(\frac{x}{0.6}\right)^{0.625} \left(\frac{1-x}{0.4}\right)^{-1.625}$$

Determine what mole fraction of benzene remains when $L = 70$. Note that it is difficult to solve this equation directly for x. Use a plot of x versus L to solve the problem.

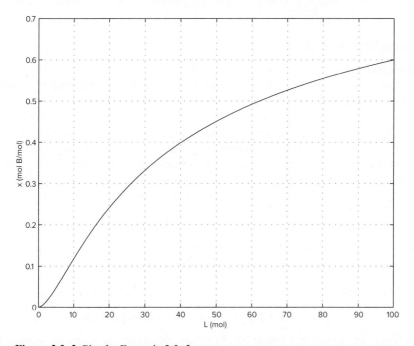

Figure 2.3–2 Plot for Example 2.3–5.

■ Solution

This equation involves both array multiplication and array exponentiation. Note that MATLAB enables us to use decimal exponents to evaluate L. It is clear that L must be in the range $0 \le L \le 100$; however, we do not know the range of x, except that $x \ge 0$. Therefore, we must make a few guesses for the range of x, using a session like the following. We find that $L > 100$ if $x > 0.6$, so we choose $x = 0:0.001:0.6$. We use the `ginput` function to find the value of x corresponding to $L = 70$.

```
>>x = 0:0.001:0.6;
>>L = 100*(x/0.6).^(0.625).*((1-x)/0.4).^(-1.625);
>>plot (L,x),grid,xlabel('L (mol)'),ylabel('x (mol B/mol)'),...
   [L,x] = ginput(1)
```

The plot is shown in Figure 2.3–2. The answer is $x = 0.52$ if $L = 70$. The plot shows that the remaining liquid becomes leaner in benzene as the liquid amount becomes smaller. Just before the still is empty ($L = 0$), the liquid is pure toluene.

2.4 Matrix Operations

Matrix addition and subtraction are identical to element-by-element addition and subtraction. The corresponding matrix elements are summed or subtracted. However, matrix multiplication and division are not the same as element-by-element multiplication and division.

Multiplication of Vectors

Recall that vectors are simply matrices with one row or one column. Thus matrix multiplication and division procedures apply to vectors as well, and we will introduce matrix multiplication by considering the vector case first.

The *vector dot product* $\mathbf{u} \cdot \mathbf{w}$ of the vectors $\mathbf{u}$ and $\mathbf{w}$ is a scalar and can be thought of as the perpendicular projection of $\mathbf{u}$ onto $\mathbf{w}$. It can be computed from $|\mathbf{u}||\mathbf{w}| \cos \theta$, where θ is the angle between the two vectors and $|\mathbf{u}|$, $|\mathbf{w}|$ are the magnitudes of the vectors. Thus if the vectors are parallel and in the same direction, $\theta = 0$ and $\mathbf{u} \cdot \mathbf{w} = |\mathbf{u}||\mathbf{w}|$. If the vectors are perpendicular, $\theta = 90°$ and thus $\mathbf{u} \cdot \mathbf{w} = 0$. Because the unit vectors $\mathbf{i}$, $\mathbf{j}$, and $\mathbf{k}$ have unit length,

$$\mathbf{i} \cdot \mathbf{i} = \mathbf{j} \cdot \mathbf{j} = \mathbf{k} \cdot \mathbf{k} = 1 \qquad (2.4\text{--}1)$$

Because the unit vectors are perpendicular,

$$\mathbf{i} \cdot \mathbf{j} = \mathbf{i} \cdot \mathbf{k} = \mathbf{j} \cdot \mathbf{k} = 0 \qquad (2.4\text{--}2)$$

Thus the vector dot product can be expressed in terms of unit vectors as

$$\mathbf{u} \cdot \mathbf{w} = (u_1 \mathbf{i} + u_2 \mathbf{j} + u_3 \mathbf{k}) \cdot (w_1 \mathbf{i} + w_2 \mathbf{j} + w_3 \mathbf{k})$$

Carrying out the multiplication algebraically and using the properties given by (2.4–1) and (2.4–2), we obtain

$$\mathbf{u} \cdot \mathbf{w} = u_1 w_1 + u_2 w_2 + u_3 w_3$$

The *matrix* product of a *row* vector **u** with a *column* vector **w** is defined in the same way as the vector dot product; the result is a scalar that is the sum of the products of the corresponding vector elements; that is,

$$[u_1 \quad u_2 \quad u_3] \begin{bmatrix} w_1 \\ w_2 \\ w_3 \end{bmatrix} = u_1 w_1 + u_2 w_2 + u_3 w_3$$

if each vector has three elements. Thus the result of multiplying a 1×3 vector by a 3×1 vector is a 1×1 array, that is, a scalar. This definition applies to vectors having any number of elements, as long as both vectors have the same number of elements.

Thus the result of multiplying a $1 \times n$ vector by an $n \times 1$ vector is a 1×1 array, that is, a scalar.

Miles Traveled

EXAMPLE 2.4–1

Table 2.4–1 gives the speed of an aircraft on each leg of a certain trip and the time spent on each leg. Compute the miles traveled on each leg and the total miles traveled.

Table 2.4–1 Aircraft speeds and times per leg

	Leg			
	1	2	3	4
Speed (mi/hr)	200	250	400	300
Time (hr)	2	5	3	4

■ Solution

We can define a row vector s containing the speeds and a row vector t containing the times for each leg. Thus s = [200, 250, 400, 300] and t = [2, 5, 3, 4].

To find the miles traveled on each leg, we multiply the speed by the time. To do so, we use the MATLAB symbol . *, which *specifies* the multiplication s . *t to produce the row vector whose elements are the products of the corresponding elements in s and t:

$$s.*t = [200(2), 250(5), 400(3), 300(4)] = [400, 1250, 1200, 1200]$$

This vector contains the miles traveled by the aircraft on each leg of the trip.

To find the total miles traveled, we use the matrix product, denoted by s*t′. In this definition the product is the *sum* of the individual element products; that is,

$$s*t′ = [200(2) + 250(5) + 400(3) + 300(4)] = 4050$$

These two examples illustrate the difference between *array* multiplication s . *t and *matrix* multiplication s*t′.

Vector-Matrix Multiplication

Not all matrix products are scalars. To generalize the preceding multiplication to a column vector multiplied by a matrix, think of the matrix as being composed of row vectors. The scalar result of each row-column multiplication forms an element in the result, which is a column vector. For example:

$$\begin{bmatrix} 2 & 7 \\ 6 & -5 \end{bmatrix}\begin{bmatrix} 3 \\ 9 \end{bmatrix} = \begin{bmatrix} 2(3) + 7(9) \\ 6(3) - 5(9) \end{bmatrix} = \begin{bmatrix} 69 \\ -27 \end{bmatrix} \tag{2.4-3}$$

Thus the result of multiplying a 2×2 matrix by a 2×1 vector is a 2×1 array, that is, a column vector. Note that the definition of multiplication requires that the number of columns in the matrix be equal to the number of rows in the vector. In general, the product $\mathbf{Ax}$, where $\mathbf{A}$ has p columns, is defined only if $\mathbf{x}$ has p rows. If $\mathbf{A}$ has m rows and $\mathbf{x}$ is a column vector, the result of $\mathbf{Ax}$ is a column vector with m rows.

Matrix-Matrix Multiplication

We can expand this definition of multiplication to include the product of two matrices $\mathbf{AB}$. The number of columns in $\mathbf{A}$ must equal the number of rows in $\mathbf{B}$. The row-column multiplications form column vectors, and these column vectors form the matrix result. The product $\mathbf{AB}$ has the same number of rows as $\mathbf{A}$ and the same number of columns as $\mathbf{B}$. For example,

$$\begin{bmatrix} 6 & -2 \\ 10 & 3 \\ 4 & 7 \end{bmatrix}\begin{bmatrix} 9 & 8 \\ -5 & 12 \end{bmatrix} = \begin{bmatrix} (6)(9) + (-2)(-5) & (6)(8) + (-2)(12) \\ (10)(9) + (3)(-5) & (10)(8) + (3)(12) \\ (4)(9) + (7)(-5) & (4)(8) + (7)(12) \end{bmatrix}$$

$$= \begin{bmatrix} 64 & 24 \\ 75 & 116 \\ 1 & 116 \end{bmatrix} \tag{2.4-4}$$

Use the operator * to perform matrix multiplication in MATLAB. The following MATLAB session shows how to perform the matrix multiplication shown in (2.4-4).

```
>>A = [6, -2; 10, 3; 4, 7];
>>B = [9, 8; -5, 12];
>>A*B
```

Element-by-element multiplication is defined for the following product:

$$[3 \quad 1 \quad 7][4 \quad 6 \quad 5] = [12 \quad 6 \quad 35]$$

However, this product is *not* defined for *matrix* multiplication, because the first matrix has three columns, but the second matrix does not have three rows. Thus if we were to type [3, 1, 7]*[4, 6, 5] in MATLAB, we would receive an error message.

The following product is defined in matrix multiplication and gives the result shown:

$$\begin{bmatrix} x_1 \\ x_2 \\ x_3 \end{bmatrix} [y_1 \quad y_2 \quad y_3] = \begin{bmatrix} x_1 y_1 & x_1 y_2 & x_1 y_3 \\ x_2 y_1 & x_2 y_2 & x_2 y_3 \\ x_3 y_1 & x_3 y_2 & x_3 y_3 \end{bmatrix}$$

The following product is also defined:

$$[10 \quad 6]\begin{bmatrix} 7 & 4 \\ 5 & 2 \end{bmatrix} = [10(7) + 6(5) \quad 10(4) + 6(2)] = [100 \quad 52]$$

Test Your Understanding

T2.4–1 Given the vectors

$$\mathbf{x} = \begin{bmatrix} 6 \\ 5 \\ 3 \end{bmatrix} \qquad \mathbf{y} = [2 \quad 8 \quad 7]$$

do the following by hand, then check your answer using MATLAB.

a. Find the matrix product $\mathbf{w} = \mathbf{x}*\mathbf{y}$.

b. Find the matrix product $\mathbf{z} = \mathbf{y}*\mathbf{x}$. Is $\mathbf{z} = \mathbf{w}$?

(Answers: a. [12, 48, 42; 10, 40, 35; 6, 24, 21]   b. 73, obviously no!)

T2.4–2 Use MATLAB to compute the dot product of the following vectors:

$$\mathbf{u} = 6\mathbf{i} - 8\mathbf{j} + 3\mathbf{k}$$
$$\mathbf{w} = 5\mathbf{i} + 3\mathbf{j} - 4\mathbf{k}$$

Check your answer by hand. (Answer: −6.)

T2.4–3 Use MATLAB to show that

$$\begin{bmatrix} 7 & 4 \\ -3 & 2 \\ 5 & 9 \end{bmatrix}\begin{bmatrix} 1 & 8 \\ 7 & 6 \end{bmatrix} = \begin{bmatrix} 35 & 80 \\ 11 & -12 \\ 68 & 94 \end{bmatrix}$$

Evaluating Multivariable Functions

To evaluate a function of two variables, say, $z = f(x, y)$, for the values $x = x_1$, $x_2,\ldots, x_m$ and $y = y_1, y_2,\ldots, y_n$, define the $m \times n$ matrices:

$$\mathbf{x} = \begin{bmatrix} x_1 & x_1 & \cdots & x_1 \\ x_2 & x_2 & \cdots & x_2 \\ \vdots & \vdots & \vdots & \vdots \\ x_m & x_m & \cdots & x_m \end{bmatrix} \qquad \mathbf{y} = \begin{bmatrix} y_1 & y_1 & \cdots & y_n \\ y_1 & y_2 & \cdots & y_n \\ \vdots & \vdots & \vdots & \vdots \\ y_1 & y_2 & \cdots & y_n \end{bmatrix}$$

When the function $z = f(x, y)$ is evaluated in MATLAB using array operations, the resulting $m \times n$ matrix $\mathbf{z}$ has the elements $z_{ij} = f(x_i, y_j)$. We can extend this technique to functions of more than two variables by using multidimensional arrays.

EXAMPLE 2.4–2	Height versus Velocity

The maximum height h achieved by an object thrown with a speed v at an angle θ to the horizontal, neglecting drag, is

$$h = \frac{v^2 \sin^2 \theta}{2g}$$

Create a table showing the maximum height for the following values of v and θ:

$$v = 10, 12, 14, 16, 18, 20 \text{ m/s} \qquad \theta = 50°, 60°, 70°, 80°$$

The rows in the table should correspond to the speed values, and the columns should correspond to the angles.

■ **Solution**

The program is shown below.

```
g = 9.8; v = 10:2:20;
theta = 50:10:80;
h = (v'.^2)*(sind(theta).^2)/(2*g);
table = [0, theta; v', h]
```

The arrays v and theta contain the given velocities and angles. The array v is 1×6 and the array theta is 1×4. Thus the term v'.^2 is a 6×1 array, and the term sind(theta).^2 is a 1×4 array. The product of these two arrays, h, is a matrix product and is a $(6 \times 1)(1 \times 4) = (6 \times 4)$ matrix.

The array [0, theta] is 1×5 and the array [v', h] is 6×5, so the matrix table is 7×5. The following table shows the matrix table rounded to one decimal place. From this table we can see that the maximum height is 8.8 m if $v = 14$ m/s and $\theta = 70°$.

0	50	60	70	80
10	3.0	3.8	4.5	4.9
12	4.3	5.5	6.5	7.1
14	5.9	7.5	8.8	9.7
16	7.7	9.8	11.5	12.7
18	9.7	12.4	14.6	16.0
20	12.0	15.3	18.0	19.8

EXAMPLE 2.4–3	Manufacturing Cost Analysis

Table 2.4–2 shows the hourly cost of four types of manufacturing processes. It also shows the number of hours required of each process to produce three different products. Use matrices and MATLAB to solve the following. (a) Determine the cost of each process to produce 1 unit of product 1. (b) Determine the cost to make 1 unit of each product. (c) Suppose we produce 10 units of product 1, 5 units of product 2, and 7 units of product 3. Compute the total cost.

Table 2.4–2 Cost and time data for manufacturing processes

		Hours required to produce one unit		
Process	**Hourly cost ($)**	**Product 1**	**Product 2**	**Product 3**
Lathe	10	6	5	4
Grinding	12	2	3	1
Milling	14	3	2	5
Welding	9	4	0	3

■ **Solution**

(a) The basic principle we can use here is that cost equals the hourly cost times the number of hours required. For example, the cost of using the lathe for product 1 is ($10/hr)(6 hr) = $60, and so forth for the other three processes. If we define the row vector of hourly costs to be `hourly_costs` and define the row vector of hours required for product 1 to be `hours_1`, then we can compute the costs of each process for product 1 using *element-by-element* multiplication. In MATLAB the session is

```
>>hourly_cost = [10, 12, 14, 9];
>>hours_1 = [6, 2, 3, 4];
>>process_cost_1 = hourly_cost.*hours_1
process_cost_1 =
    60    24    42    36
```

These are the costs of each of the four processes to produce 1 unit of product 1.

(b) To compute the total cost of 1 unit of product 1, we can use the vectors `hourly_costs` and `hours_1` but apply *matrix* multiplication instead of element-by-element multiplication, because matrix multiplication sums the individual products. The matrix multiplication gives

$$[10 \quad 12 \quad 14 \quad 9]\begin{bmatrix}6\\2\\3\\4\end{bmatrix} = 10(6) + 12(2) + 14(3) + 9(4) = 162$$

We can perform similar multiplication for products 2 and 3, using the data in the table. For product 2:

$$[10 \quad 12 \quad 14 \quad 9]\begin{bmatrix}5\\3\\2\\0\end{bmatrix} = 10(5) + 12(2) + 14(3) + 9(0) = 114$$

For product 3:

$$[10 \quad 12 \quad 14 \quad 9]\begin{bmatrix}4\\1\\5\\3\end{bmatrix} = 10(4) + 12(1) + 14(5) + 9(3) = 149$$

These three operations could have been accomplished in one operation by defining a matrix whose columns are formed by the data in the last three columns of the table:

$$[10 \quad 12 \quad 14 \quad 9] \begin{bmatrix} 6 & 5 & 4 \\ 2 & 3 & 1 \\ 3 & 2 & 5 \\ 4 & 0 & 3 \end{bmatrix} = \begin{bmatrix} 60 & + & 24 & + & 42 & + & 36 \\ 50 & + & 36 & + & 28 & + & 0 \\ 40 & + & 12 & + & 70 & + & 27 \end{bmatrix} = [162 \quad 114 \quad 149]$$

In MATLAB the session continues as follows. Remember that we must use the transpose operation to convert the row vectors into column vectors.

```
>>hours_2 = [5, 3, 2, 0];
>>hours_3 = [4, 1, 5, 3];
>>unit_cost = hourly_cost*[hours_1', hours_2', hours_3']
unit_cost =
    162    114    149
```

Thus the costs to produce 1 unit each of products 1, 2, and 3 are $162, $114, and $149, respectively.

(c) To find the total cost to produce 10, 5, and 7 units, respectively, we can use matrix multiplication:

$$[10 \quad 5 \quad 7] \begin{bmatrix} 162 \\ 114 \\ 149 \end{bmatrix} = 1620 + 570 + 1043 = 3233$$

In MATLAB the session continues as follows. Note the use of the transpose operator on the vector `unit_cost`.

```
>>units = [10, 5, 7];
>>total_cost = units*unit_cost'
total_cost =
    3233
```

The total cost is $3233.

The General Matrix Multiplication Case

We can state the general result for matrix multiplication as follows: Suppose **A** has dimension $m \times p$ and **B** has dimension $p \times q$. If **C** is the product **AB**, then **C** has dimension $m \times q$ and its elements are given by

$$c_{ij} = \sum_{k=1}^{p} a_{ik} b_{kj} \tag{2.4–5}$$

for all $i = 1, 2,\ldots, m$ and $j = 1,2,\ldots, q$. For the product to be defined, the matrices **A** and **B** must be *conformable*; that is, the number of *rows* in **B** must equal the number of *columns* in **A**. The product has the same number of rows as **A** and the same number of columns as **B**.

Matrix multiplication does not have the commutative property; that is, in general, $\mathbf{AB} \neq \mathbf{BA}$. Reversing the order of matrix multiplication is a common and easily made mistake.

The associative and distributive properties hold for matrix multiplication. The associative property states that

$$\mathbf{A}(\mathbf{B} + \mathbf{C}) = \mathbf{AB} + \mathbf{AC} \qquad (2.4\text{--}6)$$

The distributive property states that

$$(\mathbf{AB})\mathbf{C} = \mathbf{A}(\mathbf{BC}) \qquad (2.4\text{--}7)$$

Applications to Cost Analysis

Project cost data stored in tables must often be analyzed in several ways. The elements in MATLAB matrices are similar to the cells in a spreadsheet, and MATLAB can perform many spreadsheet-type calculations for analyzing such tables.

Product Cost Analysis

EXAMPLE 2.4–4

Table 2.4–3 shows the costs associated with a certain product, and Table 2.4–4 shows the production volume for the four quarters of the business year. Use MATLAB to find the quarterly costs for materials, labor, and transportation; the total material, labor, and transportation costs for the year; and the total quarterly costs.

Table 2.4–3 Product costs

Product	Unit costs ($\times 10^3$)		
	Materials	Labor	Transportation
1	6	2	1
2	2	5	4
3	4	3	2
4	9	7	3

Table 2.4–4 Quarterly production volume

Product	Quarter 1	Quarter 2	Quarter 3	Quarter 4
1	10	12	13	15
2	8	7	6	4
3	12	10	13	9
4	6	4	11	5

■ Solution

The costs are the product of the unit cost and the production volume. Thus we define two matrices: U contains the unit costs in Table 2.4–3 in thousands of dollars, and P contains the quarterly production data in Table 2.4–4.

```
>>U = [6, 2, 1; 2, 5, 4; 4, 3, 2; 9, 7, 3];
>>P = [10, 12, 13, 15; 8, 7, 6, 4; 12, 10, 13, 9; 6, 4, 11, 5];
```

Note that if we multiply the first column in U by the first column in P, we obtain the total materials cost for the first quarter. Similarly, multiplying the first column in U by the second column in P gives the total materials cost for the *second* quarter. Also, multiplying the second column in U by the first column in P gives the total *labor* cost for the first quarter, and so on. Extending this pattern, we can see that we must multiply the *transpose* of U by P. This multiplication gives the cost matrix C.

```
>>C = U'*P
```

The result is

$$C = \begin{bmatrix} 178 & 162 & 241 & 179 \\ 138 & 117 & 172 & 112 \\ 84 & 72 & 96 & 64 \end{bmatrix}$$

Each column in **C** represents one quarter. The total first-quarter cost is the sum of the elements in the first column, the second-quarter cost is the sum of the second column, and so on. Thus because the sum command sums the columns of a matrix, the quarterly costs are obtained by typing

```
>>Quarterly_Costs = sum(C)
```

The resulting vector, containing the quarterly costs in thousands of dollars, is [400 351 509 355]. Thus the total costs in each quarter are $400,000; $351,000; $509,000; and $355,000.

The elements in the first row of **C** are the material costs for each quarter; the elements in the second row are the labor costs, and those in the third row are the transportation costs. Thus to find the total material costs, we must sum across the first row of **C**. Similarly, the total labor and total transportation costs are the sums across the second and third rows of **C**. Because the sum command sums *columns*, we must use the transpose of **C**. Thus we type the following:

```
>>Category_Costs = sum(C')
```

The resulting vector, containing the category costs in thousands of dollars, is [760 539 316]. Thus the total material costs for the year are $760,000; the labor costs are $539,000; and the transportation costs are $316,000.

We displayed the matrix **C** only to interpret its structure. If we need not display **C**, the entire analysis would consist of only four command lines.

```
>>U = [6, 2, 1; 2, 5, 4; 4, 3, 2; 9, 7, 3];
>>P = [10, 12, 13, 15; 8, 7, 6, 4; 12, 10, 13, 9; 6, 4, 11, 5];
>>Quarterly_Costs = sum(U'*P)
Quarterly_Costs
    400    351    509    355
>>Category_Costs = sum((U'*P)')
Category_Costs =
    760    539    316
```

This example illustrates the compactness of MATLAB commands.

Special Matrices

Two exceptions to the noncommutative property are the *null matrix*, denoted by **0**, and the identity, or *unity, matrix*, denoted by **I**. The null matrix contains all zeros and is not the same as the empty matrix [], which has no elements. The *identity matrix* is a square matrix whose diagonal elements are all equal to 1, with the remaining elements equal to 0. For example, the 2×2 identity matrix is

$$\mathbf{I} = \begin{bmatrix} 1 & 0 \\ 0 & 1 \end{bmatrix}$$

These matrices have the following properties:

$$\mathbf{0A = A0 = 0}$$
$$\mathbf{IA = AI = A}$$

MATLAB has specific commands to create several special matrices. Type `help specmat` to see the list of special matrix commands; also check Table 2.4–5. The identity matrix **I** can be created with the `eye(n)` command, where n is the desired dimension of the matrix. To create the 2×2 identity matrix, you type `eye(2)`. Typing `eye(size(A))` creates an identity matrix having the same dimension as the matrix **A**.

Sometimes we want to initialize a matrix to have all zero elements. The `zeros` command creates a matrix of all zeros. Typing `zeros(n)` creates an $n \times n$ matrix of zeros, whereas typing `zeros(m, n)` creates an $m \times n$ matrix of zeros, as will typing `A(m, n) = 0`. Typing `zeros(size(A))` creates a matrix of all zeros having the same dimension as the matrix **A**. This type of matrix can be useful for applications in which we do not know the required dimension ahead of time. The syntax of the `ones` command is the same, except that it creates arrays filled with 1s.

For example, to create and plot the function

$$f(x) = \begin{cases} 10 & 0 \le x \le 2 \\ 0 & 2 < x < 5 \\ -3 & 5 \le x \le 7 \end{cases}$$

Table 2.4–5 Special matrices

Command	Description
`eye(n)`	Creates an $n \times n$ identity matrix.
`eye(size(A))`	Creates an identity matrix the same size as the matrix **A**.
`ones(n)`	Creates an $n \times n$ matrix of 1s.
`ones(m,n)`	Creates an $m \times n$ array of 1s.
`ones(size(A))`	Creates an array of 1s the same size as the array **A**.
`zeros(n)`	Creates an $n \times n$ matrix of 0s.
`zeros(m,n)`	Creates an $m \times n$ array of 0s.
`zeros(size(A))`	Creates an array of 0s the same size as the array **A**.

the script file is

```
x1 = 0:0.01:2;
f1 = 10*ones(size(x1));
x2 = 2.01:0.01:4.99;
f2 = zeros(size(x2));
x3 = 5:0.01:7;
f3 = -3*ones(size(x3));
f = [f1, f2, f3];
x = [x1, x2, x3];
plot(x, f),xlabel('x'),ylabel('y')
```

(Consider what the plot would look like if the command plot(x, f) were replaced with the command plot(x1, f1, x2, f2, x3, f3).)

Matrix Division and Linear Algebraic Equations

Matrix division uses both the right and left division operators, / and \, for various applications, a principal one being the solution of sets of linear algebraic equations. Chapter 8 covers a related topic, the matrix inverse.

You can use the left division operator (\) in MATLAB to solve sets of linear algebraic equations. For example, consider the set

$$6x + 12y + 4z = 70$$
$$7x - 2y + 3z = 5$$
$$2x + 8y - 9z = 64$$

To solve such sets in MATLAB you must create two arrays; we will call them A and B. The array A has as many rows as there are equations and as many columns as there are variables. The rows of A must contain the coefficients of x, y, and z in that order. In this example, the first row of A must be 6, 12, 4; the second row must be 7, −2, 3; and the third row must be 2, 8, −9. The array B contains the constants on the right-hand side of the equation; it has one column and as many rows as there are equations. In this example, the first row of B is 70, the second is 5, and the third is 64. The solution is obtained by typing A\B. The session is

```
>>A = [6, 12, 4; 7, -2, 3; 2, 8, -9];
>>B = [70; 5; 64];
>>Solution = A\B
Solution =
     3
     5
    -2
```

The solution is $x = 3$, $y = 5$, and $z = -2$.

LEFT DIVISION METHOD   The *left division method* works fine when the equation set has a unique solution. To learn how to deal with problems having a nonunique solution (or perhaps no solution at all!), see Chapter 8.

Test Your Understanding

T2.4–4 Use MATLAB to solve the following set of equations.

$$4x + 3y = 23$$
$$8x - 2y = 6$$

(Answer: $x = 2$, $y = 5$.)

T2.4–5 Use MATLAB to solve the following set of equations.

$$4x - 2y = 16$$
$$3x + 5y = -1$$

(Answer: $x = 3$, $y = -2$.)

T2.4–6 Use MATLAB to solve the following set of equations.

$$6x - 4y + 8z = 112$$
$$-5x - 3y + 7z = 75$$
$$14x + 9y - 5z = -67$$

(Answer: $x = 2$, $y = -5$, $z = 10$.)

Matrix Exponentiation

Raising a matrix to a power is equivalent to repeatedly multiplying the matrix by itself, for example, $\mathbf{A}^2 = \mathbf{AA}$. This process requires the matrix to have the same number of rows as columns; that is, it must be a *square* matrix. MATLAB uses the symbol ^ for matrix exponentiation. To find $\mathbf{A}^2$, type A^2.

We can raise a scalar n to a matrix power $\mathbf{A}$, if $\mathbf{A}$ is square, by typing n^A, but the applications for such a procedure are in advanced courses. However, raising a matrix to a matrix power—that is, $\mathbf{A}^\mathbf{B}$—is not defined, even if $\mathbf{A}$ and $\mathbf{B}$ are square.

Special Products

Many applications in physics and engineering use the cross product and dot product; for example, calculations to compute moments and force components use these special products. If $\mathbf{A}$ and $\mathbf{B}$ are vectors with three elements, the cross product command cross(A, B) computes the three-element vector that is the cross product $\mathbf{A} \times \mathbf{B}$. If $\mathbf{A}$ and $\mathbf{B}$ are $3 \times n$ matrices, cross(A, B) returns a $3 \times n$ array whose columns are the cross products of the corresponding columns in the $3 \times n$ arrays $\mathbf{A}$ and $\mathbf{B}$. For example, the moment $\mathbf{M}$ with respect to a reference point O due to the force $\mathbf{F}$ is given by $\mathbf{M} = \mathbf{r} \times \mathbf{F}$, where $\mathbf{r}$ is the position vector from the point O to the point where the force $\mathbf{F}$ is applied. To find the moment in MATLAB, you type M = cross(r, F).

The dot product command dot(A, B) computes a row vector of length n whose elements are the dot products of the corresponding columns of the $m \times n$ arrays **A** and **B**. To compute the component of the force **F** along the direction given by the vector **r**, you type dot(F, r).

2.5 Polynomial Operations Using Arrays

MATLAB has some convenient tools for working with polynomials. Type help polyfun for more information on this category of commands. We will use the following notation to describe a polynomial:

$$f(x) = a_1 x^n + a_2 x^{n-1} + a_3 x^{n-2} + \cdots + a_{n-1} x^2 + a_n x + a_{n+1}$$

We can describe a polynomial in MATLAB with a row vector whose elements are the polynomial's coefficients, *starting with the coefficient of the highest power of x.* This vector is $[a_1, a_2, a_3,\ldots, a_{n-1}, a_n, a_{n+1}]$. For example, the vector [4, -8, 7, -5] represents the polynomial $4x^3 - 8x^2 + 7x - 5$.

    Polynomial roots can be found with the roots(a) function, where a is the array containing the polynomial coefficients. For example, to obtain the roots of $x^3 + 12x^2 + 45x + 50 = 0$, you type y = roots([1, 12, 45, 50]). The answer (y) is a *column* array containing the values $-2, -5, -5$.

    The poly(r) function computes the coefficients of the polynomial whose roots are specified by the array r. The result is a *row* array that contains the polynomial's coefficients. For example, to find the polynomial whose roots are 1 and $3 \pm 5i$, the session is

```
>>p = poly([1,3+5i, 3-5i])
P =
    1   -7   40   -34
```

Thus the polynomial is $x^3 - 7x^2 + 40x - 34$.

Polynomial Addition and Subtraction

To add two polynomials, add the arrays that describe their coefficients. If the polynomials are of different degrees, add zeros to the coefficient array of the lower-degree polynomial. For example, consider

$$f(x) = 9x^3 - 5x^2 + 3x + 7$$

whose coefficient array is f = [9, -5, 3, 7] and

$$g(x) = 6x^2 - x + 2$$

whose coefficient array is g = [6, -1, 2]. The degree of $g(x)$ is 1 less that of $f(x)$. Therefore, to add $f(x)$ and $g(x)$, we append one zero to g to "fool" MATLAB into thinking $g(x)$ is a third-degree polynomial. That is, we type g = [0 g] to obtain [0, 6, -1, 2] for g. This vector represents $g(x) = 0x^3 + 6x^2 - x + 2$.

To add the polynomials, type h = f+g. The result is h = [9, 1, 2, 9], which corresponds to $h(x) = 9x^3 + x^2 + 2x + 9$. Subtraction is done in a similar way.

Polynomial Multiplication and Division

To multiply a polynomial by a scalar, simply multiply the coefficient array by that scalar. For example, $5h(x)$ is represented by [45, 5, 10, 45].

Multiplication and division of polynomials are easily done with MATLAB. Use the conv function (it stands for "convolve") to multiply polynomials and use the deconv function (deconv stands for "deconvolve") to perform synthetic division. Table 2.5–1 summarizes these functions, as well as the poly, polyval, and roots functions.

The product of the polynomials $f(x)$ and $g(x)$ is

$$f(x)g(x) = (9x^3 - 5x^2 + 3x + 7)(6x^2 - x + 2)$$
$$= 54x^5 - 39x^4 + 41x^3 + 29x^2 - x + 14$$

Dividing $f(x)$ by $g(x)$ using synthetic division gives a quotient of

$$\frac{f(x)}{g(x)} = \frac{9x^3 - 5x^2 + 3x + 7}{6x^2 - x + 2} = 1.5x - 0.5833$$

with a remainder of −0.5833 +8.1667. Here is the MATLAB session to perform these operations.

```
>>f = [9,-5,3,7];
>>g = [6,-1,2];
>>product = conv(f,g)
product =
    54   -39   41   29   -1   14
```

Table 2.5–1 Polynomial functions

Command	Description
conv(a,b)	Computes the product of the two polynomials described by the coefficient arrays a and b. The two polynomials need not be of the same degree. The result is the coefficient array of the product polynomial.
[q,r] = deconv(num,den)	Computes the result of dividing a numerator polynomial, whose coefficient array is num, by a denominator polynomial represented by the coefficient array den. The quotient polynomial is given by the coefficient array q, and the remainder polynomial is given by the coefficient array r.
poly(r)	Computes the coefficients of the polynomial whose roots are specified by the vector r. The result is a *row* vector that contains the polynomial's coefficients arranged in descending order of power.
polyval(a,x)	Evaluates a polynomial at specified values of its independent variable x, which can be a matrix or a vector. The polynomial's coefficients of descending powers are stored in the array a. The result is the same size as x.
roots(a)	Computes the roots of a polynomial specified by the coefficient array a. The result is a *column* vector that contains the polynomial's roots.

```
>>[quotient, remainder] = deconv(f,g)
quotient =
    1.5    -0.5833
remainder =
    0    0    -0.5833    8.1667
```
The conv and deconv functions do not require that the polynomials be of the same degree, so we did not have to fool MATLAB as we did when adding the polynomials.

Plotting Polynomials

The polyval(a, x) function evaluates a polynomial at specified values of its independent variable x, which can be a matrix or a vector. The polynomial's coefficient array is a. The result is the same size as x. For example, to evaluate the polynomial $f(x) = 9x^3 - 5x^2 + 3x + 7$ at the points $x = 0, 2,4,…, 10$, type

```
>>f = polyval([9,-5,3,7], [0:2:10]);
```

The resulting vector f contains six values that correspond to $f(0), f(2), f(4),…, f(10)$.

The polyval function is very useful for plotting polynomials. To do this, you should define an array that contains many values of the independent variable x in order to obtain a smooth plot. For example, to plot the polynomial $f(x) = 9x^3 - 5x^2 + 3x + 7$ for $-2 \le x \le 5$, you type

```
>>x = -2:0.01:5;
>>polyval([9,-5,3,7], x);
>>plot(x, f),xlabel('x'),ylabel('f(x)'),grid
```

Polynomial derivatives and integrals are covered in Chapter 9.

Test Your Understanding

T2.5–1 Use MATLAB to obtain the roots of

$$x^3 + 13x^2 + 52x + 6 = 0$$

Use the poly function to confirm your answer.

T2.5–2 Use MATLAB to confirm that

$$(20x^3 - 7x^2 + 5x + 10)(4x^2 + 12x - 3)$$
$$= 80x^5 + 212x^4 - 124x^3 + 121x^2 + 105x - 30$$

T2.5–3 Use MATLAB to confirm that

$$\frac{12x^3 + 5x^2 - 2x + 3}{3x^2 - 7x + 4} = 4x + 11$$

with a remainder of $59x - 41$.

T2.5–4 Use MATLAB to confirm that

$$\frac{6x^3 + 4x^2 - 5}{12x^3 - 7x^2 + 3x + 9} = 0.7108$$

when $x = 2$.

T2.5–5 Plot the polynomial

$$y = x^3 + 13x^2 + 52x + 6$$

over the range $-7 \le x \le 1$.

Earthquake-Resistant Building Design

EXAMPLE 2.5–1

Buildings designed to withstand earthquakes must have natural frequencies of vibration that are not close to the oscillation frequency of the ground motion. A building's natural frequencies are determined primarily by the masses of its floors and by the lateral stiffness of its supporting columns (which act as horizontal springs). We can find these frequencies by solving for the roots of a polynomial called the structure's *characteristic polynomial* (characteristic polynomials are discussed further in Chapter 9). Figure 2.5–1 shows the exaggerated motion of the floors of a three-story building. For such a building, if each floor has a mass m and the columns have stiffness k, the polynomial is

$$(\alpha - f^2)\left[(2\alpha - f^2)^2 - \alpha^2\right] + \alpha^2 f^2 - 2\alpha^3$$

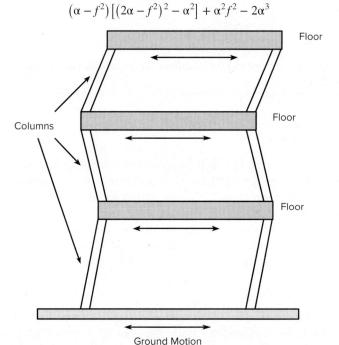

Figure 2.5–1 Simple vibration model of a building subjected to ground motion.

where $\alpha = k/4m\pi^2$ (models such as these are discussed in greater detail in [Palm, 2010]). The building's natural frequencies in cycles per second are the positive roots of this equation. Find the building's natural frequencies in cycles per second for the case where $m = 1000$ kg and $k = 5 \times 10^6$ N/m.

■ Solution

The characteristic polynomial consists of sums and products of lower-degree polynomials. We can use this fact to have MATLAB do the algebra for us. The characteristic polynomial has the form

$$p_1\left(p_2^2 - \alpha^2\right) + p_3 = 0$$

where

$$p_1 = \alpha - f^2 \qquad p_2 = 2\alpha - f^2 \qquad p_3 = \alpha^2 f^2 - 2\alpha^3$$

The MATLAB script file is

```
k = 5e+6;m = 1000;
alpha = k/(4*m*pi^2);
p1 = [-1,0,alpha];
p2 = [-1,0,2*alpha];
p3 = [alpha^2,0,-2*alpha^3];
p4 = conv(p2,p2) - (0, 0, 0,0,alpha^2];
p5 = conv(p1,p4);
p6 = p5 + [0,0,0,0,p3];
r = roots(p6)
```

The resulting positive roots and thus the frequencies, rounded to the nearest integer, are 20, 14, and 5 Hz.

2.6 Cell Arrays

The *cell array* is an array in which each element is a *bin*, or *cell*, which can contain an array. You can store different classes of arrays in a cell array, and you can group data sets that are related but have different dimensions. You access cell arrays using the same indexing operations used with ordinary arrays.

This is the only section in the text that uses cell arrays. Coverage of this section is therefore optional. Some more advanced MATLAB applications, such as those found in some of the toolboxes, do use cell arrays.

Creating Cell Arrays

CELL INDEXING

CONTENT INDEXING

You can create a cell array by using assignment statements or by using the cell function. You can assign data to the cells by using either *cell indexing* or *content indexing*. To use cell indexing, enclose in parentheses the cell subscripts on the left side of the assignment statement and use the standard array notation. Enclose the cell contents on the right side of the assignment statement in braces { }.

An Environment Database

EXAMPLE 2.6–1

Suppose you want to create a 2×2 cell array **A**, whose cells contain the location, the date, the air temperature (measured at 8 A.M., 12 noon, and 5 P.M.), and the water temperatures measured at the same time in three different points in a pond. The cell array looks like the following.

Walden Pond	June 13, 2016
[60 72 65]	$\begin{bmatrix} 55 & 57 & 56 \\ 54 & 56 & 55 \\ 52 & 55 & 53 \end{bmatrix}$

■ **Solution**
You can create this array by typing the following either in interactive mode or in a script file and running it.

```
A(1,1) = {'Walden Pond'};
A(1,2) = {'June 13, 2016'};
A(2,1) = {[60,72,65]};
A(2,2) = {[55,57,56;54,56,55;52,55,53]};
```

If you do not yet have contents for a particular cell, you can type a pair of empty braces { } to denote an empty cell, just as a pair of empty brackets [] denotes an empty numeric array. This notation creates the cell but does not store any contents in it.

To use content indexing, enclose in braces the cell subscripts on the left side, using the standard array notation. Then specify the cell contents on the right side of the assignment operator. For example:

```
A{1,1} = 'Walden Pond';
A{1,2} = 'June 13, 2016';
A{2,1} = [60,72,65];
A{2,2} = [55,57,56;54,56,55;52,55,53];
```

Type **A** at the command line. You will see

```
A =
    'Walden Pond'    'June 13, 2016'
    [1x3 double]     [3x3 double]
```

You can use the `celldisp` function to display the full contents. For example, typing `celldisp(A)` displays

```
A{1,1} =
    Walden Pond
A{2,1} =
    60 72 65

    :
    :

    etc.
```

The `cellplot` function produces a graphical display of the cell array's contents in the form of a grid. Type `cellplot(A)` to see this display for the cell array

A. Use commas or spaces with braces to indicate columns of cells and use semi-colons to indicate rows of cells (just as with numeric arrays). For example, typing

`B = {[2,4], [6,-9;3,5]; [7,2], 10};`

creates the following 2 × 2 cell array:

[2 4]	$\begin{bmatrix} 6 & -9 \\ 3 & 5 \end{bmatrix}$
[7 2]	10

You can preallocate empty cell arrays of a specified size by using the `cell` function. For example, type `C = cell(3, 5)` to create the 3 × 5 cell array C and fill it with empty matrices. Once the array has been defined in this way, you can use assignment statements to enter the contents of the cells. For example, type `C(2, 4) = {[6, -3, 7]}` to put the 1 × 3 array in cell (2,4) and type `C(1,5) = {1:10}` to put the numbers from 1 to 10 in cell (1,5). Type `C(3, 4) = {'30 mph'}` to put the string in cell (3, 4).

Accessing Cell Arrays

You can access the contents of a cell array by using either cell indexing or content indexing. To use cell indexing to place the contents of cell (3,4) of the array C in the new variable Speed, type `Speed = C(3, 4)`. To place the contents of the cells in rows 1 to 3, columns 2 to 5 in the new cell array D, type `D = C(1:3, 2:5)`. The new cell array D will have three rows, four columns, and 12 arrays. To use content indexing to access some of or all the contents in a *single cell*, enclose the cell index expression in braces to indicate that you are assigning the contents, not the cells themselves, to a new variable. For example, typing `Speed = C{3, 4}` assigns the contents `'30 mph'` in cell (3,4) to the variable Speed. You cannot use content indexing to retrieve the contents of more than one cell at a time. For example, the statements `G = C{1, :}` and `C{1, :} = var`, where var is some variable, are both invalid.

   You can access subsets of a cell's contents. For example, to obtain the second element in the 1 × 3-row vector in the (2, 4) cell of array C and assign it to the variable r, you type `r = C{2,4}(1,2)`. The result is `r = -3`.

2.7 Structure Arrays

Structure arrays are composed of *structures*. This class of arrays enables you to store dissimilar arrays together. The elements in structures are accessed using *named fields*. This feature distinguishes them from cell arrays, which are accessed using the standard array indexing operations.

   Structure arrays are used in this text only in this section. Some MATLAB toolboxes do use structure arrays.

   A specific example is the best way to introduce the terminology of structures. Suppose you want to create a database of students in a course, and you want to

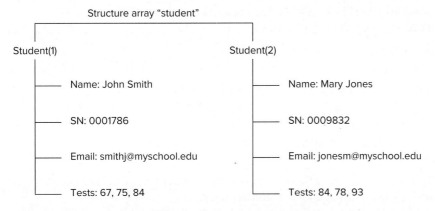

Figure 2.7–1 Arrangement of data in the structure array student.

include each student's name, student number (SN), email address, and test scores. Figure 2.7–1 shows a diagram of this data structure. Each type of data (name, student number, and so on) is a *field*, and its name is the *field name*. Thus our database has four fields. The first three fields each contain a text string, while the last field (the test scores) contains a vector having numerical elements. A *structure* consists of all this information for a single student. A *structure array* is an array of such structures for different students. The array shown in Figure 2.7–1 has two structures arranged in one row and two columns.

FIELD

Creating Structures

You can create a structure array by using assignment statements or by using the `struct` function. The following example uses assignment statements to build a structure. Structure arrays use the dot notation (.) to specify and to access the fields. You can type the commands either in the interactive mode or in a script file.

A Student Database

EXAMPLE 2.7–1

Create a structure array to contain the following types of student data:

■ Student name.
■ Student number.
■ Email address.
■ Test scores.

Enter the data shown in Figure 2.7–1 into the database.

■ Solution
You can create the structure array by typing the following either in the interactive mode or in a script file. Start with the data for the first student.

```
student.name = 'John Smith' ;
student.SN = '0001786' ;
```

```
student.email = 'smithj@myschool.edu' ;
student.tests = [67,75,84];
```

If you then type

```
>>student
```

at the command line, you will see the following response:

```
name: 'John Smith'
SN: '0001786'
email: 'smithj@myschool.edu'
tests: [67 75 84]
```

To determine the size of the array, type `size (student)`. The result is `ans = 1 1`, which indicates that it is a 1 × 1 structure array.

To add a second student to the database, use a subscript 2 enclosed in parentheses after the structure array's name and enter the new information. For example, type

```
student(2).name = 'Mary Jones' ;
student(2).SN = '0009832' ;
student(2).email = 'jonesm@myschool.edu' ;
student(2).tests = [84,78,93];
```

This process "expands" the array. Before we entered the data for the second student, the *dimension* of the structure array was 1 × 1 (it was a single structure). Now it is a 1 × 2 array consisting of two structures, arranged in one row and two columns. You can confirm this information by typing `size(student)`, which returns `ans = 1 2`. If you now type `length(student)`, you will get the result `ans = 2`, which indicates that the array has two elements (two structures). When a structure array has more than one structure, MATLAB does not display the individual field contents when you type the structure array's name. For example, if you now type `student`, MATLAB displays

```
>>student =
1x2 struct array with fields:
    name
    SN
    email
    tests
```

You can also obtain information about the fields by using the `fieldnames` function (see Table 2.7–1). For example:

```
>>fieldnames(student)
ans =
    'name'
    'SN'
    'email'
    'tests'
```

As you fill in more student information, MATLAB assigns the same number of fields and the same field names to each element. If you do not enter some information—for example, suppose you do not know someone's email address— MATLAB assigns an empty matrix to that field for that student.

Table 2.7–1 Structure functions

Function	Description
`names = fieldnames(S)`	Returns the field names associated with the structure array S as `names`, a cell array of strings.
`isfield(S,'field')`	Returns 1 if `'field'` is the name of a field in the structure array S and 0 otherwise.
`isstruct(S)`	Returns 1 if the array S is a structure array and 0 otherwise.
`S = rmfield(S,'field')`	Removes the field `'field'` from the structure array S.
`S = struct('f1','v1','f2','v2' ...)`	Creates a structure array with the fields `'f1'`, `'f2'`, ... having the values `'v1'`, `'v2'`, ...

The fields can be different sizes. For example, each name field can contain a different number of characters, and the arrays containing the test scores can be different sizes, as would be the case if a certain student did not take the second test.

In addition to the assignment statement, you can build structures using the `struct` function, which lets you "preallocate" a structure array. To build a structure array named sa_1, the syntax is

```
sa_1 = struct('field1', 'values1', 'field2',values2', ...)
```

where the arguments are the field names and their values. The values arrays `values1, values2, ...` must all be arrays of the same size, scalar cells, or single values. The elements of the values arrays are inserted into the corresponding elements of the structure array. The resulting structure array has the same size as the values arrays, or is 1 × 1 if none of the values arrays is a cell. For example, to preallocate a 1 × 1 structure array for the student database, you type

```
student = struct('name','John Smith', 'SN', ...'
0001786','email','smithj@myschool.edu', ...'
tests', [67,75,84])
```

Accessing Structure Arrays

To access the contents of a particular field, type a period after the structure array name, followed by the field name. For example, typing `student(2).name` displays the value `'Mary Jones'`. Of course, we can assign the result to a variable in the usual way. For example, typing `name2 = student(2).name` assigns the value `'Mary Jones'` to the variable name2. To access elements within a field, for example, John Smith's second test score, type `student(1).tests(2)`. This entry returns the value 75. In general, if a field contains an array, you use the array's subscripts to access its elements. In this example the statement `student (1).tests(2)` is equivalent to `student(1,1).tests(2)` because student has one row.

To store all the information for a particular structure—say, all the information about Mary Jones—in another structure array named M, you type `M = student(2)`. You can also assign or change values of field elements. For example, typing `student (2).tests(2) = 81` changes Mary Jones's second test score from 78 to 81.

Modifying Structures

Suppose you want to add phone numbers to the database. You can do this by typing the first student's phone number as follows:

```
student(1).phone = '555-1653'
```

All the other structures in the array will now have a phone field, but these fields will contain the empty array until you give them values.

To delete a field from every structure in the array, use the `rmfield` function. Its basic syntax is

```
new_struc = rmfield(array, 'field');
```

where `array` is the structure array to be modified, `'field'` is the field to be removed, and `new_struc` is the name of the new structure array so created by the removal of the field. For example, to remove the student number field and call the new structure array `new_student`, type

```
new_student = rmfield(student,'SN');
```

Using Operators and Functions with Structures

You can apply the MATLAB operators to structures in the usual way. For example, to find the maximum test score of the second student, you type `max (student(2).tests)`. The answer is 93.

The `isfield` function determines whether a structure array contains a particular field. Its syntax is `isfield(S,'field')`. It returns a value of 1 (which means "true") if `'field'` is the name of a field in the structure array S. For example, typing `isfield(student,'name')` returns the result `ans = 1`.

The `isstruct` function determines whether an array is a structure array. Its syntax is `isstruct(S)`. It returns a value of 1 if S is a structure array and 0 otherwise. For example, typing `isstruct(student)` returns the result `ans = 1`, which is equivalent to "true."

Test Your Understanding

T2.7–1   Create the structure array `student` shown in Figure 2.7–1 and add the following information about a third student: name: Alfred E. Newman; SN: 0003456; e-mail: newmana@myschool.edu; tests: 55, 45, 58.

T2.7–2   Edit your structure array to change Newman's second test score from 45 to 53.

T2.7–3   Edit your structure array to remove the SN field.

2.8 Summary

You should now be able to perform basic operations and use arrays in MATLAB. For example, you should be able to

- Create, address, and edit arrays.
- Perform array operations including addition, subtraction, multiplication, division, and exponentiation.

Table 2.8–1 Guide to commands introduced in Chapter 2

Special characters	Use
'	Transposes a matrix, creating complex conjugate elements.
.'	Transposes a matrix without creating complex conjugate elements.
;	Suppresses screen printing; also denotes a new row in an array.
:	Represents an entire row or column of an array.

	Tables	
	Array functions	Table 2.1–1
	Element-by-element operations	Table 2.3–1
	Special matrices	Table 2.4–5
	Polynomial functions	Table 2.5–1
	Structure functions	Table 2.7–1

■ Perform matrix operations including addition, subtraction, multiplication, division, and exponentiation.

■ Perform polynomial algebra.

■ Create databases using cell and structure arrays.

Table 2.8–1 is a reference guide to all the MATLAB commands introduced in this chapter.

Key Terms

Absolute value,   59
Array addressing,   55
Array operations,   62
Array size,   60
Cell array,   88
Cell indexing,   88
Column vector,   52
Content indexing,   88
Empty array,   56
Field,   91

Identity matrix,   81
Left division method,   82
Length,   59
Magnitude,   59
Matrix,   54
Matrix operations,   62
Null matrix,   81
Row vector,   52
Structure arrays,   90
Transpose,   53

Problems

You can find the answers to problems marked with an asterisk at the end of the text.

Section 2.1

1.   *a.* Use two methods to create the vector **x** having 100 regularly spaced values starting at 5 and ending at 28.

    b. Use two methods to create the vector **x** having a regular spacing of 0.2 starting at 2 and ending at 14.

    c. Use two methods to create the vector *x* having 50 regularly spaced values starting at −2 and ending at 5.

2.   *a.* Create the vector **x** having 50 logarithmically spaced values starting at 10 and ending at 1000.

    b. Create the vector **x** having 20 logarithmically spaced values starting at 10 and ending at 1000.

3.*   Use MATLAB to create a vector **x** having six values between 0 and 10 (including the endpoints 0 and 10). Create an array **A** whose first row contains the values 3*x* and whose second row contains the values 5*x* −20.

4.   Repeat Problem 3 but make the first column of **A** contain the values 3*x* and the second column contain the values 5*x* − 20.

5.   Type this matrix in MATLAB and use MATLAB to carry out the following instructions.

$$\mathbf{A} = \begin{bmatrix} 3 & 7 & -4 & 12 \\ -5 & 9 & 10 & 2 \\ 6 & 13 & 8 & 11 \\ 15 & 5 & 4 & 1 \end{bmatrix}$$

    a. Create a vector **v** consisting of the elements in the second column of **A**.

    b. Create a vector **w** consisting of the elements in the second row of **A**.

6.   Type this matrix in MATLAB and use MATLAB to carry out the following instructions.

$$\mathbf{A} = \begin{bmatrix} 3 & 7 & -4 & 12 \\ -5 & 9 & 10 & 2 \\ 6 & 13 & 8 & 11 \\ 15 & 5 & 4 & 1 \end{bmatrix}$$

    a. Create a 4 × 3 array **B** consisting of all elements in the second through fourth columns of **A**.

    b. Create a 3 × 4 array **C** consisting of all elements in the second through fourth rows of **A**.

    c. Create a 2 × 3 array **D** consisting of all elements in the first two rows and the last three columns of **A**.

7.*   Compute the length and absolute value of the following vectors:

    a. **x** = [2, 4, 7]

    b. **y** = [2, −4, 7]

    c. **z** = [5 + 3*i*, −3 + 4*i*, 2 − 7*i*]

8. Given the matrix

$$A = \begin{bmatrix} 3 & 7 & -4 & 12 \\ -5 & 9 & 10 & 2 \\ 6 & 13 & 8 & 11 \\ 15 & 5 & 4 & 1 \end{bmatrix}$$

a. Find the maximum and minimum values in each column.

b. Find the maximum and minimum values in each row.

9. Given the matrix

$$A = \begin{bmatrix} 3 & 7 & -4 & 12 \\ -5 & 9 & 10 & 2 \\ 6 & 13 & 8 & 11 \\ 15 & 5 & 4 & 1 \end{bmatrix}$$

a. Sort each column and store the result in an array **B**.

b. Sort each row and store the result in an array **C**.

c. Add each column and store the result in an array **D**.

d. Add each row and store the result in an array **E**.

10. Consider the following arrays.

$$A = \begin{bmatrix} 1 & 4 & 2 \\ 2 & 4 & 100 \\ 7 & 9 & 7 \\ 3 & \pi & 42 \end{bmatrix} \qquad B = \ln(A)$$

Write MATLAB expressions to do the following.

a. Select just the second row of **B**.

b. Evaluate the sum of the second row of **B**.

c. Multiply the second column of **B** and the first column of **A** element by element.

d. Evaluate the maximum value in the vector resulting from element-by-element multiplication of the second column of **B** with the first column of **A**.

e. Use element-by-element division to divide the first row of **A** by the first three elements of the third column of **B**. Evaluate the sum of the elements of the resulting vector.

Section 2.2

11.* *a.* Create a three-dimensional array **D** whose three "layers" are these matrices:

$$A = \begin{bmatrix} 3 & -2 & 1 \\ 6 & 8 & -5 \\ 7 & 9 & 10 \end{bmatrix} \qquad B = \begin{bmatrix} 6 & 9 & -4 \\ 7 & 5 & 3 \\ -8 & 2 & 1 \end{bmatrix} \qquad C = \begin{bmatrix} -7 & -5 & 2 \\ 10 & 6 & 1 \\ 3 & -9 & 8 \end{bmatrix}$$

b. Use MATLAB to find the largest element in each layer of **D** and the largest element in **D**.

Section 2.3

12. Given the vectors

$$\mathbf{x} = [5 \quad 9 \quad -3] \qquad\qquad \mathbf{y} = [7 \quad 4 \quad 2]$$

do the following by hand, then check your answer using MATLAB.
a. Find the sum of **x** and **y**.
b. Find the array product w = x.*y.
c. Find the array product z = y.*x. Is z = w?

13. Given the matrices

$$A = \begin{bmatrix} 9 & 6 \\ 2 & 7 \end{bmatrix} \qquad\qquad B = \begin{bmatrix} 8 & 9 \\ 6 & 2 \end{bmatrix}$$

do the following by hand, then check your answer using MATLAB.
a. Find the sum of **A** and **B**.
b. Find the array product w = A.*B.
c. Find the array product z = B.*A. Is z = w?

14. Given the vectors

$$\mathbf{x} = [10 \quad 8 \quad 3] \qquad\qquad \mathbf{y} = [9 \quad 2 \quad 6]$$

do the following by hand, then check your answer using MATLAB.
a. Find the array quotient w = x./y.
b. Find the array quotient z = y./x.

15.* Given the matrices

$$A = \begin{bmatrix} -7 & 11 \\ 4 & 9 \end{bmatrix} \qquad B = \begin{bmatrix} 4 & -5 \\ 12 & -2 \end{bmatrix} \qquad C = \begin{bmatrix} -3 & -9 \\ 7 & 8 \end{bmatrix}$$

Use MATLAB to
a. Find **A** + **B** + **C**.
b. Find **A** − **B** + **C**.

c. Verify the associative law

$$(A + B) + C = A + (B + C)$$

d. Verify the commutative law

$$A + B + C = B + C + A = A + C + B$$

16.   Given the matrices

$$A = \begin{bmatrix} 5 & 9 \\ 6 & 2 \end{bmatrix} \qquad B = \begin{bmatrix} 4 & 7 \\ 2 & 8 \end{bmatrix}$$

do the following by hand, then check your answer using MATLAB.
a. Find the array quotient C = A./B.
b. Find the array quotient D = B./A.
c. Find the array quotient E = A.\B.
d. Find the array quotient F = B.\A.
e. Are any of C, D, E, or F equal?

17.* Given the matrices

$$A = \begin{bmatrix} 56 & 32 \\ 24 & -16 \end{bmatrix} \qquad B = \begin{bmatrix} 14 & -4 \\ 6 & -2 \end{bmatrix}$$

Use MATLAB to
a. Find the result of **A** times **B** using the array product.
b. Find the result of **A** divided by **B** using array right division.
c. Find **B** raised to the third power element by element.

18.   The xy trajectory of a projectile having an initial speed v_0 at an angle A with the horizontal is described by the following equations, where $x(0) = y(0) = 0$:

$$x = (v_0 \cos A)t \qquad\qquad y = (v_0 \sin A)t - \frac{1}{2}gt^2$$

Use the values $v_0 = 100$ m/s, $A = 35$ degrees, and $g = 9.81$ m/s^2. Note that we do not know the time of flight t_{hit} (the time it takes for the projectile to hit the ground at $y = 0$).
a. Write a MATLAB program to compute t_{hit} and the maximum height y_{max} reached by the projectile. Hint: because the trajectory is symmetric, t_{hit} is twice the time to reach y_{max}.
b. Expand your program from part (a) to plot the trajectory y vs. x for $0 \le t \le t_{hit}$.

19.   Plot the following function for x over the interval $-2 \le x \le 16$

$$f(x) = \frac{4 \cos x}{x + e^{-0.75x}}$$

Use enough points to get a smooth curve.

20. Plot the following function for x over the interval $-2\pi \le x \le 2\pi$.

$$f(x) = 3x\cos^2 x - 2x$$

Use enough points to get a smooth curve.

21. Plot the following function for x over the interval $-3.5 \le x \le 10$.

$$f(x) = 2.5^{0.5x}\sin 5x$$

Use enough points to get a smooth curve.

22. A ship travels on a straight line course described by $y = 2x - 10$, where distances are measured in kilometers. The ship starts when $x = -10$ and ends when $x = 30$. Calculate the distance at closest approach to a lighthouse located at the coordinate origin (0, 0). Do not solve this using a plot.

23.* The mechanical work W done in using a force F to push a block through a distance D is $W = FD$. The following table gives data on the amount of force used to push a block through the given distance over five segments of a certain path. The force varies because of the differing friction properties of the surface.

	Path segment				
	1	**2**	**3**	**4**	**5**
Force (N)	400	550	700	500	600
Distance (m)	3	0.5	0.75	1.5	5

Use MATLAB to find (*a*) the work done on each segment of the path and (*b*) the total work done over the entire path.

24. Plane A is heading southwest at 300 mi/hr, while plane B is heading west at 150 mi/hr. What are the velocity and the speed of plane A relative to plane B?

25. The following table shows the hourly wages, hours worked, and output (number of widgets produced) in one week for five widget makers.

	Worker				
	1	**2**	**3**	**4**	**5**
Hourly wage ($)	5	5.50	6.50	6	6.25
Hours worked	40	43	37	50	45
Output (widgets)	1000	1100	1000	1200	1100

Use MATLAB to answer these questions:

a. How much did each worker earn in the week?

b. What is the total salary amount paid out?

c. How many widgets were made?

d. What is the average cost to produce one widget?

e. How many hours does it take to produce one widget on average?

f. Assuming that the output of each worker has the same quality, which worker is the most efficient? Which is the least efficient?

26. Two divers start at the surface and establish the following coordinate system: x is to the west, y is to the north, and z is down. Diver 1 swims 60 ft east, then 25 ft south, and then dives 30 ft. At the same time, diver 2 dives 20 ft, swims east 30 ft, and then south 55 ft.

    a. Compute the distance between diver 1 and the starting point.

    b. How far in each direction must diver 1 swim to reach diver 2?

    c. How far in a straight line must diver 1 swim to reach diver 2?

27. The potential energy stored in a spring is $kx^2/2$, where k is the spring constant and x is the compression in the spring. The force required to compress the spring is kx. The following table gives the data for five springs:

	Spring				
	1	**2**	**3**	**4**	**5**
Force (N)	11	7	8	10	9
Spring constant k (N/m)	1000	600	900	1300	700

    Use MATLAB to find (*a*) the compression x in each spring and (*b*) the potential energy stored in each spring.

28. A company must purchase five kinds of material. The following table gives the price the company pays per ton for each material, along with the number of tons purchased in the months of May, June, and July:

		Quantity purchased (tons)		
Material	**Price ($/ton)**	**May**	**June**	**July**
1	300	5	4	6
2	550	3	2	4
3	400	6	5	3
4	250	3	5	4
5	500	2	4	3

    Use MATLAB to answer these questions:

    a. Create a 5 × 3 matrix containing the amounts spent on each item for each month.

    b. What is the total spent in May? in June? in July?

    c. What is the total spent on each material in the three-month period?

    d. What is the total spent on all materials in the three-month period?

29. A fenced enclosure consists of a rectangle of length L and width $2R$, and a semicircle of radius R, as shown in Figure P29. The enclosure is to be

built to have an area A of 1600 ft^2. The cost of the fence is \$40/ft for the curved portion and \$30/ft for the straight sides. Use the `min` function to determine with a resolution of 0.01 ft the values of R and L required to minimize the total cost of the fence. Also compute the minimum cost.

Figure P29

30. A water tank consists of a cylindrical part of radius r and height h, and a hemispherical top. The tank is to be constructed to hold 500 m^3 of fluid when filled. The surface area of the cylindrical part is $2\pi rh$, and its volume is $\pi r^2 h$. The surface area of the hemispherical top is given by $2\pi r^2$, and its volume is given by $2\pi r^3/3$. The cost to construct the cylindrical part of the tank is \$300/m^2 of surface area; the hemispherical part costs \$400/m^2. Plot the cost versus r for $2 \le r \le 10$ m, and determine the radius that results in the least cost. Compute the corresponding height h.

31. Write a MATLAB assignment statement for each of the following functions, assuming that w, x, y, and z are row vectors of equal length and that c and d are scalars.

$$f = \frac{1}{\sqrt{2\pi c/x}} \qquad E = \frac{x + w/(y + z)}{x + w/(y - z)}$$

$$A = \frac{e^{-c/(2x)}}{(\ln y)\sqrt{dz}} \qquad S = \frac{x(2.15 + 0.35y)^{1.8}}{z(1 - x)^y}$$

32. *a.* After a dose, the concentration of medication in the blood declines due to metabolic processes. The *half-life* of a medication is the time required after an initial dosage for the concentration to be reduced by one-half. A common model for this process is

$$C(t) = C(0)e^{-kt}$$

where $C(0)$ is the initial concentration, t is time (in hours), and k is called the *elimination rate constant*, which varies among individuals. For a particular bronchodilator, k has been estimated to be in the range $0.047 \le k \le 0.107$ per hour. Find an expression for the half-life in terms of k, and obtain a plot of the half-life versus k for the indicated range.

b. If the concentration is initially zero and a constant delivery rate is started and maintained, the concentration as a function of time is described by

$$C(t) = \frac{a}{k}\left(1 - e^{-kt}\right)$$

where a is a constant that depends on the delivery rate. Plot the concentration after 1 hr, $C(1)$, versus k for the case where $a = 1$ and k is in the range $0.047 \le k \le 0.107$ per hour.

33. A cable of length L_c supports a beam of length L_b, so that it is horizontal when the weight W is attached at the beam end. The principles of statics can be used to show that the tension force T in the cable is given by

$$T = \frac{L_b L_c W}{D \sqrt{L_b^2 - D^2}}$$

where D is the distance of the cable attachment point to the beam pivot. See Figure P33.

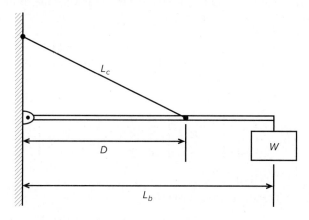

Figure P33

   a. For the case where $W = 400$ N, $L_b = 3$ m, and $L_c = 5$ m, use element-by-element operations and the min function to compute the value of D that minimizes the tension T. Compute the minimum tension value.

   b. Check the sensitivity of the solution by plotting T versus D. How much can D vary from its optimal value before the tension T increases 10 percent above its minimum value?

34. Given the vectors

$$\mathbf{x} = \begin{bmatrix} 3 \\ 7 \\ 2 \end{bmatrix} \qquad \mathbf{y} = [4 \quad 9 \quad 5]$$

do the following by hand, then check your answer using MATLAB.
   a. Find the matrix product w = x*y.
   b. Find the matrix product z = y*x. Is z = w?

35. Given

$$\mathbf{x} = \begin{bmatrix} 3 \\ 7 \\ 2 \end{bmatrix} \qquad \mathbf{A} = \begin{bmatrix} 2 & 6 & 5 \\ 3 & 7 & 4 \\ 8 & 10 & 9 \end{bmatrix}$$

do the following by hand, then check your answer using MATLAB.
a. Find the product **Ax.**
b. Find the product **xA.** Explain the result.

Section 2.4

36.* Use MATLAB to find the products **AB** and **BA** for the following matrices:

$$\mathbf{A} = \begin{bmatrix} 11 & 5 \\ -9 & -4 \end{bmatrix} \qquad \mathbf{B} = \begin{bmatrix} -7 & -8 \\ 6 & 2 \end{bmatrix}$$

37. Given the matrices

$$\mathbf{A} = \begin{bmatrix} 4 & -2 & 1 \\ 6 & 8 & -5 \\ 7 & 9 & 10 \end{bmatrix} \quad \mathbf{B} = \begin{bmatrix} 6 & 9 & -4 \\ 7 & 5 & 3 \\ -8 & 2 & 1 \end{bmatrix} \quad \mathbf{C} = \begin{bmatrix} -4 & -5 & 2 \\ 10 & 6 & 1 \\ 3 & -9 & 8 \end{bmatrix}$$

Use MATLAB to
a. Verify the associative property

$$\mathbf{A}(\mathbf{B} + \mathbf{C}) = \mathbf{A}\mathbf{B} + \mathbf{A}\mathbf{C}$$

b. Verify the distributive property

$$(\mathbf{A}\mathbf{B})\mathbf{C} = \mathbf{A}(\mathbf{B}\mathbf{C})$$

38. The following tables show the costs associated with a certain product and the production volume for the four quarters of the business year. Use MATLAB to find (*a*) the quarterly costs for materials, labor, and transportation; (*b*) the total material, labor, and transportation costs for the year; and (*c*) the total quarterly costs.

	Unit product costs ($\$ \times 10^3$)		
Product	Materials	Labor	Transportation
1	7	3	2
2	3	1	3
3	9	4	5
4	2	5	4
5	6	2	1

Product	Quarterly production volume			
	Quarter 1	**Quarter 2**	**Quarter 3**	**Quarter 4**
1	16	14	10	12
2	12	15	11	13
3	8	9	7	11
4	14	13	15	17
5	13	16	12	18

39. * Aluminum alloys are made by adding other elements to aluminum to improve its properties, such as hardness or tensile strength. The following table shows the composition of five commonly used alloys, which are known by their alloy numbers (2024, 6061, and so on) [Kutz, 1999]. Obtain a matrix algorithm to compute the amounts of raw materials needed to produce a given amount of each alloy. Use MATLAB to determine how much raw material of each type is needed to produce 1000 tons of each alloy.

Alloy	Composition of aluminum alloys				
	%Cu	**%Mg**	**%Mn**	**%Si**	**%Zn**
2024	4.4	1.5	0.6	0	0
6061	0	1	0	0.6	0
7005	0	1.4	0	0	4.5
7075	1.6	2.5	0	0	5.6
356.0	0	0.3	0	7	0

40. Redo Example 2.4–4 as a script file to allow the user to examine the effects of labor costs. Allow the user to input the four labor costs in the following table. When you run the file, it should display the quarterly costs and the category costs. Run the file for the case where the unit labor costs are $3000, $7000, $4000, and $8000, respectively.

Product costs

Product	Unit costs ($ × 103)		
	Materials	**Labor**	**Transportation**
1	6	2	1
2	2	5	4
3	4	3	2
4	9	7	3

Quarterly production volume

Product	Quarter 1	Quarter 2	Quarter 3	Quarter 4
1	10	12	13	15
2	8	7	6	4
3	12	10	13	9
4	6	4	11	5

41. Vectors with three elements can represent position, velocity, and acceleration. A mass of 5 kg, which is 3 m away from the x axis, starts at $x = 2$ m and moves with a speed of 10 m/s parallel to the y axis. Its velocity is thus described by $\mathbf{v} = [0, 10, 0]$, and its position is described by $\mathbf{r} = [2, 10t + 3, 0]$. Its angular momentum vector $\mathbf{L}$ is found from $\mathbf{L} = m(\mathbf{r} \times \mathbf{v})$, where m is the mass. Use MATLAB to

   a. Compute a matrix $\mathbf{P}$ whose 11 rows are the values of the position vector $\mathbf{r}$ evaluated at the times $t = 0, 0.5, 1, 1.5,\ldots, 5$ s.

   b. What is the location of the mass when $t = 5$ s?

   c. Compute the angular momentum vector $\mathbf{L}$. What is its direction?

42. * The *scalar triple product* computes the magnitude M of the moment of a force vector $\mathbf{F}$ about a specified line. It is $M = (\mathbf{r} \times \mathbf{F}) \cdot \mathbf{n}$, where $\mathbf{r}$ is the position vector from the line to the point of application of the force and $\mathbf{n}$ is a unit vector in the direction of the line.

Use MATLAB to compute the magnitude M for the case where $\mathbf{F} = [12, -5, 4]$ N, $\mathbf{r} = [-3, 5, 2]$ m, and $\mathbf{n} = [6, 5, -7]$.

43. Verify the identity

$$\mathbf{A} \times (\mathbf{B} \times \mathbf{C}) = \mathbf{B}(\mathbf{A} \cdot \mathbf{C}) - \mathbf{C}(\mathbf{A} \cdot \mathbf{B})$$

for the vectors $\mathbf{A} = 7\mathbf{i} - 3\mathbf{j} + 7\mathbf{k}$, $\mathbf{B} = -6\mathbf{i} + 2\mathbf{j} + 3\mathbf{k}$, and $\mathbf{C} = 2\mathbf{i} + 8\mathbf{j} - 8\mathbf{k}$.

44. The area of a parallelogram can be computed from $|\mathbf{A} \times \mathbf{B}|$, where $\mathbf{A}$ and $\mathbf{B}$ define two sides of the parallelogram (see Figure P44). Compute the area of a parallelogram defined by $\mathbf{A} = 5\mathbf{i}$ and $\mathbf{B} = \mathbf{i} + 3\mathbf{j}$.

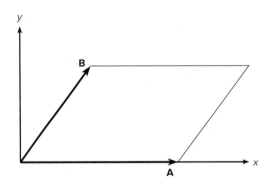

Figure P44

45. The volume of a parallelepiped can be computed from $|\mathbf{A} \cdot (\mathbf{B} \times \mathbf{C})|$, where $\mathbf{A}, \mathbf{B}$, and $\mathbf{C}$ define three sides of the parallelepiped (see Figure P45). Compute the volume of a parallelepiped defined by $\mathbf{A} = 5\mathbf{i}$, $\mathbf{B} = 2\mathbf{i} + 4\mathbf{j}$, and $\mathbf{C} = 3\mathbf{i} - 2\mathbf{k}$.

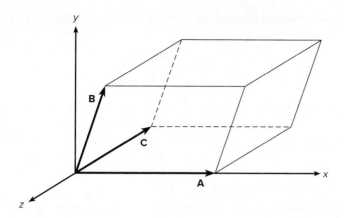

Figure P45

Section 2.5

46. Use MATLAB to plot the polynomials $y = 3x^4 - 6x^3 + 8x^2 + 4x + 90$ and $z = 3x^3 + 5x^2 - 8x + 70$ over the interval $-3 \le x \le 3$. Properly label the plot and each curve. The variables y and z represent current in milliamperes; the variable x represents voltage in volts.

47. Use MATLAB to plot the polynomial $y = 3x^4 - 5x^3 - 28x^2 - 5x + 200$ on the interval $-1 \le x \le 1$. Put a grid on the plot and use the `ginput` function to determine the coordinates of the peak of the curve.

48. Use MATLAB to find the following product:

$$\left(10x^3 - 9x^2 - 6x + 12\right)\left(5x^3 - 4x^2 - 12x + 8\right)$$

49.* Use MATLAB to find the quotient and remainder of

$$\frac{14x^3 - 6x^2 + 3x + 9}{5x^2 + 7x - 4}$$

50.* Use MATLAB to evaluate

$$\frac{24x^3 - 9x^2 - 7}{24x^3 + 5x^2 - 3x - 7}$$

at $x = 5$.

51. The *ideal gas law* provides one way to estimate the pressures and volumes of a gas in a container. The law is

$$P = \frac{RT}{\hat{V}}$$

More accurate estimates can be made with the *van der Waals* equation

$$P = \frac{RT}{\hat{V} - b} - \frac{a}{\hat{V}^2}$$

where the term b is a correction for the volume of the molecules and the term $a/\hat{V}^2$ is a correction for molecular attractions. The values of a and b depend on the type of gas. The gas constant is R, the *absolute* temperature is T, and the gas specific volume is $\hat{V}$. If 1 mol of an ideal gas were confined to a volume of 22.41 L at 0°C (273.2 K), it would exert a pressure of 1 atm. In these units, $R = 0.08206$.

For chlorine (Cl_2), $a = 6.49$ and $b = 0.0562$. Compare the specific volume estimates $\hat{V}$ given by the ideal gas law and the van der Waals equation for 1 mol of Cl_2 at 300 K and a pressure of 0.95 atm.

52. Aircraft A is flying east at 320 mi/hr, while aircraft B is flying south at 160 mi/hr. At 1:00 P.M. the aircraft are located as shown in Figure P52.

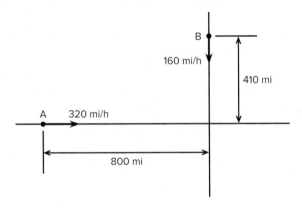

Figure P52

a. Obtain the expression for the distance D between the aircraft as a function of time. Plot D versus time until D reaches its minimum value.

b. Use the `roots` function to compute the time when the aircraft are first within 30 mi of each other.

53. The function

$$y = \frac{3x^2 - 12x + 20}{x^2 - 7x + 10}$$

approaches ∞ as $x \to 2$ and as $x \to 5$. Plot this function over the range $0 \le x \le 7$. Choose an appropriate range for the y axis.

54. The following formulas are commonly used by engineers to predict the lift and drag of an airfoil:

$$L = \frac{1}{2}\rho C_L SV^2$$

$$D = \frac{1}{2}\rho C_D SV^2$$

where L and D are the lift and drag forces, V is the airspeed, S is the wing span, ρ is the air density, and C_L and C_D are the *lift* and *drag* coefficients. Both C_L and C_D depend on α, the angle of attack, the angle between the relative air velocity and the airfoil's chord line.

Wind tunnel experiments for a particular airfoil have resulted in the following formulas.

$$C_L = 4.47 \times 10^{-5}\alpha^3 + 1.15 \times 10^{-3}\alpha^2 + 6.66 \times 10^{-2}\alpha + 1.02 \times 10^{-1}$$
$$C_D = 5.75 \times 10^{-6}\alpha^3 + 5.09 \times 10^{-4}\alpha^2 + 1.8 \times 10^{-4}\alpha + 1.25 \times 10^{-2}$$

where α is in degrees.

Plot the lift and drag of this airfoil versus V for $0 \le V \le 150$ mi/hr (you must convert V to ft/sec; there is 5280 ft/mi). Use the values $\rho = 0.002378$ slug/ft^3 (air density at sea level), $\alpha = 10°$, and $S = 36$ ft. The resulting values of L and D will be in pounds.

55. The lift-to-drag ratio is an indication of the effectiveness of an airfoil. Referring to Problem 54, the equations for lift and drag are

$$L = \frac{1}{2}\rho C_L SV^2$$

$$D = \frac{1}{2}\rho C_D SV^2$$

where, for a particular airfoil, the lift and drag coefficients versus angle of attack α are given by

$$C_L = 4.47 \times 10^{-5}\alpha^3 + 1.15 \times 10^{-3}\alpha^2 + 6.66 \times 10^{-2}\alpha + 1.02 \times 10^{-1}$$
$$C_D = 5.75 \times 10^{-6}\alpha^3 + 5.09 \times 10^{-4}\alpha^2 + 1.8 \times 10^{-4}\alpha + 1.25 \times 10^{-2}$$

Using the first two equations, we see that the lift-to-drag ratio is given simply by the ratio C_L/C_D.

$$\frac{L}{D} = \frac{\frac{1}{2}\rho C_L SV^2}{\frac{1}{2}\rho C_D SV^2} = \frac{C_L}{C_D}$$

Plot L/D versus α for $-2° \le \alpha \le 22°$. Determine the angle of attack that maximizes L/D.

56. The xy trajectory of a projectile having an initial speed v_0 at an angle A with the horizontal is described by the following equations, where $x(0) = y(0) = 0$:

$$x = (v_0 \cos A)t \qquad\qquad y = (v_0 \sin A)t - \frac{1}{2}gt^2$$

Use the values $v_0 = 100$ m/s, $A = 35$ degrees, and $g = 9.81$ m/s^2. Note that we do not know the time of flight t_{hit} (the time it takes for the projectile to hit the ground at $y = 0$).

  a. Write a MATLAB program to compute t_{hit} by solving the y equation for $y = 0$.

  b. Expand your program from part (a) to determine whether or not the projectile reaches a certain height yd, and find the time to reach this height. Do this by solving the y equation for $y = yd$. Does the projectile reach 100 m? Does it reach 200 m? How long does it take it each case?

Section 2.6

57. a. Use both cell indexing and content indexing to create the following 2 × 2 cell array.

Motor 28C	Test ID 6
$\begin{bmatrix} 3 & 9 \\ 7 & 2 \end{bmatrix}$	[6 5 1]

  b. What are the contents of the (1,1) element in the (2,1) cell in this array?

58. The capacitance of two parallel conductors of length L and radius r, separated by a distance d in air, is given by

$$C = \frac{\pi \varepsilon L}{\ln[(d - r)/r]}$$

where ε is the permittitivity of air ($\varepsilon = 8.854 \times 10^{-12}$ F/m). Create a cell array of capacitance values versus d, L, and r for $d = 0.003, 0.004, 0.005$, and 0.01 m; $L = 1, 2, 3$ m; and $r = 0.001, 0.002, 0.003$ m. Use MATLAB to determine the capacitance value for $d = 0.005$, $L = 2$, and $r = 0.001$.

Section 2.7

59. a. Create a structure array that contains the conversion factors for converting units of mass, force, and distance between the metric SI system and the British Engineering System.

  b. Use your array to compute the following:
  - The number of meters in 48 ft.
  - The number of feet in 130 m.
  - The number of pounds equivalent to 36 N.

- The number of newtons equivalent to 10 lb.
- The number of kilograms in 12 slugs.
- The number of slugs in 30 kg.

60. Create a structure array that contains the following information fields concerning the road bridges in a town: bridge location, maximum load (tons), year built, year due for maintenance. Then enter the following data into the array:

Location	Max. load	Year built	Due for maintenance
Smith St.	80	1928	2011
Hope Ave.	90	1950	2013
Clark St.	85	1933	2012
North Rd.	100	1960	2012

61. Edit the structure array created in Problem 60 to change the maintenance data for the Clark St. bridge from 2012 to 2018.

62. Add the following bridge to the structure array created in Problem 60.

Location	Max. load	Year built	Due for maintenance
Shore Rd.	85	1997	2014

Engineering in the 21st Century...

Robot-Assisted Surgery

Many advances in medicine are really engineering achievements, and many engineers will be needed in this area.

Robot-assisted surgery is now often used for hip and knee replacement. One of the challenges in such surgery is achieving proper alignment of the artificial joint. A CAT scan of the patient's hip or knee is used to create a geometric model of the patient's anatomy. A suite of sensors provides information about the patient's position during surgery, and this information, when compared with the geometric model, enables the surgeon to align the joint properly. It also enables the robot arm's controller to prevent the surgeon from cutting outside the desired region.

Robot-assisted surgery is already common in some applications that require precise, steady motions, such as prostate surgery, in which the robot filters out tremors often present in the human surgeon's hand. The next step is to develop *haptic feedback,* or a sense of touch, which enables the surgeon to indirectly feel the tissues being manipulated. Haptic feedback will be important for *telesurgery,* in which a surgeon remotely guides a surgical robot. This technology would allow delivery of medical services to remote areas.

Surgery simulators use 3D graphics and motion sensors to simulate procedures to train surgeons without the need of a patient, cadaver, or animal. They are best suited for developing eye-hand coordination and the skill to perform three-dimensional actions using a two-dimensional screen as a guide.

Designing such devices requires geometric analysis, control system design, and image processing. The MATLAB Image Processing toolbox and the several MATLAB toolboxes dealing with control system design are useful for such applications. ■

Functions

OUTLINE

3.1   Elementary Mathematical Functions

3.2   User-Defined Functions

3.3   Additional Function Types

3.4   File Functions

3.5   Summary

Problems

MATLAB has many built-in functions, including trigonometric, logarithmic, and hyperbolic functions, as well as functions for processing arrays. These functions are summarized in Section 3.1. In addition, you can define your own functions with a *function* file, and you can use them just as conveniently as the built-in functions. We explain this technique in Section 3.2. Section 3.3 covers additional topics in function programming, including function handles, anonymous functions, subfunctions, and nested functions. Data files and spreadsheet files are useful with MATLAB. The functions useful for importing and exporting such files are covered in Section 3.4.

Sections 3.1 and 3.2 contain essential topics and must be covered. The material in Section 3.3 is useful for creating large programs. The material in Section 3.4 is useful for readers who must work with large data sets or spreadsheets.

3.1 Elementary Mathematical Functions

You can use the `lookfor` command to find functions that are relevant to your application. For example, type `lookfor imaginary` to get a list of the functions that deal with imaginary numbers. You will see listed

```
imag Complex imaginary part
i    Imaginary unit
j    Imaginary unit
```

Note that `imaginary` is not a MATLAB function, but the word is found in the Help descriptions of the MATLAB function `imag` and the special symbols i and j. Their names and brief descriptions are displayed when you type `lookfor imaginary`. If you know the correct spelling of a MATLAB function, for example, `disp`, you can type `help disp` to obtain a description of the function.

Some of the functions, such as `sqrt` and `sin`, are built in. These are stored as image files and are not M-files. They are part of the MATLAB core so they are very efficient, but the computational details are not readily accessible. Some functions are implemented in M-files. You can see the code and even modify it, although this is not recommended.

Exponential and Logarithmic Functions

Table 3.1–1 summarizes some of the common mathematical functions. An example is the square root function `sqrt`. To compute $\sqrt{9}$, you type `sqrt(9)` at the command line. When you press **Enter,** you see the result ans = 3. You can use functions with variables. For example, consider the session

```
>>x = -9; y = sqrt(x)
y =
    0 + 3.0000i
```

FUNCTION ARGUMENT

Note that the `sqrt` function returns the positive root only, so `sqrt(4)` returns 2 and not −2.

The power of MATLAB is due to its ability to handle *vectorized functions*, which means that the *function argument* can be a vector. For example, if

Table 3.1–1 Some common mathematical functions

Exponential	
`exp(x)`	Exponential; e^x.
`sqrt(x)`	Square root; $\sqrt{x}$.
Logarithmic	
`log(x)`	Natural logarithm; ln x.
`log10(x)`	Common (base-10) logarithm; log $x = \log_{10}x$.
Complex	
`abs(x)`	Absolute value; x.
`angle(x)`	Angle of a complex number x.
`conj(x)`	Complex conjugate.
`imag(x)`	Imaginary part of a complex number x.
`real(x)`	Real part of a complex number x.
Numeric	
`ceil(x)`	Round to the nearest integer toward ∞.
`fix(x)`	Round to the nearest integer toward zero.
`floor(x)`	Round to the nearest integer toward $-\infty$.
`round(x)`	Round toward the nearest integer.
`sign(x)`	Signum function: $+1$ if $x > 0$; 0 if $x = 0$; -1 if $x < 0$.

$x = [4, 9, 16]$, typing `sqrt(x)` give us the vector $[2, 3, 4]$. For some MATLAB functions the argument is not restricted to being a vector; it can be a general array. For example, if $A = [4, 9, 16; 25, 36, 49]$, typing `sqrt(A)` give us the matrix $[2, 3, 4; 5, 6, 7]$. Note that the `sqrt` function returns the positive root only.

One of the strengths of MATLAB is that it will treat a variable as an array automatically. For example, to compute the square roots of 5, 7, and 15, type

```
>>x = [5,7,15]; y = sqrt(x)
y =
    2.2361    2.6358    3.8730
```

The square root function operates on every element in the array `x`.

Similarly, we can type `exp(2)` to obtain $e^2 = 7.3891$, where e is the base of the natural logarithms. Typing `exp(1)` gives 2.7183, which is e. Note that in mathematics text, ln x denotes the *natural* logarithm, where $x = e^y$ implies that

$$\ln x = \ln(e^y) = y \ln e = y$$

because ln $e = 1$. However, this notation has not been carried over into MATLAB, which uses `log(x)` to represent ln x.

The *common* (base-10) logarithm is denoted in text by log x or $\log_{10} x$. It is defined by the relation $x = 10^y$; that is,

$$\log_{10} x = \log_{10} 10^y = y \log_{10} 10 = y$$

because $\log_{10} 10 = 1$. The MATLAB common logarithm function is `log10(x)`. A common mistake is to type `log(x)`, instead of `log10(x)`.

Another common error is to forget to use the array multiplication operator `.*`. Note that in the MATLAB expression `y = exp(x).*log(x)`, we need to use the operator `.*` if `x` is an array because both `exp(x)` and `log(x)` will be arrays.

Complex Number Functions

Chapter 1 explained how MATLAB easily handles complex number arithmetic. In the *rectangular* representation the number $a + ib$ represents a point in the xy plane. The number's real part a is the x coordinate of the point, and the imaginary part b is the y coordinate. The *polar* representation uses the distance M of the point from the origin, which is the length of the hypotenuse, and the angle θ the hypotenuse makes with the *positive real* axis. The pair (M, θ) is simply the polar coordinates of the point. From the Pythagorean theorem, the length of the hypotenuse is given by $M = \sqrt{a^2 + b^2}$, which is called the *magnitude* of the number. The angle θ can be found from the trigonometry of the right triangle. It is $\theta = \arctan(b/a)$.

Adding and subtracting complex numbers by hand is easy when they are in the rectangular representation. However, the polar representation facilitates multiplication and division of complex numbers by hand. We must enter complex numbers in

MATLAB using the rectangular form, and its answers will be given in that form. We can obtain the rectangular representation from the polar representation as follows:

$$a = M \cos \theta \qquad b = M \sin \theta$$

The MATLAB abs(x) and angle(x) functions calculate the magnitude M and angle θ of the complex number x. The functions real(x) and imag(x) return the real and imaginary parts of x. The function conj(x) computes the complex conjugate of x.

The magnitude of the product z of two complex numbers x and y is equal to the product of their magnitudes: $|z| = |x||y|$. The angle of the product is equal to the sum of the angles: $\angle z = \angle x + \angle y$. These facts are demonstrated below:

```
>>x = -3 + 4i; y = 6 - 8i;
>>mag_x = abs(x)
mag_x =
    5.0000
>>mag_y = abs(y)
mag_y =
    10.0000
>>mag_product = abs(x*y)
    50.0000
>>angle_x = angle(x)
angle_x =
    2.2143
>>angle_y = angle(y)
angle_y =
    -0.9273
>>sum_angles = angle_x + angle_y
sum_angles =
    1.2870
>>angle_product = angle(x*y)
angle_product =
    1.2870
```

Similarly, for division, if $z = x/y$, then $|z| = |x|/|y|$ and $\angle z = \angle x - \angle y$.

Note that when x is a vector of *real* values, abs(x) does not give the geometric length of the vector. This length is given by norm(x). If x is a complex number representing a two-dimensional geometric vector, then abs(x) gives its geometric length.

Numeric Functions

Some functions have extended syntax that is not easily summarized in a table. An example of this is the round function. The round function rounds to the nearest integer. If x = [2.3,2.6,3.9], typing round(x) gives the results 2, 3, 4. There are several options for rounding numbers in addition to the basic syntax round(x). You can round numbers to a specified number of decimal places or significant digits. The syntax round(x,n), for positive integers n, rounds to n digits to the right of the decimal point. If n is zero, x is rounded to the nearest integer. If n is less than zero, x is rounded to the left of the decimal point. The number n must be a scalar integer.

To round to n significant digits you type round(x,n,'significant'). For example, round(pi) gives 3; round(pi,3) gives 3.1420; round(pi,3,'significant') gives 3.1400; and, round(13.47,-1) gives 10.

The fix function rounds to the nearest integer toward zero. If x = [2.3,2.6,3.9], typing fix(x) gives the results 2, 2, 3. The ceil function (which stands for "ceiling") rounds to the nearest integer toward ∞. Typing ceil(x) produces the answers 3, 3, 4.

Suppose y = [-2.6,-2.3,5.7]. The floor function rounds to the nearest integer toward $-\infty$. Typing floor(y) produces the result -3, -3, 5. Typing fix(y) produces the answer -2, -2, 5. The abs function computes the absolute value. Thus abs(y) produces 2.6, 2.3, 5.7.

Test Your Understanding

T3.1–1 For several values of x and y, confirm that $\ln(xy) = \ln x + \ln y$.

T3.1–2 Find the magnitude, angle, real part, and imaginary part of the number $\sqrt{2 + 6i}$. (Answers: magnitude = 2.5149, angle = 0.6245 rad, real part = 2.0402, imaginary part = 1.4705)

Correctly Specifying Function Arguments

When writing mathematics in text, we use parentheses ( ), brackets [ ], and braces { } to improve the readability of expressions, and we have much latitude over their use. For example, we can write sin 2 in text, but MATLAB requires parentheses surrounding the 2 (which is called the *function argument* or *parameter*). Thus to evaluate sin 2 in MATLAB, we type sin(2). The MATLAB function name must be followed by a pair of parentheses that surround the argument. To express in text the sine of the second element of the array x, we would type sin[x(2)]. However, in MATLAB you cannot use brackets or braces in this way, and you must type sin(x(2)).

FUNCTION ARGUMENT

You can include expressions and other functions as arguments. For example, if x is an array, to evaluate $\sin(x^2 + 5)$, you type sin(x.^2 + 5). To evaluate $\sin(\sqrt{x} + 1)$, you type sin(sqrt(x)+1). Be sure to check the order of precedence and the number and placement of parentheses when typing such expressions. Every left-facing parenthesis requires a right-facing mate. However, this condition does not guarantee that the expression is correct!

Another common mistake involves expressions such as $\sin^2 x$, which means $(\sin x)^2$. In MATLAB, if x is a scalar we write this expression as (sin(x))^2, *not* as sin^2(x), sin^2x, or sin(x^2)! If x is an array we must write (sin(x)).^2.

Trigonometric Functions

Other commonly used functions are `cos(x)`, `tan(x)`, `sec(x)`, and `csc(x)`, which return cos x, tan x, sec x, and csc x, respectively. Table 3.1–2 lists the MATLAB trigonometric functions that operate in radian mode. Thus `sin(5)` computes the sine of 5 rad, not the sine of 5°. Similarly, the inverse trigonometric functions return an answer in radians. The functions that operate in degree mode have the letter d appended to their names. For example, `sind(x)` accepts the value of `x` in degrees. To compute the inverse sine in radians, type `asin(x)`. For example, `asin(1)` returns the answer 1.5708 rad, which is $\pi/2$, while `asind(0.5)` returns 30 degrees. *Note*: In MATLAB, `sin(x)^(-1)` does not give $\sin^{-1}(x)$; it gives $1/\sin(x)$!

MATLAB has two inverse tangent functions. The function `atan(x)` computes $\arctan(x)$—the arctangent or inverse tangent—and returns an angle between $-\pi/2$ and $\pi/2$. Another correct answer is the angle that lies in the opposite quadrant. The user must be able to choose the correct answer. For example, `atan(1)` returns the answer 0.7854 rad, which corresponds to 45°. Thus tan 45° = 1. However, tan(45° + 180°) = tan 225° = 1 also. Thus arctan(1) = 225° is also correct.

MATLAB provides the functions `atan2(y,x)` and `atan2d(y,x)` to determine the arctangent unambiguously, where `x` and `y` are the coordinates of a point. The angle computed by these functions is the angle between the positive real axis and line from the origin $(0, 0)$ to the point (x, y). For example, the point $x = 1$, $y = -1$ corresponds to $-45°$ or -0.7854 rad, and the point $x = -1$, $y = 1$ corresponds to 135° or 2.3562 rad. Typing `atan2d(-1,1)` returns $-45°$, while

Table 3.1–2 Trigonometric functions

Trigonometric*	
`cos(x)`	Cosine; cos x.
`cot(x)`	Cotangent; cot x.
`csc(x)`	Cosecant; csc x.
`sec(x)`	Secant; sec x.
`sin(x)`	Sine; sin x.
`tan(x)`	Tangent; tan x.
Inverse trigonometric†	
`acos(x)`	Inverse cosine; arccos $x = \cos^{-1}x$.
`acot(x)`	Inverse cotangent; arccot $x = \cot^{-1}x$.
`acsc(x)`	Inverse cosecant; arccsc $x = \csc^{-1}x$.
`asec(x)`	Inverse secant; arcsec $x = \sec^{-1}x$.
`asin(x)`	Inverse sine; arcsin $x = \sin^{-1}x$.
`atan(x)`	Inverse tangent; arctan $x = \tan^{-1}x$.
`atan2(y,x)`	Four-quadrant inverse tangent.

*These functions accept x in radians.
†These functions return a value in radians.

typing `atan2d(1,-1)` returns 135°. These are examples of functions that have two arguments. The order of the arguments is important for such functions.

Test Your Understanding

T3.1–3 For several values of x, confirm that $e^{ix} = \cos x + i \sin x$.

T3.1–4 For several values of x in the range $0 \le x \le 2\pi$, confirm that $\sin^{-1}x + \cos^{-1}x = \pi/2$.

T3.1–5 For several values of x in the range $0 \le x \le 2\pi$, confirm that $\tan(2x) = 2 \tan x/(1 - \tan^2 x)$.

Hyperbolic Functions

The *hyperbolic functions* are the solutions of some common problems in engineering analysis. For example, the *catenary* curve, which describes the shape of a hanging cable supported at both ends, can be expressed in terms of the hyperbolic cosine, cosh x, which is defined as

$$\cosh x = \frac{e^x + e^{-x}}{2}$$

The hyperbolic sine, sinh x, is defined as

$$\sinh x = \frac{e^x - e^{-x}}{2}$$

The inverse hyperbolic sine, $\sinh^{-1}x$, is the value y that satisfies $\sinh y = x$.

Several other hyperbolic functions have been defined. Table 3.1–3 lists these hyperbolic functions and the MATLAB commands to obtain them.

Table 3.1–3 Hyperbolic functions

Hyperbolic	
`cosh(x)`	Hyperbolic cosine; $\cosh x = (e^x + e^{-x})/2$.
`coth(x)`	Hyperbolic cotangent; $\cosh x/\sinh x$.
`csch(x)`	Hyperbolic cosecant; $1/\sinh x$.
`sech(x)`	Hyperbolic secant; $1/\cosh x$.
`sinh(x)`	Hyperbolic sine; $\sinh x = (e^x - e^{-x})/2$.
`tanh(x)`	Hyperbolic tangent; $\sinh x/\cosh x$.
Inverse hyperbolic	
`acosh(x)`	Inverse hyperbolic cosine
`acoth(x)`	Inverse hyperbolic cotangent
`acsch(x)`	Inverse hyperbolic cosecant
`asech(x)`	Inverse hyperbolic secant
`asinh(x)`	Inverse hyperbolic sine
`atanh(x)`	Inverse hyperbolic tangent

T3.1–6   For several values of x in the range $0 \leq x \leq 5$, confirm that $\sin(ix) = i \sinh x$.

T3.1–7   For several values of x in the range $-10 \leq x \leq 10$, confirm that

$$\sinh^{-1}x = \ln(x + \sqrt{x^2 + 1}).$$

3.2 User-Defined Functions

FUNCTION FILE

LOCAL VARIABLE

Another type of M-file is a *function file*. Unlike a script file, all the variables in a function file are *local variables*, which means their values are available only within the function. Function files are useful when you need to repeat a set of commands several times. They are the building blocks of larger programs.

To create a function file, open the Editor as described in Chapter 1, by selecting **New** under the HOME tab on the Toolstrip, but instead of selecting Script, select **Function**. The default Editor window for creating function files will appear as shown in Figure 3.2–1. The first line in a function file must begin with a *function definition line* that has a list of inputs and outputs. This line distinguishes a function M-file from a script M-file. Its syntax is as follows:

FUNCTION DEFINITION LINE

```
function [output arguments] = function_name(input arguments)
```

Figure 3.2–1 The default Editor Window when creating a new function.

The output arguments are those variables whose values are computed by the function, using the given values of the input arguments. Note that the output arguments are enclosed in *square brackets* (which are optional if there is only one output), while the input arguments must be enclosed with *parentheses*. The function_name should be the same as the file name in which it is saved (with the .m extension). That is, if we name a function drop, it should be saved in the file drop.m. The function is "called" by typing its name (for example, drop) at the command line. The word function in the function definition line must be *lowercase*. Before naming a function, you can use the exist function to see if another function has the same name.

Although use of the end statement to terminate a function is sometimes optional, it is required for some cases (see Section 3.3), so it is wise to include it always.

Edit the default function window by entering the information for your specific function, then save it as you would any other M-file.

Figure 3.2–2 shows the Editor after a function has been created. The Command Window shows the results of running the function, as will be described shortly.

Useful Features of the Editor

Unless you will be using MATLAB mainly as a calculator, you will be using the Editor frequently, since you cannot create every type of function in the Command window. We briefly discussed the basic features of the Editor in Chapter 1, and now we point out some more of its useful features.

The Editor uses colors for different purposes, and the default colors described here can be changed in Preferences in the Environment category on the HOME tab. When you open the Editor to create a new function, note that the keyword function is in blue and the comment is in green. In general, regardless of whether you are creating a script or a function, keywords are shown in blue and comments in green. This feature is called *syntax highlighting*.

The Editor helps you avoid syntax errors by using *delimiter matching* to indicate matched and mismatched delimiters, such as parentheses, brackets, and braces. When you type a closing delimiter MATLAB briefly highlights and underlines in red the corresponding opening delimiter. If you type more closing delimiters than opening delimiters, MATLAB underlines the unmatched delimiter in red.

If you use the arrow keys to move the cursor over one delimiter, MATLAB briefly underlines both delimiters in a pair. If no corresponding delimiter exists, MATLAB underlines the unmatched delimiter in red.

The easiest way to find and replace variables or functions in the current file is to use the *automatic highlighting feature*. Variable and function highlighting indicates only references to a particular function or variable, not other occurrences, such as in comments. If you move the cursor over a variable, all occurrences of that variable are highlighted in teal. A more detailed find-and-replace capability is available in the Navigate category. These features have more advanced features that will be useful in later chapters.

Figure 3.2–2 The Editor after a function has been created. The Command Window shows the results of running the function.

Some Simple Function Examples

Functions operate on variables within their own workspace (called *local variables*), which is separate from the workspace you access at the MATLAB command prompt. Consider the following user-defined function fun.

```
function z = fun(x,y)
u = 3*x;
z = u + 6*y.^2;
end
```

Note the use of the array exponentiation operator (.^). This enables the function to accept y as an array. Notice also that we have placed semicolons at the end of the lines that calculate u and z. This prevents their values from being displayed when the function is called. If for some reason you want them displayed, remove the semicolons, but this is not usually the case. We usually want to strictly control what variables are available in the workspace. The reasons for this are discussed later in this chapter.

Now consider what happens when you call this function in various ways in the Command window. Call the function with its output argument:

```
>>x = 3; y = 7;
>>z = fun(x,y)
z =
   303
```

The function uses x = 3 and y = 7 to compute z. You could also insert the argument values directly into the function call, as follows:

```
>>z = fun(3,7)
z =
   303
```

If you call the function without its output argument and try to access its value, you see the following error message:

```
>>clear z, fun(3,7)
ans =
     303
>>z
???  Undefined function or variable 'z'.
```

You can assign the output argument to another variable, as:

```
>>q = fun(3,7)
q =
   303
```

You can suppress the output by putting a semicolon after the function call. For example, if you type q = fun(3,7); the value of q will be computed but not displayed.

The variables x and y are *local* to the function fun, so unless you assign them values outside of the function as was done in the first example, their values will be unavailable in the workspace outside the function. The variable u is also local to the function. For example,

```
>>x = 3; y = 7; q = fun(x,y);
>>u
???  Undefined function or variable 'u'.
```

Compare this to

```
>>q = fun(3,7);
>>x
???  Undefined function or variable 'x'.
>>y
???  Undefined function or variable 'y'.
```

Only the order of the arguments is important, not the names of the arguments:

```
>>a = 7;b = 3;
>>z = fun(b,a) % This is equivalent to z = fun(3,7).
z =
   303
```

You can use arrays as input arguments (provided you have allowed for array operations within the function, as we did with y.^2):

```
>>r = fun([2:4],[7:9])
r =
   300    393    498
```

A function may have more than one output. These are enclosed in square brackets. For example, the following function `circle` computes the area *A* and circumference *C* of a circle, given its radius as an input argument.

```
function [A, C] = circle(r)
A = pi*r.^2;
C = 2*pi*r;
end
```

The function is called as follows, if *r* = 4.

```
>>[A, C] = circle(4)
A =
   50.2655
C =
   25.1327
```

A function may have no input arguments and no output arguments. For example, the following user-defined function `show_date` computes and stores the date in the variable `today`, and displays the value of `today`.

```
function show_date
today = date
end
```

Note that no brackets, parentheses, or equal sign are required. A session using this function would look like:

```
>>show_date
today =
13-Nov-2016
```

Test Your Understanding

T3.2–1 Create a function called cube that computes the surface area A and volume V of a cube whose side length is L. (Do not forget to check if a file already exists by that name!) Test case: $L = 10$, $A = 600$, $V = 1000$.

T3.2–2 Create a function called cone that computes the volume V of a cone whose height is h and whose radius is r. (Do not forget to check if a file already exists by that name!) The volume is given by

$$V = \pi r^2 \frac{h}{3}$$

Test case: $h = 30$, $r = 5$, $V = 785.3982$.

Variations in the Function Line

The following examples show permissible variations in the format of the function line. The differences depend on whether there is no output, a single output, or multiple outputs.

Function definition line	File name
1. function [area_square] = square(side);	square.m
2. function area_square = square(side);	square.m
3. function volume_box = box(height,width,length);	box.m
4. function [area_circle,circumf] = circle(radius);	circle.m
5. function sqplot(side);	sqplot.m

Example 1 is a function with one input and one output. The square brackets are optional when there is only one output (see example 2). Example 3 has one output and three inputs. Example 4 has two outputs and one input. Example 5 has no output variable (for example, our function show_date or a function that generates a plot). In such cases the equal sign may be omitted.

Comment lines starting with the % sign can be placed anywhere in the function file. However, if you use help to obtain information about the function, MATLAB displays all comment lines immediately following the function definition line up to the first blank line or first executable line.

We can call both built-in and user-defined functions either with the output variables explicitly specified, as in examples 1 through 4, or without any output variables specified. For example, with Example 2, we can call the function square as square(side) if we are not interested in its output variable area_square. (The function might perform some other operation that we want to occur, such as producing a plot.) Note that if we omit the semicolon at the end of the function call statement, the first variable in the output variable list will be displayed.

Variations in Function Calls

The following function, called `drop`, computes a falling object's velocity and distance dropped. The input variables are the acceleration g, the initial velocity v_0, and the elapsed time t. Note that we must use the element-by-element operations for any operations involving function inputs that are arrays. Here we anticipate that t will be an array, so we use the element-by-element operator (`.^`).

```
function [dist,vel] = drop(g,v0,t);
% Computes the distance traveled and the
% velocity of a dropped object, as functions
% of g, the initial velocity v0, and the time t.
vel = g*t + v0;
dist = 0.5*g*t.^2 + v0*t;
end
```

The following examples show various ways to call the function `drop`:

1. The variable names used in the function definition may, but need not, be used when the function is called:

```
a = 32.2;
initial_speed = 10;
time = 5;
[feet_dropped,speed] = drop(a,initial_speed,time)
```

2. The input variables need not be assigned values outside the function prior to the function call:

```
[feet_dropped,speed] = drop(32.2,10,5)
```

3. The inputs and outputs may be arrays:

```
[feet_dropped,speed] = drop(32.2,10,0:1:5)
```

This function call produces the arrays `feet_dropped` and `speed`, each with six values corresponding to the six values of time in the array `0:1:5`.

We can use the function to plot the distance, speed, or both. For example, suppose the object is thrown *upward* at $t = 0$ with a velocity of 4 m/s. In this case, `g` is 9.81 and `v0` is −4. To plot the distance dropped over 2 seconds, we would enter:

```
t = 0:0.001:2;
[meters_dropped,speed] = drop(9.81,-4,t);
plot(t,meters_dropped)
```

More About Local Variables

The names of the input variables given in the function definition line are local to that function. This means that other variable names can be used when you call the function. All variables inside a function are erased after the function finishes executing, except when the same variable names appear in the output variable list used in the function call.

For example, when using the `drop` function in a program, we can assign a value to the variable `dist` before the function call, and its value will be unchanged after the call because its name was not used in the output list of the call statement (the variable `feet_dropped` was used in the place of `dist`). This is what is meant by the function's variables being "local" to the function. This feature allows us to write generally useful functions using variables of our choice, without being concerned that the calling program uses the same variable names for other calculations. This means that our function files are "portable" and need not be rewritten every time they are used in a different program.

You might find the Editor to be useful for locating errors in function files. Runtime errors in functions are more difficult to locate because the function's local workspace is lost when the error forces a return to the MATLAB base workspace. The Editor provides access to the function workspace and allows you to change values. It also enables you to execute lines one at a time and to set *breakpoints,* which are specific locations in the file where execution is temporarily halted. The applications in this text will probably not require use of these features of the Editor, which are useful mainly for very large programs. For more information, see Chapter 4.

Global Variables

The `global` command declares certain variables global, and therefore their values are available to the basic workspace and to other functions that declare these variables global. The syntax to declare the variables `A`, `X`, and `Q` is `global A X Q`. Use a space, not a comma, to separate the variables. Any assignment to those variables, in any function or in the base workspace, is available to all the other functions declaring them global. If the global variable doesn't exist the first time you issue the `global` statement, it will be initialized to the empty matrix. If a variable with the same name as the global variable already exists in the current workspace, MATLAB issues a warning and changes the value of that variable to match the global.

In a user-defined function, make the global command the first executable line. Place the same command in the calling program. It is customary, but not required, to capitalize the names of *global variables* and to use long names, to make them easily recognizable.

The decision to declare a variable global is not always clear-cut. It is recommended to avoid using global variables. This can often be done by using anonymous and nested functions, as discussed in Section 3.3.

Persistent Variables

There may be applications (but perhaps not many) where you would want to preserve the value of a variable that is local to a function but whose value is not passed through the function output. You can use the `persistent` function to declare such a variable as *persistent*, which means that their values are retained in memory between calls to that function. The syntax `persistent x y` defines `x` and `y` as persistent variables, and is placed within the function. Note that there is no *function form* of the `persistent` command, which means you cannot use parentheses or quotes to indicate the variable names. If you put this statement before the variables are created, they will be initialized to the empty matrix.

Persistent variables differ from global variables in that persistent variables are known only to the function in which they are declared. This means their values cannot be changed by other functions or from the MATLAB command line. The `clear` function can be used to clear *functions* as well as variables. Whenever you clear or modify a function that is in memory, all persistent variables declared by that function are also cleared. To prevent this, use the `mlock` function. If you declare a variable persistent and a variable with the same name exists in the current workspace, an error message will appear.

Function Handles

A *function handle* is a way to reference a given function. First introduced in MATLAB 6.0, function handles have become widely used and frequently appear in examples throughout the MATLAB documentation. You can create a function handle to any function by using the @ sign before the function name. You can then give the handle a name, if you wish, and you can use the handle to reference the function.

For example, consider the following user-defined function, which computes $y = x + 2e^{-x} - 3$.

```
function y = f1(x)
y = x + 2*exp(-x) - 3;
end
```

To create a handle to this function and name the handle `fh1`, you type `fh1 = @f1`.

Function Functions

Some MATLAB functions act on functions. These commands are called *function functions*. If the function acted upon is not a simple function, it is more convenient to define the function in an M-file. You can pass the function to the calling function by using a function handle.

Finding the Zeros of a Function   You can use the `fzero` function to find the zero of a function of a single variable, which is denoted by `x`. Its basic syntax is

```
fzero(@function, x0)
```

where `@function` is a function handle and `x0` is a user-supplied guess for the zero. The `fzero` function returns a value of `x` that is near `x0`. It identifies only points where the function crosses the `x` axis, not points where the function just touches the axis. For example, `fzero(@cos,2)` returns the value $x = 1.5708$. As another example, $y = x^2$ is a parabola that touches the x axis at $x = 0$. Because the function never crosses the x axis, however, no zero will be found.

The function `fzero(@function,x0)` tries to find a zero of `function` near `x0`, if `x0` is a scalar. The value returned by `fzero` is near a point where `function` changes sign, or `NaN` if the search fails. In this case, the search terminates when the search interval is expanded until an `Inf`, `NaN`, or a complex value is found (`fzero` cannot find complex zeros). If `x0` is a *vector* of length 2, `fzero` assumes that `x0` is an interval where the sign of `function(x0(1))` differs from the sign of `function(x0(2))`. An error occurs if this is not true. Calling `fzero` with such an interval guarantees that `fzero` will return a value near a point where `function` changes sign. Plotting the function first is a good way to get a value for the vector `x0`. If the function is not continuous, `fzero` might return values that are discontinuous points instead of zeros. For example, `x = fzero(@tan,1)` returns `x = 1.5708`, a discontinuous point in $\tan(x)$.

Functions can have more than one zero, so it helps to plot the function first and then use `fzero` to obtain an answer that is more accurate than the answer read off the plot. Figure 3.2–3 shows the plot of the function $y = x + 2e^{-x} - 3$, which has two zeros, one near $x = -0.5$ and one near $x = 3$. Using the function file `f1` created earlier to find the zero near $x = -0.5$, type `x = fzero(@f1,-0.5)`. The answer is $x = -0.5831$. To find the zero near $x = 3$, type `x = fzero(@f1,3)`. The answer is $x = 2.8887$.

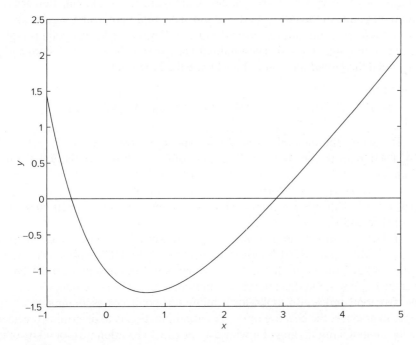

Figure 3.2–3 Plot of the function $y = x + 2e^{-x} - 3$.

The syntax `fzero(@f1,-0.5)` is preferred to the older syntax `fzero('f1', -0.5)`.

Minimizing a Function of One Variable   The `fminbnd` function finds the minimum of a function of a single variable, which is denoted by x. Its basic syntax is

```
fminbnd(@function, x1, x2)
```

where `@function` is a function handle. The `fminbnd` function returns a value of x that minimizes the function in the interval x1 ≤ x ≤ x2. For example, `fminbnd(@cos,0,4)` returns the value $x = 3.1416$.

However, to use this function to find the minimum of more-complicated functions, it is more convenient to define the function in a function file. For example, if $y = 1 - xe^{-x}$, define the following function file:

```
function y = f2(x)
y = 1 - x.*exp(-x);
end
```

To find the value of x that gives a minimum of y for $0 \le x \le 5$, type `x = fminbnd(@f2,0,5)`. The answer is $x = 1$. To find the minimum value of y, type `y = f2(x)`. The result is `y = 0.6321`.

Whenever we use a minimization technique, we should check that the solution is a true minimum. For example, consider the polynomial $y = 0.025 x^5 - 0.0625 x^4 - 0.333 x^3 + x^2$. Its plot is shown in Figure 3.2–4. The function has two minimum points in the interval $-1 < x < 4$. The minimum near $x = 3$ is called a *relative* or *local* minimum because it forms a valley whose lowest point is higher than the minimum at $x = 0$. The minimum at $x = 0$ is the true minimum and is also called the *global* minimum. First create the function file:

```
function y = f3(x)
y = polyval([0.025, -0.0625, -0.333, 1, 0, 0], x);
end
```

To specify the interval $-1 \le x \le 4$, type `x = fminbnd(@f3, -1, 4)`. MATLAB gives the answer `x = 2.0438e-006`, which is essentially 0, the true minimum point. If we specify the interval $0.1 \le x \le 2.5$, MATLAB gives the answer `x = 0.1001`, which corresponds to the minimum value of y on the interval $0.1 \le x \le 2.5$. Thus we will miss the true minimum point if our specified interval does not include it.

In fact, `fminbnd` can give misleading answers. If we specify the interval $1 \le x \le 4$, MATLAB (R2017b) gives the answer `x = 2.8236`, which corresponds to the "valley" shown in the plot, but which is not the minimum point on the interval $1 \le x \le 4$. On this interval the minimum point is at the boundary $x = 1$. The `fminbnd` procedure looks for a minimum point corresponding to a zero slope. In practice, the best use of the `fminbnd` function is to determine precisely the location of a minimum point whose approximate location was found by other means, such as by plotting the function.

Figure 3.2–4 Plot of the function $y = 0.025x^5 - 0.0625x^4 - 0.333x^3 + x^2$.

Minimizing a Function of Several Variables To find the minimum of a function of more than one variable, use the fminsearch function. Its basic syntax is

fminsearch(@function, x0)

where @function is a function handle. The vector x0 is a guess that must be supplied by the user. For example, to use the function $f = xe^{-x^2-y^2}$, first define it in an M-file, using the vector x whose elements are x(1) = x and x(2) = y.

```
function f = f4(x)
f = x(1).*exp(-x(1).^2-x(2).^2);
end
```

Suppose we guess that the minimum is near $x = y = 0$. The session is

```
>>fminsearch(@f4, [0, 0])
ans =
   -0.7071 0.000
```

Thus the minimum occurs at $x = -0.7071$, $y = 0$.

The fminsearch function can often handle discontinuities, particularly if they do not occur near the solution. The fminsearch function might give local solutions only, and it minimizes over the real numbers only; that is, x must consist of real variables only, and the function must return real

Table 3.2–1 Minimization and root-finding functions

Function	Description
`fminbnd(@function,x1,x2)`	Returns a value of x in the interval $x1 \le x \le x2$ that corresponds to a minimum of the single-variable function described by the handle `@function`.
`fminsearch(@function,x0)`	Uses the starting vector $x0$ to find a minimum of the multivariable function described by the handle `@function`.
`fzero(@function,x0)`	Uses the starting value $x0$ to find a zero of the single-variable function described by the handle `@function`.

numbers only. When x has complex values, they must be split into real and imaginary parts.

Table 3.2–1 summarizes the basic syntax of the `fminbnd`, `fminsearch`, and `fzero` commands.

These functions have extended syntax not described here. With these forms you can specify the accuracy required for the solution as well as the number of steps to use before stopping. Use the `help` facility to find out more about these functions.

EXAMPLE 3.2–1

Optimization of an Irrigation Channel

Figure 3.2–5 shows the cross section of an irrigation channel. A preliminary analysis has shown that the cross-sectional area of the channel should be 100 ft^2 to carry the desired water flow rate. To minimize the cost of concrete used to line the channel, we want to minimize the length of the channel's perimeter. Find the values of d, b, and θ that minimize this length.

■ Solution

The perimeter length L can be written in terms of the base b, depth d, and angle θ as follows:

$$L = b + \frac{2d}{\sin \theta}$$

The area of the trapezoidal cross section is

$$100 = db + \frac{d^2}{\tan \theta}$$

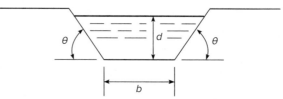

Figure 3.2–5 Cross section of an irrigation channel.

The variables to be selected are b, d, and θ. We can reduce the number of variables by solving the latter equation for b to obtain

$$b = \frac{1}{d}\left(100 - \frac{d^2}{\tan \theta}\right)$$

Substitute this expression into the equation for L. The result is

$$L = \frac{100}{d} - \frac{d}{\tan \theta} + \frac{2d}{\sin \theta}$$

We must now find the values of d and θ to minimize L.

First define the function file for the perimeter length. Let the vector x be $[d\ \theta]$.

```
function L = channel(x)
L = 100./x(1) - x(1)./tan(x(2)) + 2*x(1)./sin(x(2));
end
```

Then use the `fminsearch` function. Using a guess of $d = 20$ and $\theta = 1$ rad, the session is

```
>>x = fminsearch(@channel,[20,1])
x =
    7.5984    1.0472
```

Thus the minimum perimeter length is obtained with $d = 7.5984$ ft and $\theta = 1.0472$ rad, or $\theta = 60°$. Using a different guess, $d = 1$, $\theta = 0.1$, produces the same answer. The value of the base b corresponding to these values is $b = 8.7738$.

However, using the guess $d = 20$, $\theta = 0.1$ produces the physically meaningless result $d = -781$, $\theta = 3.1416$. The guess $d = 1$, $\theta = 1.5$ produces the physically meaningless result $d = 3.6058$, $\theta = -3.1416$.

The equation for L is a function of the two variables d and θ, and it forms a surface when L is plotted versus d and θ on a three-dimensional coordinate system. This surface might have multiple peaks, multiple valleys, and "mountain passes" called saddle points that can fool a minimization technique. Different initial guesses for the solution vector can cause the minimization technique to find different valleys and thus report different results. We can use the surface-plotting functions covered in Chapter 5 to look for multiple valleys, or we can use a large number of initial values for d and θ, say, over the physically realistic ranges $0 < d < 30$ and $0 < \theta < \pi/2$. If all the physically meaningful answers are identical, then we can be reasonably sure that we have found the minimum.

Test Your Understanding

T3.2–3 The equation $e^{-0.2x} \sin(x + 2) = 0.1$ has three solutions in the interval $0 < x < 10$. Find these three solutions. (Answers: $x = 1.0187$, 4.5334, 7.0066)

T3.2–4 The function $y = 1 + e^{-0.2x} \sin(x + 2)$ has two minimum points in the interval $0 < x < 10$. Find the values of x and y at each minimum. (Answers: $(x, y) = (2.5150, 0.4070)$, $(9.0001, 0.8347)$)

T3.2–5 Find the depth d and angle θ to minimize the perimeter length of the channel shown in Figure 3.2–5 to provide an area of 200 ft^2. (Answer: $d = 10.7457$ ft, $\theta = 60°$.)

3.3 Additional Function Types

In addition to function handles, *anonymous functions, subfunctions, nested functions,* and *private functions* some other types of user-created functions in MATLAB. This section covers the basic features of these functions. This is a more advanced topic, so unless you will be creating large, complex programs, you will probably not need to use these function types. The remainder of the text does not depend on knowledge of this topic.

Methods for Calling Functions

There are four ways to invoke, or "call," a function into action:

1. As a character string identifying the appropriate function M-file
2. As a function handle
3. As an "inline" function object
4. As a string expression

Examples of these ways follow for the `fzero` function used with the user-defined function `fun1`, which computes $y = x^2 - 4$.

1. As a character string identifying the appropriate function M-file, which is

   ```
   function y = fun1(x)
       y = x.^2-4;
   end
   ```

   The function may be called as follows, to compute the zero over the range $0 \le x \le 3$:

   ```
   >>x = fzero('fun1',[0, 3])
   ```

2. As a function handle to an existing function M-file:

   ```
   >>x = fzero(@fun1,[0, 3])
   ```

3. As an "inline" function object:

   ```
   >>fun1 = 'x.^2-4';
   >>fun_inline = inline(fun1);
   >>x = fzero(fun_inline,[0, 3])
   ```

4. As a string expression:

   ```
   >>fun1 = 'x.^2-4';
   >>x = fzero(fun1,[0, 3])
   ```

   or as

   ```
   >>x = fzero('x.^2-4',[0, 3])
   ```

Method 2 was not available prior to MATLAB 6.0, and it is now preferred over method 1. The third method is not discussed in this text because it is a slower method than the first two. The third and fourth methods are equivalent because they both utilize the `inline` function; the only difference is that with the fourth method MATLAB determines that the first argument of `fzero` is a string variable and calls `inline` to convert the string variable to an inline function object. The function handle method (method 2) is the fastest method, followed by method 1.

In addition to speed improvement, another advantage of using a function handle is that it provides access to subfunctions, which are normally not visible outside of their defining M-file. This is discussed later in this section.

Types of Functions

At this point it is helpful to review the types of functions provided for in MATLAB. MATLAB provides built-in functions, such as `clear`, `sin`, and `plot`, which are not M-files, and some functions that are M-files, such as the function `mean`. In addition, the following types of *user-defined* functions can be created in MATLAB.

- The *primary function* is the first function in an M-file and typically contains the main program. Following the primary function in the same file can be any number of subfunctions, which can serve as subroutines to the primary function. Usually the primary function is the only function in an M-file that you can call from the MATLAB command line or from another M-file function. You invoke this function by using the name of the M-file in which it is defined. We normally use the same name for the function and its file, but if the function name differs from the file name, you must use the file name to invoke the function.

  PRIMARY FUNCTION

- *Anonymous functions* enable you to create a simple function without needing to create an M-file for it. You can construct an anonymous function either at the MATLAB command line or from within another function or script. Thus, anonymous functions provide a quick way of making a function from any MATLAB expression without the need to create, name, and save a file.

  ANONYMOUS FUNCTIONS

- *Subfunctions* are placed in the primary function and are called by the primary function. You can use multiple functions within a single primary function M-file.

  SUBFUNCTIONS

- *Nested functions* are functions defined within another function. They can help to improve the readability of your program and also give you more flexible access to variables in the M-file. The difference between nested functions and subfunctions is that subfunctions normally cannot be accessed outside of their primary function file.

  NESTED FUNCTIONS

- *Overloaded* functions are functions that respond differently to different types of input arguments. They are similar to overloaded functions in any object-oriented language. For example, an overloaded function can be created to treat integer inputs differently than inputs of class double.

- *Private functions* enable you to restrict access to a function. They can be called only from an M-file function in the parent directory.

  PRIVATE FUNCTIONS

Anonymous Functions

Anonymous functions enable you to create a simple function without needing to create an M-file for it. You can construct an anonymous function either at the MATLAB command line or from within another function or script. The syntax for creating an anonymous function from an expression is

```
fhandle = @(arglist) expr
```

where `arglist` is a comma-separated list of input arguments to be passed to the function and `expr` is any single, valid MATLAB expression. This syntax creates the function handle `fhandle`, which enables you to invoke the function. Note that this syntax is different from that used to create other function handles, `fhandle = @functionname`. The handle is also useful for passing the anonymous function in a call to some other function in the same way as any other function handle.

For example, to create a simple function called `sq` to calculate the square of a number, type

```
sq = @(x) x.^2;
```

To improve readability, you may enclose the expression in parentheses, as `sq = @(x) (x.^2);`. To execute the function, type the name of the function handle, followed by any input arguments enclosed in parentheses. For example,

```
>>sq(5)
ans =
    25
>>sq([5,7])
ans =
    25    49
```

You might think that this particular anonymous function will not save you any work because typing `sq([5,7])` requires nine keystrokes, one more than is required to type `[5,7].^2`. Here, however, the anonymous function protects you from forgetting to type the period (.) required for array exponentiation. Anonymous functions are useful, however, for more complicated functions involving numerous keystrokes.

You can pass the handle of an anonymous function to other functions. For example, to find the minimum of the polynomial $4x^2 - 50x + 5$ over the interval $[-10, 10]$, you type

```
>>poly1 = @(x) 4*x.^2 - 50*x + 5;
>>fminbnd(poly1, -10, 10)
ans =
    6.2500
```

If you are not going to use that polynomial again, you can omit the handle definition line and type instead

```
>>fminbnd(@(x) 4*x.^2 - 50*x + 5, -10, 10)
```

Multiple-Input Arguments   You can create anonymous functions having more than one input. For example, to define the function $\sqrt{x^2 + y^2}$, type

```
>>sqrtsum = @(x,y) sqrt(x.^2 + y.^2);
```

Then

```
>>sqrtsum(3, 4)
ans =
     5
```

As another example, consider the function $z = Ax + By$ defining a plane. The scalar variables A and B must be assigned values before you create the function handle. For example,

```
>>A = 6; B = 4:
>>plane = @(x,y) A*x + B*y;
>>z = plane(2,8)
z =
    44
```

No-Input Arguments   To construct a handle for an anonymous function that has no input arguments, use empty parentheses for the input argument list, as shown by the following: `d = @() date;`.
Use empty parentheses when invoking the function, as follows:

```
>>d()
ans =
    12-Jul-2016
```

You must include the parentheses. If you do not, MATLAB just identifies the handle; it does not execute the function.

Calling One Function within Another   One anonymous function can call another to implement function composition. Consider the function $5 \sin(x^3)$. It is composed of the functions $g(y) = 5 \sin(y)$ and $f(x) = x^3$. In the following session the function whose handle is h calls the functions whose handles are f and g to compute $5 \sin(2^3)$.

```
>>f = @(x) x.^3;
>>g = @(x) 5*sin(x);
>>h = @(x) g(f(x));
>>h(2)
ans =
     4.9468
```

To preserve an anonymous function from one MATLAB session to the next, save the function handle to a MAT-file. For example, to save the function associated with the handle h, type `save anon.mat h`. To recover it in a later session, type `load anon.mat h`.

Variables and Anonymous Functions   Variables can appear in anonymous functions in two ways:

- As variables specified in the argument list, such as `f = @(x) x.^3;`.
- As variables specified in the body of the expression, such as with the variables A and B in `plane = @(x,y) A*x + B*y`. In this case, when the function is created, MATLAB captures the values of these variables and retains those values for the lifetime of the function handle. In this example, if the values of A or B are changed after the handle is created, their values associated with the handle do *not* change. This feature has both advantages and disadvantages, so you must keep it in mind. If you later decide to change the value of A or B, you must redefine the anonymous function using the new value.

Subfunctions

A function M-file may contain more than one user-defined function. The first defined function in the file is called the *primary function,* whose name is the same as the M-file name. All other functions in the file are called *subfunctions,* also called *local functions.* Subfunctions are normally "visible" only to the primary function and to other subfunctions in the same file; that is, they normally cannot be called by programs or functions outside the file. However, this limitation can be removed with the use of function handles, as we will see later in this section.

Create the primary function first with a function definition line and its defining code, and name the file with this function name as usual. Then create each subfunction with its own function definition line and defining code. The order of the subfunctions does not matter, but function names must be unique within the M-file.

The order in which MATLAB checks for functions is very important. When a function is called from within an M-file, MATLAB first checks to see if the function is a built-in function such as `sin`. If not, it checks to see if it is a *subfunction* in the file, then checks to see if it is a *private* function (which is a function M-file residing in the `private` subdirectory of the calling function). Then MATLAB checks for a standard M-file on your search path. Thus, because MATLAB checks for a subfunction before checking for private and standard M-file functions, you may use subfunctions with the same name as another existing M-file. This feature allows you to name subfunctions without being concerned about whether another function exists with the *same name,* so you need not choose long function names to avoid conflict. This feature also protects you from using another function unintentionally.

Note that you may even supercede a MATLAB M-function in this way. The following example shows how the MATLAB M-function `mean` can be superceded by our own definition of the mean, one which gives the root-mean-square value. The function `mean` is a subfunction. The function `subfun_demo` is the primary function.

```
function y = subfun_demo(a)
    y = a - mean(a);
    function w = mean(x)
        w = sqrt(sum(x.^2))/length(x);
    end
end
```

A sample session follows.

```
>>y = subfun_demo([4, -4])
y =
    1.1716    -6.8284
```

If we had used the MATLAB M-function mean, we would have obtained a different answer, that is,

```
>>a = [4,-4];
>>b = a - mean(a)
b =
    4    -4
```

Thus the use of subfunctions enables you to reduce the number of files that define your functions. For example, if it were not for the subfunction mean in the previous example, we would have had to define a separate M-file for our mean function and give it a different name so as not to confuse it with the MATLAB function of the same name.

Subfunctions are normally visible only to the primary function and other subfunctions in the same file. However, we can use a function handle to allow access to the subfunction from outside the M-file, as the following example shows. Create the following M-file with the primary function fn_demo1(range) and the subfunction testfun(x) to compute the zeros of the function $(x^2 - 4)$ cos x over the range specified in the input variable range. Note the use of a function handle in the second line.

```
function yzero = fun_demo1(range)
    fun = @testfun;
    [yzero,value] = fzero(fun,range);
%
    function y = testfun(x)
        y = (x.^2-4).*cos(x);
    end
end
```

A test session gives the following results.

```
>>yzero = fun_demo1([3, 6])
yzero =
    4.7124
```

So the zero of $(x^2 - 4)$ cos x over $3 \leq x \leq 6$ occurs at $x = 4.7124$.

Nested Functions

Starting with MATLAB 7 you can place the definitions of one or more functions within another function. Functions so defined are said to be *nested* within the main function. You can also nest functions within other nested functions. Like any M-file function, a nested function contains the usual components of an M-file function. You must, however, always terminate a nested function with an end statement. In fact, if an M-file contains at least one nested function, you must terminate *all* functions, including subfunctions, in the file with an end statement, whether or not they contain nested functions.

The following example constructs a function handle for a nested function p(x) and then passes the handle to the MATLAB function fminbnd to find the minimum point on the parabola. The parabola function constructs and returns a function handle f for the nested function p that evaluates the parabola $ax^2 + bx + c$. This handle gets passed to fminbnd.

```
function f = parabola(a, b, c)
f = @p;
    % Nested function
    function y = p(x)
       y = polyval([a,b,c],x);
    end
end
```

In the Command window type

```
>>f = parabola(4, -50, 5);
>>fminbnd(f, -10, 10)
ans =
     6.2500
```

Note that the function p(x) can see the variables a, b, and c in the calling function's workspace.

Contrast this approach to that required using global variables. First create the function p(x).

```
function y = p(x)
    global a b c
    y = polyval([a, b, c], x);
end
```

Then, in the Command window, type

```
>>global a b c
>>a = 4; b = -50; c = 5;
>> fminbnd(@p, -10, 10)
```

Nested functions might seem to be the same as subfunctions, but they are not. Nested functions have two unique properties:

1. A nested function can access the workspaces of all functions inside of which it is nested. So, for example, a variable that has a value assigned to it by the

primary function can be read or overwritten by a function nested at any level within the main function. In addition, a variable assigned in a nested function can be read or overwritten by any of the functions containing that function.

2. If you construct a function handle for a nested function, the handle not only stores the information needed to access the nested function, but also stores the values of all variables shared between the nested function and those functions that contain it. This means that these variables persist in memory between calls made by means of the function handle.

Consider the following representation of some functions named A, B, . . . , E.

```
function A(x, y)    % The primary function
    B(x, y);
    D(y);
    function B(x, y)    % Nested in A
        C(x);
        D(y);
        function C(x)    % Nested in B
            D(x);
        end    % This terminates C
    end    % This terminates B
    function D(x)    % Nested in A
        E(x);
        function E    % Nested in D
         . . .
        end    % This terminates E
    end    % This terminates D
end    % This terminates A
```

You call a nested function in several ways.

1. You can call it from the level immediately above it. (In the previous code, function A can call B or D, but not C or E.)
2. You can call it from a function nested at the same level within the same parent function. (Function B can call D, and D can call B.)
3. You can call it from a function at any lower level. (Function C can call B or D, but not E.)
4. If you construct a function handle for a nested function, you can call the nested function from any MATLAB function that has access to the handle.

You can call a subfunction from any nested function in the same M-file.

Private Functions

Private functions reside in subdirectories with the special name `private`, and in the future they will be visible only to functions or scripts in the directory immediately above the private directory. Assume the directory `rsmith` is on the MATLAB search path. A subdirectory of `rsmith` called `private` may contain functions that only the functions in `rsmith` can call. Because private functions are invisible outside the parent directory `rsmith`, they can use the same names as functions in other directories. This is useful if the main directory is used by several individuals including R. Smith, but R. Smith wants to create a personal version of a particular function while retaining the original in the main directory. Because MATLAB looks for private functions before standard M-file functions, it will find a private function named, say, `cylinder.m` before a nonprivate M-file named `cylinder.m`.

Primary functions and subfunctions can be implemented as private functions. Create a private directory by creating a subdirectory called `private` using the standard procedure for creating a directory or a folder on your computer, but do not place the private directory on your path.

3.4 File Functions

There are two kinds of computer files normally of interest to us: binary files and text files, which are commonly called ASCII files. In a binary file eight bits are used to represent each character, so each position can hold one of 256 different binary numbers. Binary files require special treatment and we will not discuss them further. In an ASCII file each byte represents one character according to the *ASCII* code. ASCII files use only seven bits of the byte, which restricts them to 128 combinations. The ASCII character set contains the characters on an English-language keyboard, plus some special characters. It is the most common format for English-language text files.

Word processors can store information in ASCII files. In spreadsheet and database files, binary codes describing the file structure are in a "header" and are interspersed throughout the file. However, the text information in the file (names, phone numbers, addresses, etc.) is ASCII. So in this section we will limit ourselves to ASCII files, with special attention paid to spreadsheet files. ASCII files usually have the extension .txt or .dat, except for spread sheet files, which have their own extension, such as .xlsx for Excel files.

An ASCII file may have one or more lines of text, called the header, at the beginning. These might be comments that describe what the data represent, the date they were created, and who created the data, for example. One or more lines of data, arranged in rows and columns, follow the header. The numbers in each row might be separated by spaces or by commas.

If it is inconvenient to edit the data file, the MATLAB environment provides many ways to bring data created by other applications into the MATLAB workspace, a process called *importing data,* and to package workspace variables so that they can be exported to other applications.

Creating and Importing ASCII Files

As we will see in the next example, we can create a data file by opening a new script in the MATLAB Editor, typing in the data (make sure the data has the same number of entries on each line), and saving it as a .dat file (Note: be sure not to save it as the default M-file type). Once the file is created, you can use the load command to load the data into a variable whose name you choose.

Typing load file_name loads data from the file called file_name. As we saw in Chapter 1, if file_name is a MAT-file, then load file_name loads variables in the MAT-File into the MATLAB workspace. If file_name is an ASCII file, then load file_name creates a matrix containing the data in the file. Since the data is stored in a matrix, the data must have the same number of entries in each line.

Importing Spreadsheet Files

The command

```
xlswrite(file_name,array_name,sheet_number,range)
```

writes the array array_name to the Excel file specified by file_name. The array is stored in the Excel sheet specified by sheet_number in the range specified with the syntax 'C1:C2', where C1 and C2 are opposing corners of the region. Optionally, just the upper-left corner can be specified. Unless otherwise specified, the default extension is .xls.

The command A = xlsread('filename') imports the Microsoft Excel workbook file filename.xls into the array A. The command [A, B] = xlsread('filename') imports all numeric data into the array A and all text data into the cell array B.

Creating a Data File and Loading It into a Variable

EXAMPLE 3.4–1

Create a file containing the following data, load the data into MATLAB, and plot it.

Time (s)	1	2	3	4	5
Speed (m/s)	12	14	16	21	27

■ Solution
Open a new script in the MATLAB Editor, create the file, separating the items by a space, and save it as speed_data.dat (note: be sure not to save it as the default M-file type).

```
% speed_data.dat
% speed vs. time
1, 2, 3, 4, 5;
12, 14, 16, 21, 27;
```

Then in the MATLAB Command window type

```
>>load speed_data.dat
>>time = speed_data(1,:)
```

```
>>speed = speed_data(2,:)
>>plot(time,speed,'o'),xlabel('time(s)'), ...
    ylabel('speed(m/s)')
```

Note that the comments in the file are not saved, only the numerical values.

For example, to write mixed text and numeric data to an Excel .xlsx file starting at cell C1 of sheet 3, the session is

```
>>file_name = 'speed_data.xlsx';
>>A = {'Time(s)','Speed(m/s)';1,12;2,14;3,16;4,21;5,27};
>>sheet = 3;
>>range = 'C1';
>>xlswrite(file_name,A,sheet,range)
```

MATLAB provides several other ways to import data. These are described in the help files, which are comprehensive. You can use the Import Tool, which provides a graphical interface to import data into an array. Type `uiimport` or select **Import Data** on the Toolstrip. The command `importdata` can import other types of files besides text and data files, such as graphics files. New as of R2016a, the `readtable` command creates a table from a file.

3.5 Summary

In Section 3.1 we introduced just some of the most commonly used mathematical functions. You should now be able to use the MATLAB Help to find other functions you need. If necessary, you can create your own functions, using the methods of Section 3.2. This section also covered function handles and their use with function functions.

Anonymous functions, subfunctions, and nested functions extend the capabilities of MATLAB. These topics were treated in Section 3.3. In addition to function files, data files are useful for many applications. Section 3.4 shows how to import and export such files in MATLAB.

Key Terms

Anonymous functions,   135

Function argument,   117

Function definition line,   120

Function file,   120

Function handle,   128

Global variables,   127

Local variable,   120

Nested functions,   135

Primary function,   135

Private functions,   135

Subfunctions,   135

Problems

You can find the answers to problems marked with an asterisk at the end of the text.

Section 3.1

1.* Suppose that $y = -3 + ix$. For $x = 0$, 1, and 2, use MATLAB to compute the following expressions. Hand-check the answers.

a. $|y|$ b. $\sqrt{y}$

c. $(-5 - 7i)y$ d. $\dfrac{y}{6 - 3i}$

2.* Let $x = -5 - 8i$ and $y = 10 - 5i$. Use MATLAB to compute the following expressions. Hand-check the answers.

a. The magnitude and angle of xy.

b. The magnitude and angle of $\frac{x}{y}$.

3.* Use MATLAB to find the angles corresponding to the following coordinates. Hand-check the answers.

a. $(x, y) = (5, 8)$ b. $(x, y) = (-5, 8)$

c. $(x, y) = (5, -8)$ d. $(x, y) = (-5, -8)$

4. For several values of x, use MATLAB to confirm that $\sinh x = (e^x - e^{-x})/2$.

5. For several values of x, use MATLAB to confirm that $\cosh^{-1} x = \ln (x + \sqrt{x^2 - 1})$, $x \geq 1$

6. The capacitance of two parallel conductors of length L and radius r, separated by a distance d in air, is given by

$$C = \frac{\pi \varepsilon L}{\ln [(d - r)/r]}$$

where ε is the permittivity of air ($\varepsilon = 8.854 \times 10^{-12}$ F/m).

Write a script file that accepts user input for d, L, and r and computes and displays C. Test the file with the values $L = 1$ m, $r = 0.001$ m, and $d = 0.004$ m.

7.* When a belt is wrapped around a cylinder, the relation between the belt forces on each side of the cylinder is

$$F_1 = F_2 e^{\mu \beta}$$

where β is the angle of wrap of the belt and μ is the friction coefficient. Write a script file that first prompts a user to specify β, μ, and F_2 and then computes the force F_1. Test your program with the values $\beta = 130°$, $\mu = 0.3$, and $F_2 = 100$ N. (*Hint:* Be careful with β!)

Section 3.2

8. Write a function that accepts temperature in degrees Fahrenheit (°F) and computes the corresponding value in degrees Celsius (°C). The relation between the two is

$$T\,°C = \frac{5}{9}(T\,°F - 32)$$

Be sure to test your function.

9.* An object thrown vertically with a speed v_0 reaches a height h at time t, where

$$h = v_0 t - \frac{1}{2}gt^2$$

Write and test a function that computes the time t required to reach a specified height h, for a given value of v_0. The function's inputs should be h, v_0, and g. Test your function for the case where $h = 100$ m, $v_0 = 50$ m/s, and $g = 9.81$ m/s^2. Interpret both answers.

10. A water tank consists of a cylindrical part of radius r and height h and a hemispherical top. The tank is to be constructed to hold 600 m^3 when filled. The surface area of the cylindrical part is $2\pi rh$, and its volume is $\pi r^2 h$. The surface area of the hemispherical top is given by $2\pi r^2$, and its volume is given by $2\pi r^3/3$. The cost to construct the cylindrical part of the tank is \$400 per square meter of surface area; the hemispherical part costs \$600 per square meter. Use the `fminbnd` function to compute the radius that results in the least cost. Compute the corresponding height h.

11. A fence around a field is shaped as shown in Figure P11. It consists of a rectangle of length L and width W, and a right triangle that is symmetrical about the central horizontal axis of the rectangle. Suppose the width W is known (in meters) and the enclosed area A is known (in square meters). Write a user-defined function file with W and A as inputs. The outputs are the length L required so that the enclosed area is A and the total length of fence required. Test your function for the values $W = 6$ m and $A = 80$ m^2.

Figure P11

12. A fence around a field is shaped as shown in Figure P11. It consists of a rectangle of length L and width W and a right triangle that is symmetric about the central horizontal axis of the rectangle. Suppose the desired enclosed area A is given. Note that the length L can be expressed as a function of A and W, so that the perimeter P can be expressed solely as a function of A and W.

    a. Write a MATLAB function using the `min` function to compute the width W required to minimize the fence perimeter P and to calculate the corresponding values of L and P. The function should create a vector of guessed values for W, whose minimum and maximum values are *W1* and *W2*, with spacing d. The function inputs should be the desired area A, guesses *W1* and *W2*, and spacing d. The function outputs should be the solution for W and the corresponding values of L and P. Test your function for the value $A = 80$ m^2.

    b. Write a MATLAB function to use with the `fminbnd` function to compute the width W required to minimize the fence perimeter P and to calculate the corresponding values of L and P. The function input should be the desired area A. The function outputs should be the solution for W and the corresponding values of L and P. Test your function for the value $A = 80$ m^2.

13. A fenced enclosure consists of a rectangle of length L and width $2R$, and a semicircle of radius R, as shown in Figure P13. The enclosure is to be built to have an area A of 1600 ft^2. The cost of the fence is \$40/ft for the curved portion and \$30/ft for the straight sides. Use the `min` function to determine with a resolution of 0.01 ft the values of R and L required to minimize the total cost of the fence. Also compute the minimum cost.

Figure P13

14. A fenced enclosure consists of a rectangle of length L and width $2R$ and a semicircle of radius R, as shown in Figure P13. The enclosure is to be built to have an area A of 2000 ft^2. The cost of the fence is \$50 per foot for the curved portion and \$40 per foot for the straight sides. Use the `fminbnd` function to determine with a resolution of 0.01 ft the values of R and L required to minimize the total cost of the fence. Also compute the minimum cost.

15. Using estimates of rainfall, evaporation, and water consumption, the town engineer developed the following model of the water volume in the reservoir as a function of time

$$V(t) = 10^9 + 10^8(1 - e^{-t/100}) - rt$$

where V is the water volume in liters, t is time in days, and r is the town's consumption rate in liters per day. Write two user-defined functions. The first function should define the function $V(t)$ for use with the `fzero` function. The second function should use `fzero` to compute how long it will take for the water volume to decrease to x percent of its initial value of 10^9 L. The inputs to the second function should be x and r. Test your functions for the case where $x = 50$ percent and $r = 10^7$ L/day.

16. The volume V and paper surface area A of a conical paper cup are given by

$$V = \frac{1}{3}\pi r^2 h \qquad A = \pi r \sqrt{r^2 + h^2}$$

where r is the radius of the base of the cone and h is the height of the cone.

a. By eliminating h, obtain the expression for A as a function of r and V.

b. Create a user-defined function that accepts R as the only argument and computes A for a given value of V. Declare V to be global within the function.

c. For $V = 10$ in.^{3}, use the function with the `fminbnd` function to compute the value of r that minimizes the area A. What is the corresponding value of the height h? Investigate the sensitivity of the solution by plotting V versus r. How much can R vary about its optimal value before the area increases 10 percent above its minimum value?

17. A torus is shaped like a doughnut. If its inner radius is a and its outer radius is b, its volume and surface area are given by

$$V = \frac{1}{4}\pi^2(a + b)(b - a)^2 \qquad A = \pi^2(b^2 - a^2)$$

a. Create a user-defined function that computes V and A from the arguments a and b.

b. Suppose that the outer radius is constrained to be 2 in. greater than the inner radius. Write a script file that uses your function to plot A and V versus a for $0.25 \leq a \leq 4$ in.

18. Suppose it is known that the graph of the function $y = ax^3 + bx^2 + cx + d$ passes through four given points (x_i, y_i), $i = 1, 2, 3, 4$. Write a user-defined function that accepts these four points as input and computes the coefficients a, b, c, and d. The function should solve four linear equations in terms of the four unknowns a, b, c, and d. Test your function for the

case where $(x_i, y_i) = (-2, -20), (0, 4), (2, 68),$ and $(4, 508),$ whose answer is $a = 7, b = 5, c = -6,$ and $d = 4.$

19. Create a function called `savings_balance` that determines the balance in a savings account at the end of every year for the first n years, where n is an input. The account has an initial investment A (to be provided as input; for example, enter $10,000 as 10000) and an annually compounded interest rate of r% (to be provided as input; for example, enter 3.5% as 3.5). Display the information on screen in a table in which the first column is Year and the second is Balance ($). (Test case: $n = 10, A = 10000, r = 3.5.$ After 10 years the balance is $14,105.99.)

With an initial investment of A and interest rate r, the balance B after n years is given by:

$$B = A(1 + r/100)^n$$

20. Planets and planetary satellites move in elliptical orbits. The general equation for an ellipse centered at the origin, whose major and minor axes lie along the x and y axes, is

$$\frac{x^2}{a^2} + \frac{y^2}{b^2} = 1$$

This can be solved for y as follows:

$$y = \pm b\sqrt{1 - \frac{x^2}{a^2}}$$

Create a function that will plot the entire ellipse, given the inputs a and b. Obtain the plot for the case $a = 1, b = 2.$

21. Solve the following equation for positive values of x.

$$1 - 3xe^{-x} = 0$$

a. First plot the left-hand side to see how many roots there may be.

b. Then use the `fzero` function to find all the roots.

22. Two models of population growth are the exponential growth model

$$p(t) = p(0)e^{rt}$$

and the logistic growth model

$$p(t) = \frac{Kp(0)}{p(0) + [K - p(0)]e^{-rt}}$$

where $p(t)$ is the population size as a function of time t, and $p(0)$ is the initial population size at $t = 0$. The constant r is the growth rate, and the constant K is called the carrying capacity of the environment. As $t \to \infty$ the exponential predicts that $p(t) \to \infty$ but the logistic model predicts that

$p(t) \to K$. Both models have been used extensively to model a number of different populations, including bacteria, animals, fish, and human populations.

If $p(0)$ and r are the same for both models, it is easy to see that the exponential model will predict a larger population for all $t > 0$. But suppose that $p(0)$ is the same for both models but the r values are different. In particular, let $r = 0.1$ for the exponential model, $r = 1$ and $K = 10$ for the logistic model, and $p(0) = 10$ for both models. Then the two models will predict the same population at time t if

$$\frac{50}{10 + 40e^{-t}} = e^{0.1t}$$

This equation cannot be solved analytically, so we must use a numerical method. Use the `fzero` function to solve this equation for t, and calculate the population at that time.

Section 3.3

23. Create an anonymous function for $10e^{-2x}$ and use it to plot the function over the range $0 \le x \le 2$.

24. Create an anonymous function for $20x^2 - 200x + 3$ and use it

   a. To plot the function to determine the approximate location of its minimum

   b. With the `fminbnd` function to precisely determine the location of the minimum

25. Create four anonymous functions to represent the function $6e^{3\cos x^2}$, which is composed of the functions $h(z) = 6e^z$, $g(y) = 3 \cos y$, and $f(x) = x^2$. Use the anonymous functions to plot $6e^{3\cos x^2}$ over the range $0 \le x \le 4$.

26. Use a primary function with a subfunction to compute the zeros of the function $3x^3 - 12x^2 - 33x + 80$ over the range $-10 \le x \le 10$.

27. Create a primary function that uses a function handle with a nested function to compute the minimum of the function $20x^2 - 200x + 12$ over the range $0 \le x \le 10$.

Section 3.4

28. Use a text editor to create a file containing the following data. Then use the `load` function to load the file into MATLAB, and use the `mean` function to compute the mean value of each column.

55	42	98
49	39	95
63	51	92
58	45	90

29. Enter and save the data given in Problem 28 in a spreadsheet. Then import the spreadsheet file into the MATLAB variable A. Use MATLAB to compute the sum of each column.

30. Use a text editor to create a file from the data given in Problem 28, but separate each number with a semicolon. Then use the Import Wizard to load and save the data in the MATLAB variable A.

Engineering in the
21st Century. . .

Nanotechnology

Many of the engineering challenges and opportunities in the 21st century will involve the development of extremely small devices and even the manipulation of individual atoms. This technology is called *nanotechnology* because it involves processing materials whose size is about 1 nanometer (nm), which is 10^{-9} m, or 1/1 000 000 mm. The distance between atoms in single-crystal silicon is 0.5 nm.

Nanotechnology is in its infancy, although some working devices have been created. One class of such devices is called *lab-on-a-chip* (LOC), such as the one shown in the photo. Using lithography to create nanoscale channel structures on the surface of metals and semiconductors, and microfluidics to control the flow of droplets, LOC technology enables several laboratory functions to be carried out on a chip whose size ranges from a few millimeters to a few square centimeters. The goal is to achieve rapid and inexpensive screening for a number of diseases using drop-size samples of blood or saliva.

Another application of nanotechnology is in the creation of micromechanical machines (MEMS). MEMS are widely used in vehicle systems such as airbag sensors, accelerometers, and gyroscopes for detecting yaw to achieve electronic stability control. At such small sizes, simple dynamics and heat transfer principles are not always sufficient for design. MEMS have a large surface area to volume ratio, so surface effects such as electrostatics, surface tension, and wetting have more influence than volume effects such as inertia or heat capacity.

To design and apply these devices, engineers must first model the appropriate mechanical, fluid, and electrical properties. The features of MATLAB provide excellent support for such analyses. ■

Programming with MATLAB

The MATLAB interactive mode is very useful for simple problems, but more-complex problems require a script file. Such a file can be called a *computer program,* and writing such a file is called *programming.* Section 4.1 presents a general and efficient approach to the design and development of programs.

The usefulness of MATLAB is greatly increased by the use of decision-making functions in its programs. These functions enable you to write programs whose operations depend on the results of calculations made by the program. Sections 4.2, 4.3, and 4.4 deal with these decision-making functions.

MATLAB can also repeat calculations a specified number of times or until some condition is satisfied. This feature enables engineers to solve problems of great complexity or requiring numerous calculations. These "loop" structures are covered in Sections 4.5 and 4.6.

The `switch` structure enhances the MATLAB decision-making capabilities. This topic is covered in Section 4.7. Use of the MATLAB Editor/Debugger for debugging programs is covered in Section 4.8.

Section 4.9 discusses "simulation," a major application of MATLAB programs that enables us to study the operation of complicated systems, processes, and organizations. Tables summarizing the MATLAB commands introduced in this chapter appear throughout the chapter, and Table 4.10–1 will help you locate the information you need.

4.1 Program Design and Development

In this chapter we introduce *relational operators,* such as > and ==, and the two types of loops used in MATLAB, the `for` loop and the `while` loop. These features, plus MATLAB functions and the *logical operators* to be introduced in Section 4.3, form the basis for constructing MATLAB programs to solve complex problems. Design of computer programs to solve complex problems needs to be done in a systematic manner from the start to avoid time-consuming and frustrating difficulties later in the process. In this section we show how to structure and manage the program design process.

Algorithms and Control Structures

An *algorithm* is an ordered sequence of precisely defined instructions that performs some task in a finite amount of time. An ordered sequence means that the instructions can be numbered, but an algorithm often must have the ability to alter the order of its instructions using what is called a *control structure.* There are three categories of algorithmic operations:

Sequential operations.   These instructions are executed in order.

Conditional operations.   These control structures first ask a question to be answered with a true/false answer and then select the next instruction based on the answer.

Iterative operations (loops).   These control structures repeat the execution of a block of instructions.

Not every problem can be solved with an algorithm, and some potential algorithmic solutions can fail because they take too long to find a solution.

Structured Programming

Structured programming is a technique for designing programs in which a hierarchy of *modules* is used, each having a single entry and a single exit point, and in which control is passed downward through the structure without unconditional branches to higher levels of the structure. In MATLAB these modules can be built-in or user-defined functions.

Control of the program flow uses the same three types of control structures used in algorithms: sequential, conditional, and iterative. In general, any computer program can be written with these three structures. This realization led to the development of structured programming. Languages suitable for structured programming, such as MATLAB, thus do not have an equivalent to the goto statement that you might have seen in the BASIC and FORTRAN languages. An unfortunate result of the goto statement was confusing code, called *spaghetti code,* composed of a complex tangle of branches.

Structured programming, if used properly, results in programs that are easy to write, understand, and modify. The advantages of structured programming are as follows:

1. Structured programs are easier to write because the programmer can study the overall problem first and deal with the details later.
2. Modules (functions) written for one application can be used for other applications (this is called *reusable code*).
3. Structured programs are easier to debug because each module is designed to perform just one task, and thus it can be tested separately from the other modules.
4. Structured programming is effective in a teamwork environment because several people can work on a common program, each person developing one or more modules.
5. Structured programs are easier to understand and modify, especially if meaningful names are chosen for the modules and if the documentation clearly identifies the module's task.

Top-Down Design and Program Documentation

A method for creating structured programs is *top-down design*, which aims to describe a program's intended purpose at a very high level initially and then partition the problem repeatedly into more detailed levels, one level at a time, until enough is understood about the program structure to enable it to be coded. Table 4.1–1, which is repeated from Chapter 1, summarizes the process of

Table 4.1–1 Steps for developing a computer solution

1. State the problem concisely.
2. Specify the data to be used by the program. This is the *input*.
3. Specify the information to be generated by the program. This is the *output*.
4. Work through the solution steps by hand or with a calculator; use a simpler set of data, if necessary.
5. Write and run the program.
6. Check the output of the program with your hand solution.
7. Run the program with your input data and perform a "reality check" on the output. Does it make sense? Estimate the range of the expected result and compare it with you answer.
8. If you will use the program as a general tool in the future, test it by running it for a range of reasonable data values; perform a reality check on the results.

top-down design. In step 4 you create the algorithms used to obtain the solution. Note that step 5, Write and run the program, is only part of the top-down design process. In this step you create the necessary modules and test them separately.

Two types of charts aid in developing structured programs and in documenting them. These are *structure charts* and *flowcharts*. A structure chart is a graphical description showing how the different parts of the program are connected. This type of diagram is particularly useful in the initial stages of top-down design.

STRUCTURE CHART

A structure chart displays the organization of a program without showing the details of the calculations and decision processes. For example, we can create program modules using function files that do specific, readily identifiable tasks. Larger programs are usually composed of a main program that calls on the modules to do their specialized tasks as needed. A structure chart shows the connection between the main program and the modules.

For example, suppose you want to write a program that plays a game, say, Tic-Tac-Toe. You need a module to allow the human player to input a move, a module to update and display the game grid, and a module that contains the computer's strategy for selecting its moves. Figure 4.1–1 shows the structure chart of such a program.

FLOWCHARTS

Flowcharts are useful for developing and documenting programs that contain conditional statements, because they can display the various paths (called *branches*) that a program can take, depending on how the conditional statements are executed. The flowchart representation of the verbal description of the `if` statement (covered in Section 4.3) is shown in Figure 4.1–2. Flowcharts use the diamond symbol to indicate decision points.

The usefulness of structure charts and flowcharts is limited by their size. For large, more complicated programs, it might be impractical to draw such charts. Nevertheless, for smaller projects, sketching a flowchart and/or a structure chart might help you organize your thoughts before you begin to write the specific MATLAB code. Because of the space required for such charts, we do not use them in this text. You are encouraged, however, to use them when solving problems.

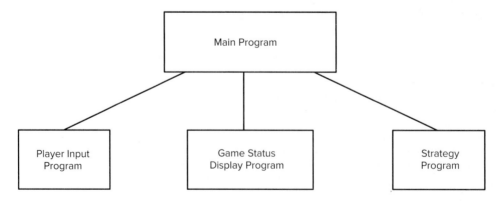

Figure 4.1–1 Structure chart of a game program.

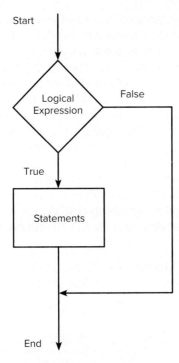

Figure 4.1–2 Flowchart representation of the verbal description of the `if` statement.

Documenting programs properly is very important, even if you never give your programs to other people. If you need to modify one of your programs, you will find that it is often very difficult to recall how it operates if you have not used it for some time. Effective documentation can be accomplished with the use of

1. Proper selection of variable names to reflect the quantities they represent.
2. Comments within the program.
3. Structure charts.
4. Flowcharts.
5. A verbal description of the program, often in *pseudocode*.

The advantage of using suitable variable names and comments is that they reside with the program; anyone who gets a copy of the program will see such documentation. However, they often do not provide enough of an overview of the program. The latter three elements can provide such an overview.

Pseudocode

Use of natural language, such as English, to describe algorithms often results in a description that is too verbose and is subject to misinterpretation. To avoid dealing immediately with the possibly complicated syntax of the programming language, we can instead use *pseudocode*, in which natural language and mathematical expressions are used to construct statements that look like computer statements but without detailed syntax. Pseudocode may also use some simple MATLAB syntax to explain the operation of the program.

As its name implies, pseudocode is an imitation of the actual computer code. The pseudocode can provide the basis for comments within the program. In addition to providing documentation, pseudocode is useful for outlining a program before writing the detailed code, which takes longer to write because it must conform to the strict rules of MATLAB.

Each pseudocode instruction may be numbered, but should be unambiguous and computable. Note that MATLAB does not use line numbers except in the Debugger. Each of the following examples illustrates how pseudocode can document each of the control structures used in algorithms: sequential, conditional, and iterative operations.

Example 1. Sequential Operations   Compute the perimeter p and the area A of a triangle whose sides are a, b, and c. The formulas are

$$p = a + b + c \qquad s = \frac{p}{2} \qquad A = \sqrt{s(s - a)(s - b)(s - c)}$$

1. Enter the side lengths a, b, and c.
2. Compute the perimeter p.

$$p = a + b + c$$

3. Compute the semiperimeter s.

$$s = \frac{p}{2}$$

4. Compute the area A.

$$A = \sqrt{s(s - a)(s - b)(s - c)}$$

5. Display the results p and A.
6. Stop.

The program is

```
a = input('Enter the value of side a: ');
b = input('Enter the value of side b: ');
c = input('Enter the value of side c: ');
p = a + b + c;
s = p/2;
A = sqrt(s*(s-a)*(s-b)*(s-c));
disp('The perimeter is:')
p
disp('The area is:')
A
```

Example 2. Conditional Operations   Given the (x, y) coordinates of a point, compute its polar coordinates (r, θ), where

$$r = \sqrt{x^2 + y^2} \quad \theta = \tan^{-1}\left(\frac{y}{x}\right)$$

1.   Enter the coordinates x and y.
2.   Compute the hypotenuse r.

```
r = sqrt(x^2+y^2)
```

3.   Compute the angle θ.

   3.1   If $x \geq 0$

```
        theta = atan(y/x)
```

   3.2   Else

```
        theta = atan(y/x) + pi
```

4.   Convert the angle to degrees.

```
   theta = theta*(180/pi)
```

5.   Display the results r and theta.
6.   Stop.

Note the use of the numbering scheme 3.1 and 3.2 to indicate subordinate clauses. Note also that MATLAB syntax may be used for clarity where needed. The following program implements the pseudocode using some of the MATLAB features to be introduced in this chapter. It uses the relational operator >=, which means "greater than or equal to" (in Section 4.2). The program also uses the "if-else-end" construct, which is covered in Section 4.3.

```
x = input('Enter the value of x: ');
y = input('Enter the value of y: ');
r = sqrt(x^2+y^2);
if x >= 0
    theta = atan(y/x);
else
    theta = atan(y/x) + pi;
end
disp('The hypotenuse is:')
disp(r)
theta = theta*(180/pi);
disp('The angle is degrees is:')
disp(theta)
```

Example 3. Iterative Operations   Determine how many terms are required for the sum of the series $10k^2 - 4k + 2$, $k = 1, 2, 3, \ldots$ to exceed 20,000. What is the sum for this many terms?

Because we do not know how many times we must evaluate the expression $10k^2 - 4k + 2$, we use a while loop, which is covered in Section 4.6.

1. Initialize the total to zero.
2. Initialize the counter to zero.
3. While the total is less than 20,000 compute the total.
    3.1   Increment the counter by 1.
          ```
          k = k + 1
          ```
    3.2   Update the total.
          ```
          total = 10*k^2 - 4*k + 2 + total
          ```
4. Display the current value of the counter.
5. Display the value of the total.
6. Stop.

The following program implements the pseudocode. The statements in the `while` loop are executed until the variable `total` equals or exceeds 2×10^4.

```
total = 0;
k = 0;
while total < 2e+4
   k = k+1;
   total = 10*k^2 - 4*k + 2 + total;
end
disp('The number of terms is:')
disp(k)
disp('The sum is:')
disp(total)
```

Finding Bugs

Debugging a program is the process of finding and removing the "bugs," or errors, in a program. Such errors usually fall into one of the following categories.

1. Syntax errors such as omitting a parenthesis or comma, or spelling a command name incorrectly. MATLAB usually detects the more obvious errors and displays a message describing the error and its location.
2. Errors due to an incorrect mathematical procedure. These are called *runtime errors*. They do not necessarily occur every time the program is executed; their occurrence often depends on the particular input data. A common example is division by zero.

The MATLAB error messages usually enable you to find syntax errors. However, runtime errors are more difficult to locate. To locate such an error, try the following:

1. Always test your program with a simple version of the problem, whose answers can be checked by hand calculations.
2. Display any intermediate calculations by removing semicolons at the end of statements.

Table 4.2–1 Relational operators

Relational operator	Meaning
<	Less than.
<=	Less than or equal to.
>	Greater than.
>=	Greater than or equal to.
==	Equal to.
~=	Not equal to.

3. To test user-defined functions, try commenting out the `function` line and running the file as a script.

4. Use the debugging features of the Editor, which are discussed in Section 4.8.

4.2 Relational Operators and Logical Variables

MATLAB has six *relational operators* to make comparisons between arrays. These operators are shown in Table 4.2–1. Note that the *equal to* operator consists of two = signs, not a single = sign as you might expect. The single = sign is the *assignment,* or *replacement,* operator in MATLAB.

The result of a comparison using the relational operators is either 0 (if the comparison is *false*) or 1 (if the comparison is *true*), and the result can be used as a variable. For example, if x = 2 and y = 5, typing z = x < y returns the value z = 1 and typing u = x == y returns the value u = 0. To make the statements more readable, we can group the logical operations using parentheses. For example, z = (x < y) and u = (x == y).

When used to compare arrays, the relational operators compare the arrays on an element-by-element basis. The arrays being compared must have the same dimension. The only exception occurs when we compare an array to a scalar. In that case all the elements of the array are compared to the scalar. For example, suppose that x = [6,3,9] and y = [14,2,9]. The following MATLAB session shows some examples.

```
>>z = (x < y)
z =
    1   0   0
>>z = (x ~= y)
z =
    1   1   0
>>z = (x > 8)
z =
    0   0   1
```

The relational operators can be used for array addressing. For example, with x = [6,3,9] and y = [14,2,9], typing z = x(x < y) finds all the elements in x that are less than the corresponding elements in y. The result is z = 6.

The arithmetic operators +, −, *, /, and \ have precedence over the relational operators. Thus the statement z = 5 > 2 + 7 is equivalent to z = 5 >(2+7) and returns the result z = 0. We can use parentheses to change the order of precedence; for example, z = (5 > 2) + 7 evaluates to z = 8.

The relational operators have equal precedence among themselves, and MATLAB evaluates them in order from left to right. Thus the statement

```
z = 5 > 3 ~= 1
```

is equivalent to

```
z = (5 > 3) ~= 1
```

Both statements return the result z = 0.

With relational operators that consist of more than one character, such as == or >=, be careful not to put a space between the characters.

The **logical** Class

When the relational operators are used, such as x = (5 > 2), they create a logical variable, in this case, x. Prior to MATLAB 6.5, logical was an attribute of any numeric data type. Now logical is a first-class data type and a MATLAB class, and so logical is now equivalent to other first-class types such as character and cell arrays. Logical variables may have only the values 1 (true) and 0 (false).

Just because an array contains only 0s and 1s, however, it is not necessarily a logical array. For example, in the following session k and w appear the same, but k is a logical array and w is a numeric array, and thus an error message is issued.

```
>>x = -2:2
x =
    -2    -1    0    1    2
>>k = (abs(x)>1)
k =
    1    0    0    0    1
>>z = x(k)
z =
    -2    2
>>w = [1,0,0,0,1];
>>v = x(w)
??? Subscript indices must either be real positive. . . integers
    or logicals.
```

The **logical** Function

Logical arrays can be created with the relational and logical operators and with the logical function. The logical function returns an array that can be used for logical indexing and logical tests. Typing B = logical(A), where A is a numeric array, returns the logical array B. So to correct the error in the previous session, you may type instead w = logical([1,0,0,0,1]) before typing v = x(w).

When a finite, real value other than 1 or 0 is assigned to a logical variable, the value is converted to logical 1 and a warning message is issued. For example, when you type y = logical(9), y will be assigned the value logical 1 and a warning will be issued. You may use the double function to convert a logical array to an array of class double. For example, x = (5 > 3); y = double(x);. Some arithmetic operations convert a logical array to a double array. For example, if we add zero to each element of B by typing B = B + 0, then B will be converted to a numeric (double) array. However, not all mathematical operations are defined for logical variables. For example, typing

```
>>x = ([2, 3] > [1, 6]);
>>y = sin(x)
```

will generate an error message. This is not an important issue because it hardly makes sense to compute the sine of logical data or logical variables.

Accessing Arrays Using Logical Arrays

When a logical array is used to address another array, it extracts from that array the elements in the locations where the logical array has 1s. So typing A(B), where B is a logical array of the same size as A, returns the values of A at the indices where B is 1.

Given A = [5,6,7;8,9,10;11,12,13] and B = logical (eye(3)), we can extract the diagonal elements of A by typing C = A(B) to obtain C = [5;9;13]. Specifying array subscripts with logical arrays extracts the elements that correspond to the true (1) elements in the logical array.

Note, however, that using the *numeric* array eye(3), as C = A(eye(3)), results in an error message because the elements of eye(3) do not correspond to locations in A. If the numeric array values correspond to valid locations, you may use a numeric array to extract the elements. For example, to extract the diagonal elements of A with a numeric array, type C = A([1,5,9]).

MATLAB data types are preserved when indexed assignment is used. So now that logical is a MATLAB data type, if A is a logical array, for example, A = logical(eye(4)), then typing A(3,4) = 1 does not change A to a double array. However, typing A(3,4) = 5 will set A(3,4) to logical 1 and cause a warning to be issued.

4.3 Logical Operators and Functions

MATLAB has five *logical operators,* which are sometimes called *Boolean* operators (see Table 4.3–1). These operators perform element-by-element operations. With the exception of the NOT operator (~), they have a lower precedence than the arithmetic and relational operators (see Table 4.3–2). To see a very detailed order of precedence, type help precedence in the Command window. The NOT symbol is called the *tilde.*

Table 4.3–1 Logical operators

Operator	Name	Definition
~	NOT	~A returns an array of the same dimension as A; the new array has 1s where A is 0 and 0s where A is nonzero.
&	AND	A & B returns an array of the same dimension as A and B; the new array has 1s where both A and B have nonzero elements and 0s where either A or B is 0.
|	OR	A | B returns an array of the same dimension as A and B; the new array has 1s where at least one element in A or B is nonzero and 0s where A and B are both 0.
&&	Short-Circuit AND	Operator for scalar logical expressions. A && B returns true if both A and B evaluate to true, and false if they do not.
||	Short-Circuit OR	Operator for scalar logical expressions. A || B returns true if either A or B or both evaluate to true, and false if they do not.

Table 4.3–2 Order of precedence for operator types

Precedence	Operator type
First	Parentheses; evaluated starting with the innermost pair.
Second	Arithmetic operators and logical NOT (~); evaluated from left to right.
Third	Relational operators; evaluated from left to right.
Fourth	Logical AND.
Fifth	Logical OR.

The NOT operation ~A returns an array of the same dimension as A; the new array has 1s where A is 0 and 0s where A is nonzero. If A is logical, then ~A replaces 1s with 0s and 0s with 1s. For example, if x = [0,3,9] and y = [14,-2,9], then z = ~x returns the array z = [1,0,0] and the statement u = ~x > y returns the result u = [0,1,0]. This expression is equivalent to u = (~x) > y, whereas v = ~(x > y) gives the result v = [1,0,1]. This expression is equivalent to v = (x <= y).

The & and | operators compare two arrays of the same dimension. The only exception, as with the relational operators, is that an array can be compared to a scalar. The AND operation A & B returns 1s where both A and B have nonzero elements and 0s where any element of A or B is 0. The expression z = 0 & 3 returns z = 0; z = 2 & 3 returns z = 1; z = 0 & 0 returns z = 0, and z = [5,-3,0,0] & [2,4,0,5] returns z = [1,1,0,0]. Because of operator precedence, z = 1 & 2 + 3 is equivalent to z = 1 & (2 + 3), which returns z = 1. Similarly, z = 5 < 6 & 1 is equivalent to z = (5 < 6) & 1, which returns z = 1.

Let x = [6,3,9] and y = [14,2,9] and let a = [4,3,12]. The expression

z = (x > y) & a

gives z = [0,1,0], and

z = (x > y) & (x > a)

returns the result z = [0,0,0]. This is equivalent to

```
z = x > y & x > a
```

which is much less readable.

Be careful when using the logical operators with inequalities. For example, note that ~(x > y) is equivalent to x <= y. It is *not* equivalent to x < y. As another example, the relation $5 < x < 10$ must be written as

```
(5 < x) & (x < 10)
```

in MATLAB.

The OR operation A|B returns 1s where at least one of A and B has nonzero elements and 0s where both A and B are 0. The expression z = 0|3 returns z = 1; the expression z = 0|0 returns z = 0; and

```
z = [5,-3,0,0]|[2,4,0,5]
```

returns z = [1,1,0,1]. Because of operator precedence,

```
z = 3 < 5|4 == 7
```

is equivalent to

```
z = (3 < 5)|(4 == 7)
```

which returns z = 1. Similarly, z = 1|0 & 1 is equivalent to z = (1|0) & 1, which returns z = 1, while z = 1|0 & 0 returns z = 0, and z = 0 & 0|1 returns z = 1.

Because of the precedence of the NOT operator, the statement

```
z = ~3 == 7|4 == 6
```

returns the result z = 0, which is equivalent to

```
z = ((~3) == 7)|(4 == 6)
```

The exclusive OR function xor(A,B) returns 0s where A and B are either both nonzero or both 0, and 1s where either A or B is nonzero, *but not both.* The function is defined in terms of the AND, OR, and NOT operators as follows.

```
function z = xor(A,B)
z = (A|B) & ~(A & B);
```

The expression

```
z = xor([3,0,6],[5,0,0])
```

returns z = [0,0,1], whereas

```
z = [3, 0, 6]|[5,0,0]
```

returns z = [1,0,1].

Table 4.3–3 is a *truth table* that defines the operations of the logical operators and the function xor. Until you acquire more experience with the

TRUTH TABLE

Table 4.3–3 Truth table

x	y	~x	x|y	x & y	xor(x,y)
true	true	false	true	true	false
true	false	false	true	false	true
false	true	true	true	false	true
false	false	true	false	false	false

logical operators, you should use this table to check your statements. Remember that *true* is equivalent to logical 1, and *false* is equivalent to logical 0. We can test the truth table by building its numerical equivalent as follows. Let x and y represent the first two columns of the truth table in terms of 1s and 0s.

The following MATLAB session generates the truth table in terms of 1s and 0s.

```
>>x = [1,1,0,0]';
>>y = [1,0,1,0]';
>>Truth_Table = [x,y,~x,x|y,x & y,xor(x,y)]
Truth_Table =
    1 1 0 1 1 0
    1 0 0 1 0 1
    0 1 1 1 0 1
    0 0 1 0 0 0
```

Starting with MATLAB 6, the AND operator (&) was given a higher precedence than the OR operator (|). This was not true in earlier versions of MATLAB, so if you are using code created in an earlier version, you should make the necessary changes before using it in MATLAB 6 or higher. For example, now the statement y = 1|5 & 0 is evaluated as y = 1|(5 & 0), yielding the result y = 1, whereas in MATLAB 5.3 and earlier, the statement would have been evaluated as y = (1|5) & 0, yielding the result y = 0. To avoid potential problems due to precedence, it is important to use parentheses in statements containing arithmetic, relational, or logical operators, even where parentheses are optional.

Short-Circuit Operators

The following operators perform AND and OR operations on logical expressions containing *scalar* values only. They are called short-circuit operators because they evaluate their second operand only when the result is not fully determined by the first operand. They are defined as follows in terms of the two logical variables A and B.

A && B Returns true (logical 1) if both A and B evaluate to true, and false (logical 0) if they do not.

A||B Returns true (logical 1) if either A or B or both evaluate to true, and false (logical 0) if they do not.

Thus in the statement A && B, if A equals logical zero, then the entire expression will evaluate to false, regardless of the value of B, and therefore there is no need to evaluate B.

For A||B, if A is true, regardless of the value of B, the statement will evaluate to true.

Table 4.3–4 lists several useful logical functions.

Table 4.3–4 Logical functions

Logical function	Definition
all(x)	Returns a scalar, which is 1 if all the elements in the vector x are nonzero and 0 otherwise.
all(A)	Returns a row vector having the same number of columns as the matrix A and containing 1s and 0s, depending on whether the corresponding column of A has all nonzero elements.
any(x)	Returns a scalar, which is 1 if any of the elements in the vector x is nonzero and 0 otherwise.
any(A)	Returns a row vector having the same number of columns as A and containing 1s and 0s, depending on whether the corresponding column of the matrix A contains any nonzero elements.
find(A)	Computes an array containing the indices of the nonzero elements of the array A.
[u,v,w] = find(A)	Computes the arrays u and v containing the row and column indices of the nonzero elements of the array A and computes the array w containing the values of the nonzero elements. The array w may be omitted.
finite(A)	Returns an array of the same dimension as A with 1s where the elements of are finite and 0s elsewhere.
ischar(A)	Returns a 1 if A is a character array and 0 otherwise.
isempty(A)	Returns a 1 if A is an empty matrix and 0 otherwise.
isinf(A)	Returns an array of the same dimension as A, with 1s where A has 'inf' and 0s elsewhere.
isnan(A)	Returns an array of the same dimension as A with 1s where A has 'NaN' and 0s elsewhere. ('NaN' stands for "not a number," which means an undefined result.)
isnumeric(A)	Returns a 1 if A is a numeric array and 0 otherwise.
isreal(A)	Returns a 1 if A has no elements with imaginary parts and 0 otherwise.
logical(A)	Converts the elements of the array A into logical values.
xor(A,B)	Returns an array the same dimension as A and B; the new array has 1s where either A or B is nonzero, but not both, and 0s where A and B are either both nonzero or both zero.

Logical Operators and the find Function

The find function is very useful for creating decision-making programs, especially when combined with the relational or logical operators. The function find(x) computes an array containing the indices of the nonzero elements of the array x. For example, consider the session

```
>>x = [-2, 0, 4];
>>y = find(x)
y =
    1    3
```

The resulting array y = [1, 3] indicates that the first and third elements of x are nonzero. Note that the find function returns the *indices*, not the *values*. In the following session, note the difference between the result obtained by x(x < y) and the result obtained by find(x < y).

```
>>x = [6, 3, 9, 11];y = [14, 2, 9, 13];
>>values = x(x < y)
values =
    6     11
>>how_many = length (values)
how_many =
    2
>>indices = find(x < y)
indices =
    1    4
```

Thus two values in the array x are less than the *corresponding* values in the array y. They are the first and fourth values, 6 and 11. To find out how many, we could also have typed length(indices).

The find function is also useful when combined with the logical operators. For example, consider the session

```
>>x = [5, -3, 0, 0, 8]; y = [2, 4, 0, 5, 7];
>>z = find(x & y)
z =
    1    2    5
```

The resulting array z = [1, 2, 5] indicates that the first, second, and fifth elements of both x and y are nonzero. Note that the find function returns the *indices*, and not the *values*. In the following session, note the difference between the result obtained by y(x & y) and the result obtained by find(x & y) above.

```
>>x = [5, -3, 0, 0, 8];y = [2, 4, 0, 5, 7];
>>values = y(x & y)
values =
    2    4    7
>>how_many = length(values)
how_many =
    3
```

Thus there are three nonzero values in the array y that correspond to nonzero values in the array x. They are the first, second, and fifth values, which are 2, 4, and 7.

In the above examples, there were only a few numbers in the arrays x and y, and thus we could have obtained the answers by visual inspection. However, these MATLAB methods are very useful either where there are so many data that visual inspection would be very time-consuming, or where the values are generated internally in a program.

Test Your Understanding

T4.3–1 If x = [5,-3,18,4] and y = [-9,13,7,4], what will be the result of the following operations? Use MATLAB to check your answer.

a. z = ~y > x

b. z = x & y

c. z = x | y

d. z = xor(x,y)

T4.3–2 Suppose that x = [-9, -6, 0, 2, 5] and y = [-10,-6, 2, 4, 6]. What is the result of the following operations? Determine the answers by hand, and then use MATLAB to check your answers.

a. z = (x < y)

b. z = (x > y)

c. z = (x ~= y)

d. z = (x == y)

e. z = (x > 2)

T4.3–3 Suppose that x = [-4, -1, 0, 2, 10] and y = [-5, -2, 2, 5, 9]. Use MATLAB to find the values and the indices of the elements in x that are greater than the corresponding elements in y.

Height and Speed of a Projectile

EXAMPLE 4.3–1

The height and speed of a projectile (such as a thrown ball) launched with a speed of v_0 at an angle A to the horizontal are given by

$$h(t) = v_0 t \sin A - 0.5 g t^2$$

$$v(t) = \sqrt{v_0^2 - 2 v_0 g t \sin A + g^2 t^2}$$

where g is the acceleration due to gravity. The projectile will strike the ground when $h(t) = 0$, which gives the time to hit $t_{\text{hit}} = 2(v_0/g)\sin A$. Suppose that $A = 40°$, $v_0 = 20$ m/s, and $g = 9.81 \text{m/s}^2$. Use the MATLAB relational and logical operators to find the times when the height is no less than 6 m and the speed is simultaneously no greater than 16 m/s. In addition, discuss another approach to obtaining a solution.

■ **Solution**

The key to solving this problem with relational and logical operators is to use the `find` command to determine the times at which the logical expression `(h >= 6) & (v <= 16)` is true. First we must generate the vectors h and v corresponding to times t_1 and t_2 between $0 \le t \le t_{hit}$, using a spacing for time t that is small enough to achieve sufficient accuracy for our purposes. We will choose a spacing of $t_{hit}/100$, which provides 101 values of time. The program follows. When computing the times t_1 and t_2, we must subtract 1 from `u(1)` and from `length(u)` because the first element in the array t corresponds to $t = 0$ (that is, `t(1)` is 0).

```
% Set the values for initial speed, gravity, and angle.
v0 = 20; g = 9.81; A = 40*pi/180;
% Compute the time to hit.
t_hit = 2*v0*sin(A)/g;
% Compute the arrays containing time, height, and speed.
t = 0:t_hit/100:t_hit;
h = v0*t*sin(A) - 0.5*g*t.^2;
v = sqrt(v0^2 - 2*v0*g*sin(A)*t + g^2*t.^2);
% Determine when the height is no less than 6
% and the speed is no greater than 16.
u = find(h >= 6 & v <= 16);
% Compute the corresponding times.
t_1 = (u(1) - 1)*(t_hit/100)
t_2 = u(length(u) - 1)*(t_hit/100)
```

The results are $t_1 = 0.8649$ and $t_2 = 1.7560$. Between these two times $h \ge 6$ m and $v \le 16$ m/s.

We could have solved this problem by plotting $h(t)$ and $v(t)$, but the accuracy of the results would be limited by our ability to pick points off the graph; in addition, if we had to solve many such problems, the graphical method would be more time-consuming.

Test Your Understanding

T4.3–4 Consider the problem given in Example 4.3–1. Use relational and logical operators to find the times for which either the projectile's height is less than 4 m or the speed is greater than 17 m/s. Plot $h(t)$ and $v(t)$ to confirm your answer.

4.4 Conditional Statements

In everyday language we describe our decision making by using conditional phrases such as "If I get a raise, I will buy a new car." If the statement "I get a raise" is true, the action indicated (buy a new car) will be executed. Here is another example: "If I get at least a $100 per week raise, I will buy a new car; else, I will put the raise into savings." A slightly more involved example is: "If I get at least a $100 per week raise, I will buy a new car; else, if the raise is greater than $50, I will buy a new sound system; otherwise, I will put the raise into savings."

We can illustrate the logic of the first example as follows:

```
If I get a raise,
      I will buy a new car
. (period)
```

Note how the period marks the end of the statement.

The second example can be illustrated as follows:

```
If I get at least a $100 per week raise,
    I will buy a new car;
else,
    I will put the raise into savings
. (period)
```

The third example follows:

```
If I get at least a $100 per week raise,
    I will buy a new car;
else, if the raise is greater than $50,
    I will buy a new sound system;
otherwise,
    I will put the raise into savings
. (period)
```

The MATLAB *conditional statements* enable us to write programs that make decisions. Conditional statements contain one or more of the if, else, and elseif statements. The end statement denotes the end of a conditional statement, just as the period was used in the preceding examples. These conditional statements have a form similar to the examples, and they read somewhat like their English-language equivalents.

The if Statement

The if statement's basic form is

if *logical expression*
    statements
end

Every if statement must have an accompanying end statement. The end statement marks the end of the *statements* that are to be executed if the *logical expression* is true. A space is required between the if and the *logical expression,* which may be a scalar, a vector, or a matrix.

For example, suppose that x is a scalar and that we want to compute $y = \sqrt{x}$ only if $x \geq 0$. In English, we could specify this procedure as follows: If x is greater than or equal to zero, compute y from $y = \sqrt{x}$. The following if statement implements this procedure in MATLAB, assuming x already has a scalar value.

```
if x >= 0
    y = sqrt(x)
end
```

If x is negative, the program takes no action. The *logical expression* here is x >= 0, and the *statement* is the single line y = sqrt(x).

The if structure may be written on a single line; for example,

```
if x >= 0, y = sqrt(x), end
```

However, this form is less readable than the previous form. The usual practice is to indent the *statements* to clarify which statements belong to the if and its corresponding end and thereby improve readability.

The *logical expression* may be a compound expression; the *statements* may be a single command or a series of commands separated by commas or semicolons or on separate lines. For example, if *x* and *y* have scalar values,

```
z = 0;w = 0;
if (x >= 0)&(y >= 0)
    z = sqrt(x) + sqrt(y)
    w = sqrt(x*y)
end
```

The new values of z and w are computed only if both x and y are nonnegative. Otherwise, *z* and *w* retain their values of zero. The flowchart is shown in Figure 4.4–1.

We may "nest" if statements, as shown by the following example.

if *logical expression 1*
    statement group 1
    if *logical expression 2*
        statement group 2
    end
end

Note that each if statement has an accompanying end statement.

For example, assuming x and y have already been assigned scalar values,

```
if x >= 0
  % Calculate new value for y.
  y = 2 - log(x);
  if y >= 0
      z = log(x);
  end
end
```

The else Statement

When more than one action can occur as a result of a decision, we can use the else and elseif statements along with the if statement. The basic structure for the use of the else statement is

if *logical expression*
    statement group 1

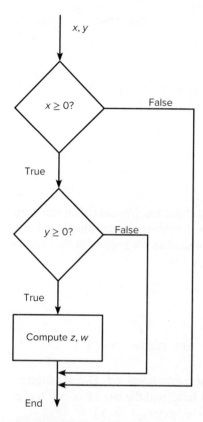

Figure 4.4–1 Flowchart illustrating two logical tests.

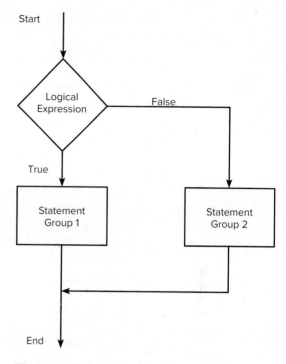

Figure 4.4–2 Flowchart of the else structure.

```
else
        statement group 2
end
```

Figure 4.4–2 shows the flowchart of this structure.

For example, suppose that $y = \sqrt{x}$ for $x \geq 0$ and that $y = e^x - 1$ for $x < 0$. The following statements will calculate y, assuming that x already has a scalar value.

```
if x >= 0
   y = sqrt(x)
else
   y = exp(x) - 1
end
```

When the test, if *logical expression,* is performed, where the logical expression may be an *array,* the test returns a value of true only if *all* the elements of the logical expression are true! For example, if we fail to recognize how the test works, the following statements do not perform the way we might expect.

```
x = [4,-9,25];
if x < 0
    disp('Some of the elements of x are negative.')
else
    y = sqrt(x)
end
```

When this program is run, it gives the result

```
y =
    2      0 + 3.000i      5
```

The program does not test each element in x in sequence. Instead it tests the truth of the vector relation x < 0. The test if x < 0 returns a false value because it generates the vector [0,1,0]. Compare the preceding program with the following program.

```
x = [4,-9,25];
if x >= 0
    y = sqrt(x)
else
    disp('Some of the elements of x are negative.')
end
```

When executed, it produces the following result: Some of the elements of x are negative. The test if x < 0 is false, and the test if x >= 0 also returns a false value because x >= 0 returns the vector [1,0,1].

We sometimes must choose between a program that is concise, but perhaps more difficult to understand, and one that uses more statements than is necessary. For example, the statements

if *logical expression 1*
    if *logical expression 2*
       statements
    end
end

can be replaced with the more concise program

if *logical expression 1* & *logical expression 2*
       statements
end

The **elseif** Statement

The general form of the if statement is

if *logical expression 1*
    statement group 1
elseif *logical expression 2*
    statement group 2

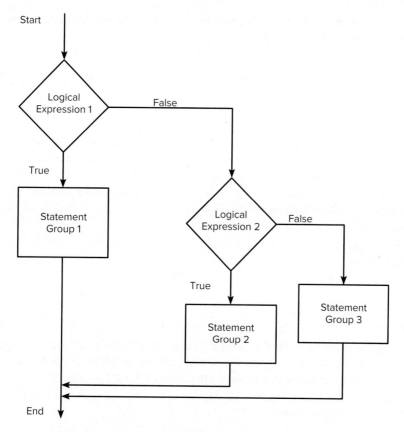

Figure 4.4–3 Flowchart for the general if structure.

```
else
    statement group 3
end
```

The else and elseif statements may be omitted if not required. However, if both are used, the else statement must come after the elseif statement to take care of all conditions that might be unaccounted for. Figure 4.4–3 is the flowchart for the general if structure.

For example, suppose that $y = \ln x$ if $x \geq 5$ and that $y = \sqrt{x}$ if $0 \leq x < 5$. The following statements will compute y if x has a scalar value.

```
if x >= 5
    y = log(x)
else
    if x >= 0
        y = sqrt(x)
    end
end
```

If $x = -2$, for example, no action will be taken. If we use an `elseif`, we need fewer statements. For example,

```
if x >= 5
    y = log(x)
elseif x >= 0
    y = sqrt(x)
end
```

Note that the `elseif` statement does not require a separate end statement.

The `else` statement can be used with `elseif` to create detailed decision-making programs. For example, suppose that $y = \ln x$ for $x > 10$, $y = \sqrt{x}$ for $0 \le x \le 10$, and $y = e^x - 1$ for $x < 0$. The following statements will compute y if x already has a scalar value.

```
if x > 10
    y = log(x)
elseif x >= 0
    y = sqrt(x)
else
    y = exp(x) - 1
end
```

Decision structures may be *nested*; that is, one structure can contain another structure, which in turn can contain another, and so on. Indentations are used to emphasize the statement groups associated with each end statement.

Test Your Understanding

T4.4–1 Given a number x and the quadrant q ($q = 1, 2, 3, 4$), write a program to compute $\sin^{-1}(x)$ in degrees, taking into account the quadrant. The program should display an error message if $|x| > 1$.

Checking the Number of Input and Output Arguments

Sometimes you will want to have a function act differently depending on how many inputs it has. You can use the function `nargin`, which stands for "number of input arguments." Within the function you can use conditional statements to direct the flow of the computation depending on how many input arguments there are. For example, suppose you want to compute the square root of the input if there is only one, but compute the square root of the average if there are two inputs. The following function does this.

```
function z = sqrtfun(x, y)
if (nargin == 1)
    z = sqrt(x);
elseif (nargin == 2)
    z = sqrt((x + y)/2);
end
```

The `nargout` function can be used to determine the number of output arguments.

Strings and Conditional Statements

A string is a variable that contains characters. Strings are useful for creating input prompts and messages and for storing and operating on data such as names and addresses. To create a string variable, enclose the characters in single quotes. For example, the string variable `name` is created as follows:

```
>>name = 'Leslie Student'
name =
    Leslie Student
```

The following string, called `number`,

```
>>number = '123'
number =
    123
```

is *not* the same as the variable `number` created by typing `number = 123`.

Strings are stored as row vectors in which each column represents a character. For example, the variable `name` has 1 row and 14 columns (each blank space occupies one column). We can access any column the way we access any other vector. For example, the letter S in the name Leslie Student occupies the eighth column in the vector `name`. It can be accessed by typing `name(8)`.

One of the most important applications for strings is to create input prompts and output messages. The following prompt program uses the `isempty(x)` function, which returns a 1 if the array `x` is empty and 0 otherwise. It also uses the `input` function, whose syntax is

```
x = input('prompt', 'string')
```

This function displays the string *prompt* on the screen, waits for input from the keyboard, and returns the entered value in the string variable `x`. The function returns an empty matrix if you press the **Enter** key without typing anything.

The following prompt program is a script file that allows the user to answer Yes by typing either Y or y or by pressing the **Enter** key. Any other response is treated as a No answer.

```
response = input('Do you want to continue? Y/N [Y]: ','s');
if (isempty(response))|(response == 'Y')|(response == 'y')
    response = 'Y'
else
    response = 'N'
end
```

Many more string functions are available in MATLAB. Type `help strfun` to obtain information on these.

The following function demonstrates the `elseif` structure and the use of a string variable. A *certificate of deposit* (CD) is one type of investment whose interest rate depends on the length of the term. Suppose a bank offers CDs with terms from 0.5 to 5 years. The following function displays the rate offered as a function of the term. Note how the function tests for an improper input (a term outside the range of 0.5 to 5 years).

```
function r = CD(t);
% Displays CD rate r as a function of the term t.
if t >= 0.5 & t <= 5
    if t >= 4, r = '3.5%';
    elseif t >= 3, r = '3%';
    elseif t >= 2, r = '2.5%';
    elseif t >= 1, r = '2%';
    else r = '1.5%';
    end
else
    disp('An incorrect term was entered')
end
```

Here are some common mistakes when using logical operators, strings, and the elseif clause.

- Typing if t >= 0.5 & <= 5 in the previous code instead of if t >= 0.5 & t <= 5.
- Typing and in the previous code instead of &.
- Typing else if instead of elseif.
- Failing to put quotes around a string variable, as typing r = 3% instead of r = '3%'.
- Typing = instead of == to test equality.

4.5 for Loops

A *loop* is a structure for repeating a calculation a number of times. Each repetition of the loop is a *pass*. MATLAB uses two types of explicit loops: the for loop, when the number of passes is known ahead of time, and the while loop, when the looping process must terminate when a specified condition is satisfied and thus the number of passes is not known in advance.

A simple example of a for loop is

```
for k = 5:10:35
    x = k^2
end
```

The *loop variable* k is initially assigned the value 5, and x is calculated from x = k^2. Each successive pass through the loop increments k by 10 and calculates x until k exceeds 35. Thus k takes on the values 5, 15, 25, and 35; and x takes on the values 25, 225, 625, and 1225. The program then continues to execute any statements following the end statement.

The typical structure of a for loop is

for *loop variable* = *m:s:n*
    statements
end

The expression m:s:n assigns an initial value of m to the loop variable, which is incremented by the value s, called the *step value* or *incremental value*. The *statements* are executed once during each pass, using the current value of the loop variable. The looping continues until the loop variable exceeds the *terminating value* n. For example, in the expression for k = 5:10:36, the final value of k is 35. Note that we need not place a semicolon after the for m:s:n statement to suppress printing k. Figure 4.5–1 shows the flowchart of a for loop.

Note that a for statement needs an accompanying end statement. The end statement marks the end of the *statements* that are to be executed. A space is required between the for and the *loop variable,* which may be a scalar, a vector, or a matrix, although the scalar case is by far the most common.

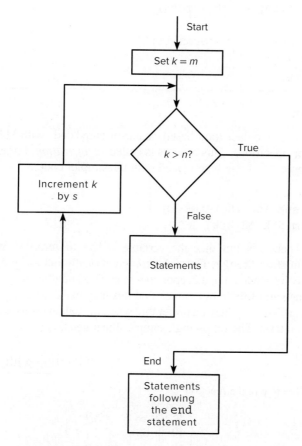

Figure 4.5–1 Flowchart of a for loop.

The for loop may be written on a single line; for example,

```
for x = 0:2:10, y = sqrt(x), end
```

However, this form is less readable than the previous form. The usual practice is to indent the *statements* to clarify which statements belong to the for and its corresponding end and thereby improve readability.

<table>
<tr><td>EXAMPLE 4.5–1</td><td style="text-align:right">Series Calculation with a for Loop</td></tr>
</table>

Write a script file to compute the sum of the first 15 terms in the series $5k^2 - 2k$, $k = 1$, $2, 3, \ldots , 15$.

■ **Solution**

Because we know how many times we must evaluate the expression $5k^2 - 2k$, we can use a for loop. The script file is the following:

```
total = 0;
for k = 1:15
    total = 5*k^2 - 2*k + total;
end
disp ('The sum for 15 terms is:')
disp (total)
```

The answer is 5960.

Vectorization

Sometimes you can replace loop-based, scalar-oriented code with MATLAB matrix and vector operations. This process is called *vectorization*. For example, the code in Example 4.5–1 can be replaced with the simpler code:

```
k = [1:15];
disp('The sum for 15 terms is:')
total = sum(5*k.^2-2*k)
```

Note that we need not initialize the variable total to zero, but we must use array exponentiation (k.^2). For more computationally intense programs, such efficiency may be needed, but a deeper understanding and greater comfort level is required on the part of the programmer, and this may make mistakes more likely.

However, a for loop may be preferable when the calculations depend on one or more logical tests. The following example illustrates this.

<table>
<tr><td>EXAMPLE 4.5–2</td><td style="text-align:right">Plotting with a for Loop</td></tr>
</table>

Write a script file to plot the function

$$y = \begin{cases} 15 \sqrt{4x} + 10 & x \geq 9 \\ 10x + 10 & 0 \leq x < 9 \\ 10 & x < 0 \end{cases}$$

for $-5 \leq x \leq 30$.

■ **Solution**

We choose a spacing $dx = 35/300$ to obtain 301 points, which is sufficient to obtain a smooth plot. The script file is the following:

```
dx = 35/300;
x = -5:dx:30;
for k = 1:length(x)
   if x(k) >= 9
     y(k) = 15*sqrt(4*x(k)) + 10;
   elseif x(k) >= 0
     y(k) = 10*x(k) + 10;
   else
     y(k) = 10;
   end
end
plot(x,y), xlabel('x'), ylabel('y')
```

Note that we must use the index k to refer to x within the loop, as x(k).

We may *nest loops* and conditional statements, as shown by the following example. (Note that each for and if statement needs an accompanying end statement.)

NESTED LOOPS

Suppose we want to create a special square matrix that has 1s in the first row and first column, and whose remaining elements are the sum of two elements, the element above and the element to the left, if the sum is less than 20. Otherwise, the element is the maximum of those two element values. The following function creates this matrix. The row index is r; the column index is c. Note how indenting improves the readability.

```
function A = specmat(n)
A = ones(n);
for r = 1:n
    for c = 1:n
        if (r > 1) & (c > 1)
            s = A(r-1,c) + A(r,c-1);
            if s < 20
                A(r,c) = s;
            else
                A(r,c) = max(A(r-1,c),A(r,c-1));
            end
        end
    end
end
end
```

Typing `specmat(5)` produces the following matrix:

$$
\begin{bmatrix}
1 & 1 & 1 & 1 & 1 \\
1 & 2 & 3 & 4 & 5 \\
1 & 3 & 6 & 10 & 15 \\
1 & 4 & 10 & 10 & 15 \\
1 & 5 & 15 & 15 & 15
\end{bmatrix}
$$

Test Your Understanding

T4.5–1 Write a script file using conditional statements to evaluate the following function, assuming that the scalar variable x has a value. The function is $y = \sqrt{x^2 + 1}$ for $x < 0$, $y = 3x + 1$ for $0 \leq x < 10$, and $y = 9 \sin(5x - 50) + 31$ for $x \geq 10$. Use your file to evaluate y for $x = -5$, $x = 5$, and $x = 15$, and check the results by hand.

T4.5–2 Use a `for` loop to determine the sum of the first 20 terms in the series $3k^2$, $k = 1, 2, 3, \ldots, 20$. (Answer: 8610.)

T4.5–3 Write a program to produce the following matrix:

$$
\mathbf{A} =
\begin{bmatrix}
4 & 8 & 12 \\
10 & 14 & 18 \\
16 & 20 & 24 \\
22 & 26 & 30
\end{bmatrix}
$$

Note the following rules when using `for` loops with the loop variable expression `k = m:s:n`:

- The step value s may be negative. For example, `k = 10:-2:4` produces $k = 10, 8, 6, 4$.
- If s is omitted, the step value defaults to 1.
- If s is positive, the loop will not be executed if m is greater than n.
- If s is negative, the loop will not be executed if m is less than n.
- If m equals n, the loop will be executed only once.
- If the step value s is not an integer, round-off errors can cause the loop to execute a different number of passes than intended.

When the loop is completed, k retains its last value. You should not alter the value of the loop variable k within the *statements*. Doing so can cause unpredictable results.

A common practice in traditional programming languages such as BASIC and FORTRAN is to use the symbols i and j as loop variables. However, this convention is not good practice in MATLAB, which uses these symbols for the imaginary unit $\sqrt{-1}$.

The break and continue Statements

It is permissible to use an if statement to "jump" out of the loop before the loop variable reaches its terminating value. The break command, which terminates the loop but does not stop the entire program, can be used for this purpose. For example,

```
for k = 1:10
    x = 50 - k^2;
    if x < 0
        break
    end
    y = sqrt(x)
end
% The program execution jumps to here
% if the break command is executed.
```

However, it is usually possible to write the code to avoid using the break command. This can often be done with a while loop as explained in the next section.

The break statement stops the execution of the loop. There can be applications where we want to not execute the case producing an error but continue executing the loop for the remaining passes. We can use the continue statement to do this. The continue statement passes control to the next iteration of the for or while loop in which it appears, skipping any remaining statements in the body of the loop. In nested loops, continue passes control to the next iteration of the for or while loop enclosing it.

For example, the following code uses a continue statement to avoid computing the logarithm of a negative number.

```
x = [10,1000,-10,100];
y = NaN*x;
for k = 1:length(x)
    if x(k) < 0
        continue
    end
    kvalue(k) = k;
    y(k) = log10(x(k));
end
kvalue
y
```

The results are k = 1, 2, 0, 4 and y = 1, 3, NaN, 2.

Using an Array as a Loop Index

It is permissible to use a matrix expression to specify the number of passes. In this case the loop variable is a vector that is set equal to the successive columns of the matrix expression during each pass. For example,

```
A = [1,2,3;4,5,6];
for v = A
    disp(v)
end
```

is equivalent to

```
A = [1,2,3;4,5,6];
n = 3;
for k = 1:n
    v = A(:,k)
end
```

The common expression k = $m:s:n$ is a special case of a matrix expression in which the columns of the expression are scalars, not vectors.

For example, suppose we want to compute the distance from the origin to a set of three points specified by their xy coordinates (3, 7), (6, 6), and (2, 8). We can arrange the coordinates in the array coord as follows.

$$\begin{bmatrix} 3 & 6 & 2 \\ 7 & 6 & 8 \end{bmatrix}$$

Then coord = [3,6,2;7,6,8]. The following program computes the distance and determines which point is farthest from the origin. The first time through the loop the index coord is [3, 7]′. The second time the index is [6, 6]′, and during the final pass it is [2, 8]′.

```
k = 0;
for coord = [3,6,2;7,6,8]
    k = k + 1;
    distance(k) = sqrt(coord'*coord)
end
[max_distance,farthest] = max(distance)
```

The previous program illustrates the use of an array index, but the problem can be solved more concisely with the following program, which uses the diag function to extract the diagonal elements of an array.

```
coord = [3,6,2;7,6,8];
distance = sqrt(diag(coord'*coord))
[max_distance,farthest] = max(distance)
```

Implied Loops

Many MATLAB commands contain *implied loops*. For example, consider these statements.

```
x = [0:5:100];
y = cos(x);
```

To achieve the same result using a for loop, we would type

```
for k = 1:21
    x = (k - 1)*5;
    y(k) = cos(x);
end
```

The find command is another example of an implied loop. The statement `y = find(x>0)` is equivalent to

```
m = 0;
for k = 1:length(x)
    if x(k) > 0
        m = m + 1;
        y(m) = k;
    end
end
```

If you are familiar with a traditional programming language such as FORTRAN or BASIC, you might be inclined to solve problems in MATLAB using loops, instead of using the powerful MATLAB commands such as find. To use these commands and to maximize the power of MATLAB, you might need to adopt a new approach to problem solving. As the preceding example shows, you often can save many lines of code by using MATLAB commands, instead of using loops. Your programs will also run faster because MATLAB was designed for high-speed vector computations.

Test Your Understanding

T4.5–4 Write a for loop that is equivalent to the command sum(A), where A is a matrix.

Data Sorting

EXAMPLE 4.5–3

A vector x has been obtained from measurements. Suppose we want to consider any data value in the range $-0.1 < x < 0.1$ as being erroneous. We want to remove all such elements and replace them with zeros at the end of the array. Develop two ways of doing this. An example is given in the following table.

	Before	**After**
x(1)	1.92	1.92
x(2)	0.05	−2.43
x(3)	−2.43	0.85
x(4)	−0.02	0
x(5)	0.09	0
x(6)	0.85	0
x(7)	−0.06	0

■ **Solution**

The following script file uses a for loop with conditional statements. Note how the null array [ ] is used.

```
x = [1.92,0.05,-2.43,-0.02,0.09,0.85,-0.06];
y = [];z = [];
for k = 1:length(x)
    if abs(x(k)) >= 0.1
        y = [y,x(k)];
    else
        z = [z,x(k)];
    end
end
xnew = [y,zeros(size(z))]
```

The next script file uses the `find` function.

```
x = [1.92,0.05,-2.43,-0.02,0.09,0.85,-0.06];
y = x(find(abs(x) >= 0.1));
z = zeros(size(find(abs(x) < 0.1)));
xnew = [y,z]
```

Use of Logical Arrays as Masks

Consider the array **A**.

$$\mathbf{A} = \begin{bmatrix} 0 & -1 & 4 \\ 9 & -14 & 25 \\ -34 & 49 & 64 \end{bmatrix}$$

The following program computes the array **B** by computing the square roots of all the elements of **A** whose value is no less than 0 and adding 50 to each element that is negative.

```
A = [0, -1, 4; 9, -14, 25; -34, 49, 64];
for m = 1:size(A,1)
    for n = 1:size(A,2)
        if A(m,n) >= 0
            B(m,n) = sqrt(A(m,n));
        else
            B(m,n) = A(m,n) + 50;
        end
    end
end
B
```

The result is

$$\mathbf{B} = \begin{bmatrix} 0 & 49 & 2 \\ 3 & 36 & 5 \\ 16 & 7 & 8 \end{bmatrix}$$

When a logical array is used to address another array, it extracts from that array the elements in the locations where the logical array has 1s. We can often avoid the use of loops and branching and thus create simpler and faster programs by using a logical array as a *mask* that selects elements of another array. Any elements not selected will remain unchanged.

MASK

The following session creates the logical array C from the numeric array A given previously.

```
>>A = [0, -1, 4; 9, -14, 25; -34, 49, 64];
>>C = (A >= 0);
```

The result is

$$C = \begin{bmatrix} 1 & 0 & 1 \\ 1 & 0 & 1 \\ 0 & 1 & 1 \end{bmatrix}$$

We can use this technique to compute the square root of only those elements of A given in the previous program that are no less than 0 and add 50 to those elements that are negative. The program is

```
A = [0, -1, 4; 9, -14, 25; -34, 49, 64];
C = (A >= 0);
A(C) = sqrt(A(C))
A(~C) = A(~C) + 50
```

The result after the third line is executed is

$$A = \begin{bmatrix} 0 & -1 & 2 \\ 3 & -14 & 25 \\ -34 & 49 & 64 \end{bmatrix}$$

The result after the last line is executed is

$$A = \begin{bmatrix} 0 & 49 & 2 \\ 3 & 36 & 5 \\ 16 & 7 & 8 \end{bmatrix}$$

Flight of an Instrumented Rocket

EXAMPLE 4.5–4

All rockets lose weight as they burn fuel; thus the mass of the system is variable. The following equations describe the speed v and height h of a rocket launched vertically, neglecting air resistance. They can be derived from Newton's law.

$$v(t) = u \ln\frac{m_0}{m_0 - qt} - gt \tag{4.5-1}$$

$$h(t) = \frac{u}{q}(m_0 - qt)\ln(m_0 - qt)$$

$$+ u(\ln m_0 + 1)t - \frac{gt^2}{2} - \frac{m_0 u}{q}\ln m_0 \tag{4.5-2}$$

where m_0 is the rocket's initial mass, q is the rate at which the rocket burns fuel mass, u is the exhaust velocity of the burned fuel relative to the rocket, and g is the acceleration due to gravity. Let b be the *burn time*, after which all the fuel is consumed. Thus the rocket's mass without fuel is $m_e = m_0 - qb$.

For $t > b$ the rocket engine no longer produces thrust, and the speed and height are given by

$$v(t) = v(b) - g(t - b) \tag{4.5-3}$$

$$h(t) = h(b) + v(b)(t - b) - \frac{g(t - b)^2}{2} \tag{4.5-4}$$

The time t_p to reach the peak height is found by setting $v(t) = 0$. The result is $t_p = b + v(b)/g$. Substituting this expression into the expression (4.5–4) for $h(t)$ gives the following expression for the peak height: $h_p = h(b) + v^2(b)/(2g)$. The time at which the rocket hits the ground is $t_{\text{hit}} = t_p + \sqrt{2h_p/g}$.

Suppose the rocket is carrying instruments to study the upper atmosphere, and we need to determine the amount of time spent above 50,000 ft as a function of the burn time b (and thus as a function of the fuel mass qb). Assume that we are given the following values: $m_e = 100$ slugs, $q = 1$ slugs/sec, $u = 8000$ ft/sec, and $g = 32.2$ ft/sec^2. If the rocket's maximum fuel load is 100 slugs, the maximum value of b is $100/q = 100$. Write a MATLAB program to solve this problem.

■ Solution

Pseudocode for developing the program appears in Table 4.5–1. A `for` loop is a logical choice to solve this problem because we know the burn time b and t_{hit}, the time it takes to hit the ground. A MATLAB program to solve this problem appears in Table 4.5–2. It has two

Table 4.5–1 Pseudocode for Example 4.5–4

Enter data.
Increment burn time from 0 to 100. For each burn time value:
    Compute m_0, v_b, h_b, h_p.
    If $h_p \geq h_{\text{desired}}$,
        Compute t_p, t_{hit}.
        Increment time from 0 to t_{hit}.
            Compute height as a function of time, using
            the appropriate equation, depending on whether
            burnout has occurred.
        Compute the duration above desired height.
        End of the time loop.
    If $h_p < h_{\text{desired}}$, set duration equal to zero.
End of the burn time loop.
Plot the results.

nested for loops. The inner loop is over time and evaluates the equations of motion at times spaced 1/10 sec apart. This loop calculates the duration above 50,000 ft for a specific value of the burn time b. We can obtain greater accuracy by using a smaller value of the time increment dt. The outer loop varies the burn time in integer values from $b = 1$ to $b = 100$. The final result is the vector of durations for the various burn times. Figure 4.5–2 gives the resulting plot.

Table 4.5–2 MATLAB program for Example 4.5–4

```
% Script file rocket1.m
% Computes flight duration as a function of burn time.
% Basic data values.
m_e = 100; q = 1; u = 8000; g = 32.2;
dt = 0.1; h_desired = 50000;
for b = 1:100 % Loop over burn time.
    burn_time(b) = b;
    % The following lines implement the formulas in the text.
    m_0 = m_e + q*b; v_b = u*log(m_0/m_e) - g*b;
    h_b = ((u*m_e)/q)*log(m_e/(m_e+q*b))+u*b - 0.5*g*b^2;
    h_p = h_b + v b^2/(2*g);
    if h_p >= h_desired
    % Calculate only if peak height > desired height.
        t_p = b + v_b/g; % Compute peak time.
        t_hit = t_p + sqrt(2*h_p/g); % Compute time to hit.
        for p = 0:t_hit/dt
            % Use a loop to compute the height vector.
            k = p + 1; t = p*dt; time(k) = t;
            if t <= b
                % Burnout has not yet occurred.
                h(k) = (u/q)*(m_0 - q*t)*log(m_0 - q*t). . .
                    + u*(log(m_0) + 1)*t - 0.5*g*t^2 . . .
                    - (m_0*u/q)*log(m_0);
            else
                % Burnout has occurred.
                h(k) = h_b - 0.5*g*(t - b)^2 + v_b*(t - b);
            end
        end
        % Compute the duration.
        duration(b) = length(find(h >= h_desired))*dt;
    else
        % Rocket did not reach the desired height.
        duration(b) = 0;
    end
end % Plot the results.
plot(burn_time,duration),xlabel('Burn Time (sec)'), . . .
ylabel('Duration (sec)'),title('Duration Above 50,000 Feet')
```

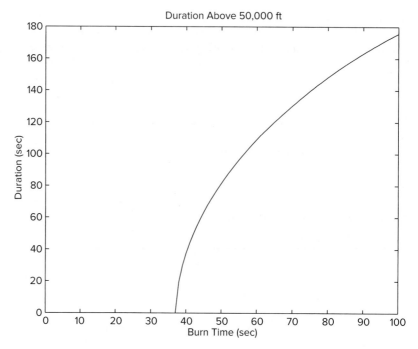

Figure 4.5–2 Duration above 50,000 ft as a function of the burn time.

4.6 while Loops

The while loop is used when the looping process terminates because a specified condition is satisfied, and thus the number of passes is not known in advance. A simple example of a while loop is

```
x = 5;
while x < 25
    disp(x)
    x = 2*x - 1;
end
```

The results displayed by the disp statement are 5, 9, and 17. The *loop variable* x is initially assigned the value 5, and it has this value until the statement x = 2*x - 1 is encountered the first time. The value then changes to 9. Before each pass through the loop, x is checked to see whether its value is less than 25. If so, the pass is made. If not, the loop is skipped and the program continues to execute any statements following the end statement.

A principal application of while loops is when we want the loop to continue as long as a certain statement is true. Such a task is often more difficult to do with a for loop. The typical structure of a while loop follows.

while *logical expression*
    statements
end

MATLAB first tests the truth of the *logical expression.* A loop variable must be included in the *logical expression.* For example, x is the loop variable in the statement while x < 25. If the *logical expression* is true, the *statements* are executed. For the while loop to function properly, the following two conditions must occur:

1. The loop variable must have a value before the while statement is executed.
2. The loop variable must be changed somehow by the *statements.*

The *statements* are executed once during each pass, using the current value of the loop variable. The looping continues until the *logical expression* is false. Figure 4.6–1 shows the flowchart of the while loop.

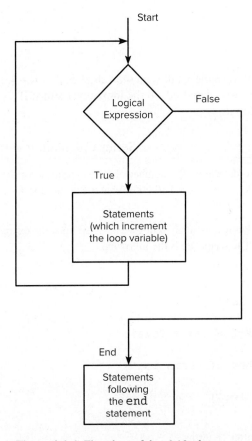

Figure 4.6–1 Flowchart of the while loop.

Each `while` statement must be matched by an accompanying `end`. As with for loops, the *statements* should be indented to improve readability. You may nest `while` loops, and you may nest them with `for` loops and `if` statements.

Always make sure that the loop variable has a value assigned to it before the start of the loop. For example, the following loop can give unintended results if `x` has an overlooked previous value.

```
while x < 10
    x = x + 1;
    y = 2*x;
end
```

If `x` has not been assigned a value prior to the loop, an error message will occur. If we intend `x` to start at zero, then we should place the statement `x = 0;` before the `while` statement.

It is possible to create an *infinite loop,* which is a loop that never ends. For example,

```
x = 8;
while x ~= 0
    x = x - 3;
end
```

Within the loop the variable `x` takes on the values 5, 2, −1, −4, . . . , and the condition `x ~= 0` is always satisfied, so the loop never stops. If such a loop occurs, press **Ctrl-C** to stop it.

| EXAMPLE 4.6–1 | Series Calculation with a `while` Loop |

Write a script file to determine the number of terms required for the sum of the series $5k^2 - 2k, k = 1, 2, 3, . . . ,$ to exceed 10,000. What is the sum for this many terms?

■ Solution

Because we do not know how many times we must evaluate the expression $5k^2 - 2k$, we use a `while` loop. The script file is the following:

```
total = 0;
k = 0;
while total < 1e+4
    k = k + 1;
    total = 5*k^2 - 2*k + total;
end
disp('The number of terms is:')
disp(k)
disp('The sum is:')
disp(total)
```

The sum is 10,203 after 18 terms.

Growth of a Bank Account

EXAMPLE 4.6–2

Determine how long it will take to accumulate at least $10,000 in a bank account if you deposit $500 initially and $500 at the end of each year, if the account pays 5 percent annual interest.

■ Solution

Because we do not know how many years it will take, a `while` loop should be used. The script file is the following.

```
amount = 500;
k=0;
while amount < 10000
    k = k+1;
    amount = amount*1.05 + 500;
end
amount
k
```

The final results are `amount = 1.0789e+004`, or $10,789, and `k = 14`, or 14 years.

Time to Reach a Specified Height

EXAMPLE 4.6–3

Consider the variable-mass rocket treated in Example 4.5–4. Write a program to determine how long it takes for the rocket to reach 40,000 ft if the burn time is 50 sec.

■ Solution

The pseudocode appears in Table 4.6–1. Because we do not know the time required, a `while` loop is convenient to use. The program in Table 4.6–2 performs the task and is a modification of the program in Table 4.5–2. Note that the new program allows for the possibility that the rocket might not reach 40,000 ft. It is important to write your programs to handle all such foreseeable circumstances. The answer given by the program is 53 sec.

Table 4.6–1 Pseudocode for Example 4.6–3

Enter data.
Compute m_0, v_b, h_b, h_p.
If $h_p \geq h_{desired}$,
    Use a `while` loop to increment time and compute height until desired height is reached.
        Compute height as a function of time, using the appropriate equation, depending on whether burnout has occurred.
    End of the time loop.
    Display the results.
If $h_p < h_{desired}$, rocket cannot reach desired height.

Table 4.6–2 MATLAB program for Example 4.6–3

```
% Script file rocket2.m
% Computes time to reach desired height.
% Set the data values.
h_desired = 40000; m_e = 100; q = 1;
u = 8000; g = 32.2; dt = 0.1; b = 50;
% Compute values at burnout, peak time, and height.
m_0 = m_e + q*b; v_b = u*log(m_0/m_e) - g*b;
h_b = ((u*m e)/q)*log(m e/(m e+q*b))+u*b - 0.5*g*b^2;
t_p = b + v b/g;
h_p = h_b + v_b^2/(2*g);
% If h_p > h_desired, compute time to reached h_desired.
if h_p > h_desired
    h - 0; k - 0;
    while h < h_desired % Compute h until h = h_desired.
        t = k*dt; k = k + 1;
        if t <= b
            % Burnout has not yet occurred.
            h = (u/q)*(m_0 - q*t)*log(m_0 - q*t)...
                + u*(log(m_0) + 1)*t - 0.5*g*t^2 ...
                - (m_0*u/q)*log(m_0);
        else
            % Burnout has occurred.
            h = h_b - 0.5*g*(t - b)^2 + v_b*(t - b);
        end
    end
    % Display the results.
    disp('The time to reach the desired height is:')
    disp(t)
else
    disp('Rocket cannot achieve the desired height.')
end
```

Test Your Understanding

T4.6–1   Use a `while` loop to determine how many terms in the series $3k^2$, $k = 1, 2, 3, \ldots$, are required for the sum of the terms to exceed 2000. What is the sum for this number of terms? (Answer: 13 terms, with a sum of 2457.)

T4.6–2   Rewrite the following code, using a `while` loop to avoid using the `break` command.

```
for k = 1:10
    x = 50 - k^2;
    if x < 0
        break
    end
    y = sqrt(x)
end
```

T4.6–3   Find to two decimal places the largest value of x before the error in the series approximation $e^x \approx 1 + x + x^2/2 + x^3/6$ exceeds 1 percent. (Answer: $x = 0.83$.)

4.7 The switch Structure

The switch structure provides an alternative to using the if, elseif, and else commands. Anything programmed using switch can also be programmed using if structures. However, for some applications the switch structure is more readable than code using the if structure. The syntax is

```
switch input expression (scalar or string)
    case value1
        statement group 1
    case value2
        statement group 2
    .
    .
    .
    otherwise
        statement group n
end
```

The *input expression* is compared to each case value. If they are the same, then the statements following that case statement are executed and processing continues with any statements after the end statement. If the *input expression* is a string, then it is equal to the case *value* if strcmp returns a value of 1 (true). Only the *first* matching case is executed. If no match occurs, the statements following the otherwise statement are executed. However, the otherwise statement is optional. If it is absent, execution continues with the statements following the end statement if no match exists. Each case *value* statement must be on a single line.

For example, suppose the variable angle has an integer value that represents an angle measured in degrees from North. The following switch block displays the point on the compass that corresponds to that angle.

```
switch angle
    case 45
        disp('Northeast')
    case 135
        disp('Southeast')
    case 225
        disp('Southwest')
    case 315
        disp('Northwest')
    otherwise
        disp('Direction Unknown')
end
```

The use of a string variable for the *input expression* can result in very readable programs. For example, in the following code the numeric vector x has values, and the user enters the value of the string variable response; its intended values are min, max, or sum. The code then either finds the minimum or maximum value of x or sums the elements of x, as directed by the user.

```
t = [0:100]; x = exp(-t).*sin(t);
response = input('Type min, max, or sum.','s')
response = lower('response');
switch response
   case min
      minimum = min(x)
   case max
      maximum = max(x)
   case sum
      total = sum(x)
   otherwise
      disp('You have not entered a proper choice.')
end
```

The switch statement can handle multiple conditions in a single case statement by enclosing the case *value* in a cell array. For example, the following switch block displays the corresponding point on the compass, given the integer angle measured from North.

```
switch angle
   case {0,360}
      disp('North')
   case {-180,180}
      disp('South')
   case {-270,90}
      disp('East')
   case {-90,270}
      disp('West')
   otherwise
      disp('Direction Unknown')
end
```

Test Your Understanding

T4.7–1   Write a program using the switch structure to input one angle, whose value may be 45, −45, 135, or −135°, and display the quadrant (1, 2, 3, or 4) containing the angle.

Using the `switch` Structure for Calendar Calculations

EXAMPLE 4.7–1

Use the `switch` structure to compute the total elapsed days in a year, given the number (1–12) of the month, the day, and an indication of whether the year is a leap year.

■ Solution

Note that February has an extra day if the year is a leap year. The following function computes the total elapsed number of days in a year, given the month, the day of the month, and the value of `extra_day`, which is 1 for a leap year and 0 otherwise.

```
function total_days = total(month,day,extra_day)
total_days = day;
for k = 1:month - 1
   switch k
     case {1,3,5,7,8,10,12}
         total_days = total_days + 31;
     case {4,6,9,11}
         total_days = total_days + 30;
     case 2
         total_days = total_days + 28 + extra_day;
   end
end
```

The function can be used as shown in the following program.

```
month = input('Enter month (1 - 12): ');
day = input('Enter day (1 - 31): ');
extra_day = input('Enter 1 for leap year; 0 otherwise: ');
total_days = total(month,day,extra_day)
```

One of the chapter problems for Section 4.4 (Problem 19) asks you to write a program to determine whether a given year is a leap year.

4.8 Debugging MATLAB Programs

Use of the MATLAB Editor as an M-file *editor* was discussed in earlier chapters. Here we discuss its use as a *debugger*. Figure 4.8–1 shows the Editor containing a program to be analyzed. Before you use the Editor as a debugger, try to debug your program using the commonsense guidelines presented under **Debugging Script Files** in Section 1.4. MATLAB programs are often short because of the power of the commands, and you may not need to use the Editor as a debugger unless you are writing large programs. However, the cell mode discussed in this section is useful even for short programs. In Chapters 1 and 3 we have already discussed the leftmost items on the menu bar in the FILE section of the EDITOR tab, and their functions are obvious.

The middle group of items, under the NAVIGATE and EDIT tabs, is useful mainly for large programs. The forward and back arrows, and the Find item

Figure 4.8–1 The Editor containing a program to be analyzed.

enable you to navigate through the program. The Insert item lets you insert a new section, a function from a list, or fixed point data. The Comment item lets you insert, remove, or wrap comments. Click *anywhere* in a previously typed line, and then click Comment. This makes the entire line a comment. A multi-line comment can be indicated by inserting %{ before the first line of the comment and %} after the last line. To turn a commented line into an executable line, click *anywhere* in the line, and then click **Uncomment**. The Indent item lets you increase or decrease the amount of indenting, or turn on smart indenting.

The most important features for debugging with the Editor are the five items in the BREAKPOINTS and RUN sections.

Cell Mode

The cell mode can be used to debug programs. It can also be used to generate a report. See the end of Section 5.2 for a discussion of the latter usage. A *code cell* is a group of commands. (Such a cell should not be confused with the cell array data type covered in Section 2.6.) Insert two percent signs (%%), called a *cell divider,* to mark the beginning of a new cell, or click on Insert Section in the EDIT section. The cell toolbar is shown in Figure 4.8–2.

Consider the following simple program that plots either a quadratic or a cubic function.

```
%% Evaluate a quadratic and a cubic.
clear, clc
x = linspace(0, 10, 300);
%% Quadratic
y1 = polyval([1, -8, 6], x); plot(x,y1)
%% Cubic
y2 = polyval([1, -11, 9, 9], x); plot(x,y2)
```

After entering and saving the program, place the cursor in one of the sections and click the **Run Section** icon (for this program, you should obviously

Figure 4.8–2 The cell mode of the Editor.

run the first section initially to establish values for **x**). You can also click on **Run and Advance** or **Advance**. These enable you to evaluate the current single cell (where the cursor is currently), to evaluate the current cell and advance to the next cell, or to evaluate the entire program. These features are obviously more useful in larger programs.

Clicking **Run and Time** starts the Profiler, which is a user interface that uses the results returned by the `profile` function. They provide a way to measure which program code requires the most time, so you can evaluate that code for possible performance improvements. They can be also be used to determine what lines of code do not run for particular input values. You can then develop test cases that exercise that specific code to see if it is causing a problem.

A useful feature of cell mode is that it enables you to evaluate quickly the results of changing a parameter without saving the program. For example, in Figure 4.8–2, if you have already run the program and the quadratic plot is on the screen, delete the minus sign before the number −8 and click **Run Section**. Watch the plot change. You need not first save the program.

Breakpoints

Breakpoints are points in the file where execution stops temporarily so that you can examine the values of the variables up to that point. The drop-down menu under Breakpoints lets you set and clear breakpoints, set conditions, and specify how to handle errors. You set a breakpoint on a line by placing the cursor on the line and selecting **Set/Clear** on the menu. You clear a breakpoint by repeating this process.

You need not use the menus to debug a program; you can use the Command window. Type `help debug` to see a list of MATLAB debugging functions, or type "debugging" in the Search window. They all begin with db (for debug). The commonly used ones are `dbstop` (to set a breakpoint), `dbclear` (to remove a breakpoint), dbcont (to resume execution), dbstep (to execute one or more lines), and dbquit (to quit debug mode).

When a breakpoint is hit, MATLAB goes into debug mode, the debugger window becomes active, and the prompt changes to a K>>. Any MATLAB command is allowed at the prompt. To resume program execution, use dbcont or dbstep. To exit from the debug mode, type dbquit.

Consider the following function test3(x) shown in Figure 4.8–1. If we type test3(10) we get the message No solution, which is correct if *y* is negative. Since *y* is local to the function we do not know its value. An advantage of debug mode is that it lets us see the values of local variables. Notice that the Editor uses line numbers. To check the value of *y* we set a breakpoint at line 5 by typing dbstop test3 5 in the Command window. A red dot will appear at line 5 between the line number and the code. This strip is the *breakpoint alley*. (This dot and strip are shown in Figure 4.8–1.) Now type test3(10) in the Command window. You will see

```
>> test3(10)
5 if y < 0
K>>
```

This is the prompt in debug mode (the K stands for "keyboard"). After this prompt, type y. You will see

```
y =
-2.2023e+04
```

thus confirming that *y* is negative. In the debug mode type dbcont to continue the execution or dbstep to step through the program one executable line at a time. Type dbclear test3 5 to clear the breakpoint at line 5. Type dbclear test3 to clear all the breakpoints. Type dbquit to exit the debug mode.

Personal experience has shown that is often is easier to use a combination of the drop-down menu under Breakpoints and the Command window. For example, it is easier to set a breakpoint with the cursor than to use the dbstop command, and it is easier to clear breakpoints with the drop-down menu.

Another way to trace the execution of a program is to use the echo function, which displays every line (including comments) as the execution proceeds, including the results, and omits any lines that are skipped. To trace a script file, simply type echo on in the Command window. For functions, the syntax is echo function name on. To turn off tracing type echo off or echo function name off.

4.9 Applications to Simulation

Simulation is the process of building and analyzing the output of computer programs that describe the operations of an organization, process, or physical system. Such a program is called a *computer model*. Simulation is often used in *operations research*, which is the quantitative study of an organization in action, to find ways to improve the functioning of the organization. Simulation enables

OPERATIONS RESEARCH

engineers to study the past, present, and future actions of the organization for this purpose. Operations research techniques are useful in all engineering fields. Common examples include airline scheduling, traffic flow studies, and production lines. The MATLAB logical operators and loops are excellent tools for building simulation programs.

A College Enrollment Model: Part I

EXAMPLE 4.9–1

As an example of how simulation can be used for operations research, consider the following college enrollment model. A certain college wants to analyze the effect of admissions and freshman retention rate on the college's enrollment so that it can predict the future need for instructors and other resources. Assume that the college has estimates of the percentages of students repeating a grade or leaving school before graduating. Develop a matrix equation on which to base a simulation model that can help in this analysis.

■ Solution

Suppose that the current freshman enrollment is 500 students and the college decides to admit 1000 freshmen per year from now on. The college estimates that 10 percent of the freshman class will repeat the year. The number of freshmen in the following year will be $0.1(500) + 1000 = 1050$, then it will be $0.1(1050) + 1000 = 1105$, and so on. Let $x_1(k)$ be the number of freshmen in year k, where $k = 1, 2, 3, 4, 5, 6, \ldots$. Then in year $k + 1$, the number of freshmen is given by

$$
\begin{aligned}
x_1(k + 1) = \; & 10 \text{ percent of previous freshman class} \\
& \text{repeating freshman year} \\
& + 1000 \text{ new freshmen} \\
= \; & 0.1x_1(k) + 1000
\end{aligned}
\tag{4.9–1}
$$

Because we know the number of freshmen in the first year of our analysis (which is 500), we can solve this equation step by step to predict the number of freshmen in the future.

Let $x_2(k)$ be the number of sophomores in year k. Suppose that 15 percent of the freshmen do not return and that 10 percent repeat freshman year. Thus 75 percent of the freshman class returns as sophomores. Suppose also 5 percent of the sophomores repeat the sophomore year and that 200 sophomores each year transfer from other schools. Then in year $k + 1$, the number of sophomores is given by

$$
x_2(k + 1) = 0.75x_1(k) + 0.05x_2(k) + 200
$$

To solve this equation, we need to solve the "freshman" equation (4.9–1) at the same time, which is easy to do with MATLAB. Before we solve these equations, let us develop the rest of the model.

Let $x_3(k)$ and $x_4(k)$ be the number of juniors and seniors, respectively, in year k. Suppose that 5 percent of the sophomores and juniors leave school and that 5 percent of the sophomores, juniors, and seniors repeat the grade. Thus 90 percent of the sophomores and juniors return and advance in grade. The models for the juniors and seniors are

$$
\begin{aligned}
x_3(k + 1) &= 0.9x_2(k) + 0.05x_3(k) \\
x_4(k + 1) &= 0.9x_3(k) + 0.05x_4(k)
\end{aligned}
$$

These four equations can be written in the following matrix form:

$$\begin{bmatrix} x_1(k+1) \\ x_2(k+1) \\ x_3(k+1) \\ x_4(k+1) \end{bmatrix} = \begin{bmatrix} 0.1 & 0 & 0 & 0 \\ 0.75 & 0.05 & 0 & 0 \\ 0 & 0.9 & 0.05 & 0 \\ 0 & 0 & 0.9 & 0.05 \end{bmatrix} \begin{bmatrix} x_1(k) \\ x_2(k) \\ x_3(k) \\ x_4(k) \end{bmatrix} + \begin{bmatrix} 1000 \\ 200 \\ 0 \\ 0 \end{bmatrix}$$

In Example 4.9–2 we will see how to use MATLAB to solve such equations.

Test Your Understanding

T4.9–1   Suppose that 70 percent of the freshmen, instead of 75 percent, return for the sophomore year. How does the previous equation change?

| EXAMPLE 4.9–2 | A College Enrollment Model: Part II |

To study the effects of admissions and transfer policies, generalize the enrollment model in Example 4.9–1 to allow for varying admissions and transfers.

■ **Solution**

Let $a(k)$ be the number of new freshmen admitted in the spring of year k for the following year $k + 1$, and let $d(k)$ be the number of transfers into the following year's sophomore class. Then the model becomes

$$\begin{aligned} x_1(k+1) &= c_{11}x_1(k) + a(k) \\ x_2(k+1) &= c_{21}x_1(k) + c_{22}x_2(k) + d(k) \\ x_3(k+1) &= c_{32}x_2(k) + c_{33}x_3(k) \\ x_4(k+1) &= c_{43}x_3(k) + c_{44}x_4(k) \end{aligned}$$

STATE TRANSITION DIAGRAM

where we have written the coefficients c_{21}, c_{22}, and so on in symbolic, rather than numerical, form so that we can change their values if desired.

   This model can be represented graphically by a *state transition diagram*, like the one shown in Figure 4.9–1. Such diagrams are widely used to represent time-dependent and probabilistic processes. The arrows indicate how the model's calculations are updated for each new year. The enrollment at year k is described completely by the

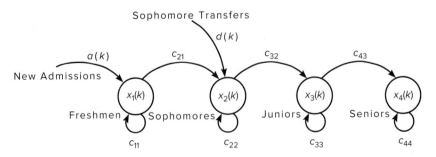

Figure 4.9–1 The state transition diagram for the college enrollment model.

values of $x_1(k)$, $x_2(k)$, $x_3(k)$, and $x_4(k)$, that is, by the vector $\mathbf{x}(k)$, which is called the *state vector*. The elements of the state vector are the *state variables*. The state transition diagram shows how the new values of the state variables depend on both the previous values and the inputs $a(k)$ and $d(k)$.

The four equations can be written in the following matrix form:

$$\begin{bmatrix} x_1(k+1) \\ x_2(k+1) \\ x_3(k+1) \\ x_4(k+1) \end{bmatrix} = \begin{bmatrix} c_{11} & 0 & 0 & 0 \\ c_{21} & c_{22} & 0 & 0 \\ 0 & c_{32} & c_{33} & 0 \\ 0 & 0 & c_{43} & c_{44} \end{bmatrix} \begin{bmatrix} x_1(k) \\ x_2(k) \\ x_3(k) \\ x_4(k) \end{bmatrix} + \begin{bmatrix} a(k) \\ d(k) \\ 0 \\ 0 \end{bmatrix}$$

or more compactly as

$$\mathbf{x}(k+1) = \mathbf{C}\mathbf{x}(k) + \mathbf{b}(k)$$

where

$$\mathbf{x}(k) = \begin{bmatrix} x_1(k) \\ x_2(k) \\ x_3(k) \\ x_4(k) \end{bmatrix} \quad \mathbf{b}(k) = \begin{bmatrix} a(k) \\ d(k) \\ 0 \\ 0 \end{bmatrix}$$

and

$$\mathbf{C} = \begin{bmatrix} c_{11} & 0 & 0 & 0 \\ c_{21} & c_{22} & 0 & 0 \\ 0 & c_{32} & c_{33} & 0 \\ 0 & 0 & c_{43} & c_{44} \end{bmatrix}$$

Suppose that the initial total enrollment of 1480 consists of 500 freshmen, 400 sophomores, 300 juniors, and 280 seniors. The college wants to study, over a 10-year period, the effects of increasing admissions by 100 each year and transfers by 50 each year until the total enrollment reaches 4000; then admissions and transfers will be held constant. Thus the admissions and transfers for the next 10 years are given by

$$a(k) = 900 + 100k$$
$$d(k) = 150 + 50k$$

for $k = 1, 2, 3, \ldots$ until the college's total enrollment reaches 4000; then admissions and transfers are held constant at the previous year's levels. We cannot determine when this event will occur without doing a simulation. Table 4.9–1 gives the pseudocode for solving this problem. The enrollment matrix $\mathbf{E}$ is a 4×10 matrix whose columns represent the enrollment in each year.

Because we know the length of the study (10 years), a `for` loop is a natural choice. We use an `if` statement to determine when to switch from the increasing admissions and transfer schedule to the constant schedule. A MATLAB script file to predict the enrollment for the next 10 years appears in Table 4.9–2. Figure 4.9–2 shows the resulting plot. Note

Table 4.9–1 Pseudocode for Example 4.9–2

Enter the coefficient matrix **C** and the initial enrollment vector **x**.
Enter the initial admissions and transfers, $a(1)$ and $d(1)$.
Set the first column of the enrollment matrix **E** equal to **x**.
Loop over years 2 to 10.
    If the total enrollment is ≤ 4000, increase admissions by 100 and transfers
    by 50 each year.
    If the total enrollment is > 4000, hold admissions and transfers constant.
    Update the vector x, using $\mathbf{x} = \mathbf{Cx} + \mathbf{b}$.
    Update the enrollment matrix **E** by adding another column composed of **x**.
End of the loop over years 2 to 10.
Plot the results.

Table 4.9–2 College enrollment model

```
% Script file enroll1.m. Computes college enrollment.
% Model's coefficients.
C = [0.1,0,0,0;0.75,0.05,0,0;0,0.9,0.05,0;0,0,0.9,0.05];
% Initial enrollment vector.
x = [500;400;300;280];
% Initial admissions and transfers.
a(1) = 1000; d(1) = 200;
% E is the 4 x 10 enrollment matrix.
E(:,1) = x;
% Loop over years 2 to 10.
for k = 2:10
    % The following describes the admissions
    % and transfer policies.
    if sum(x) <= 4000
        % Increase admissions and transfers.
        a(k) = 900+100*k;
        d(k) = 150+50*k;
    else
        % Hold admissions and transfers constant.
        a(k) = a(k-1);
        d(k) = d(k-1);
    end
        % Update enrollment matrix.
        b = [a(k);d(k);0;0];
        x = C*x+b;
        E(:,k) = x;
end
% Plot the results.
plot(E'),hold,plot(E(1,:),'o'),plot(E(2,:),'+'),plot(E(3,:),'*'), . . .
plot(E(4,:),'x'),xlabel('Year'),ylabel('Number of Students'), . . .
gtext('Frosh'),gtext('Soph'),gtext('Jr'),gtext('Sr'), . . .
title('Enrollment as a Function of Time')
```

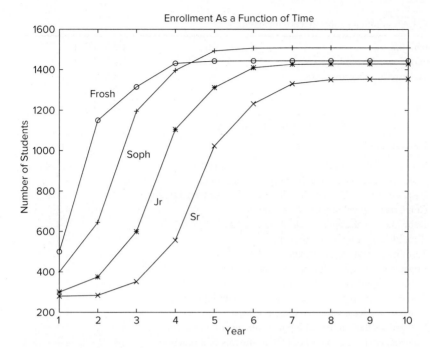

Figure 4.9–2 Class enrollments versus time.

that after year 4 there are more sophomores than freshmen. The reason is that the increasing transfer rate eventually overcomes the effect of the increasing admission rate.

In actual practice this program would be run many times to analyze the effects of different admissions and transfer policies and to examine what happens if different values are used for the coefficients in the matrix **C** (indicating different dropout and repeat rates).

Test Your Understanding

T4.9–2 In the program in Table 4.9–2, lines 16 and 17 compute the values of a(k) and d(k). These lines are repeated here:

```
a(k) = 900 + 100 * k
d(k) = 150 + 50 * k;
```

Why does the program contain the line a(1)=1000; d(1)=200;?

Table 4.10–1 Guide to MATLAB commands introduced in Chapter 4

Relational operators	Table 4.2–1
Logical operators	Table 4.3–1
Order of precedence for operator types	Table 4.3–2
Truth table	Table 4.3–3
Logical functions	Table 4.3–4

Miscellaneous commands

Command	Description	Section
break	Terminates the execution of a for or a while loop.	4.5, 4.6
case	Used with switch to direct program execution.	4.7
continue	Passes control to the next iteration of a for or while loop.	4.5, 4.6
double	Converts a logical array to class double.	4.2
else	Delineates an alternate block of statements.	4.4
elseif	Conditionally executes statements.	4.4
end	Terminates for, while, and if statements.	4.4, 4.5, 4.6
for	Repeats statements a specific number of times.	4.5
if	Executes statements conditionally.	4.4
input('s1', 's')	Display the prompt string s1 and stores user input as a string.	4.4
logical	Converts numeric values to logical values.	4.2
nargin	Determines the number of input arguments of a function.	4.4
nargout	Determines the number of output arguments of a function.	4.4
switch	Directs program execution by comparing the input expression with the associated case expressions.	4.7
while	Repeats statements an indefinite number of times.	4.6
xor	Exclusive OR function.	4.3

4.10 Summary

Now that you have finished this chapter, you should be able to write programs that can perform decision-making procedures; that is, the program's operations depend on results of the program's calculations or on input from the user. Sections 4.2, 4.3, and 4.4 covered the necessary functions: the relational operators, the logical operators and functions, and the conditional statements.

You should also be able to use MATLAB loop structures to write programs that repeat calculations a specified number of times or until some condition is satisfied. This feature enables engineers to solve problems of great complexity or requiring numerous calculations. The for loop and while loop structures were covered in Sections 4.5 and 4.6. Section 4.7 covered the switch structure.

Section 4.8 gave an overview and an example of how to debug programs using the Editor/Debugger. Section 4.9 presented an application of these methods to simulation, which enables engineers to study the operation of complicated systems, processes, and organizations.

Tables summarizing the MATLAB commands introduced in this chapter are located throughout the chapter. Table 4.10–1 will help you locate these tables. It also summarizes those commands not found in the other tables.

Key Terms

Problems

You can find answers to problems marked with an asterisk at the end of the text.

Section 4.1

1.   The volume V and surface area A of a sphere of radius r are given by

$$V = \frac{4}{3}\pi r^3 \quad A = 4\pi r^2$$

a. Develop a pseudocode description of a program to compute V and A for $0 \le r \le 3$ and to plot V versus A.

b. Write and run the program described in part *a*.

2.   The roots of the quadratic equation $ax^2 + bx + c = 0$ are given by

$$x = \frac{-b \pm \sqrt{b^2 - 4ac}}{2a}$$

a. Develop a pseudocode description of a program to compute both roots given the values of a, b, and c. Be sure to identify the real and imaginary parts.

b. Write the program described in part *a* and test it for the following cases:

1. $a = 2$, $b = 10$, $c = 12$
2. $a = 3$, $b = 24$, $c = 48$
3. $a = 4$, $b = 24$, $c = 100$

3.   It is desired to compute the sum of the first 10 terms of the series

$$14k^3 - 20k^2 + 5k, \, k = 1, 2, 3, \ldots$$

a. Develop a pseudocode description of the required program.

b. Write and run the program described in part *a*.

Section 4.2

4.* Suppose that x = 6. Find the results of the following operations by hand and use MATLAB to check your results.

a. z = (x < 10)
b. z = (x == 10)
c. z = (x >= 4)
d. z = (x ~= 7)

5.* Find the results of the following operations by hand and use MATLAB to check your results.

a. z = 6 > 3 + 8
b. z = 6 + 3 > 8
c. z = 4 > (2 + 9)
d. z = (4 < 7) + 3
e. z = 4 < 7 + 3
f. z = (4 < 7)*5
g. z = 4 < (7*5)
h. z = 2/5 >= 5

6.* Suppose that x = [10, -2, 6, 5, -3] and y = [9, -3, 2, 5, -1]. Find the results of the following operations by hand and use MATLAB to check your results.

a. z = (x < 6)
b. z = (x <= y)
c. z = (x == y)
d. z = (x ~= y)

7. For the arrays x and y given below, use MATLAB to find all the elements in x that are greater than the corresponding elements in y.

x = [-3, 0, 0, 2, 6, 8] y = [-5, -2, 0, 3, 4, 10]

8. The array price given below contains the price in dollars of a certain stock over 10 days. Use MATLAB to determine how many days the price was above $20.

price = [19, 18, 22, 21, 25, 19, 17, 21, 27, 29]

9. The arrays price_A and price_B given below contain the price in dollars of two stocks over 10 days. Use MATLAB to determine how many days the price of stock A was above the price of stock B.

price_A = [19, 18, 22, 21, 25, 19, 17, 21, 27, 29]
price_B = [22, 17, 20, 19, 24, 18, 16, 25, 28, 27]

10. The arrays price_A, price_B, and price_C given below contain the price in dollars of three stocks over 10 days.

a. Use MATLAB to determine how many days the price of stock A was above both the price of stock B and the price of stock C.

b. Use MATLAB to determine how many days the price of stock A was above either the price of stock B or the price of stock C.

c. Use MATLAB to determine how many days the price of stock A was above either the price of stock B or the price of stock C, but not both.

```
price_A = [19, 18, 22, 21, 25, 19, 17, 21, 27, 29]
price_B = [22, 17, 20, 19, 24, 18, 16, 25, 28, 27]
price_C = [17, 13, 22, 23, 19, 17, 20, 21, 24, 28]
```

Section 4.3

11.* Suppose that x = [-3, 0, 0, 2, 5, 8] and y = [-5, -2, 0, 3, 4, 10]. Find the results of the following operations by hand and use MATLAB to check your results.

a. z = y <~ x

b. z = x & y

c. z = x | y

d. z = xor(x,y)

12. The height and speed of a projectile (such as a thrown ball) launched with a speed of v_0 at an angle A to the horizontal are given by

$$h(t) = v_0 t \sin A - 0.5\, g t^2$$

$$v(t) = \sqrt{v_0^2 - 2 v_0 g t \sin A + g^2 t^2}$$

where g is the acceleration due to gravity. The projectile will strike the ground when $h(t) = 0$, which gives the time to hit $t_{hit} = 2(v_0/g)\sin A$.

Suppose that $A = 30°$, $v_0 = 40$ m/s, and $g = 9.81$ m/s². Use the MATLAB relational and logical operators to find the times when

a. The height is no less than 15 m.

b. The height is no less than 15 m and the speed is simultaneously no greater than 36 m/s.

c. The height is less than 5 m or the speed is greater than 35 m/s.

13.* The price, in dollars, of a certain stock over a 10-day period is given in the following array.

```
price = [19, 18, 22, 21, 25, 19, 17, 21, 27, 29]
```

Suppose you owned 1000 shares at the start of the 10-day period, and you bought 100 shares every day the price was below $20 and sold 100 shares every day the price was above $25. Use MATLAB to compute (*a*) the amount you spent in buying shares, (*b*) the amount you received from the sale of shares, (*c*) the total number of shares you own after the 10th day, and (*d*) the net increase in the worth of your portfolio.

14. Let e1 and e2 be logical expressions. DeMorgan's laws for logical expressions state that

    NOT(e1 AND e2) implies that (NOT e1) OR (NOT e2)

    and

    NOT(e1 OR e2) implies that (NOT e1) AND (NOT e2)

    Use these laws to find an equivalent expression for each of the following expressions and use MATLAB to verify the equivalence.

    a. ~((x < 10) & (x >= 6))
    b. ~((x == 2) | (x > 5))

15. Are these following expressions equivalent? Use MATLAB to check your answer for specific values of *a*, *b*, *c*, and *d*.

    a. 1. (a == b) & ((b == c)|(a == c))
       2. (a == b)|((b == c)&(a == c))
    b. 1. (a < b) & ((a > c)|(a > d))
       2. (a < b) & (a > c)|((a < b)&(a > d))

16. Write a script file using conditional statements to evaluate the following function, assuming that the scalar variable x has a value. The function is $y = e^{x+1}$ for $x < -1$, $y = 2 + \cos(\pi x)$ for $-1 \le x < 5$, and $y = 10(x - 5) + 1$ for $x \ge 5$. Use your file to evaluate y for $x = -5$, $x = 3$, and $x = 15$, and check the results by hand.

Section 4.4

17. Rewrite the following statements to use only one if statement.

```
if x < y
   if z < 10
      w = x*y*z
   end
end
```

18. Write a program that accepts a numerical value *x* from 0 to 100 as input and computes and displays the corresponding letter grade given by the following table.

    A $x \ge 90$
    B $80 \le x \le 89$
    C $70 \le x \le 79$
    D $60 \le x \le 69$
    F $x < 60$

    a. Use nested if statements in your program (do not use elseif).
    b. Use only elseif clauses in your program.

19. Write a program that accepts a year and determines whether the year is a leap year. Use the mod function. The output should be the variable extra_day, which should be 1 if the year is a leap year and 0 otherwise. The rules for determining leap years in the Gregorian calendar are as follows:

   1. All years evenly divisible by 400 are leap years.
   2. Years evenly divisible by 100 but not by 400 are not leap years.
   3. Years divisible by 4 but not by 100 are leap years.
   4. All other years are not leap years.

   For example, the years 1800, 1900, 2100, 2300, and 2500 are not leap years, but 2400 is a leap year.

20. Figure P20 shows a mass-spring model of the type used to design packaging systems and vehicle suspensions, for example. The springs exert a force that is proportional to their compression, and the proportionality constant is the spring constant k. The two side springs provide additional resistance if the weight W is too heavy for the center spring. When the weight W is gently placed, it moves through a distance x before coming to rest. From statics, the weight force must balance the spring forces at this new position. Thus

$$W = k_1 x \qquad\qquad \text{if } x < d$$
$$W = k_1 x + 2 k_2 (x - d) \quad \text{if } x \geq d$$

   These relations can be used to generate the plot of x versus W.

   a. Create a function file that computes the distance x, using the input parameters W, k_1, k_2, and d. Test your function for the following two cases, using the values $k_1 = 10^4$ N/m; $k_2 = 1.5 \times 10^4$ N/m; $d = 0.1$ m.

   $$W = 500 \text{ N}$$
   $$W = 2000 \text{ N}$$

   b. Use your function to plot x versus W for $0 \leq W \leq 3000$ N for the values of k_1, k_2, and d given in part a.

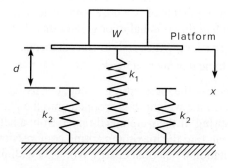

Figure P20

21. Create a MATLAB function called `fxy` to evaluate the function $f(x,y)$ defined as follows:

$$f(x,y) = \begin{cases} xy & \text{if } x \geq 0 \text{ and } y \geq 0 \\ xy^2 & \text{if } x \geq 0 \text{ and } y \leq 0 \\ x^2y & \text{if } x < 0 \text{ and } y \geq 0 \\ x^2y^2 & \text{if } x < 0 \text{ and } y < 0 \end{cases}$$

Test your function for all four cases.

Section 4.5

22. Use a `for` loop to plot the function given in Problem 16 over the interval $-2 \leq x \leq 6$. Properly label the plot. The variable y represents height in kilometers, and the variable x represents time in seconds.

23. Use a `for` loop to determine the sum of the first 10 terms in the series $5k^3$, $k = 1, 2, 3, \ldots, 10$.

24. The (x, y) coordinates of a certain object as a function of time t are given by

$$x(t) = 5t - 10 \qquad y(t) = 25t^2 - 120t + 144$$

for $0 \leq t \leq 4$. Write a program to determine the time at which the object is the closest to the origin at $(0, 0)$. Determine also the minimum distance. Do this in two ways:

a. By using a `for` loop.

b. By not using a `for` loop.

25. Consider the array **A**.

$$A = \begin{bmatrix} 3 & 5 & -4 \\ -8 & -1 & 33 \\ -17 & 6 & -9 \end{bmatrix}$$

Write a program that computes the array **B** by computing the natural logarithm of all the elements of **A** whose value is no less than 1, and adding 20 to each element that is equal to or greater than 1. Do this in two ways:

a. By using a `for` loop with conditional statements.

b. By using a logical array as a mask.

26. We want to analyze the mass-spring system discussed in Problem 20 for the case in which the weight W is dropped onto the platform attached to the center spring. If the weight is dropped from a height h above the platform, we can find the maximum spring compression x by equating the weight's gravitational potential energy $W(h + x)$ with the potential energy stored in the springs. Thus

$$W(h + x) = \frac{1}{2}k_1 x^2 \qquad \text{if } x < d$$

which can be solved for x as

$$x = \frac{W \pm \sqrt{W^2 + 2k_1 Wh}}{k_1} \qquad \text{if } x < d$$

and

$$W(h + x) = \frac{1}{2}k_1 x^2 + \frac{1}{2}(2k_2)(x - d)^2 \qquad \text{if } x \geq d$$

which gives the following quadratic equation to solve for x:

$$(k_1 + 2k_2)x^2 - (4k_2 d + 2W)x + 2k_2 d^2 - 2Wh = 0 \qquad \text{if } x \geq d$$

a. Create a function file that computes the maximum compression x due to the falling weight. The function's input parameters are k_1, k_2, d, W, and h. Test your function for the following two cases, using the values $k_1 = 10^4$ N/m; $k_2 = 1.5 \times 10^4$ N/m; and $d = 0.1$ m.

$$W = 100 \text{ N} \qquad h = 0.5 \text{ m}$$

$$W = 2000 \text{ N} \qquad h = 0.5 \text{ m}$$

b. Use your function file to generate a plot of x versus h for $0 \leq h \leq 2$ m. Use $W = 100$ N and the preceding values for k_1, k_2, and d.

27. Electrical resistors are said to be connected "in series" if the same current passes through each and "in parallel" if the same voltage is applied across each. If in series, they are equivalent to a single resistor whose resistance is given by

$$R = R_1 + R_2 + R_3 + \cdots + R_n$$

If in parallel, their equivalent resistance is given by

$$\frac{1}{R} = \frac{1}{R_1} + \frac{1}{R_2} + \frac{1}{R_3} + \cdots + \frac{1}{R_n}$$

Write an M-file that prompts the user for the type of connection (series or parallel) and the number of resistors n and then computes the equivalent resistance.

28. a. An *ideal* diode blocks the flow of current in the direction opposite that of the diode's arrow symbol. It can be used to make a *half-wave rectifier*, as shown in Figure P28a. For the ideal diode, the voltage v_L across the load R_L is given by

$$v_L = \begin{cases} v_S & \text{if } v_S > 0 \\ 0 & \text{if } v_S \leq 0 \end{cases}$$

Suppose the supply voltage is

$$v_S(t) = 3e^{-t/3}\sin(\pi t) \text{ V}$$

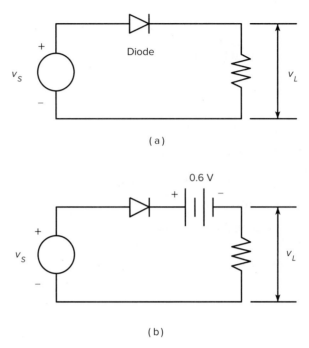

(a)

(b)

Figure P28

where time t is in seconds. Write a MATLAB program to plot the voltage v_L versus t for $0 \leq t \leq 10$.

b. A more accurate model of the diode's behavior is given by the *offset diode* model, which accounts for the offset voltage inherent in semiconductor diodes. The offset model contains an ideal diode and a battery whose voltage equals the offset voltage (which is approximately 0.6 V for silicon diodes) [Rizzoni, 2007]. The half-wave rectifier using this model is shown in Figure P28b. For this circuit,

$$v_L = \begin{cases} v_S - 0.6 & \text{if } v_S > 0.6 \\ 0 & \text{if } v_S \leq 0.6 \end{cases}$$

Using the same supply voltage given in part a, plot the voltage v_L versus t for $0 \leq t \leq 10$; then compare the results with the plot obtained in part a.

29.* A company wants to locate a distribution center that will serve six of its major customers in a 30×30 mi area. The locations of the customers relative to the southwest corner of the area are given in the following table in terms of (x, y) coordinates (the x direction is east; the y direction is north) (see Figure P29). Also given is the volume in tons per week that must be delivered from the distribution center to each customer. The

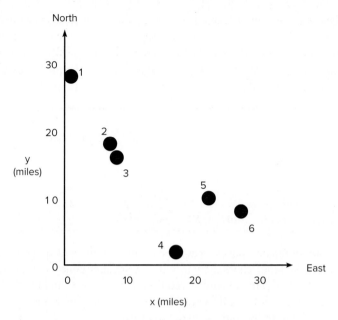

Figure P29

Customer	x location (mi)	y location (mi)	Volume (tons/week)
1	1	28	3
2	7	18	7
3	8	16	4
4	17	2	5
5	22	10	2
6	27	8	6

weekly delivery cost c_i for customer i depends on the volume V_i and the distance d_i from the distribution center. For simplicity we will assume that this distance is the straight-line distance. (This assumes that the road network is dense.) The weekly cost is given by $c_i = 0.5\, d_i\, V_i$, $i = 1, \ldots, 6$. Find the location of the distribution center (to the nearest mile) that minimizes the total weekly cost to service all six customers.

30. A company has the choice of producing up to four different products with its machinery, which consists of lathes, grinders, and milling machines. The number of hours on each machine required to produce a product is given in the following table, along with the number of hours available per week on each type of machine. Assume that the company can sell everything it produces. The profit per item for each product appears in the last line of the table.

   a. Determine how many units of each product the company should make to maximize its total profit, and then compute this profit. Remember, the company cannot make fractional units, so your answer must be in integers. (*Hint:* First estimate the upper limits on the number of products that can be produced without exceeding the available capacity.)

b. How sensitive is your answer? How much does the profit decrease if you make one more or one less item than the optimum?

	Product				
	1	2	3	4	Hours available
Hours required					
Lathe	1	2	0.5	3	40
Grinder	0	2	4	1	30
Milling	3	1	5	2	45
Unit profit ($)	100	150	90	120	

31. A certain company makes televisions, stereo units, and speakers. Its parts inventory includes chassis, picture tubes, speaker cones, power supplies, and electronics. The inventory, required components, and profit for each product appear in the following table. Determine how many of each product to make in order to maximize the profit.

	Product			
	Television	Stereo unit	Speaker unit	Inventory
Requirements				
Chassis	1	1	0	450
Picture tube	1	0	0	250
Speaker cone	2	2	1	800
Power supply	1	1	0	450
Electronics	2	2	1	600
Unit profit ($)	80	50	40	

Section 4.6

32. Plot the function $y = 10(1 - e^{-x/4})$ over the interval $0 \le x \le x_{max}$, using a `while` loop to determine the value of x_{max} such that $y(x_{max}) = 9.8$. Properly label the plot. The variable y represents force in newtons, and the variable x represents time in seconds.

33. Use a `while` loop to determine how many terms in the series 2^k, $k = 1$, $2, 3, \ldots$, are required for the sum of the terms to exceed 2000. What is the sum for this number of terms?

34. One bank pays 5.5 percent annual interest, while a second bank pays 4.5 percent annual interest. Determine how much longer it will take to accumulate at least $50,000 in the second bank account if you deposit $1000 initially and $1000 at the end of each year.

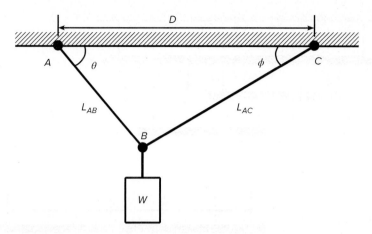

Figure P36

35.* Use a loop in MATLAB to determine how long it will take to accumulate $1,000,000 in a bank account if you deposit $10,000 initially and $10,000 at the end of each year; the account pays 6 percent annual interest.

36. A weight W is supported by two cables anchored a distance D apart (see Figure P36). The cable length L_{AB} is given, but the length L_{AC} is to be selected. Each cable can support a maximum tension force equal to W. For the weight to remain stationary, the total horizontal force and total vertical force must each be zero. This principle gives the equations

$$-T_{AB}\cos\theta + T_{AC}\cos\phi = 0$$
$$T_{AB}\sin\theta + T_{AC}\sin\phi = W$$

We can solve these equations for the tension forces T_{AB} and T_{AC} if we know the angles θ and ϕ. From the law of cosines

$$\theta = \cos^{-1}\left(\frac{D^2 + L_{AB}^2 - L_{AC}^2}{2DL_{AB}}\right)$$

From the law of sines

$$\phi = \sin^{-1}\left(\frac{L_{AB}\sin\theta}{L_{AC}}\right)$$

For the given values $D = 6$ ft, $L_{AB} = 3$ ft, and $W = 2000$ lb, use a loop in MATLAB to find L_{ACmin}, the shortest length L_{AC} we can use without T_{AB} or T_{AC} exceeding 2000 lb. Note that the largest L_{AC} can be is 6.7 ft (which corresponds to $\theta = 90°$). Plot the tension forces T_{AB} and T_{AC} on the same graph versus L_{AC} for $L_{ACmin} \le L_{AC} \le 6.7$.

37.* In the structure in Figure P37a, six wires support three beams. Wires 1 and 2 can support no more than 1200 N each, wires 3 and 4 can support no more than 400 N each, and wires 5 and 6 can support no more than

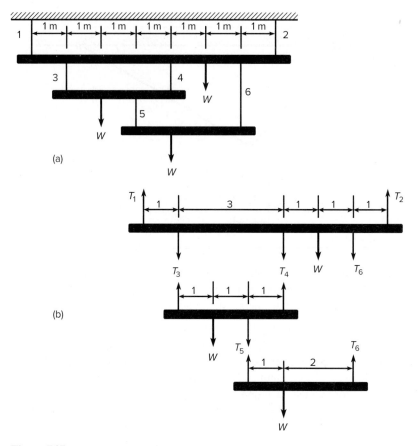

Figure P37

200 N each. Three equal weights W are attached at the points shown. Assuming that the structure is stationary and that the weights of the wires and the beams are very small compared to W, the principles of statics applied to a particular beam state that the sum of vertical forces is zero and that the sum of moments about any point is also zero. Applying these principles to each beam using the free-body diagrams shown in Figure P37b, we obtain the following equations. Let the tension force in wire i be T_i. For beam 1

$$T_1 + T_2 = T_3 + T_4 + W + T_6$$

$$-T_3 - 4T_4 - 5W - 6T_6 + 7T_2 = 0$$

For beam 2

$$T_3 + T_4 = W + T_5$$

$$-W - 2T_5 + 3T_4 = 0$$

Figure P38

For beam 3

$$T_5 + T_6 = W$$

$$-W + 3T_6 = 0$$

Find the maximum value of the weight W the structure can support. Remember that the wires cannot support compression, so T_i must be nonnegative.

38. The equations describing the circuit shown in Figure P38 are

$$-v_1 + R_1 i_1 + R_4 i_4 = 0$$
$$-R_4 i_4 + R_2 i_2 + R_5 i_5 = 0$$
$$-R_5 i_5 + R_3 i_3 + v_2 = 0$$

$$i_1 = i_2 + i_4$$
$$i_2 = i_3 + i_5$$

a. The given values of the resistances and the voltage v_1 are $R_1 = 5$, $R_2 = 100$, $R_3 = 200$, $R_4 = 150$, $R_5 = 250$ kΩ, and $v_1 = 100$ V. (Note that 1 k$\Omega = 1000 \ \Omega$.) Suppose that each resistance is rated to carry a current of no more than 1 mA ($= 0.001$ A). Determine the allowable range of positive values for the voltage v_2.

b. Suppose we want to investigate how the resistance R_3 limits the allowable range for v_2. Obtain a plot of the allowable limit on v_2 as a function of R_3 for $150 \leq R_3 \leq 250$ kΩ.

39. Many applications require us to know the temperature distribution in an object. For example, this information is important for controlling the material properties, such as hardness, when cooling an object formed from molten metal. In a heat-transfer course, the following description of the temperature distribution in a flat, rectangular metal plate is often derived. The temperature is held constant at T_1 on three sides and at T_2 on the

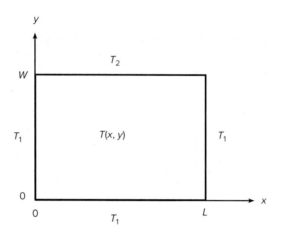

Figure P39

fourth side (see Figure P39). The temperature $T(x, y)$ as a function of the xy coordinates shown is given by

$$T(x, y) = (T_2 - T_1)w(x, y) + T_1$$

where

$$w(x,y) = \frac{2}{\pi} \sum_{n \text{ odd}}^{\infty} \frac{2}{n} \sin\left(\frac{n\pi x}{L}\right) \frac{\sinh(n\pi y/L)}{\sinh(n\pi W/L)}$$

Use the following data: $T_1 = 70°\,\text{F}$, $T_2 = 200°\text{F}$, and $W = L = 2$ ft.

a. The terms in the preceding series become smaller in magnitude as n increases. Write a MATLAB program to verify this fact for $n = 1, \ldots, 19$ for the center of the plate ($x = y = 1$).

b. Using $x = y = 1$, write a MATLAB program to determine how many terms are required in the series to produce a temperature calculation that is accurate to within 1 percent. (That is, for what value of n will the addition of the next term in the series produce a change in T of less than 1 percent?) Use your physical insight to determine whether this answer gives the correct temperature at the center of the plate.

c. Modify the program from part b to compute the temperatures in the plate; use a spacing of 0.2 for both x and y.

40. Consider the following script file. Fill in the lines of the following table with the values that would be displayed immediately after the while statement if you ran the script file. Write in the values the variables have each time the while statement is executed. You might need more or fewer lines in the table. Then type in the file, and run it to check your answers.

```
k = 1;b = -2;x = -1;y = -2;
while k <= 3
```

```
   k, b, x, y
   y = x^2 - 3;
   if y < b
       b = y;
   end
   x = x + 1;
   k = k + 1;
end
```

Pass	k	b	x	y
First				
Second				
Third				
Fourth				
Fifth				

41. Assume that the human player makes the first move against the computer in a game of Tic-Tac-Toe, which has a 3 × 3 grid. Write a MATLAB function that lets the computer respond to that move. The function's input argument should be the cell location of the human player's move. The function's output should be the cell location of the computer's first move. Label the cells as 1, 2, 3 across the top row; 4, 5, 6 across the middle row; and 7, 8, 9 across the bottom row.

Section 4.7

42. The following table gives the approximate values of the static coefficient of friction μ for various materials.

Materials	μ
Metal on metal	0.20
Wood on wood	0.35
Metal on wood	0.40
Rubber on concrete	0.70

To start a weight W moving on a horizontal surface, you must push with a force F, where $F = \mu W$. Write a MATLAB program that uses the `switch` structure to compute the force F. The program should accept as input the value of W and the type of materials.

43. The height and speed of a projectile (such as a thrown ball) launched with a speed of v_0 at an angle A to the horizontal are given by

$$h(t) = v_0 t \sin A - 0.5g t^2$$

$$v(t) = \sqrt{v_0^2 - 2v_0 gt \sin A + g^2 t^2}$$

where g is the acceleration due to gravity. The projectile will strike the ground when $h(t) = 0$, which gives the time to hit $t_{hit} = 2(v_0/g)\sin A$.

Use the switch structure to write a MATLAB program to compute the maximum height reached by the projectile, the total horizontal distance traveled, or the time to hit. The program should accept as input the user's choice of which quantity to compute and the values of v_0, A, and g. Test the program for the case where $v_0 = 40$ m/s, $A = 30°$, and $g = 9.81$ m/s^2.

44.   Use the switch structure to write a MATLAB program to compute the amount of money that accumulates in a savings account in one year. The program should accept the following input: the initial amount of money deposited in the account; the frequency of interest compounding (monthly, quarterly, semiannually, or annually); and the interest rate. Run your program for a $1000 initial deposit for each case; use a 5 percent interest rate. Compare the amounts of money that accumulate for each case.

45.   Engineers often need to estimate the pressures and volumes of a gas in a container. The *van der Waals* equation is often used for this purpose. It is

$$P = \frac{RT}{\hat{V} - b} - \frac{a}{\hat{V}^2}$$

where the term b is a correction for the volume of the molecules and the term $a/\hat{V}^2$ is a correction for molecular attractions. The gas constant is R, the *absolute* temperature is T, and the gas specific volume is $\hat{V}$. The value of R is the same for all gases; it is $R = 0.08206$ L-atm/mol-K. The values of a and b depend on the type of gas. Some values are given in the following table. Write a user-defined function using the switch structure that computes the pressure P on the basis of the van der Waals equation. The function's input arguments should be T, $\hat{V}$, and a string variable containing the name of a gas listed in the table. Test your function for chlorine (Cl_2) for $T = 300$ K and $\hat{V} = 20$ L/mol.

Gas	a (L^2-atm/mol^2)	b (L/mol)
Helium, He	0.0341	0.0237
Hydrogen, H_2	0.244	0.0266
Oxygen, O_2	1.36	0.0318
Chlorine, Cl_2	6.49	0.0562
Carbon dioxide, CO_2	3.59	0.0427

46.   Using the program developed in Problem 19, write a program that uses the switch structure to compute the number of days in a year up to a given date, given the year, the month, and the day of the month.

Section 4.9

47. Consider the college enrollment model discussed in Example 4.9–2. Suppose the college wants to limit freshman admissions to 120 percent of the current sophomore class and limit sophomore transfers to 10 percent of the current freshman class. Rewrite and run the program given in the example to examine the effects of these policies over a 10-year period. Plot the results.

48. Suppose you project that you will be able to deposit the following monthly amounts into a savings account for a period of 5 years. The account initially has no money in it.

At the end of each year in which the account balance is at least $3000, you withdraw $2000 to buy a certificate of deposit (CD), which pays 6 percent interest compounded annually.

Write a MATLAB program to compute how much money will accumulate in 5 years in the account and in any CDs you buy. Run the program for two different savings interest rates: 4 percent and 5 percent.

Year	1	2	3	4	5
Monthly deposit ($)	300	350	350	350	400

49.* A certain company manufactures and sells golf carts. At the end of each week, the company transfers the carts produced that week into storage (inventory). All carts that are sold are taken from the inventory. A simple model of this process is

$$I(k + 1) = P(k) + I(k) - S(k)$$

where

$P(k)$ = number of carts produced in week k

$I(k)$ = number of carts in inventory in week k

$S(k)$ = number of carts sold in week k

The projected weekly sales for 10 weeks are

Week	1	2	3	4	5	6	7	8	9	10
Sales	50	55	60	70	70	75	80	80	90	55

Suppose the weekly production is based on the previous week's sales so that $P(k) = S(k - 1)$. Assume that the first week's production is 50 carts; that is, $P(1) = 50$. Write a MATLAB program to compute and plot the number of carts in inventory for each of the 10 weeks or until the inventory drops below zero. Run the program for two cases: (*a*) an initial inventory of 50 carts so that $I(1) = 50$ and (*b*) an initial inventory of 30 carts so that $I(1) = 30$.

50. Redo Problem 49 with the restriction that the next week's production is set to zero if the inventory exceeds 40 carts.

Engineering in the 21st Century. . .

Small Scale Aeronautics

Although now called a *drone* in popular culture, the proper term for an aircraft without a human pilot aboard is an *unmanned aerial vehicle (UAV)*. UAVs may operate either by remote control with a human operator, or controlled continuously or intermittently by an onboard computer. Such vehicles became possible in the last few years because of miniaturization of computers and sensors such as cameras and gyroscopes.

Drones like the one shown in the photograph are inherently unstable, and the direction and speed of their motors must continuously controlled to provide stability and to achieve the desired vehicle orientation, altitude, and course. The computer code and electrical hardware required to do this is called a *feedback loop* or *control loop*. In some designs each motor can rotate about three axes (roll, pitch, and yaw), so each motor requires four loops, one for the motor speed and three for the axes. The MathWorks provides supporting software for using MATLAB and Simulink to design and implement control code for use with several popular microprocessors that are used in drones.

Not all such vehicles have the standard configuration of the drone shown in the photograph. Some depend on aerodynamics rather than motors to provide lift. One such *micro air vehicle (MAV)* is 6 in. long, carries a 2-g video camera the size of a sugar cube, and flies at about 65 km/h with a range of 10 km. Other MAVs utilize flapping wings. Air at such small scales and speeds behaves more like a viscous fluid, and one of the unexpected challenges to designing better MAVs is to improve our understanding of low-speed aeronautics.

The MATLAB advanced graphics capabilities make it useful for visualizing flow patterns, and the Optimization and Control Systems toolboxes are useful for designing such vehicles. ■

Advanced Plotting

In this chapter you will learn additional features to use to create a variety of two-dimensional plots, which are also called *xy plots,* and three-dimensional plots called *xyz plots,* or *surface plots.* Two-dimensional plots are discussed in Sections 5.1 through 5.3. Section 5.4 discusses three-dimensional plots. These plotting functions are described in the graph2d and graph3d Help categories, so typing help graph2d or help graph3d will display a list of the relevant plotting functions.

An important application of plotting is *function discovery,* the technique for using data plots to obtain a mathematical function or "mathematical model" that describes the process that generated the data. This topic is treated in Chapter 6.

5.1 *xy* Plotting Functions

The "anatomy" and nomenclature of a typical *xy* plot is shown in Figure 5.1–1, in which the plot of a data set and a curve generated from an equation appear. A plot can be made from measured data or from an equation. When data are plotted, each data point is plotted with a *data symbol,* or *point marker,* such as the small circles shown in Figure 5.1–1. An exception to this rule would be when there are so many data points that the symbols would be too densely packed. In that case,

DATA SYMBOL

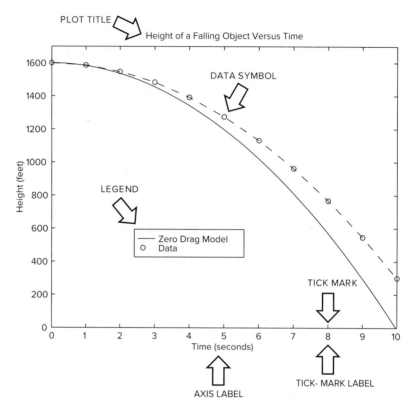

Figure 5.1–1 Nomenclature for a typical *xy* plot.

the data points should be plotted with a dot. However, when the plot is generated from a function, data symbols must *never* be used! Lines between closely spaced points are always used to plot a function.

The MATLAB basic *xy* plotting function is `plot(x,y)` as we saw in Chapter 1. If `x` and `y` are vectors, a single curve is plotted with the `x` values on the abscissa and the `y` values on the ordinate. The `xlabel` and `ylabel` commands put labels on the abscissa and the ordinate, respectively. The syntax is `xlabel('text')`, where `text` is the text of the label. Note that you must enclose the label's text in single quotes. The syntax for `ylabel` is the same. The `title` command puts a title at the top of the plot. Its syntax is `title('text')`, where `text` is the title's text.

The `plot(x,y)` function in MATLAB automatically selects a tick-mark spacing for each axis and places appropriate tick labels. This feature is called *autoscaling*. MATLAB also chooses limits for the *x* and *y* axes. The order of the `xlabel`, `ylabel`, and `title` commands does not matter, but we must place them *after* the `plot` command, either on separate lines using ellipses or on the same line separated by commas.

After the `plot` command is executed, the plot will appear in the Figure window. You can obtain a hard copy of the plot in one of several ways:

1. Use the menu system. Select **Print** on the File menu in the Figure window. A pop-up window will appear for your default printer.
2. Type `print` at the command line. This command sends the current plot directly to the default printer.
3. Save the plot to a file to be printed later or imported into another application such as a word processor. You need to know something about graphics file formats to use this file properly. See the subsection **Exporting Figures** later in this section.
4. Select **Copy** in the Edit menu in the Figure window. Then paste the figure into a word processor. This method provides a quick and easy way to include figures in reports.

Type `help print` to obtain more information.

MATLAB assigns the output of the `plot` command to Figure window number 1. When another `plot` command is executed, MATLAB overwrites the contents of the existing Figure window with the new plot. Although you can keep more than one Figure window active, we do not use this feature in this text.

When you have finished with the plot, close the Figure window by selecting **Close** from the File menu in the Figure window. If you do not close the window, it will not reappear when a new `plot` command is executed. However, the figure will still be updated.

Table 5.1–1 lists the requirements essential to producing plots that communicate effectively.

Table 5.1–1 Requirements for a correct plot

1. Each axis must be labeled with the name of the quantity being plotted *and its units!* If two or more quantities having different units are plotted (such as when in a plot of both speed and distance versus time), indicate the units in the axis label, if there is room, or in the legend or labels for each curve.
2. Each axis should have regularly spaced tick marks at convenient intervals—not too sparse, but not too dense—with a spacing that is easy to interpret and interpolate. For example, use 0.1, 0.2, and so on, rather than 0.13, 0.26, and so on.
3. If you are plotting more than one curve or data set, label each on its plot, use different line types, or use a legend to distinguish them.
4. If you are preparing multiple plots of a similar type or if the axes' labels cannot convey enough information, use a title.
5. If you are plotting measured data, plot each data point with a symbol such as a circle, square, or cross (use the same symbol for every point in the same data set). If there are many data points, plot them using the dot symbol.
6. Sometimes data symbols are connected by lines to help the viewer visualize the data, especially if there are few data points. However, connecting the data points, especially with a solid line, might be interpreted to imply knowledge of what occurs between the data points. Thus you should be careful to prevent such misinterpretation.
7. If you are plotting points generated by evaluating a function (as opposed to measured data), do *not* use a symbol to plot the points. Instead, be sure to generate many points, and connect the points with solid lines.

grid and axis **Commands**

The `grid` command displays gridlines at the tick marks corresponding to the tick labels. You can use the `axis` command to override the MATLAB selections for the *axis limits*. The basic syntax is `axis([xmin xmax ymin ymax])`. This command sets the scaling for the *x* and *y* axes to the minimum and maximum values indicated. Note that, unlike an array, this command does not use commas to separate the values.

AXIS LIMITS

Figure 5.1–2 shows a plot in which the command `axis([0 10 -2 5])` was used to override the limits chosen by auto scaling (which chose the upper limit of the ordinate to be 4).

The following list shows some of the variants of the `axis` command:

- `axis square` selects the axes' limits so that the plot will be square.
- `axis equal` selects the scale factors and tick spacing to be the same on each axis. This variation makes `plot(sin(x),cos(x))` look like a circle, instead of an oval.
- `axis auto` returns the axis scaling to its default autoscaling mode in which the best axes' limits are computed automatically.
- `axis tight` sets the axis limits to the range of the data.

Type `help axis` to see the full list of variants. Notice that sometimes axis behaves like a function with an argument, as in `axis([xmin xmax ymin ymax])`,

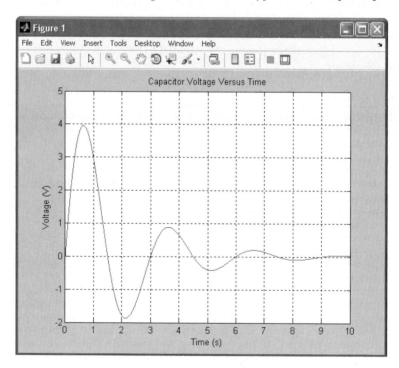

Figure 5.1–2 A sample plot shown in a Figure window.

and sometimes it behaves as a command, as in `axis equal`. The MATLAB interpreter can recognize the difference according to the context in which it is used.

Plots of Complex Numbers

With only one argument, say, `plot(y)`, the `plot` function will plot the values in the vector `y` versus their indices 1, 2, 3, . . . , and so on. If `y` is complex, `plot(y)` plots the imaginary parts versus the real parts. Thus `plot(y)` in this case is equivalent to `plot(real(y),imag(y))`. This situation is the only time when the `plot` function handles the imaginary parts; in all other variants of the `plot` function, it ignores the imaginary parts. For example, the script file

```
z = 0.1 + 0.9i;
n = 0:0.01:10;
plot(z.^n),xlabel('Real'),ylabel('Imaginary')
```

generates a spiral plot.

The Function Plot Command fplot

MATLAB has a "smart" command for plotting functions. The `fplot` command automatically analyzes the function to be plotted and decides how many plotting points to use so that the plot will show all the features of the function. Its basic syntax is `fplot(function)`, where `function` is a function handle to the function to be plotted over the default interval [-5, 5]. To specify the interval, use the syntax `fplot(function, [xmin xmax])`. For additional syntax consult the MATLAB help.

For example, the session

```
>>f = @(x) (cos(tan(x)) - tan(sin(x)));
>>fplot(f,[1 2])
```

produces the plot shown in Figure 5.1–3a. The `fplot` command automatically chooses enough plotting points to display all the variations in the function. We can achieve the same results using the `plot` command, but we need to know how many values to compute to generate the plot. For example, choosing a spacing of 0.01, and using `plot`, we obtain the plot in Figure 5.1–3b. We see that this choice of spacing misses some of the function's behavior.

Other commands can be used with the `fplot` command to enhance a plot's appearance, for example, the `title`, `xlabel`, and `ylabel` commands and the line type commands to be introduced in the next section.

Plotting Polynomials

We can plot polynomials more easily by using the `polyval` function. The function `polyval(p,x)` evaluates the polynomial p at specified values of its

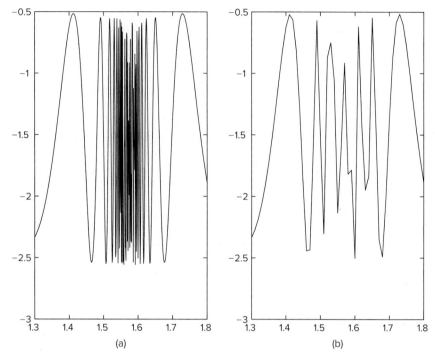

Figure 5.1–3 (a) The plot was generated with `fplot`. (b) The plot was generated with `plot` using 101 points.

independent variable x. For example, to plot the polynomial $3x^5 + 2x^4 - 100x^3 + 2x^2 - 7x + 90$ over the range $-6 \leq x \leq 6$ with a spacing of 0.01, you type

```
>>x = -6:0.01:6;
>>p = [3,2,-100,2,-7,90];
>>plot(x,polyval(p,x)),xlabel('x'),ylabel('p')
```

Table 5.1–2 summarizes the xy plotting commands discussed in this section.

Test Your Understanding

T5.1–1 Plot the equation $y = 0.4 \sqrt{1.8x}$ for $0 \leq x \leq 35$ and $0 \leq y \leq 3.5$.

T5.1–2 Use the `fplot` command to investigate the function $\tan(\cos x) - \sin(\tan x)$ for $0 \leq x \leq 2\pi$. How many values of x are needed to obtain the same plot using the `plot` command? (Answer: 292 values.)

T5.1–3 Plot the imaginary part versus the real part of the function $(0.2 + 0.8i)^n$ for $0 \leq n \leq 20$. Choose enough points to obtain a smooth curve. Label each `axis` and put a title on the plot. Use the `axis` command to change the tick-label spacing.

Table 5.1–2 Basic *xy* plotting commands

Command	Description
axis([xmin xmax ymin ymax])	Sets the minimum and maximum limits of the *x* and *y* axes.
fplot(function,[xmin xmax])	Performs intelligent plotting of functions, where function is a function handle that describes the function to be plotted and [xmin xmax] specifies the minimum and maximum values of the independent variable. The range of the dependent variable can also be specified. In this case the syntax is fplot(function, [xmin xmax ymin ymax]).
grid	Displays gridlines at the tick marks corresponding to the tick labels.
plot(x,y)	Generates a plot of the array y versus the array x on rectilinear axes.
plot(y)	Plots the values of y versus their indices if y is a vector. Plots the imaginary parts of y versus the real parts if y is a vector having complex values.
polyval(p,x)	Evaluates the polynomial p at specified values of its independent variable x.
print	Prints the plot in the Figure window.
title('text')	Puts text in a title at the top of a plot.
xlabel('text')	Adds a text label to the *x* axis (the abscissa).
ylabel('text')	Adds a text label to the *y* axis (the ordinate).

Saving Figures

When you create a plot, the Figure window appears. This window has eight menus, which are discussed in detail in Section 5.3. The File menu is used for saving and printing the figure. You can save your figure in a format that can be opened during another MATLAB session or in a format that can be used by other applications.

To save a figure that can be opened in subsequent MATLAB sessions, save it in a figure file with the .fig file name extension. To do this, select **Save** from the Figure window File menu or click the **Save** button (the disk icon) on the toolbar. If this is the first time you are saving the file, the default file type is the MATLAB Figure (*.fig). Specify the name you want assigned to the figure file. Click OK.

If you want to save the figure as another file type, such as JPEG, BMP, or PNG, select **Save As.** In the dialog box that appears you can select the desired type. These are popular types used by many applications. You can also use the saveas command from the command line.

Caution: If you might want to edit the figure later, be sure to save it first as a MATLAB Figure (*.fig) file. If you first save it as another graphics file type (JPEG, etc.) you will no longer be able to edit it using the MATLAB plot tools.

To open a figure file, select **Open** from the File menu or click the **Open** button (the opened folder icon) on the toolbar. Select the figure file you want to open and click OK. The figure file appears in a new figure window.

Exporting Figures

Exporting a figure is not the same as simply saving it. You can use the Export Setup window to customize a figure before saving it. You can change the figure size, background color, font size, and line width, and you can save the settings as an export style that you can apply to other figures before saving them:

If you want to save the figure in a format that can be used by another application, in a number of different graphics file formats, perform these steps:

1. Select **Export Setup** from the File menu. This dialog provides options you can specify for the output file, such as the figure size, fonts, line size and style, and output format.
2. Select **Export** from the Export Setup dialog. A standard Save As dialog appears.
3. Select the format from the list of formats in the Save As type menu. This selects the format of the exported file and adds the standard file name extension given to files of that type.
4. Enter the name you want to give the file, less the extension.
5. Click **Save.**

You can also export the figure from the command line, by using the `print` command. See MATLAB Help for more information about exporting figures in different formats.

You can also copy a figure to the clipboard and then paste it into another application:

1. Select **Copy Options** from the Edit menu of the Figure window. The Copying Options page of the Preferences dialog box appears.
2. Complete the fields on the Copying Options page and click **OK.**
3. Select **Copy Figure** from the Edit menu.

The figure is copied to the Windows clipboard and can be pasted into another application.

The graphics functions covered in this section and in Section 5.3 can be placed in script files that can be reused to create similar plots. This feature gives them an advantage over the interactive plotting tools discussed in Section 5.3.

When you are creating plots, keep in mind that the actions listed in Table 5.1–3, while not required, can nevertheless improve the appearance and usefulness of your plots.

The Live Editor

A *live script* is an *interactive* document that contains output, including graphics, along with the code that produced them, together in a single interactive environment called the *Live Editor*. You can also include formatted text, images, hyperlinks, and equations to produce an interactive shareable narrative. Live scripts, which were introduced in MATLAB R2016a, are stored in a file with the extension **.mlx.** You can convert the scripts to HTML or PDF files for publication.

Table 5.1–3 Hints for improving plots

1. Start scales from zero whenever possible. This technique prevents a false impression of the magnitudes of any variations shown on the plot.
2. Use sensible tick-mark spacing. For example, if the quantities are months, choose a spacing of 12 because 1/10 of a year is not a convenient division. Space tick marks as close as is useful, but no closer.
3. Minimize the number of zeros in the tick labels. For example, use a scale in millions of dollars when appropriate, instead of a scale in dollars with six zeros after every number.
4. Determine the minimum and maximum data values for each axis before plotting the data. Then set the axis limits to cover the entire data range plus an additional amount to allow convenient tick-mark spacing to be selected.

The Live Editor enables you to work more efficiently because you can write, execute, and test code without leaving the environment, and you can run blocks of code individually or the whole file. You can see the results and graphics next to the code that produced them, and you can see errors at the file location where they occur.

Two ways to open a new live script are:

■ On the Home tab, in the New drop-down menu, select **Live Script**.
■ Highlight the desired commands from the Command History, right-click, and select **Create Live Script**.

Enter your code in the Live Editor as you would in the Command window. See Figure 5.1–4 for an example. After entering the code, click at the top of the

Figure 5.1–4 Screen shot of the Live Editor with code and graphical output.

blue boundary to the left. The code will run and MATLAB will alert you to any errors. In the example shown, the plot will appear after the third line of code is executed. To the far right of the Editor there are two icons for selecting where to put the outputs (in this case the plot). Click the left-most icon to put the output inline; click the right-most to display the output on the right.

You can open existing scripts as live scripts. This creates a copy of the file, and leaves the original file untouched. Only script files can be opened as live scripts. Function files are not converted.

You can open an existing script (.m) as a live script (.mlx) by using one of the following methods:

- Open the script in the Editor, click **Save**, and select **Save As**. Then, select the **Save as type: MATLAB Live Scripts (*.mlx)** and click **Save**.
- Right-click the file in the Current Folder browser and select **Open as Live Script** from the context menu.

Note: You must use one of these methods to convert your script into a live script. Simply renaming the script with the extension .mlx does not work, and can corrupt the file.

You can insert equations as typeset mathematics. Only text lines, not code lines, can contain equations. There are three ways to insert an equation into a live script. You can build an equation interactively from a palette of symbols and structures (go to the Insert tab of the Live Editor and click on Σ**Equation**). Or you can create an equation using LaTex commands (go to the Insert tab of the Live Editor, click on **Equation**, then select **LaTex Equation**). For information about these two methods, see the Help under the topic "Insert Equations into Live Scripts." Lastly, you can use the commands from the Symbolic Math Toolbox (see Chapter 11, Section 11.3 and Figure 11.3–2 for an example).

The best way to learn more is to type Live Editor in the documentation search box in the top right of the Desktop.

5.2 Additional Commands and Plot Types

MATLAB can create figures that contain an array of plots, called *subplots*. These are useful when you want to compare the same data plotted with different axis types, for example. The MATLAB subplot command creates such figures. We frequently need to plot more than one curve or data set on a single plot. Such a plot is called an *overlay plot*. This section describes these plots and several other types of plots.

OVERLAY PLOT

Subplots

You can use the subplot command to obtain several smaller "subplots" in the same figure. The syntax is subplot(m,n,p). This command divides the Figure window into an array of rectangular panes with *m* rows and *n* columns. The variable p tells MATLAB to place the output of the plot command following the

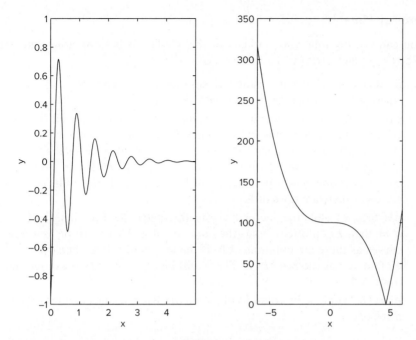

Figure 5.2–1 Application of the `subplot` command.

subplot command into the pth pane. For example, `subplot(3,2,5)` creates an array of six panes, three panes deep and two panes across, and directs the next plot to appear in the fifth pane (in the bottom left corner). The following script file created Figure 5.2–1, which shows the plots of the functions $y = e^{-1.2x}$ $\sin(10x + 5)$ for $0 \le x \le 5$ and $y = |x^3 - 100|$ for $-6 \le x \le 6$.

```
x = 0:0.01:5;
y = exp(-1.2*x).*sin(10*x+5);
subplot(1,2,1)
plot(x,y),xlabel('x'),ylabel('y'),axis([0 5 -1 1])
x = -6:0.01:6;
y = abs(x.^3-100);
subplot(1,2,2)
plot(x,y),xlabel('x'),ylabel('y'),axis([-6 6 0 350])
```

Test Your Understanding

T5.2–1 Pick a suitable spacing for t and v, and use the `subplot` command to plot the function $z = e^{-0.5t} \cos(20t - 6)$ for $0 \le t \le 8$ and the function $u = 6 \log_{10}(v^2 + 20)$ for $-8 \le v \le 8$. Label each axis.

Overlay Plots

You can use the following variants of the MATLAB basic plotting functions plot(x,y) and plot(y) to create *overlay plots*:

- plot(A) plots the columns of A versus their indices and generates *n* curves, where A is a matrix with *m* rows and *n* columns.
- plot(x,A) plots the matrix A versus the vector x, where x is either a row vector or a column vector and A is a matrix with *m* rows and *n* columns. If the length of x is *m*, then each *column* of A is plotted versus the vector x. There will be as many curves as there are columns of A. If x has length *n*, then each *row* of A is plotted versus the vector x. There will be as many curves as there are rows of A.
- plot(A,x) plots the vector x versus the matrix A. If the length of x is *m*, then x is plotted versus the *columns* of A. There will be as many curves as there are columns of A. If the length of x is *n*, then x is plotted versus the *rows* of A. There will be as many curves as there are rows of A.
- plot(A,B) plots the columns of the matrix B versus the columns of the matrix A.

Data Markers and Line Types

To plot the vector y versus the vector x and mark each point with a data marker, enclose the symbol for the marker in single quotes in the plot function. Table 5.2–1 shows the symbols for some of the available data markers. For example, to use a small circle, which is represented by the lowercase letter o, type plot(x,y,'o'). This notation results in a plot like the one on the left in Figure 5.2–2. To connect each data marker with a straight line, we must plot the data twice, by typing plot(x,y,x,y,'o'). See the plot on the right in Figure 5.2–2.

Suppose we have two curves or data sets stored in the vectors x, y, u, and v. To plot y versus x and v versus u on the same plot, type plot(x,y,u,v). Both

Table 5.2–1 Specifiers for data markers, line types, and colors

Data markers[†]		Line types		Colors	
Dot (·)	.	Solid line	-	Black	k
Asterisk (*)	*	Dashed line	--	Blue	b
Cross (×)	×	Dash-dotted line	-.	Cyan	c
Circle (o)	o	Dotted line	:	Green	g
Plus sign (+)	+			Magenta	m
Square (□)	s			Red	r
Diamond (◇)	d			White	w
Five-pointed star (★)	p			Yellow	y

[†]Other data markers are available. Search for "markers" in MATLAB Help.

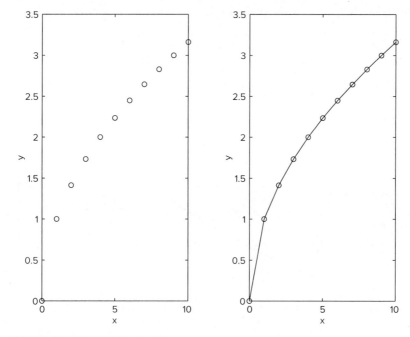

Figure 5.2–2 Use of data markers.

sets will be plotted with a solid line, which is the default line style. To distinguish the sets, we can plot them with different line types. To plot y versus x with a solid line and u versus v with a dashed line, type plot(x,y,u,v,'- -'), where the symbols '− −' represent a dashed line. Table 5.2–1 gives the symbols for other line types. To plot y versus x with asterisks (*) connected with a dotted line, you must plot the data twice by typing plot(x,y,'*',x,y,':').

You can obtain symbols and lines of different colors by using the color symbols shown in Table 5.2–1. The color symbol can be combined with the data-marker symbol and the line-type symbol. For example, to plot y versus x with green asterisks (*) connected with a red dashed line, you must plot the data twice by typing plot(x,y,'g*',x,y,'r- -'). (Do not use colors if you are going to print the plot on a black-and-white printer.)

Labeling Curves and Data

When more than one curve or data set is plotted on a graph, we must distinguish between them. If we use different data symbols or different line types, then we must either provide a legend or place a label next to each curve. To create a legend, use the legend command. The basic form of this command is legend ('string1','string2'), where string1 and string2 are text strings of your choice. The legend command automatically obtains from the plot the line type used for each data set and displays a sample of this line type in the legend

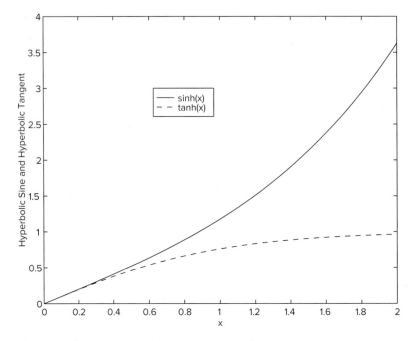

Figure 5.2–3 Application of the legend command.

box next to the string you selected. The following script file produced the plot in Figure 5.2–3.

```
x = 0:0.01:2;
y = sinh(x);
z = tanh(x);
plot(x,y,x,z,'- -'),xlabel('x'),...
    ylabel('Hyperbolic Sine and Hyperbolic Tangent'),...
    legend('sinh(x)','tanh(x)')
```

The legend command must be placed somewhere after the plot command. When the plot appears in the Figure window, use the mouse to position the legend box. (Hold down the left button on the mouse to move the box.)

Another way to distinguish curves is to place a label next to each. The label can be generated either with the gtext command, which lets you place the label by using the mouse, or with the text command, which requires you to specify the coordinates of the label. The syntax of the gtext command is gtext('string'), where string is a text string that specifies the label of your choice. When this command is executed, MATLAB waits for a mouse button or a key to be pressed while the mouse pointer is within the Figure window; the label is placed at that position of the mouse pointer. You may use more than one gtext command for a given plot. The text command text(x,y,'string') adds a text string to the plot at the location specified by the coordinates x,y.

These coordinates are in the same units as the plot's data. Of course, finding the proper coordinates to use with the `text` command usually requires some trial and error.

The hold Command

The `hold` command creates a plot that needs two or more `plot` commands. Suppose we wanted to plot $y_2 = 4 + e^{-x} \cos 6x$ versus $y_1 = 3 + e^{-x} \sin 6x$, $-1 \leq x \leq 1$ on the same plot with the complex function $z = (0.1 + 0.9i)^n$, where $0 \leq n \leq 10$. The following script file creates the plot in Figure 5.2–4.

```
x = -1:0.01:1;
y1 = 3+exp(-x).*sin(6*x);
y2 = 4+exp(-x).*cos(6*x);
plot((0.1+0.9i).^(0:0.01:10)),hold,plot(y1,y2),...
   gtext('y2 versus y1'),gtext('Imag(z) versus Real(z)')
```

When more than one `plot` command is used, do not place any of the `gtext` commands before any `plot` command. Because the scaling changes as each `plot` command is executed, the label placed by the `gtext` command might end up in the wrong position. Using the command `axis manual` freezes the scaling at the current limits, so that if `hold` is turned on, subsequent plots will use the same limits.

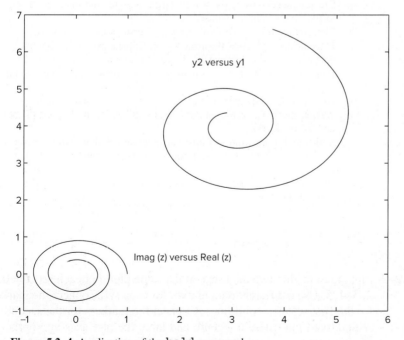

Figure 5.2–4 Application of the `hold` command.

Functions of the following form often appear in applications.

$$x(t) = e^{-0.3t}(\cos 2t + j \sin 2t)$$

If we were to attempt to plot x versus t, only the real part would be plotted, and MATLAB would issue a warning. If we want to plot both the real and imaginary parts on the same plot, we could use the hold command as shown in the following program:

```
t = 0:pi/50:2*pi;
x = exp(-0.3t).*(cos(2t)+j*sin(2*t));
plot(t,real(x));
hold on;
plot(t,imag(x),'- -');
hold off;
```

Table 5.2–2 summarizes the plot enhancement commands introduced in this section.

Table 5.2–2 Plot enhancement commands

Command	Description
gtext('text')	Places the string text in the Figure window at a point specified by the mouse.
hold	Freezes the current plot for subsequent graphics commands.
legend('leg1','leg2',...)	Creates a legend using the strings leg1, leg2, and so on and enables its placement with the mouse.
plot(x,y,u,v)	Plots, on rectilinear axes, four arrays: y versus x and v versus u.
plot(x,y,'type')	Plots the array y versus the array x on rectilinear axes, using the line type, data marker, and colors specified in the string type. See Table 5.2–1.
plot(A)	Plots the columns of the $m \times n$ array A versus their indices and generates n curves.
plot(P,Q)	Plots array Q versus array P. See the text for a description of the possible variants involving vectors and/or matrices: plot(x,A), plot(A,x), and plot(A,B).
subplot(m,n,p)	Splits the Figure window into an array of subwindows with m rows and n columns and directs the subsequent plotting commands to the pth subwindow.
text(x,y,'text')	Places the string text in the Figure window at a point specified by coordinates x, y.

Test Your Understanding

T5.2–2 Plot the following two data sets on the same plot. For each set, $x = 0, 1, 2, 3, 4, 5$. Use a different data marker for each set. Connect the markers for the first set with solid lines. Connect the markers for the second set with dashed lines. Use a legend, and label the plot axes appropriately. The first set is $y = 11, 13, 8, 7, 5, 9$. The second set is $y = 2, 4, 5, 3, 2, 4$.

T5.2–3 Plot $y = \cosh x$ and $y = 0.5e^x$ on the same plot for $0 \le x \le 2$. Use different line types and a legend to distinguish the curves. Label the plot axes appropriately.

T5.2–4 Plot $y = \sinh x$ and $y = 0.5e^x$ on the same plot for $0 \le x \le 2$. Use a solid line type for each, the `gtext` command to label the $\sinh x$ curve, and the `text` command to label the $0.5e^x$ curve. Label the plot axes appropriately.

T5.2–5 Use the `hold` command and the `plot` command twice to plot $y = \sin x$ and $y = x - x^3/3$ on the same plot for $0 \le x \le 1$. Use a solid line type for each, and use the `gtext` command to label each curve. Label the plot axes appropriately.

Annotating Plots

You can create text, titles, and labels that contain mathematical symbols, Greek letters, and other effects such as italics. The features are based on the T_EX typesetting language. For more information, including a list of the available characters, search the online Help for the "Text Properties" page. See also the "Mathematical symbols, Greek Letters, and T_EX Characters" page.

You can create a title having the mathematical function $Ae^{-t/\tau} \sin(\omega t)$ by typing

```
>>title('{\it Ae}^{-{\it t/\tau}}\sin({\it \omega t})')
```

The backslash character \ precedes all T_EX character sequences. Thus the strings \tau and \omega represent the Greek letters τ and ω. Superscripts are created by typing ^; subscripts are created by typing _. To set multiple characters as superscripts or subscripts, enclose them in braces. For example, type x_{13} to produce x_{13}. In mathematical text variables are usually set in italic, and functions, such as sin, are set in roman type. To set a character, say, x, in italic using the T_EX commands, you type {\it x}.

Logarithmic Plots

Logarithmic scales—abbreviated log scales—are widely used (1) to represent a data set that covers a wide range of values and (2) to identify certain trends in data. Certain types of functional relationships appear as straight lines when plotted using a log scale. This method makes it easier to identify the function. A *log-log* plot has log scales on both axes. A *semilog* plot has a log scale on only one axis.

Figure 5.2–5 shows a rectilinear plot and a log-log plot of the function

$$y = \sqrt{\frac{100(1 - 0.01x^2)^2 + 0.02x^2}{(1 - x^2)^2 + 0.1x^2}} \qquad 0.1 \le x \le 100 \qquad (5.2\text{–}1)$$

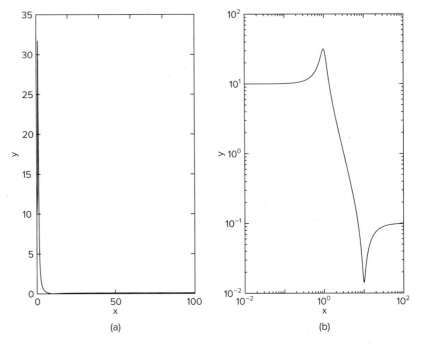

Figure 5.2–5 (a) Rectilinear plot of the function in Equation (5.2–1). (b) Log-log plot of the function. Note the wide range of values of both x and y.

Because of the wide range in values on both the abscissa and the ordinate, rectilinear scales do not reveal the important features. The following program produced Figure 5.2–5.

```
% Create the Rectilinear Plot
x1 = 0:0.01:100; u1 = x1.^2;
num1 = 100*(1-0.01*u1).^2 + 0.02*u1;
den1 = (1-u1).^2 + 0.1*u1;
y1 = sqrt(num1./den1);
subplot(1,2,1),plot(x1,y1),xlabel('x'),ylabel('y'),
% Create the Loglog Plot
x2 = logspace(-2, 2, 500); u2 = x2.^2;
num2 = 100*(1-0.01*u2).^2 + 0.02*u2;
den2 = (1-u2).^2 + 0.1*u2;
y2 = sqrt(num2./den2);
subplot(1,2,2),loglog(x2,y2),xlabel('x'),ylabel('y')
```

It is important to remember the following points when using log scales:

1. You cannot plot negative numbers on a log scale, because the logarithm of a negative number is not defined as a real number.
2. You cannot plot the number 0 on a log scale, because $\log_{10} 0 = \ln 0 = -\infty$. You must choose an appropriately small number as the lower limit on the plot.

3. The tick-mark labels on a log scale are the actual values being plotted; they are not the logarithms of the numbers. For example, the range of x values in the plot in Figure 5.2–5b is from $10^{-2} = 0.01$ to $10^2 = 100$, and the range of y values is from 10^{-2} to $10^2 = 100$.

MATLAB has three commands for generating plots having log scales. The appropriate command depends on which axis must have a log scale. Follow these rules:

1. Use the `loglog(x,y)` command to have both scales logarithmic.
2. Use the `semilogx(x,y)` command to have the x scale logarithmic and the y scale rectilinear.
3. Use the `semilogy(x,y)` command to have the y scale logarithmic and the x scale rectilinear.

Table 5.2–3 summarizes these functions. For other two-dimensional plot types, type `help specgraph`. We can plot multiple curves with these commands just as with the `plot` command. In addition, we can use the other commands, such as `grid`, `xlabel`, and `axis`, in the same manner. Figure 5.2–6 shows how these commands are applied. It was created with the following program:

```
x1 = 0:0.01:3; y1 = 25*exp(0.5*x1);
y2 = 40*(1.7.^x1);
x2 = logspace(-1,1,500); y3 = 15*x2.^(0.37);
subplot(1,2,1),semilogy(x1,y1,x1,y2, '--'),...
    legend ('y = 25e^{0.5x}', 'y = 40(1.7) ^x'),...
    xlabel('x'),ylabel('y'),grid,...
    subplot(1,2,2),loglog(x2,y3),legend('y = 15x^{0.37}'),...
    xlabel('x'),ylabel('y'),grid
```

Note that the two *exponential* functions $y = 25e^{0.5x}$ and $y = 40(1.7)^x$ both produce straight lines on a semilog plot with the y axis logarithmic. The *power* function $y = 15x^{0.37}$ produces a straight line on a log-log plot.

Table 5.2–3 Specialized plot commands

Command	Description
`bar(x,y)`	Creates a bar chart of **y** versus **x**.
`fimplicit(f)`	Plots an implicit function.
`loglog(x,y)`	Produces a log-log plot of **y** versus **x**.
`polarplot(theta,r,'type')`	Produces a polar plot from the polar coordinates **theta** and **r**, using the line type, data marker, and colors specified in the string **type**.
`semilogx(x,y)`	Produces a semilog plot of **y** versus **x** with logarithmic abscissa scale.
`semilogy(x,y)`	Produces a semilog plot of **y** versus **x** with logarithmic ordinate scale.
`stairs(x,y)`	Produces a stairs plot of **y** versus **x**.
`stem(x,y)`	Produces a stem plot of **y** versus **x**.
`yyaxis(x1,y1,x2,y2)`	Produces a plot with two y axes, **y1** on the left and **y2** on the right.

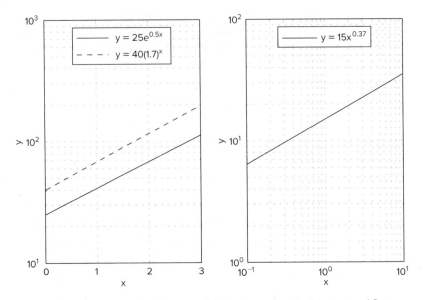

Figure 5.2–6 Two examples of exponential functions plotted with the `semilogy` function (the left-hand plot), and an example of a power function plotted with the `loglog` function (right-hand) plot.

Stem, Stairs, and Bar Plots

MATLAB has several other plot types that are related to *xy* plots. These include the stem, stairs, and bar plots. Their syntax is very simple, namely, `stem(x,y)`, `stairs(x,y)`, and `bar(x,y)`. See Table 5.2–3.

Separate *y* Axes

The `yyaxis` function (formerly called `plotyy`) generates a graph with two *y* axes. The syntax `yyaxis(x1,y1,x2,y2)` plots `y1` versus `x1` with *y* axis labeling on the left, and plots `y2` versus `x2` with *y* axis labeling on the right. The syntax `yyaxis(x1,y1,x2,y2,'type1','type2')` generates a `'type1'` plot of `y1` versus `x1` with *y* axis labeling on the left, and generates a `'type2'` plot of `y2` versus `x2` with *y* axis labeling on the right. For example, `yyaxis(x1,y1,x2, y2,'plot','stem')` uses `plot(x1,y1)` to generate a plot for the left axis, and `stem(x2,y2)` to generate a plot for the right axis. To see other variations of the `yyaxis` function, type `help yyaxis`.

Polar Plots

Polar plots are two-dimensional plots made using polar coordinates. If the polar coordinates are (θ, r), where θ is the angular coordinate and r is the radial coordinate of a point, then the command `polarplot(theta,r)` will produce the polar plot (formerly called `polar`). A grid is automatically overlaid on a polar plot. This grid consists of concentric circles and radial lines every 30°. The `title` and `gtext` commands can be used to place a title and text. The variant command `polarplot(theta,r,'type')` can be used to specify the line type or data marker, just as with the `plot` command.

Plotting Orbits

EXAMPLE 5.2–1

The equation

$$r = \frac{p}{1 - \in \cos \theta}$$

describes the polar coordinates of an orbit measured from one of the orbit's two focal points. For objects in orbit around the sun, the sun is at one of the focal points. Thus r is the distance of the object from the sun. The parameters p and $\in$ determine the size of the orbit and its eccentricity, respectively. Obtain the polar plot that represents an orbit having $\in = 0.5$ and $p = 2$ AU (AU stands for "astronomical unit"; 1 AU is the mean distance from the sun to Earth). How far away does the orbiting object get from the sun? How close does it approach Earth's orbit?

■ Solution

Figure 5.2–7 shows the polar plot of the orbit. The plot was generated by the following session:

```
>>theta = 0:pi/90:2*pi;
>>r = 2./(1-0.5*cos(theta));
>>polarplot(theta,r),title('Orbital Eccentricity = 0.5')
```

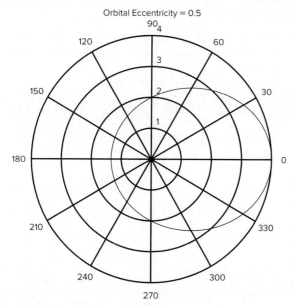

Figure 5.2–7 A polar plot showing an orbit having an eccentricity of 0.5.

The sun is at the origin, and the plot's concentric circular grid enables us to determine that the closest and farthest distances the object is from the sun are approximately 1.3 and 4 AU. Earth's orbit, which is nearly circular, is represented by the innermost circle. Thus the closest the object gets to Earth's orbit is approximately 0.3 AU. The radial grid lines allow us to determine that when $\theta = 90°$ and $270°$, the object is 2 AU from the sun.

Error Bar Plots

Experimental data are often represented with plots containing error bars. The bars show the estimated or calculated errors for each data point. They can also be used to display the error in an approximate formula. The basic syntax errorbar(x,y,e) plots y versus x with symmetric vertical error bars $2e(i)$ long. The arrays x, y, and e must be the same size. When they are vectors, each error bar is a distance of $e(i)$ above and below the point defined by $(x(i),y(i))$. When they are matrices, each error bar is a distance of $e(i,j)$ above and below the point defined by $(x(i,j),y(i,j))$.

For example, keeping two terms in the Taylor series expansion of $\cos x$ about $x = 0$ gives $\cos x \approx 1 - x^2/2$. The following program creates the plot shown in Figure 5.2–8:

```
% errorbar example
x = linspace(0.1, pi, 20);
approx = 1 - x.^2/2;
error = approx - cos(x);
errorbar(x, cos(x), error), legend('cos(x)'),...
    title('Approximation = 1 - x^2/2')
```

There are more than 20 two-dimensional plot functions available in MATLAB. We have shown the most important ones for engineering applications.

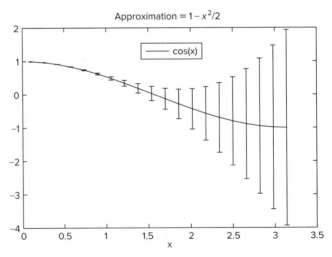

Figure 5.2–8 Error bars for the approximation $\cos x \approx 1 - x^2/2$.

Plotting Implicit Functions

An *implicit function* with two variables, say x and y, is a function in which we cannot isolate one variable in terms of the other. Fortunately, MATLAB provides the function `fimplicit(f)` to plot the implicit function defined by the equation $f(x,y) = 0$ over the default interval $[-5\ 5]$ for x and y. For example, to plot the hyperbola defined by $x^2 - y^2 - 1 = 0$ over the default interval of $[-5\ 5]$, you type

```
>>fimplicit(@(x,y) x.^2 - y.^2 - 1)
```

You can specify the interval with the syntax `fimplicit(f,interval)`. The equation for an ellipse centered at the origin has the form

$$\frac{x^2}{a^2} + \frac{y^2}{b^2} = 1$$

This equation is technically not an implicit function because we can isolate the variable y as follows:

$$y = \pm b \sqrt{1 - \frac{x^2}{a^2}}$$

However, the $\pm$ sign forces us to consider both possibilities when evaluating y. It is thus easier to use the `fimplicit` function. To plot the specific ellipse given by $a = 2$ and $b = 4$, the full ellipse will be displayed if the limits for x are $[-2\ 2]$ and the limits for y are $[-4\ 4]$. You would type

```
>>fimplicit(@(x,y) x.^2/4 + y.^2/16 - 1,[-2 2 -4 4])
```

Test Your Understanding

T5.2–6 Plot the following functions using axes that will produce a straight-line plot. The power function is $y = 2x^{-0.5}$, and the exponential function is $y = 10^{1-x}$.

T5.2–7 Plot the function $y = 8x^3$ for $-1 \le x \le 1$ with a tick spacing of 0.25 on the x axis and 2 on the y axis.

T5.2–8 The *spiral of Archimedes* is described by the polar coordinates (θ, r), where $r = a\theta$. Obtain a polar plot of this spiral for $0 \le \theta \le 4\pi$, with the parameter $a = 2$.

T5.2–9 Obtain the plot of the following implicit function, known as the Ampersand curve. Use the `axis equal` command.

$$(y^2 - x^2)(x - 1)(2x - 3) = 4(x^2 + y^2 - 2x)^2$$

Publishing Reports Containing Graphics

The `publish` function is available for creating reports, which may have embedded graphics. Reports generated by the `publish` function may be exported to a variety of common formats including HTML (Hyper Text Markup Language), which is used for Web-based reports, MS Word, PowerPoint, and L^{A}T$_E$X. To publish a report, do the following:

1. Open the Editor, type in the M-file that forms the basis of the report, and save it. Use the double percent character (%%) to indicate a section heading in the report. This character marks the beginning of a new cell, which is a group of commands. (Such a cell should not be confused with the cell array data type covered in Section 2.6.) Enter any blank lines you wish to appear in the report. Consider, as a very simple example, the following sample file polyplot.m.

```
%% Example of Report Publishing:
% Plotting the cubic y = x^3 - 6x^2 + 10x+4.
%% Create the independent variable.
x = linspace(0, 4, 300); % Use 300 points between 0 and 4.
%% Define the cubic from its coefficients.
p = [1, -6, 10, 4]; % p contains the coefficients.
%% Plot the cubic
plot(x,polyval(p,x)), xlabel('x'), ylabel('y')
```

2. Run the file to check it for errors. (To do this for a larger file, you may use the cell mode of the Editor to execute its each cell one at a time; see Section 4.7.)

3. Use the publish and open functions to create the report in the desired format. Using our sample file, we can obtain a report in HTML format by typing

```
>>publish ('polyplot','html')
>>open html/polyplot.html
```

You should see a report like the one shown in Figure 5.2–9.

Instead of using the publish and open functions, you may use the menu items under the PUBLISH tab of the toolstrip.

Once it is published in HTML, you may click on a section heading in the Contents to go to that section. This is useful for larger reports.

If you want the equation to look professionally typeset, you may edit the resulting report in the appropriate editor (say, MS Word or L^{A}T$_E$X). For example, to set the cubic polynomial in the resulting L^{A}T$_E$X file, use the commands presented earlier in this section to replace the equation in the second line of the report with

```
y = {\it x}^3 - 6{\it x}^2 + 10{\it x} + 4
```

You can also use the Live Editor to obtain your equations in standard mathematical form (see Section 5.1 of this chapter).

Example of Report Publishing:

Plotting the cubic $y = x^3 - 6x^2 + 10x + 4$.

Contents

- Create the independent variable.
- Define the cubic.
- Plot the cubic.

Create the independent variable.

```
x = linspace (0,4,300); % Use 300 points between 0 and 4.
```

Define the cubic.

```
p = [1,-6,10,4]; % p contains the coefficients.
```

Plot the cubic.

```
plot(x,polyval(p,x)),xlabel('x'),ylabel('y')
```

Figure 5.2–9 A sample report published from MATLAB.

5.3 Interactive Plotting in MATLAB

The interactive plotting environment in MATLAB is a set of tools for

- Creating different types of graphs,
- Selecting variables to plot directly from the Workspace Browser,
- Creating and editing subplots,
- Adding annotations such as lines, arrows, text, rectangles, and ellipses, and
- Editing properties of graphics objects, such as their color, line weight, and font.

The Plot Tools interface includes the following three panels associated with a given figure.

- **The Figure Palette:** Use this to create and arrange subplots, to view and plot workspace variables, and to add annotations.
- **The Plot Browser:** Use this to select and control the visibility of the axes or graphics objects plotted in the figure, and to add data for plotting.
- **The Property Editor:** Use this to set basic properties of the selected object and to obtain access to all properties through the Property Inspector.

The Figure Window

When you create a plot, the Figure window appears with the Figure toolbar visible (see Figure 5.3–1). This window has eight menus.

The File Menu   The File menu is used for saving and printing the figure. This menu was discussed in Section 5.1 under **Saving Figures** and **Exporting Figures.**

The Edit Menu   You can use the Edit menu to cut, copy, and paste items, such as legend or title text, that appear in the figure. Click on **Figure Properties** to open the Property Editor—Figure dialog box to change certain properties of the figure.

Three items on the Edit menu are very useful for editing the figure. Clicking the **Axes Properties** item brings up the Property Editor—Axes dialog box. Double-clicking on any axis also brings up this box. You can change the scale type (linear, log, etc.), the labels, and the tick marks by selecting the tab for the desired axis or the font to be edited.

The Current Object Properties item enables you to change the properties of an object in the figure. To do this, first click on the object, such as a plotted line,

Figure 5.3–1 The Figure toolbar displayed.

then click on **Current Object Properties** in the Edit menu. You will see the Property Editor—Line series dialog box that lets you change properties such as line weight and color, data-marker type, and plot type.

Clicking on any text, such as that placed with the `title`, `xlabel`, `ylabel`, `legend`, or `gtext` commands, and then selecting **Current Object Properties** in the Edit menu bring up the Property Editor—Text dialog box, which enables you to edit the text.

The View Menu   The items on the View menu are the three toolbars (Figure Toolbar, Plot Edit Toolbar, and Camera Toolbar), the Figure Palette, the Plot Browser, and the Property Editor. These will be discussed later in this section.

The Insert Menu   The Insert menu enables you to insert labels, legends, titles, text, and drawing objects, rather than using the relevant commands from the Command window. To insert a label on the y axis, for example, click on the **Y Label** item on the menu; a box will appear on the y axis. Type the label in this box, and then click outside the box to finish.

The Insert menu also enables you to insert arrows, lines, text, rectangles, and ellipses in the figure. To insert an arrow, for example, click on the **Arrow** item; the mouse cursor changes to a crosshair style. Then click the mouse button, and move the cursor to create the arrow. The arrowhead will appear at the point where you release the mouse button. Be sure to add arrows, lines, and other annotations only after you are finished moving or resizing your axes, because these objects are not anchored to the axes. (They can be anchored to the plot by *pinning;* see the MATLAB Help under "Add Annotation to Graph Interactively.")

To delete or move a line or arrow, click on it, then press the **Delete** key to delete it, or press the mouse button and move it to the desired location. The Axes item lets you use the mouse to place a new set of axes within the existing plot. Click on the new axes, and a box will surround them. Any further plot commands issued from the Command window will direct the output to these axes.

The Light item applies to three-dimensional plots.

The Tools Menu   The Tools menu includes items for adjusting the view (by zooming and panning) and the alignment of objects on the plot. The Edit Plot item starts the plot editing mode, which can also be started by clicking on the northwest-facing arrow on the Figure toolbar. The Tools menu also gives access to the Data Cursor, which is discussed later in this section. The last two items, Basic Fitting and Data Statistics, will be discussed in Sections 6.3 and 7.1, respectively.

Other Menus   The Desktop menu enables you to dock the Figure window within the desktop. The Window menu lets you switch between the Command window and any other Figure windows. The Help menu accesses the general MATLAB Help System as well as Help features specific to plotting.

There are three toolbars available in the Figure window: the Figure toolbar, the Plot Edit toolbar, and the Camera toolbar. The View menu lets you select which ones you want to appear. We will discuss the Figure toolbar and the Plot

Edit toolbar in this section. The Camera toolbar is useful for three-dimensional plots, which are discussed at the end of this chapter.

The Figure Toolbar

To activate the Figure toolbar, select it from the View menu (see Figure 5.3–1). The four leftmost buttons are for opening, saving, and printing the figure. Clicking on the northwest-facing arrow button toggles the plot edit mode on and off.

The Zoom-in and Zoom-out buttons let you obtain a close-up or faraway view of the figure. The Pan and Rotate 3D buttons are used for three-dimensional plots.

The Data Cursor button enables you to read data directly from a graph by displaying the values of points you select on plotted lines, surfaces, images, and so on.

The Insert Colorbar button inserts a color map strip in the graph and is useful for three-dimensional surface plots. The Insert Legend button enables you to insert a legend in the plot. The last two buttons hide or show the plot tools and dock the figure if it is undocked.

The Plot Edit Toolbar

Once a plot is in the window, you can display the Plot Edit toolbar from the View menu. This toolbar is shown in Figure 5.3–2. You can enable plot editing by clicking on the northwest-facing arrow on the Figure toolbar. Then double-click on an axis, a plotted line, or a label to activate the appropriate property editor. To add text that is not a label, title, or legend, click the button labeled **T,** move the cursor to the desired location for the text, click the mouse button, and type the text. When finished, click outside the text box and note that the nine leftmost buttons become highlighted and available. These enable you to modify the color, font, and other attributes of the text.

To insert arrows, lines, rectangles, and ellipses, click on the appropriate button and follow the instructions given previously for the Insert menu.

The Plot Tools

Once a figure has been created, you can display any of or all three Plot Tools (Figure Palette, Plot Browser, and Property Editor) by selecting them from the View menu. You can also start the environment by first creating a plot and then clicking on the **Show Plot Tools** icon at the far right of the Figure toolbar

Figure 5.3–2 The Figure and Plot Edit toolbars displayed.

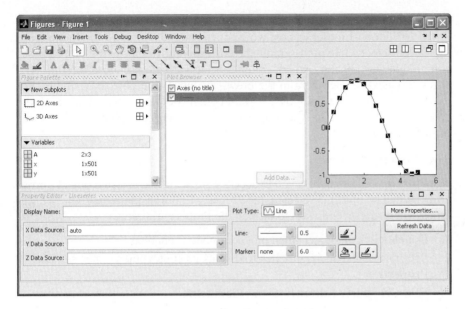

Figure 5.3–3 The Figure window with the Plot Tools activated.

(see Figure 5.3–3), or by creating a figure with the plotting tools attached by placing the `plottools` command after the `plot` function. Remove the tools by clicking on the **Hide Tools** icon, which is second from the left.

Figure 5.3–3 shows the result of clicking on the plotted line after clicking the **Show Plot Tools** icon. The plotting interface then displays the Property Editor—Lineseries.

The Figure Palette

The Figure Palette contains three panels, which are selected and expanded by clicking the appropriate button. Click on the grid icon in the **New Subplots** panel to display the selector grid that enables you to specify the layout of the subplots. In the Variables panel you can select a graphics function to plot the variable by selecting the variable and right-clicking to display the context menu. This menu contains a list of possible plot types based on the type of variable you select. You can also drag the variable into an axes set, and MATLAB will select an appropriate plot type.

Click on the **Annotations** panel to display a menu of objects such as lines, arrows, etc. Click on the desired object, and use the mouse to position and size it.

The Plot Browser

The Plot Browser provides a legend of all the graphs in the figure. For example, if you plot an array with multiple rows and columns, the browser lists each axis and the objects (lines, surfaces, etc.) used to create the graph. To set the properties of an individual line, double-click on the line. Its properties are displayed in the Property Editor—Lineseries box, which opens on the bottom of the figure.

If you select a line in the graph, the corresponding entry in the Plot Browser is highlighted, indicating which column in the variable produced the line. The check box next to each item in the browser controls the object's visibility. For example, if you want to plot only certain columns of data, you can uncheck the columns not wanted. The graph updates as you uncheck each box and rescales the axes as required.

The Property Editor

The Property Editor enables you to access a subset of the selected object's properties. When no object is selected, the Property Editor displays the figure's properties. There are several ways to display the Property Editor.

1. Double-click an object when plot edit mode is enabled.
2. Select an object and right-click to display its context menu, then select **Properties.**
3. Select **Property Editor** from the View menu.
4. Use the `propertyeditor` command.

The Property Editor enables you to change the most commonly used object properties. If you want to access all object properties, use the Property Inspector. To display the Property Inspector, click on the **More Properties** button. Use of this feature requires detailed knowledge of object properties and handle graphics, and thus will not be covered here.

Recreating Graphs from M-Files

Once your graph is finished, you can generate MATLAB code to reproduce the graph by selecting **Generate Code** from the File menu. MATLAB creates a function that recreates the graph and opens the generated M-File in the editor. This feature is particularly useful for capturing property settings and other modifications made in the plot editor.

Adding Data to Axes

The Plot Browser provides the mechanism by which you add data to axes. The procedure is as follows:

1. Select a two-dimensional or three-dimensional axis from the New Subplots subpanel.
2. After creating the axis, select it in the Plot Browser panel to enable the **Add Data** button at the bottom of the panel.
3. Click the **Add Data** button to display the Add Data to Axes dialog box. The Add Data to Axes dialog enables you to select a plot type and specify the workspace variables to pass to the plotting function. You can also specify a MATLAB expression, which is evaluated to produce the data to plot.

5.4 Three-Dimensional Plots

MATLAB provides many functions for creating three-dimensional plots. Here we will summarize the basic functions to create three types of plots: line plots, surface plots, and contour plots. The extended syntax for all the functions treated in this section is extensive. This syntax enables you to customize your plots with colors, spacing, labels, and shading. The nature of three-dimensional plots itself is quite involved, because the viewer's point of view can affect how much information and understanding can be obtained from the graph. Thus the Camera Toolbar in the View menu of the Figure window is helpful for determining the proper point of view. Information about these features and functions is available in MATLAB Help (categories `graph3d` and `specgraph`).

Three-Dimensional Line Plots

Lines in three-dimensional space can be plotted with the `plot3` function. Its syntax is `plot3(x,y,z)`. For example, the following equations generate a three-dimensional curve as the parameter t is varied over some range:

$$
\begin{aligned}
x &= e^{-0.05t}\sin t \\
y &= e^{-0.05t}\cos t \\
z &= t
\end{aligned}
$$

If we let t vary from $t = 0$ to $t = 10\pi$, the sine and cosine functions will vary through five cycles, while the absolute values of x and y become smaller as t increases. This process results in the spiral curve shown in Figure 5.4–1, which was produced with the following session.

```
>>t = 0:pi/50:10*pi;
>>plot3(exp(-0.05*t).*sin(t),exp(-0.05*t).*cos(t),t),...
        xlabel('x'),ylabel('y'),zlabel('z'),grid
```

Note that the `grid` and `label` functions work with the `plot3` function, and that we can label the z axis by using the `zlabel` function, which we have seen for the first time. Similarly, we can use the other plot enhancement functions discussed in Sections 5.1 and 5.2 to add a title and text and to specify line type and color.

The `plot3(x,y,z)` function generates a three-dimensional plot of a set of data points where `x,y,z` are vectors or matrices, by plotting lines in three-dimensional space through the points whose coordinates are the elements of `x, y, z`. The `fplot3` function, introduced in MATLAB release R2016a, supplements the `plot3` function. Its syntax `fplot3(fx,fy,fz,t_interval)` plots the parametric curve defined by the functions $x = fx(t)$, $y = fy(t)$, and $z = fz(t)$ over the interval `t_interval` for t.

For example, the plot in Figure 5.4–1, produced with `plot3`, can also be created with `fplot3` as follows:

```
>>fx = @(t)exp(-0.05*t).*sin(t)
>>fy = @(t)exp(-0.05*t).*cos(t)
```

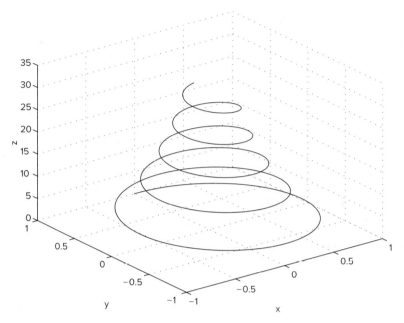

Figure 5.4–1 The curve $x = e^{-0.05t} \sin t$, $y = e^{-0.05t} \cos t$, $z = t$ plotted with the plot3 function.

```
>>fz = @(t)t
>>fplot3(fx,fy,fz,[0,10*pi]),xlabel('x'),...
    ylabel('y'),zlabel('z'),grid on
```

or

```
>>fplot3(@(t)exp(-0.05*t).*sin(t),...
    @(t)exp(-0.05*t).*cos(t),@(t)t,0,10*pi]),
    xlabel('x'),ylabel('y'),zlabel('z'),grid on
```

Test Your Understanding

T5.4-1   Use plot3 and fplot3 to plot the 3-D line plot described by $x = \sin(t)$, $y = \cos(t)$, $z = \ln(t)$ for t between 0 and 30.

Surface Mesh Plots

The function $z = f(x, y)$ represents a surface when plotted on xyz axes, and the mesh function provides the means to generate a *surface mesh plot*. Before you can use this function, you must generate a grid of points in the xy plane and then evaluate the function $f(x, y)$ at these points. The meshgrid function generates the grid. Its syntax is [X,Y] = meshgrid(x,y). If x = xmin:xspacing:xmax and y = ymin:yspacing:ymax, then this function will generate the coordinates of a rectangular grid with one corner at (*xmin, ymin*) and the opposite corner at (*xmax, ymax*). Each rectangular panel in the grid will have a width equal to *xspacing* and a depth equal to *yspacing*. The resulting matrices X and Y contain the coordinate pairs of every point in the grid. These pairs are then used to evaluate the function.

The function [X,Y] = meshgrid(x) is equivalent to [X,Y] = meshgrid (x,x) and can be used if *x* and *y* have the same minimum values, the same maximum values, and the same spacing. Using this form, you can type [X,Y] = meshgrid(min:spacing:max), where min and max specify the minimum and maximum values of both *x* and *y* and spacing is the desired spacing of the *x* and *y* values.

After the grid is computed, you create the surface plot with the mesh function. Its syntax is mesh(x,y,z). The grid, label, and text functions can be used with the mesh function. The following session shows how to generate the surface mesh plot of the function $z = xe^{-[(x - y^2)^2 - y^2]}$, for $-2 \leq x \leq 2$ and $-2 \leq y \leq 2$, with a spacing of 0.1. This plot appears in Figure 5.4–2.

```
>>[X,Y] = meshgrid(-2:0.1:2);
>>Z = X.*exp(-((X-Y.^2).^2 - Y.^2));
>>mesh(X,Y,Z),xlabel('x'),ylabel('y'),zlabel('z')
```

Be careful not to select too small a spacing for the *x* and *y* values for two reasons: (1) Small spacing creates small grid panels, which make the surface difficult to visualize, and (2) the matrices X and Y can become too large.

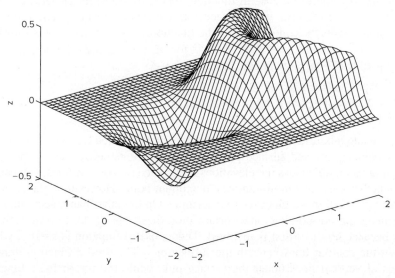

Figure 5.4–2 A plot of the surface $z = xe^{-[(x - y^2)^2 - y^2]}$ created with the mesh function.

The fmesh(f,xy_interval) function generates a surface plot of a function $f(x,y)$. This function was introduced in MATLAB release R2016a, and supplements the mesh function. To use the same interval for both x and y, specify xy_interval as a two-element vector of the form [min max]. To use different intervals, specify a four-element vector of the form [xmin xmax ymin ymax].

For example, the plot in Figure 5.4–2, produced with mesh, can also be created with fmesh as follows:

```
>>fmesh(@(x,y) x.*exp(-(x-y.^2).^2-y.^2),[-2,2]),...
    xlabel('x'),ylabel('y'),zlabel('z')
```

The surf and surfc functions are similar to mesh and meshc except that the former create a shaded surface plot. You can use the Camera toolbar and some menu items in the Figure window to change the view and lighting of the figure.

The fsurf(f,xy_interval) function generates a shaded surface plot of a function $f(x,y)$. This function was introduced in MATLAB release R2016a, and supplements the surf function. To use the same interval for both x and y, specify xy_interval as a two-element vector of the form [min max]. To use different intervals, specify a four-element vector of the form [xmin xmax ymin ymax].

At present there is no fmeshc or fsurfc function.

Contour Plots

Topographic plots show the contours of the land by means of constant elevation lines. These lines are also called *contour lines,* and such a plot is called a *contour plot.* If you walk along a contour line, you remain at the same elevation. Contour plots can help you visualize the shape of a function. They can be created with the contour function, whose syntax is contour(X,Y,Z). You use this function the same way you use the mesh function; that is, first use the meshgrid function to generate the grid and then generate the function values. The following session generates the contour plot of the function whose surface plot is shown in Figure 5.4–2, namely, $z = xe^{-[(x-y^2)^2 + y^2]}$, for $-2 \le x \le 2$ and $-2 \le y \le 2$, with a spacing of 0.1. This plot appears in Figure 5.4–3.

```
>>[X,Y] = meshgrid(-2:0.1:2);
>>Z = X.*exp(-((X-Y.^2).^2+Y.^2));
>>contour(X,Y,Z),xlabel('x'),ylabel('y')
```

You can add labels to the contour lines. Type help clabel.

Contour plots and surface plots can be used together to clarify the function. For example, unless the elevations are labeled on contour lines, you cannot tell whether there is a minimum or a maximum point. However, a glance at the surface plot will make this easy to determine. On the other hand, accurate measurements are not possible on a surface plot; these can be done on the contour plot because no distortion is involved. Thus a useful function is meshc, which shows the contour lines beneath the surface plot. The meshz function draws a series of vertical lines under the surface plot, while the waterfall function draws mesh lines in one direction only. The results of these functions are shown in Figure 5.4–4 for the function $z = xe^{-(x^2 + y^2)}$.

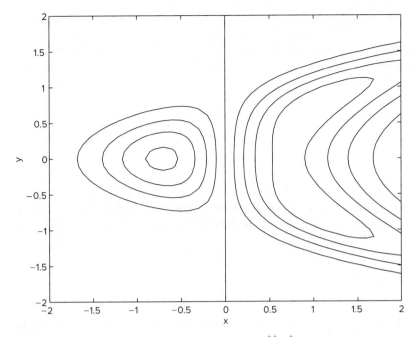

Figure 5.4–3 A contour plot of the surface $z = xe^{-[(x-y^2)^2 + y^2]}$ created with the contour function.

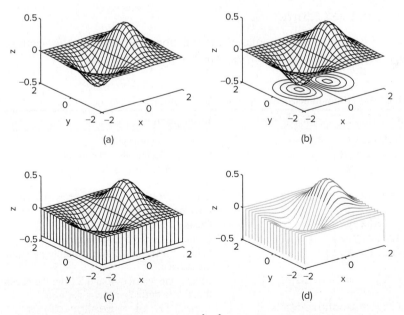

Figure 5.4–4 Plots of the surface $z = xe^{-(x^2 + y^2)}$ created with the mesh function and its variant forms: meshc, meshz, and waterfall. (a) mesh, (b) meshc, (c) meshz, (d) waterfall.

The fcontour(f) function plots the contour lines of the function $z = f(x,y)$ for constant levels of z over the default interval $[-5\ 5]$ for x and y. This function was introduced in MATLAB release R2016a, and supplements the contour function. The extended syntax is fcontour(f,xy_interval). To use the same interval for both x and y, specify xy_interval as a two-element vector of the form [min max]. To use different intervals, specify a four-element vector of the form [xmin xmax ymin ymax].

Surface Plots of Implicit Functions

In Section 5.2 we saw that an *implicit function* is a function in which we cannot isolate one variable in terms of the other. Fortunately, MATLAB provides the function fimplicit3(f) to plot the three-dimensional implicit function defined by the equation $f(x, y, z) = 0$ over the default interval $[-5\ 5]$ for x, y, and z. You can specify the interval with the syntax fimplicit3(f,interval). For example, to plot the hyperboloid $x^2 + y^2 - z^2 = 0$ over the default interval $[-5\ 5]$, you type

```
>>f = @(x,y,z) x.^2 + y.^2 - z.^2;
>>fimplicit3(f)
```

To plot the upper half of the hyperboloid $x^2 + y^2 - z^2 = 0$ you specify the interval as for z as $[0\ 5]$, and for x and y, use the default interval $[-5\ 5]$, as follows.

```
>>f = @(x,y,z) x.^2 + y.^2 - z.^2;
>>interval = [-5 5 -5 5 0 5];
>>fimplicit3(f,interval)
```

Tables 5.4–1 and 5.4–2 summarize the functions introduced in this section. For other three-dimensional plot types, type help specgraph.

Table 5.4–1 Three-dimensional plotting functions using array inputs

Function	Description
contour(x,y,z)	Creates a contour plot.
mesh(x,y,z)	Creates a 3-D mesh surface plot.
meshc(x,y,z)	Same as mesh but draws a contour plot under the surface.
meshz(x,y,z)	Same as mesh but draws a series of vertical reference lines under the surface.
plot3(x,y,z)	Creates a 3-D line plot.
surf(x,y,z)	Creates a shaded 3-D surface plot.
surfc(x,y,z)	Same as surf but draws a contour plot under the surface.
[X,Y] = meshgrid(x,y)	Creates the matrices X and Y from the vectors x and y to define a rectangular grid.
[X,Y] = meshgrid(x)	Same as [X,Y] = meshgrid(x,x).
waterfall(x,y,z)	Same as mesh but draws mesh lines in one direction only.

Table 5.4–2 Three-dimensional plotting functions using function inputs

Function	Description
fcontour(f)	Creates a contour plot.
fimplicit3(f)	Plots an implicit 3-D function.
fmesh(f)	Creates a 3-D surface plot.
fplot3(fx,fy,fz)	Creates a 3-D line plot.
fsurf(f)	Creates a shaded 3-D surface plot.

Test Your Understanding

T5.4–2 Use mesh, fmesh, contour, and fcontour to create a surface plot and a contour plot of the function $z = (x - 2)^2 + 2xy + y^2$.

T5.4–3 Use the fimplicit3 function to create a surface plot of the function

$$x^2 - y^2 - z^2 = 0$$

5.5 Summary

This chapter explained how to use the powerful MATLAB commands to create effective and pleasing two-dimensional and three-dimensional plots. The following guidelines will help you create plots that effectively convey the desired information.

■ Label each axis with the name of the quantity being plotted *and its units!*
■ Use regularly spaced tick marks at convenient intervals along each axis.
■ If you are plotting more than one curve or data set, label each on its plot or use a legend to distinguish them.
■ If you are preparing multiple plots of a similar type or if the axes' labels cannot convey enough information, use a title.
■ If you are plotting measured data, plot each data point in a given set with the same symbol, such as a circle, square, or cross.
■ If you are plotting points generated by evaluating a function (as opposed to measured data), do *not* use a symbol to plot the points. Instead, connect the points with solid lines.

Key Terms

Axis limits, 228
Contour plot, 258
Data symbol, 225
Overlay plot, 234

Polar plots, 244
Subplots, 234
Surface mesh plot, 256

Problems

You can find the answers to problems marked with an asterisk at the end of the text.

Sections 5.1, 5.2, and 5.3

1.* *Breakeven analysis* determines the production volume at which the total production cost is equal to the total revenue. At the breakeven point, there is neither profit nor loss. In general, production costs consist of fixed costs and variable costs. Fixed costs include salaries of those not directly involved with production, factory maintenance costs, insurance costs, and so on. Variable costs depend on production volume and include material costs, labor costs, and energy costs. In the following analysis, assume that we produce only what we can sell; thus the production quantity equals the sales. Let the production quantity be Q, in gallons per year.

   Consider the following costs for a certain chemical product:

   Fixed cost: $3 million per year.

   Variable cost: 2.5 cents per gallon of product.

   The selling price is 5.5 cents per gallon.

   Use these data to plot the total cost and the revenue versus Q, and graphically determine the breakeven point. Fully label the plot and mark the breakeven point. For what range of Q is production profitable? For what value of Q is profit a maximum?

2. Consider the following costs for a certain chemical product:

   Fixed cost: $2.045 million/year.

   Variable costs:

   Material cost: 62 cents per gallon of product.

   Energy cost: 24 cents per gallon of product.

   Labor cost: 16 cents per gallon of product.

   Assume that we produce only what we sell. Let P be the selling price in dollars per gallon. Suppose that the selling price and the sales quantity Q are interrelated as follows: $Q = 6 \times 10^6 - 1.1 \times 10^6 P$. Accordingly, if we raise the price, the product becomes less competitive and sales drop.

   Use this information to plot the fixed and total variable costs versus Q, and graphically determine the breakeven point(s). Fully label the plot and mark the breakeven points. For what range of Q is production profitable? For what value of Q is profit a maximum?

3*. *a.* Estimate the roots of the equation

$$x^3 - 3x^2 + 5x \sin\left(\frac{\pi x}{4} - \frac{5\pi}{4}\right) + 3 = 0$$

   by plotting the equation.

   b. Use the estimates found in part *a* to find the roots more accurately with the `fzero` function.

4. To compute the forces in structures, sometimes we must solve equations similar to the following. Use the `fplot` function to find all the positive roots of this equation:

$$x \tan(x) = 9$$

5*. Cables are used to suspend bridge decks and other structures. If a heavy uniform cable hangs suspended from its two endpoints, it takes the shape of a *catenary* curve whose equation is

$$y = a \cosh\left(\frac{x}{a}\right)$$

where a is the height of the lowest point on the chain above some horizontal reference line, x is the horizontal coordinate measured to the right from the lowest point, and y is the vertical coordinate measured up from the reference line.

     Let $a = 10$ m. Plot the catenary curve for $-20 \le x \le 30$ m. How high is each endpoint?

6. Using estimates of rainfall, evaporation, and water consumption, the town engineer developed the following model of the water volume in the reservoir as a function of time

$$V(t) = 10^9 + 10^8(1 - e^{-t/100}) - 10^7 t$$

where V is the water volume in liters and t is time in days. Plot $V(t)$ versus t. Use the plot to estimate how many days it will take before the water volume in the reservoir is 50 percent of its initial volume of 10^9 L.

7. It is known that the following Leibniz series converges to the value $\pi/4$ as $n \to \infty$.

$$S(n) = \sum_{k=0}^{n} (-1)^k \frac{1}{2k + 1}$$

Plot the difference between $\pi/4$ and the sum $S(n)$ versus n for $0 \le n \le 200$.

8. A certain fishing vessel is initially located in a horizontal plane at $x = 0$ and $y = 10$ mi. It moves on a path for 10 hr such that $x = t$ and $y = 0.5t^2 + 10$, where t is in hours. An international fishing boundary is described by the line $y = 2x + 6$.

a. Plot and label the path of the vessel and the boundary.

b. The perpendicular distance of the point (x_1, y_1) from the line $Ax + By + C = 0$ is given by

$$d = \frac{Ax_1 + By_1 + C}{\pm\sqrt{A^2 + B^2}}$$

where the sign is chosen to make $d \ge 0$. Use this result to plot the distance of the fishing vessel from the fishing boundary as a function of time for $0 \le t \le 10$ hr.

9. Plot columns 2 and 3 of the following matrix **A** versus column 1. The data in column 1 are time (seconds). The data in columns 2 and 3 are force (newtons).

$$\mathbf{A} = \begin{bmatrix} 0 & -7 & 6 \\ 5 & -4 & 3 \\ 10 & -1 & 9 \\ 15 & 1 & 0 \\ 20 & 2 & -1 \end{bmatrix}$$

10*. Many applications use the following "small angle" approximation for the sine to obtain a simpler model that is easy to understand and analyze. This approximation states that $\sin x \approx x$, where x must be in radians. Investigate the accuracy of this approximation by creating three plots. For the first, plot $\sin x$ and x versus x for $0 \le x \le 1$. For the second, plot the approximation error $\sin x - x$ versus x for $0 \le x \le 1$. For the third, plot the relative error $[\sin(x) - x]/\sin(x)$ versus x for $0 \le x \le 1$. How small must x be for the approximation to be accurate within 5 percent?

11. You can use trigonometric identities to simplify the equations that appear in many applications. Confirm the identity $\tan(2x) = 2\tan x/(1 - \tan^2 x)$ by plotting both the left and the right sides versus x over the range $0 \le x \le 2\pi$.

12. The complex number identity $e^{ix} = \cos x + i \sin x$ is often used to convert the solutions of equations into a form that is relatively easy to visualize. Confirm this identity by plotting the imaginary part versus the real part for both the left and right sides over the range $0 \le x \le 2\pi$.

13. Use a plot over the range $0 \le x \le 5$ to confirm that $\sin(ix) = i \sinh x$.

14*. The function $y(t) = 1 - e^{-bt}$, where t is time and $b > 0$, describes many processes, such as the height of liquid in a tank as it is being filled and the temperature of an object being heated. Investigate the effect of the parameter b on $y(t)$. To do this, plot y versus t for several values of b on the same plot. How long will it take for $y(t)$ to reach 98 percent of its steady-state value?

15. The following functions describe the oscillations in electric circuits and the vibrations of machines and structures. Plot these functions on the same plot. Because they are similar, decide how best to plot and label them to avoid confusion.

$$x(t) = 10e^{-0.5t}\sin(3t + 2)$$
$$y(t) = 7e^{-0.4t}\cos(5t - 3)$$

16. In certain kinds of structural vibrations, a periodic force acting on the structure will cause the vibration amplitude to repeatedly increase and decrease with time. This phenomenon, called *beating,* also occurs in musical sounds. A particular structure's displacement is described by

$$y(t) = \frac{1}{f_1^2 - f_2^2}[\cos(f_2 t) - \cos(f_1 t)]$$

where y is the displacement in inches and t is the time in seconds. Plot y versus t over the range $0 \le t \le 20$ for $f_1 = 8$ rad/sec and $f_2 = 1$ rad/sec. Be sure to choose enough points to obtain an accurate plot.

17*. The height $h(t)$ and horizontal distance $x(t)$ traveled by a ball thrown at an angle A with a speed v are given by

$$h(t) = vt \sin A - \frac{1}{2}gt^2$$

$$x(t) = vt \cos A$$

At Earth's surface the acceleration due to gravity is $g = 9.81$ m/s^2.

a. Suppose the ball is thrown with a velocity $v = 10$ m/s at an angle of $35°$. Use MATLAB to compute how high the ball will go, how far it will go, and how long it will take to hit the ground.

b. Use the values of v and A given in part a to plot the ball's *trajectory;* that is, plot h versus x for positive values of h.

c. Plot the trajectories for $v = 10$ m/s corresponding to five values of the angle A: $20°$, $30°$, $45°$, $60°$, and $70°$.

d. Plot the trajectories for $A = 45°$ corresponding to five values of the initial velocity v: 10, 12, 14, 16, and 18 m/s.

18. The perfect gas law relates the pressure p, absolute temperature T, mass m, and volume V of a gas. It states that

$$pV = mRT$$

The constant R is the *gas constant.* The value of R for air is 286.7 (N · m)/ (kg · K). Suppose air is contained in a chamber at room temperature ($20°C = 293$ K). Create a plot having three curves of the gas pressure in N/m^2 versus the container volume V in m^3 for $20 \le V \le 100$. The three curves correspond to the following masses of air in the container: $m = 1$ kg, $m = 3$ kg, and $m = 7$ kg.

19. Oscillations in mechanical structures and electric circuits can often be described by the function

$$y(t) = e^{-t/\tau} \sin(\omega t + \phi)$$

where t is time and ω is the oscillation frequency in radians per unit time. The oscillations have a period of $2\pi/\omega$, and their amplitudes decay in time at a rate determined by τ, which is called the *time constant.* The smaller τ is, the faster the oscillations die out.

a. Use these facts to develop a criterion for choosing the spacing of the t values and the upper limit on t to obtain an accurate plot of $y(t)$. (*Hint:* Consider two cases: $4\tau > 2\pi/\omega$ and $4\tau < 2\pi/\omega$.)

b. Apply your criterion, and plot $y(t)$ for $\tau = 10$, $\omega = \pi$, and $\phi = 2$.

c. Apply your criterion, and plot $y(t)$ for $\tau = 0.1$, $\omega = 8\pi$, and $\phi = 2$.

20. When a constant voltage was applied to a certain motor initially at rest, its rotational speed $s(t)$ versus time was measured. The data appear in the following table:

Time (sec)	1	2	3	4	5	6	7	8	10
Speed (rpm)	1210	1866	2301	2564	2724	2881	2879	2915	3010

Determine whether the following function can describe the data. If so, find the values of the constants b and c.

$$s(t) = b(1 - e^{ct})$$

21. The following table shows the average temperature for each year in a certain city. Plot the data as a stem plot, a bar plot, and a stairs plot.

Year	2000	2001	2002	2003	2004
Temperature (°C)	21	18	19	20	17

22. A sum of $10,000 invested at 4 percent interest compounded annually will grow according to the formula

$$y(k) = 10^4 (1.04)^k$$

where k is the number of years ($k = 0, 1, 2, \ldots$). Plot the amount of money in the account for a 10-year period. Do this problem with four types of plots: the xy plot, the stem plot, the stairs plot, and the bar plot.

23. The volume V and surface area A of a sphere of radius r are given by

$$V = \frac{4}{3}\pi r^3 \qquad A = 4\pi r^2$$

a. Plot V and A versus r in two subplots, for $0.1 \leq r \leq 100$ m. Choose axes that will result in straight-line graphs for both V and A.

b. Plot V and r versus A in two subplots, for $1 \leq A \leq 10^4$ m². Choose axes that will result in straight-line graphs for both V and r.

24. The current amount A of a principal P invested in a savings account paying an annual interest rate r is given by

$$A = P\left(1 + \frac{r}{n}\right)^{nt}$$

where n is the number of times per year the interest is compounded. For continuous compounding, $A = Pe^{rt}$. Suppose $10,000 is initially invested at 3.5 percent ($r = 0.035$).

a. Plot A versus t for $0 \leq t \leq 20$ years for four cases: continuous compounding, annual compounding ($n = 1$), quarterly compounding ($n = 4$), and monthly compounding ($n = 12$). Show all four cases on the same subplot and label each curve. On a second subplot, plot the difference between the amount obtained from continuous compounding and the other three cases.

b. Redo part a, but plot A versus t on log-log and semilog plots. Which plot gives a straight line?

25. Figure P25 is a representation of an electrical system with a power supply and a load. The power supply produces the fixed voltage v_1 and supplies the current i_1 required by the load, whose voltage drop is v_2. The current–voltage relationship for a specific load is found from experiments to be

$$i_1 = 0.16(e^{0.12v_2} - 1)$$

Suppose that the supply resistance is $R_1 = 30 \, \Omega$ and the supply voltage is $v_1 = 15$ V. To select or design an adequate power supply, we need to determine how much current will be drawn from the power supply when this load is attached. Find the voltage drop v_2 as well.

Figure P25

26. The circuit shown in Figure P26 consists of a resistor and a capacitor and is thus called an RC circuit. If we apply a sinusoidal voltage v_i, called the input voltage, to the circuit as shown, then eventually the output voltage v_o will be sinusoidal also, with the same frequency but with a different amplitude and shifted in time relative to the input voltage. Specifically, if $v_i = A_i \sin \omega t$, then $v_o = A_o \sin(\omega t + \phi)$. The frequency–response plot is a plot of A_o/A_i versus frequency ω. It is usually plotted on logarithmic axes. Upper-level engineering courses explain that for the RC circuit shown, this ratio depends on ω and RC as follows:

$$\frac{A_o}{A_i} = \left| \frac{1}{RCs + 1} \right|$$

where $s = \omega i$. For $RC = 0.1$ s, obtain the log-log plot of $|A_o/A_i|$ versus ω and use it to find the range of frequencies for which the output amplitude A_o is less than 70 percent of the input amplitude A_i.

Figure P26

27. An approximation to the function sin x is $\sin x \approx x - x^3/6$. Plot the sin x function and 20 evenly spaced error bars representing the error in the approximation.

28. Consider the following functions:

$$f(x) = 3x \cos^2 x - 2x$$

$$g(x) = -6x \cos x \sin x + 3 \cos^2 x - 2$$

Plot $f(x)$ and $g(x)$ on the same plot for x in the range of $[-2\pi, 2\pi]$. Label the axes and add a grid and legend. Use red with a solid line for $f(x)$ and blue with a dashed line for $g(x)$.

29. Create a polar plot of the following function for the range $0 \le \theta \le 2\pi$.

$$r = 4 \cos^2(0.6\theta) + \theta$$

30. Given the following function

$$y = 3^{(-0.5x + 15)}$$

Plot the function with a grid over the range $[0.1, 100]$ using four types of axes: linear-linear, linear-log, log-linear, and log-log. Do not use `subplot`.

31. Write a MATLAB script that allows the user to plot one of the following functions over the range $0 \le x \le 10$. Use the `input` command to enable the user to select which function to plot.

$$f_1(x) = \cos(x)$$
$$f_2(x) = \sin(x)$$
$$f_3(x) = -x^2 + 10x$$

32. Planets and planetary satellites move in elliptical orbits. A certain ellipse centered at the origin has the equation

$$x^2 + \frac{y^2}{4} = 1$$

Another ellipse, also centered at the origin, is rotated relative to the first ellipse. Its equation is

$$0.5833x^2 - 0.2887xy + 0.4167y^2 = 1$$

We want to find all points where the ellipses intersect. Use the `fimplicit` function and the `hold` command to plot both ellipses on the same plot. Since both ellipses are centered at the origin, if they intersect they will intersect at four points, so you will need to use the `ginput` function for four points.

Section 5.4

33. The popular amusement ride known as the corkscrew has a helical shape. The parametric equations for a circular helix are

$$x = a \cos(t)$$
$$y = a \sin(t)$$
$$z = bt$$

where a is the radius of the helical path and b is a constant that determines the "tightness" of the path. In addition, if $b > 0$, the helix has the shape of a right-handed screw; if $b < 0$, the helix is left-handed.

Obtain the three-dimensional plot of the helix for the following three cases and compare their appearance with one another. Use $0 \leq t \leq 10\pi$ and $a = 1$.

a. $b = 0.1$

b. $b = 0.2$

c. $b = -0.1$

34. A robot rotates about its base at 2 rpm while lowering its arm and extending its hand. It lowers its arm at the rate of $120°$ per minute and extends its hand at the rate of 5 m/min. The arm is 0.5 m long. The xyz coordinates of the hand are given by

$$x = \left(0.5 + 5t\right)\sin\left(\tfrac{2\pi}{3}t\right)\cos\left(4\pi t\right)$$
$$y = \left(0.5 + 5t\right)\sin\left(\tfrac{2\pi}{3}t\right)\sin\left(4\pi t\right)$$
$$z = \left(0.5 + 5t\right)\cos\left(\tfrac{2\pi}{3}t\right)$$

where t is time in minutes.

Obtain the three-dimensional plot of the path of the hand for $0 \leq t \leq 0.2$ min.

35. Obtain the surface and contour plots for the function $z = x^2 - 2xy + 4y^2$, showing the minimum at $x = y = 0$.

36. Obtain the surface and contour plots for the function $z = -x^2 + 2xy + 3y^2$. This surface has the shape of a saddle. At its saddlepoint at $x = y = 0$, the surface has zero slope, but this point does not correspond to either a minimum or a maximum. What type of contour lines corresponds to a saddlepoint?

37. Obtain the surface and contour plots for the function $z = (x - y^2)(x - 3y^2)$. This surface has a singular point at $x = y = 0$, where the surface has zero slope, but this point does not correspond to either a minimum or a maximum. What type of contour lines corresponds to a singular point?

38. A square metal plate is heated to $80°C$ at the corner corresponding to $x = y = 1$. The temperature distribution in the plate is described by

$$T = 80e^{-(x-1)^2}e^{-3(y-1)^2}$$

Obtain the surface and contour plots for the temperature. Label each axis. What is the temperature at the corner corresponding to $x = y = 0$?

39. The following function describes oscillations in some mechanical structures and electric circuits.

$$z(t) = e^{-t/\tau}\sin(\omega t + \phi)$$

In this function t is time, and ω is the oscillation frequency in radians per unit time. The oscillations have a period of $2\pi/\omega$, and their amplitudes decay in time at a rate determined by τ, which is called the *time constant*. The smaller τ is, the faster the oscillations die out.

Suppose that $\phi = 0$, $\omega = 2$, and τ can have values in the range $0.5 \le \tau \le 10$ sec. Then the preceding equation becomes

$$z(t) = e^{-t/\tau}\sin(2t)$$

Obtain a surface plot and a contour plot of this function to help visualize the effect of τ for $0 \le t \le 15$ sec. Let the x variable be time t and the y variable be τ.

40. The following equation describes the temperature distribution in a flat rectangular metal plate. The temperature on three sides is held constant at T_1, and at T_2 on the fourth side (see Figure P40). The temperature $T(x, y)$ as a function of the xy coordinates shown is given by

$$T(x, y) = (T_2 - T_1)w(x, y) + T_1$$

where

$$w(x, y) = \frac{2}{\pi}\sum_{n\ \text{odd}}^{\infty}\frac{2}{n}\sin\left(\frac{n\pi x}{L}\right)\frac{\sinh(n\pi y/L)}{\sinh(n\pi W/L)}$$

The given data for this problem are $T_1 = 70°F$, $T_2 = 200°F$, and $W = L = 2$ ft.

Using a spacing of 0.2 for both x and y, generate a surface mesh plot and a contour plot of the temperature distribution.

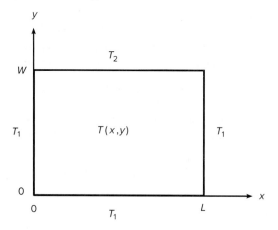

Figure P40

41. The electric potential field V at a point, due to two charged particles, is given by

$$V = \frac{1}{4\pi\epsilon_0}\left(\frac{q_1}{r_1} + \frac{q_2}{r_2}\right)$$

where q_1 and q_2 are the charges of the particles in coulombs (C), r_1 and r_2 are the distances of the charges from the point (in meters), and ϵ_0 is the permittivity of free space, whose value is

$$\epsilon_0 = 8.854 \times 10^{-12} \text{ C}^2/(\text{N} \cdot \text{m}^2)$$

Suppose the charges are $q_1 = 2 \times 10^{-10}$ C and $q_2 = 4 \times 10^{-10}$ C. Their respective locations in the xy plane are $(0.3, 0)$ and $(-0.3, 0)$ m. Plot the electric potential field on a three-dimensional surface plot with V plotted on the z axis over the ranges $-0.25 \le x \le 0.25$ and $-0.25 \le y \le 0.25$. Create the plot in two ways: (*a*) by using the surf function and (*b*) by using the meshc function.

42. Refer to Problem 26 of Chapter 4. Use the function file created for that problem to generate a surface mesh plot and a contour plot of x versus h and W for $0 \le W \le 500$ N and for $0 \le h \le 2$ m. Use the values $k_1 = 10^4$ N/m, $k_2 = 1.5 \times 10^4$ N/m, and $d = 0.1$ m.

43. Refer to Problem 29 of Chapter 4. To see how sensitive the cost is to location of the distribution center, obtain a surface plot and a contour plot of the total cost as a function of the x and y coordinates of the distribution center location. How much would the cost increase if we located the center 1 mi in any direction from the optimal location?

44. Refer to Example 3.2–1 in Chapter 3. Use a surface plot and a contour plot of the perimeter length L as a function of d and θ over the ranges $1 \le d \le 30$ ft and $0.1 \le \theta \le 1.5$ rad. Are there valleys other than the one corresponding to $d = 7.5984$ and $\theta = 1.0472$? Are there any saddle points?

45. The range of a projectile thrown with an initial velocity v at an angle A is given by

$$R = \frac{2v^2\cos A \sin A}{g}$$

Create a function range(v,A) that computes R, given A in degrees. Use the function to obtain surface plots with the functions mesh and meshc. For v on the interval [10, 25] with a spacing of 1 m/s, and A on the interval [5, 85] with a spacing of 1 degree.

46. Use the fimplicit3 function to create a surface plot of the function

$$x^2 + 30y^2 + 30z^2 = 120$$

Engineering in the 21st Century. . .

Virtual Prototyping

Virtual prototyping is a method of product development by which a design is validated before committing to making a physical prototype. It usually employs computer-aided design (CAD) software, computer-aided engineering (CAE) software, and simulation software, such as MATLAB and Simulink. The method is an extension of traditional design methods, but made more practical because of the power of modern computers and the improved accuracy of the software.

More than just computer-aided drafting, CAD and CAE include stress analysis on components and assemblies using finite element analysis (FEA), computational fluid dynamics (CFD) to calculate flow patterns and forces, multi-body dynamics, and optimization. Simulation is used to speed development, integration, and testing of microcontroller units. Engineers can use computers to determine the forces, voltages, currents, and so on that might occur in a proposed design. They can use this information to make sure the hardware can withstand the predicted forces or supply the required voltages or currents.

The normal stages in the development of a new vehicle, such as an aircraft, formerly consisted of aerodynamic testing a scale model; building a full-size wooden *mock-up* to check for pipe, cable, and structural interferences; and finally building and testing a *prototype*, the first complete vehicle.

Virtual prototyping is changing the traditional development cycle. The Boeing 777 is the first aircraft to be designed and built using virtual prototyping, without the extra time and expense of building a mock-up. The design teams responsible for the various subsystems, such as aerodynamics, structures, hydraulics, and electrical systems, all had access to the same computer database that described the aircraft. Thus when one team made a design change, the database was updated, allowing the other teams to see whether the change affected their subsystem. ■

CHAPTER 6

Model Building and Regression

OUTLINE

6.1   Function Discovery

6.2   Regression

6.3   The Basic Fitting Interface

6.4   Summary

Problems

An important application of the plotting techniques covered in Chapter 5 is *function discovery,* the technique for using data plots to obtain a mathematical function or "mathematical model" that describes the process that generated the data. This is the topic of Section 6.1. A systematic way of finding an equation that best fits the data is *regression* (also called the *least-squares method*). Regression is treated in Section 6.2. Section 6.3 introduces the MATLAB Basic Fitting interface, which supports regression.

6.1 Function Discovery

Function discovery is the process of finding, or "discovering," a function that can describe a particular set of data. The following three function types can often describe physical phenomena.

1. The *linear* function: $y(x) = mx + b$. Note that $y(0) = b$.
2. The *power* function: $y(x) = bx^m$. Note that $y(0) = 0$ if $m \geq 0$, and $y(0) = \infty$ if $m < 0$.

3. The *exponential* function: $y(x) = b(10)^{mx}$ or its equivalent form $y = be^{mx}$, where e is the base of the natural logarithm ($\ln e = 1$). Note that $y(0) = b$ for both forms.

Each function gives a straight line when plotted using a specific set of axes:

1. The linear function $y = mx + b$ gives a straight line when plotted on rectilinear axes. Its slope is m and its intercept is b.
2. The power function $y = bx^m$ gives a straight line when plotted on log-log axes.
3. The exponential function $y = b(10)^{mx}$ and its equivalent form $y = be^{mx}$ give a straight line when plotted on a semilog plot whose y axis is logarithmic.

We look for a straight line on the plot because it is relatively easy to recognize, and therefore we can easily tell whether the function will fit the data well.

Use the following procedure to find a function that describes a given set of data. We assume that one of the function types (linear, exponential, or power) can describe the data.

1. Examine the data near the origin. The exponential function can never pass through the origin (unless of course $b = 0$, which is a trivial case). (See Figure 6.1–1 for examples with $b = 1$.) The linear function can pass through the origin only if $b = 0$. The power function can pass through the origin but only if $m > 0$. (See Figure 6.1–2 for examples with $b = 1$.)

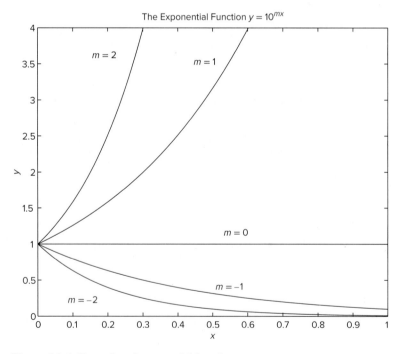

Figure 6.1–1 Examples of exponential functions.

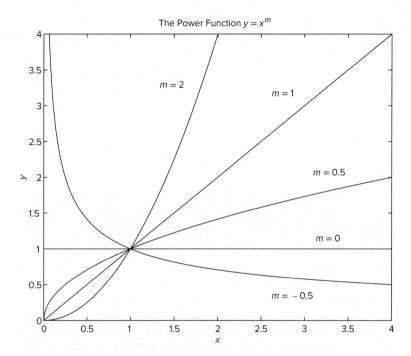

Figure 6.1–2 Examples of power functions.

2. Plot the data using rectilinear scales. If the data form a straight line, then the data can be represented by the linear function and you are finished. Otherwise, if you have data at $x = 0$, then

   a. If $y(0) = 0$, try the power function.

   b. If $y(0) \neq 0$, try the exponential function.

   If data are not given for $x = 0$, proceed to step 3.

3. If you suspect a power function, plot the data using log-log scales. Only a power function will form a straight line on a log-log plot. If you suspect an exponential function, plot the data using the semilog scales. Only an exponential function will form a straight line on a semilog plot.

4. In function discovery applications, we use the log-log and semilog plots *only* to identify the function type, but not to find the coefficients b and m. The reason is that it is difficult to interpolate on log scales.

We can find the values of b and m with the MATLAB `polyfit` function. This function finds the coefficients of a polynomial of specified degree n that best fits the data, in the so-called least-squares sense. The syntax appears in Table 6.1–1. The mathematical foundation of the least-squares method is presented in Section 6.2.

    Because we are assuming that our data will form a straight line on a rectilinear, semilog, or log-log plot, we are interested only in a polynomial that corresponds to a straight line, that is, a first-degree polynomial, which we will

Table 6.1–1 The `polyfit` function

Command	Description
`p = polyfit(x,y,n)`	Fits a polynomial of degree n to data described by the vectors x and y, where x is the independent variable. Returns a row vector p of length $n + 1$ that contains the polynomial coefficients in order of descending powers.

denote as $w = p_1 z + p_2$. Thus, referring to Table 6.1–1, we see that the vector p will be $[p_1, p_2]$ if n is 1. This polynomial has a different interpretation in each of the three cases:

- **The linear function:** $y = mx + b$. In this case the variables w and z in the polynomial $w = p_1 z + p_2$ are the original data variables x and y, and we can find the linear function that fits the data by typing `p = polyfit(x,y,1)`. The first element p_1 of the vector p will be m, and the second element p_2 will be b.

- **The power function:** $y = bx^m$. In this case $\log_{10} y = m \log_{10} x + \log_{10} b$, which has the form $w = p_1 z + p_2$, where the polynomial variables w and z are related to the original data variables x and y by $w = \log_{10} y$ and $z = \log_{10} x$. Thus we can find the power function that fits the data by typing `p = polyfit(log10(x),log10(y),1)`. The first element p_1 of the vector p will be m, and the second element p_2 will be $\log_{10} b$. We can find b from $b = 10^{p_2}$.

- **The exponential function:** $y = b(10)^{mx}$. In this case $\log_{10} y = mx + \log_{10} b$, which has the form $w = p_1 z + p_2$, where the polynomial variables w and z are related to the original data variables x and y by $w = \log_{10} y$ and $z = x$. Thus we can find the exponential function that fits the data by typing `p = polyfit(x, log10(y),1)`. The first element p_1 of the vector p will be m, and the second element p_2 will be $\log_{10} b$. We can find b from $b = 10^{p_2}$.

EXAMPLE 6.1–1   Speed Estimation from Sonar Measurements

Sonar measurements of the range of an approaching underwater vehicle are given in the following table, where the distance is measured in nautical miles (nmi). Assuming the relative speed v is constant, the range as function of time is given by $r = -vt + r_0$ where is r_0 the initial range at $t = 0$. Estimate the speed v and when the range will be zero.

Time, t (min)	0	2	4	6	8	10
Range, r (nmi)	3.8	3.5	2.7	2.1	1.2	0.7

■ **Solution**

The MATLAB program follows.

```
% Data
t = 0:2:10;
r = [3.8,3.5,2.7,2.1,1.2,0.7];
% First-order curve fit.
```

```
p = polyfit(t,r,1)
% Create plotting variable.
rp = p(1)*t+p(2);
plot(t,r,'o',t,rp),xlabel('t (min'),ylabel('r (nmi)')
% Speed calculation.
v = -p(1)*60 % speed in knots (nmi/hr)
p
```

Figure 6.1–3 shows the plot. The estimated relative speed is 0.3286 nmi/min, or 19.7 knots. The coefficients are $p(1) = -0.3286$ and $p(2) = 3.9762$, which is r_0. Thus the fitted equation is $r = -0.3286t + 3.9762$. From this we can estimate when the range will be zero: $t = 3.9762/0.3286 = 12.1$ minutes.

Figure 6.1–3 Range versus time: the sonar data and the fitted line.

Temperature Dynamics

The temperature of coffee cooling in a porcelain mug at room temperature (68°F) was measured at various times. The data follow.

Time t (sec)	Temperature T (°F)
0	145
620	130
2266	103
3482	90

Develop a model of the coffee's temperature as a function of time, and use the model to estimate how long it took the temperature to reach 120°F.

■ Solution

Because $T(0)$ is finite but nonzero, the power function cannot describe these data, so we do not bother to plot the data on log-log axes. Common sense tells us that the coffee will cool and its temperature will eventually equal the room temperature. So we subtract the room temperature from the data and plot the relative temperature, $T - 68$, versus time. If the relative temperature is a linear function of time, the model is $T - 68 = mt + b$. If the relative temperature is an exponential function of time, the model is $T - 68 = b(10)^{mt}$. Figure 6.1–4 shows the plots used to solve the problem. The following MATLAB script file generates the top two plots. The time data are entered in the array `time`, and the temperature data are entered in `temp`.

```
% Enter the data.
time = [0,620,2266,3482];
temp = [145,130,103,90];
% Subtract the room temperature.
temp = temp - 68;
% Plot the data on rectilinear scales.
```

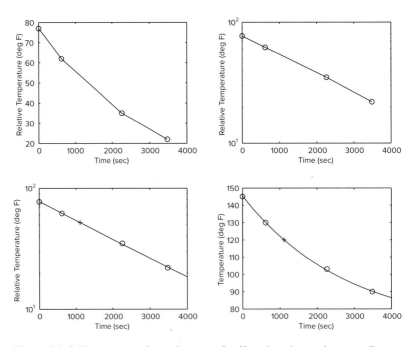

Figure 6.1–4 Temperature of a cooling cup of coffee, plotted on various coordinates.

```
subplot(2,2,1)
plot(time,temp,time,temp,'o'),xlabel('Time (sec)'),...
    ylabel('Relative Temperature (deg F)')
%
% Plot the data on semilog scales.
subplot(2,2,2)
semilogy(time,temp,time,temp,'o'),xlabel('Time (sec)'),...
    ylabel('Relative Temperature (deg F)')
```

The data form a straight line on the semilog plot only (the top right plot). Thus the data can be described with the exponential function $T = 68 + b(10)^{mt}$. Using the `polyfit` command, the following lines can be added to the script file.

```
% Fit a straight line to the transformed data.
p = polyfit(time,log10(temp),1);
m = p(1)
b = 10^p(2)
```

The computed values are $m = -1.5557 \times 10^{-4}$ and $b = 77.4469$. Thus our derived model is $T = 68 + b(10)^{mt}$. To estimate how long it will take for the coffee to cool to 120°F, we must solve the equation $120 = 68 + b(10)^{mt}$ for t. The solution is $t = [\log_{10}(120-68) - \log_{10}(b)]/m$. The MATLAB command for this calculation is shown in the following script file, which is a continuation of the previous script and produces the bottom two subplots shown in Figure 6.1–4.

```
% Compute the time to reach 120 degrees.
t_120 = (log10(120-68)-log10(b))/m
% Show derived curve and estimated point on semilog scales.
t = 0:10:4000;
T = 68+b*10.^(m*t);
subplot(2,2,3)
semilogy(t,T-68,time,temp,'o',t_120,120-68,'+'),
xlabel('Time (sec)'),...
    ylabel('Relative Temperature (deg F)')
%
% Show derived curve and estimated point on linear scales.
subplot(2,2,4)
plot(t,T,time,temp+68,'o',t_120,120,'+'),xlabel('Time (sec)'),...
    ylabel('Temperature (deg F)')
```

The computed value of `t_120` is 1112. Thus the time to reach 120°F is 1112 sec. The plot of the model, along with the data and the estimated point (1112, 120) marked with a + sign, is shown in the bottom two subplots in Figure 6.1–4. Because the graph of our model lies near the data points, we can treat its prediction of 1112 sec with some confidence.

Hydraulic Resistance

Figure 6.1–5 An experiment to verify Torricelli's principle.

A 15-cup coffee pot (see Figure 6.1–5) was placed under a water faucet and filled to the 15-cup line. With the outlet valve open, the faucet's flow rate was adjusted until the water level remained constant at 15 cups, and the time for one cup to flow out of the pot was measured. This experiment was repeated with the pot filled to the various levels shown in the following table:

Liquid volume V (cups)	Time to fill 1 cup t (sec)
15	6
12	7
9	8
6	9

(a) Use the preceding data to obtain a relation between the flow rate and the number of cups in the pot. (b) The manufacturer wants to make a 36-cup pot using the same outlet valve but is concerned that a cup will fill too quickly, causing spills. Extrapolate the relation developed in part (a) and predict how long it will take to fill one cup when the pot contains 36 cups.

■ **Solution**

(a) Torricelli's principle in hydraulics states that $f = rV^{1/2}$, where f is the flow rate through the outlet valve in cups per second, V is the volume of liquid in the pot in cups, and r is a constant whose value is to be found. We see that this relation is a power function where the exponent is 0.5. Thus if we plot $\log_{10}(f)$ versus $\log_{10}(V)$, we should obtain a straight line. The values for f are obtained from the reciprocals of the given data for t. That is, $f = 1/t$ cups per second.

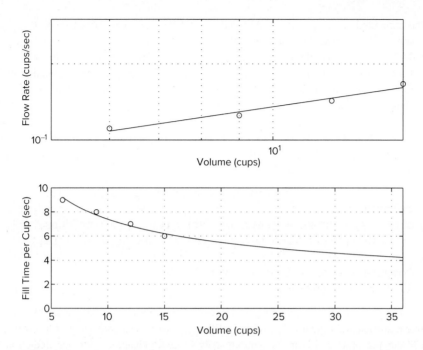

Figure 6.1–6 Flow rate and fill time for a coffee pot.

The MATLAB script file follows. The resulting plots appear in Figure 6.1–6. The volume data are entered in the array cups, and the time data are entered in meas_times.

```
% Data for the problem.
cups = [6,9,12,15];
meas_times = [9,8,7,6];
meas_flow = 1./meas_times;
%
% Fit a straight line to the transformed data.
p = polyfit(log10(cups),log10(meas_flow),1);
coeffs = [p(1),10^p(2)];
m = coeffs(1)
b = coeffs(2)
%
% Plot data and fitted line on a loglog plot to see
% how well the line fits the data.
x = 6:0.01:40;
y = b*x.^m;
subplot(2,1,1)
loglog(x,y,cups,meas_flow,'o'),grid,xlabel('Volume (cups)'),...
    ylabel('Flow Rate (cups/sec)'),axis([5 15 0.1 0.3])
```

The computed values are $m = 0.433$ and $b = 0.0499$, and our derived relation is $f = 0.0499V^{0.433}$. Because the exponent is 0.433, not 0.5, our model does not agree exactly with Torricelli's principle, but it is close. Note that the first plot in Figure 6.1–6 shows that the data points do not lie exactly on the fitted straight line. In this application it is difficult to measure the time to fill one cup with an accuracy greater than an integer second, so this inaccuracy could have caused our result to disagree with that predicted by Torricelli.

(b) Note that the fill time is $1/f$, the reciprocal of the flow rate. The remainder of the MATLAB script uses the derived flow rate relation $f = 0.0499V^{0.433}$ to plot the extrapolated fill-time curve $1/f$ versus t.

```
% Plot the fill time curve extrapolated to 36 cups.
subplot(2,1,2)
plot(x,1./y,cups,meas_times,'o'),grid,xlabel('Volume(cups)'),...
    ylabel('Fill Time per Cup (sec)'),axis([5 36 0 10])
%
% Compute the fill time for V = 36 cups.
fill_time = 1/(b*36^m)
```

The predicted fill time for 1 cup is 4.2 sec. The manufacturer must now decide if this time is sufficient for the user to avoid overfilling. (In fact, the manufacturer did construct a 36-cup pot, and the fill time is approximately 4 sec, which agrees with our prediction.)

EXAMPLE 6.1–4   A Cantilever Beam Model

The deflection of a cantilever beam is the distance its end moves in response to a vertical force applied at the end (see Figure 6.1–7). The following table gives the measured deflection x that was produced in a particular beam by the given applied force f. Is there a set of axes (rectilinear, semilog, or log-log) on which the data forms an approximate straight line? If so, use that information to obtain a functional relation between f and x.

Force f (lb)	0	100	200	300	400	500	600	700	800
Deflection x (in.)	0	0.15	0.23	0.35	0.37	0.5	0.57	0.68	0.77

■ **Solution**
The following MATLAB script file generates two plots on rectilinear axes. The data is entered in the arrays `deflection` and `force`.

```
% Enter the data.
force = 0:100:800;
deflection=[0,0.15,0.23,0.35,0.37,0.5,0.57,0.68,0.77];
%
% Plot the data on rectilinear scales.
subplot(2,1,1)
```

Figure 6.1–7 Measurement of beam deflection.

```
plot(deflection,force,'o'),...
    xlabel('Deflection (in.)'),ylabel('Force(lb)'),...
    axis([0 0.8 0 800])
```

The plot appears as the first plot of Figure 6.1–8. The points appear to lie on a straight line that can be described by an equation that is usually written as $f = kx + c$,

Figure 6.1–8 Plots for the cantilever beam example.

where k is called the beam's *spring constant*. We can determine the value of k by using the `polyfit` command as shown in the following script file, which is a continuation of the preceding script.

```
% Fit a straight line to the data.
p = polyfit(deflection, force,1);
% Here k = p(1) and c = p(2).
k = p(1)
c = p(2)
% Plot the fitted line and the data.
x = deflection;
f = k*x+c;
subplot(2,1,2)
plot(x,f,deflection,force,'o'),...
    xlabel('Deflection (in.)'),ylabel('Force (lb)'),...
    axis([0 0.8 0 800])
```

The plot appears as the second plot of Figure 6.1–8. The computed values are $k = 1082$ lb/in. and $c = -34.6592$ lb.

Many applications require a model whose form is dictated by physical principles. For example, the force-extension model of a spring must pass through the origin (0, 0) because the spring exerts no force when it is not stretched or compressed. Thus the linear spring model should be $f = kx$ where $c = 0$. In Section 6.2 we present a method for finding the spring constant k for a straight-line model that passes through the origin.

6.2 Regression

In Section 6.1 we used the MATLAB function `polyfit` to perform regression analysis with functions that are linear or could be converted to linear form by a logarithmic or other transformation. The `polyfit` function is based on the least-squares method, which is also called *regression*. We now show how to use this function to develop polynomial and other types of functions.

The Least-Squares Method

Suppose we have the three data points given in the following table, and we need to determine the coefficients of the straight line $y = mx + b$ that best fit the following data in the least-squares sense.

x	y
0	2
5	6
10	11

According to the least-squares criterion, the line that gives the best fit is the one that minimizes J, the sum of the squares of the vertical differences between the line and the data points. These differences are called the *residuals*. Here there are three data points, and J is given by

$$J = \sum_{i=1}^{3} (mx_i + b - y_i)^2$$
$$= (0m + b - 2)^2 + (5m + b - 6)^2 + (10m + b - 11)^2$$

The values of m and b that minimize J are found by setting the partial derivatives $\partial J/\partial m$ and $\partial J/\partial b$ equal to zero.

$$\frac{\partial J}{\partial m} = 250m + 30b - 280 = 0$$

$$\frac{\partial J}{\partial b} = 30m + 6b - 38 = 0$$

These conditions give two equations that must be solved for the two unknowns m and b. The solution is $m = 0.9$ and $b = 11/6$. The best straight line in the least-squares sense is $y = 0.9x + 11/6$. If we evaluate this equation at the data values $x = 0, 5$, and 10, we obtain the values $y = 1.833, 6.333, 10.8333$. These values are different from the given data values $y = 2, 6, 11$ because the line is not a perfect fit to the data. The value of J is $J = (1.833 - 2)^2 + (6.333 - 6)^2 + (10.8333 - 11)^2 = 0.16656689$. No other straight line will give a lower value of J for these data.

In general, for the polynomial $a_1x^n + a_2x^{n-1} + \cdots + a_nx + a_{n+1}$, the sum of the squares of the residuals for m data points is

$$J = \sum_{i=1}^{m} (a_1x^n + a_2x^{n-1} + \cdots + a_nx + a_{n+1} - y_i)^2$$

The values of the $n + 1$ coefficients a_i that minimize J can be found by solving a set of $n + 1$ linear equations. The `polyfit` function provides this solution. Its syntax is `p = polyfit(x,y,n)`. Table 6.2–1 summarizes the `polyfit` and `polyval` functions.

Table 6.2–1 Functions for polynomial regression

Command	Description
`p = polyfit(x,y,n)`	Fits a polynomial of degree `n` to data described by the vectors `x` and `y`, where `x` is the independent variable. Returns a row vector `p` of length `n+1` that contains the polynomial coefficients in order of descending powers.
`[p,s,mu] = polyfit(x,y,n)`	Fits a polynomial of degree `n` to data described by the vectors `x` and `y`, where `x` is the independent variable. Returns a row vector `p` of length `n+1` that contains the polynomial coefficients in order of descending powers and a structure `s` for use with `polyval` to obtain error estimates for predictions. The optional output variable `mu` is a two-element vector containing the mean and standard deviation of `x`.
`[y,delta] = polyval(p,x,s,mu)`	Uses the optional output structure s generated by `[p,s,mu] = polyfit(x,y,n)` to generate error estimates. If the errors in the data used with `polyfit` are independent and normally distributed with constant variance, at least 50 percent of the data will lie within the band `y ± delta`.

EXAMPLE 6.2–1

Effect of Polynomial Degree

Consider the data set where $x = 1, 2, 3, \ldots, 9$ and $y = 5, 6, 10, 20, 28, 33, 34, 36, 42$. Fit polynomials of first through fourth degree to this data and compare the results.

■ **Solution**

The following script file computes the coefficients of the first- through fourth-degree polynomials for these data and evaluates J for each polynomial.

```
x = 1:9;
y = [5,6,10,20,28,33,34,36,42];
for k = 1:4
   coeff = polyfit(x,y,k)
   J(k) = sum ((polyval(coeff,x)-y).^2)
end
```

The J values are, to two significant figures, 72, 57, 42, and 4.7. Thus the value of J decreases as the polynomial degree is increased, as we would expect. Figure 6.2–1 shows this data and the four polynomials. Note now the fit improves with the higher-degree polynomial.

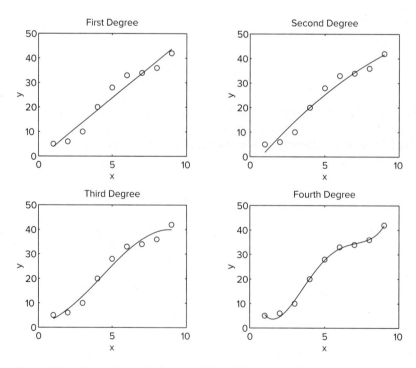

Figure 6.2–1 Regression using polynomials of first through fourth degree.

Caution: It is tempting to use a high-degree polynomial to obtain the best possible fit. However, there are two dangers in using high-degree polynomials. High-degree polynomials often exhibit large excursions between the data points and thus should be avoided if possible. Figure 6.2–2 shows an example of this phenomenon. The second danger with using high-degree polynomials is that they can produce large errors if their coefficients are not represented with a large number of significant figures. In some cases it might not be possible to fit the data with a low-degree polynomial. In such cases we might be able to use several cubic polynomials. This method, called *cubic splines*, is covered in Chapter 7.

Test Your Understanding

T6.2–1 Obtain and plot the first- through fourth-degree polynomials for the following data: $x = 0, 1, \ldots, 5$ and $y = 0, 1, 60, 40, 41,$ and 47. Find the coefficients and the J values.

(Answer: The polynomials are $9.5714x + 7.5714$; $-3.6964x^2 + 28.0536x - 4.7500$; $0.3241x^3 - 6.1270x^2 + 32.4934x - 5.7222$; and $2.5208x^4 - 24.8843x^3 + 71.2986x^2 - 39.5304x - 1.4008$. The corresponding J values are 1534, 1024, 1017, and 495, respectively.)

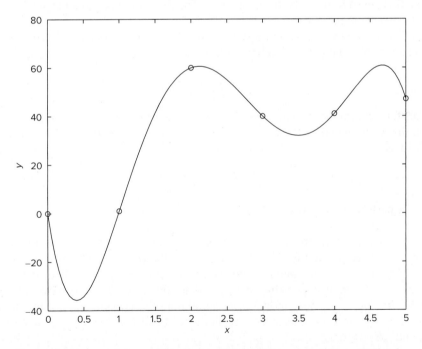

Figure 6.2–2 An example of a fifth-degree polynomial that passes through all six data points but exhibits large excursions between points.

Fitting Other Functions

Given the data (y, z), the logarithmic function $y = m \ln z + b$ can be converted to a first-degree polynomial by transforming the z values into x values by the transformation $x = \ln z$. The resulting function is $y = mx + b$.

Given the data (y, z), the function $y = b(10)^{m/z}$ can be converted to an exponential function by transforming the z values by the transformation $x = 1/z$.

Given the data (v, x), the function $v = 1/(mx + b)$ can be converted to a first-degree polynomial by transforming the v data values with the transformation $y = 1/v$. The resulting function is $y = mx + b$.

To see how to obtain a function $y = kx$ that passes through the origin, see Problem 8.

The Quality of a Curve Fit

The least-squares criterion used to fit a function $f(x)$ is the sum of the squares of the residuals J. It is defined as

$$J = \sum_{i=1}^{m} [f(x_i) - y_i]^2 \tag{6.2-1}$$

We can use the J value to compare the quality of the curve fit for two or more functions used to describe the same data. The function that gives the smallest J value gives the best fit.

We denote the sum of the squares of the deviation of the y values from their mean $\bar{y}$ by S, which can be computed from

$$S = \sum_{i=1}^{m} (y_i - \bar{y})^2 \tag{6.2-2}$$

COEFFICIENT OF DETERMINATION

This formula can be used to compute another measure of the quality of the curve fit, the *coefficient of determination*, also known as the *r-squared value*. It is defined as

$$r^2 = 1 - \frac{J}{S} \tag{6.2-3}$$

For a perfect fit, $J = 0$ and thus $r^2 = 1$. Thus the closer r^2 is to 1, the better the fit. The largest r^2 can be is 1. The value of S indicates how much the data is spread around the mean, and the value of J indicates how much of the data spread is unaccounted for by the model. Thus the ratio J/S indicates the fractional variation unaccounted for by the model. It is possible for J to be larger than S, and thus it is possible for r^2 to be negative. Such cases, however, are indicative of a very poor model that should not be used. As a rule of thumb, a good fit accounts for at least 99 percent of the data variation. This value corresponds to $r^2 \geq 0.99$.

For example, the following table gives the values of J, S, and r^2 for the first-through fourth-degree polynomials used to fit the data $x = 1, 2, 3, \ldots, 9$ and $y = 5, 6, 10, 20, 28, 33, 34, 36, 42$.

Degree n	J	S	r^2
1	72	1562	0.9542
2	57	1562	0.9637
3	42	1562	0.9732
4	4.7	1562	0.9970

Because the fourth-degree polynomial has the largest r^2 value, it represents the data better than the representation from first- through third-degree polynomials, according to the r^2 criterion.

The values of S and r^2 were calculated by adding the following lines to the end of the script file shown in Example 6.2–1.

```
mu = mean(y);
for k=1:4
   S(k) = sum((y-mu).^2);
   r2(k) = 1 - J(k)/S(k);
end
S
r2
```

Scaling the Data

The effect of computational errors in computing the coefficients can be lessened by properly scaling the x values. When the function `polyfit(x,y,n)` is executed, it will issue a warning message if the polynomial degree n is greater than or equal to the number of data points (because there will not be enough equations for MATLAB to solve for the coefficients), or if the vector x has repeated, or nearly repeated, points, or if the vector x needs centering and/or scaling. The alternate syntax `[p, s, mu] = polyfit(x,y,n)` finds the coefficients p of a polynomial of degree n in terms of the variable

$$\hat{x} = (x - \mu_x)/\sigma_x$$

The output variable mu is a two-element vector $[\mu_x, \sigma_x]$, where μ_x is the mean of x and σ_x is the standard deviation of x (the standard deviation is discussed in Chapter 7).

You can scale the data yourself before using `polyfit`. Some common scaling methods are

$$\hat{x} = x - x_{min} \quad \text{or} \quad \hat{x} = x - \mu_x$$

if the range of x is small, or

$$\hat{x} = \frac{x}{x_{max}} \quad \text{or} \quad \hat{x} = \frac{x}{x_{mean}}$$

if the range of x is large.

EXAMPLE 6.2–2

The following data give the number of vehicles (in millions) crossing a bridge each year for 10 years. Fit a cubic polynomial to the data and use the fit to estimate the flow in the year 2010.

Year	2000	2001	2002	2003	2004	2005	2006	2007	2008	2009
Vehicle flow (millions)	2.1	3.4	4.5	5.3	6.2	6.6	6.8	7	7.4	7.8

■ **Solution**

If we attempt to fit a cubic to these data, as in the following session, we get a warning message.

```
>>Year = 2000:2009;
>>Veh_Flow = [2.1,3.4,4.5,5.3,6.2,6.6,6.8,7,7.4,7.8];
>>p = polyfit(Year,Veh_Flow,3)
Warning: Polynomial is badly conditioned.
```

The problem is caused by the large values of the independent variable `Year`. Because their range is small, we can simply subtract 2000 from each value. Continue the session as follows.

```
>>x = Year-2000; y = Veh_Flow;
>>p = polyfit(x,y,3)
p =
   0.0087 -0.1851 1.5991 2.0362
>>J = sum((polyval(p,x)-y).^2);
>>S = sum((y-mean(y)).^2);
>>r2 = 1 - J/S
r2 =
   0.9972
```

Thus the polynomial fit is good because the coefficient of determination is 0.9972. The corresponding polynomial is

$$f = 0.0087(t - 2000)^3 - 0.1851(t - 2000)^2 + 1.5991(t - 2000) + 2.0362$$

where f is the traffic flow in millions of vehicles and t is the time in years measured from 0. We can use this equation to estimate the flow at the year 2010 by substituting $t = 2010$, or by typing in MATLAB `polyval(p,10)`. Rounded to one decimal place, the answer is 8.2 million vehicles.

Test Your Understanding

T6.2–2 The U.S. census data from 1790 to 1990 is stored in the file `census. dat`, which is supplied with MATLAB. Type `load census` to load this file. The first column, cdate, contains the years, and the second column, pop, contains the population in millions. First try to fit a cubic polynomial to the data. If you get a warning message, scale the data by subtracting 1790 from the years, and fit a cubic. Compute the correlation coefficient and interpolate to estimate the population in 1965.

(Answer:

$$y = 3.8550 \times 10^{-6} x^3 + 5.3845 \times 10^{-3} x^2 - 2.2203 \times 10^{-3} x + 4.2644$$

where $x =$ cdate $- 1790$. The correlation coefficient is $r^2 = 0.9988$. Estimated population in 1965 is 189 million.)

Using Residuals

We now show how to use the residuals as a guide to choosing an appropriate function to describe the data. In general, if you see a pattern in the plot of the residuals, it indicates that another function can be found to describe the data better.

Modeling Bacteria Growth   **EXAMPLE 6.2–3**

The following table gives data on the growth of a certain bacteria population with time. Fit an equation to these data.

Time (min)	Bacteria (ppm)	Time (min)	Bacteria (ppm)
0	6	10	350
1	13	11	440
2	23	12	557
3	33	13	685
4	54	14	815
5	83	15	990
6	118	16	1170
7	156	17	1350
8	210	18	1575
9	282	19	1830

■ **Solution**
We try three polynomial fits (linear, quadratic, and cubic) and an exponential fit. The script file is given below. Note that we can write the exponential form as $y = b(10)^{mt} = 10^{mt+a}$, where $b = 10^a$.

```
% Time data
x = 0:19;
% Population data
y = [6,13,23,33,54,83,118,156,210,282,...
    350,440,557,685,815,990,1170,1350,1575,1830];
```

```
% Linear fit
p1 = polyfit(x,y,1);
% Quadratic fit
p2 = polyfit(x,y,2);
% Cubic fit
p3 = polyfit(x,y,3);
% Exponential fit
p4 = polyfit(x,log10(y),1);
% Residuals
res1 = polyval(p1,x) - y;
res2 = polyval(p2,x) - y;
res3 = polyval(p3,x) - y;
res4 = 10.^polyval(p4,x) - y;
```

You can then plot the residuals as shown in Figure 6.2–3. Pay attention to the *magnitudes* of the residuals. The magnitudes of the cubic are by far the smallest. Note that there is a definite pattern in the residuals of the linear fit. This indicates that the linear function cannot match the curvature of the data. The residuals of the quadratic fit are much smaller, but there is still a pattern, with a random component. This indicates that the quadratic function also cannot match the curvature of the data. The residuals of the cubic fit are even smaller, with no strong pattern and a large random component. This indicates that a polynomial degree higher than 3 will not be able to match the data curvature any better than the cubic. The residuals for the exponential are the largest of all, and indicate a poor

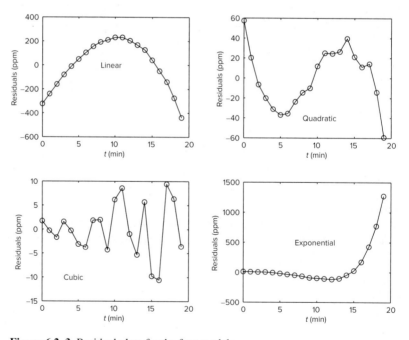

Figure 6.2–3 Residual plots for the four models.

fit. Note also how the residuals systematically increase with t, indicating that the exponential cannot describe the data's behavior after a certain time.

Thus the cubic is the best fit of the four models considered. Its coefficient of determination is $r^2 = 0.9999$. The model is

$$y = 0.1916t^3 + 1.2082t^2 + 3.607t + 7.7307$$

where y is the bacteria population in ppm and t is time in minutes.

Test Your Understanding

T6.2–3 Refer to T6.2–2. Using the scaled data, try three polynomial fits (linear, quadratic, and cubic), and an exponential fit. Then plot the residuals, and decide which is the better fit.

Multiple Linear Regression

Suppose that y is a linear function of two or more variables $x_1, x_2, \ldots$, for example, $y = a_0 + a_1 x_1 + a_2 x_2$. To find the coefficient values a_0, a_1, and a_2 to fit a set of data (y, x_1, x_2) in the least-squares sense, we can make use of the fact that the left-division method for solving linear equations uses the least-squares method when the equation set is overdetermined. To use this method, let n be the number of data points and write the linear equation in matrix form as $\mathbf{Xa} = \mathbf{y}$, where

$$\mathbf{a} = \begin{bmatrix} a_0 \\ a_1 \\ a_2 \end{bmatrix} \qquad \mathbf{X} = \begin{bmatrix} 1 & x_{11} & x_{21} \\ 1 & x_{12} & x_{22} \\ 1 & x_{13} & x_{23} \\ \cdots & \cdots & \cdots \\ 1 & x_{1n} & x_{2n} \end{bmatrix} \qquad \mathbf{y} = \begin{bmatrix} y_1 \\ y_2 \\ y_3 \\ \cdots \\ y_n \end{bmatrix}$$

where x_{1i}, x_{2i}, and y_i are the data, $i = 1, \ldots, n$. The solution for the coefficients is given by a = X\y.

Breaking Strength and Alloy Composition   EXAMPLE 6.2–4

We want to predict the strength of metal parts as a function of their alloy composition. The tension force y required to break a steel bar is a function of the percentage x_1 and x_2 of each of two alloying elements present in the metal. The following table gives some pertinent data. Obtain a linear model $y = a_0 + a_1 x_1 + a_2 x_2$ to describe the relationship.

Breaking strength (kN) y	% of element 1 x_1	% of element 2 x_2
7.1	0	5
19.2	1	7
31	2	8
45	3	11

■ **Solution**

The script file is as follows:

```
x1 = (0:3)';x2 = [5,7,8,11]';
y = [7.1,19.2,31,45]';
X = [ones(size(x1)), x1, x2];
a = X\y
yp = X*a;
Max_Percent_Error = 100*max(abs((yp-y)./y))
```

The vector `yp` is the vector of breaking strength values predicted by the model. The scalar `Max_Percent_Error` is the maximum percent error in the four predictions. The results are a = `[0.8000, 10.2429, 1.2143]'` and `Max_Percent_Error` = `3.2193`. Thus the model is $y = 0.8 + 10.2429x_1 + 1.2143x_2$. The maximum percent error of the model's predictions, as compared to the given data, is 3.2193 percent.

Test Your Understanding

T6.2–4 Obtain a linear model $y = a_0 + a_1x_1 + a_2x_2$ for the following data to describe the relationship.

y	x_1	x_2
3.8	7.5	6
5.6	12	9
6	13.5	10.5
5	16.5	18
5.8	19.5	21
5.6	21	25.5

(Answer:

$$y = 1.3153 + 0.6043\,x_1 - 0.3386\,x_2$$

Maximum percent error = 4.1058%. Maximum error = 0.2299.)

Linear-in-Parameters Regression

Sometimes we want to fit an expression that is neither a polynomial nor a function that can be converted to linear form by a logarithmic or other transformation. In some cases we can still do a least-squares fit if the function is a linear expression in terms of its parameters. The following example illustrates the method.

Response of a Biomedical Instrument

Engineers developing instrumentation often need to obtain a *response* curve that describes how fast the instrument can make measurements. The theory of instrumentation shows that often the response can be described by one of the following equations, where v is the voltage output and t is time. In both models, the voltage reaches a steady-state constant value as $t \to \infty$, and T is the time required for the voltage to equal 95 percent of the steady-state value.

$$v(t) = a_1 + a_2 e^{-3t/T} \qquad\qquad \text{(first-order model)}$$

$$v(t) = a_1 + a_2 e^{-3t/T} + a_2 t e^{-3t/T} \qquad\qquad \text{(second-order model)}$$

The following data give the output voltage of a certain device as a function of time. Obtain a function that describes these data.

t(s)	0	0.3	0.8	1.1	1.6	2.3	3
v(V)	0	0.6	1.28	1.5	1.7	1.75	1.8

■ **Solution**
Plotting the data, we estimate that it takes approximately 3 s for the voltage to become constant. Thus we estimate that $T = 3$. The first-order model written for each of the n data points results in n equations, which can be expressed as follows:

$$\begin{bmatrix} 1 & e^{-t_1} \\ 1 & e^{-t_2} \\ \cdots & \cdots \\ 1 & e^{-t_n} \end{bmatrix} \begin{bmatrix} a_1 \\ a_2 \end{bmatrix} = \begin{bmatrix} y_1 \\ y_2 \\ \cdots \\ y_n \end{bmatrix}$$

or, in matrix form,

$$\mathbf{Xa = y'}$$

which can be solved for the coefficient vector $\mathbf{a}$ using left division. The following MATLAB script solves the problem.

```
t = [0,0.3,0.8,1.1,1.6,2.3,3];
y = [0,0.6,1.28,1.5,1.7,1.75,1.8];
X = [ones(size(t));exp(-t)]';
a = X\y'
```

The answer is $a_1 = 2.0258$ and $a_2 = -1.9307$.

A similar procedure can be followed for the second-order model.

$$\begin{bmatrix} 1 & e^{-t_1} & t_1 e^{-t_1} \\ 1 & e^{-t_2} & t_2 e^{-t_2} \\ \cdots & \cdots & \cdots \\ 1 & e^{-t_n} & t_n e^{-t_n} \end{bmatrix} \begin{bmatrix} a_1 \\ a_2 \\ a_3 \end{bmatrix} = \begin{bmatrix} y_1 \\ y_2 \\ \cdots \\ y_n \end{bmatrix}$$

Continue the previous script as follows:

```
X = [ones(size(t));exp(-t);t.*exp(-t)]';
a = X\y'
```

The answer is $a_1 = 1.7496$, $a_2 = -1.7682$, and $a_3 = 0.8885$. The two models are plotted with the data in Figure 6.2–4. Clearly the second-order model gives the better fit.

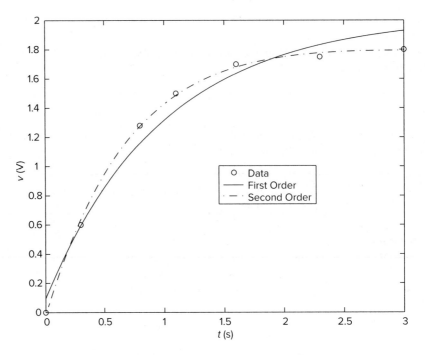

Figure 6.2–4 Comparison of first- and second-order model fits.

Test Your Understanding

T6.2–5 A mass attached to a spring and damper is displaced a distance x_0 (cm) while being given an initial velocity v_0 (cm/s). We know from physics and mathematics (see Chapter 8) that the displacement x as a function of time is given by

$$x(t) = \left(\frac{5x_0}{3} + \frac{v_0}{3}\right)e^{-2t} - \left(\frac{2x_0 + v_0}{3}\right)e^{-5t}$$

The displacement is measured every 0.2 s. The measured displacement versus time is given by

t (s)	0	0.2	0.4	0.6	0.8	1	1.2	1.4	1.6	1.8	2
x (cm)	1.9	2.1	1.7	1.2	0.9	0.6	0.4	0.3	0.2	0.1	0.1

Estimate the initial displacement and velocity.

(Answer: $x_0 = 1.9044$, $v_0 = 4.2090$.)

Constraining the Curve to Pass Through a Specified Point

Consider the cantilever beam shown in Figure 6.1–7. The deflection x of the beam is the distance its end moves in response to a force f applied at the end. Common sense tells us that there must be zero beam deflection if there is no applied force, so the equation describing the data must pass through the origin. So if the data are linearly related, the relation must be of the form $f = kx$. In this form the constant k is called the *spring constant* or *elastic constant*.

In general a linear model $y = mx + b$ sometimes must have a zero value for b. However, in general the least-squares method will give a nonzero value for b because of the scatter or measurement error that is usually present in the data. Then we cannot use the function p = polyfit(x,y,1) because in general p(2) will not be zero.

To obtain a zero-intercept model of the form $y = mx$, we take advantage of the fact that the *right*-division method uses the least-squares method to obtain a solution to a set of equations that contains more equations than unknowns. Such an equation set is said to be *overdetermined*. Solving overdetermined sets is the subject of Section 8.4.

The following program illustrates this method as applied to the cantilever beam data given in Example 6.1–4. The ten data points represent ten equations, with one unknown, k. The desired form of the fitted equation is $f = kx$, so the scalar k is found from $k = f/x$. If the data on f and x are stored as row vectors, then in vector form this equation must be written using right division as k = x'\f'. The program follows.

```
% Deflection and force data.
x = [0,0.15,0.23,0.35,0.37,0.5,0.57,0.68,0.77];
f=0:100:800;
k=x'\f'
```

The result is $k = 1017$ lb/in.

Suppose the model is required to pass through a point not at the origin, say the point (x_0, y_0) *and* that point is known to be an *exact* solution to the equation, so that $y_0 = mx_0 + b$. In that case, simply subtract x_0 from all the x values, subtract y_0 from all the y values. Specifically, let $u = x - x_0$ and $w = y - y_0$. The resulting equation will be of the form $w = mu$, and the coefficient m can be calculated using right division. In MATLAB we would write m = u'\w' where u and w are *row* vectors containing the transformed data.

6.3 The Basic Fitting Interface

MATLAB supports curve fitting through the Basic Fitting interface. Using this interface, you can quickly perform basic curve-fitting tasks within the same easy-to-use environment. The interface is designed so that you can do the following:

- Fit data using a cubic spline or a polynomial up to degree 10.
- Plot multiple fits simultaneously for a given data set.
- Plot the residuals.

■ Examine the numerical results of a fit.
■ Interpolate or extrapolate a fit.
■ Annotate the plot with the numerical fit results and the norm of residuals.
■ Save the fit and evaluated results to the MATLAB workspace.

Depending on your specific curve-fitting application, you can use the Basic Fitting interface, the command line functions, or both. *Note*: You can use the Basic Fitting interface only with two-dimensional data. However, if you plot multiple data sets as a subplot, and at least one data set is two-dimensional, then the interface is enabled.

Two panes of the Basic Fitting interface are shown in Figure 6.3–1. To reproduce this state:

1. Plot some data.
2. Select **Basic Fitting** from the Tools menu of the Figure window.
3. When the first pane of the Basic Fitting interface appears, click the right arrow button once.

Figure 6.3–1 The Basic Fitting interface.
Source: MATLAB

The third pane is used for interpolating or extrapolating a fit. It appears when you click the right arrow button a second time.

At the top of the first pane is the Select data window which contains the names of all the data sets you display in the Figure window associated with the Basic Fitting interface. Use this menu to select the data set to be fit. You can perform multiple fits for the current data set. Use the Plot Editor to change the name of a data set. The remaining items on the first pane are used as follows.

■ **Center and scale X data.** If checked, the data are centered at zero mean and scaled to unit standard deviation. You may need to center and scale your data to improve the accuracy of the subsequent numerical computations. As described in the previous section, a warning is returned to the Command window if a fit produces results that might be inaccurate.

■ **Plot fits.** This panel allows you to visually explore one or more fits to the current data set.

■ **Check to display fits on figure.** Select the fits you want to display for the current data set. You can choose as many fits for a given data set as you want. However, if your data set has n points, then you should use polynomials with at most n coefficients. If you fit using polynomials with more than n coefficients, the interface will automatically set a sufficient number of coefficients to zero during the calculation so that a solution can be obtained.

■ **Show equations.** If checked, the fit equation is displayed on the plot.

■ **Significant digits.** Select the significant digits associated with the fit coefficient display.

■ **Plot residuals.** If checked, the residuals are displayed. You can display the residuals as a bar plot, a scatter plot, a line plot using either the same figure window as the data or using a separate figure window. If you plot multiple data sets as a subplot, then residuals can be plotted only in a separate figure window. See Figure 6.3–2.

■ **Show norm of residuals.** If checked, the norm of residuals is displayed. The norm of residuals is a measure of the goodness of fit, where a smaller value indicates a better fit. The norm is the square root of the sum of the squares of the residuals.

The second pane of the Basic Fitting interface is labeled Numerical Results. This pane enables you to explore the numerical results of a single fit to the current data set without plotting the fit. It contains three items.

■ **Fit.** Use this menu to select an equation to fit to the current data set. The fit results are displayed in the box below the menu. Note that selecting an equation in this menu does not affect the state of the Plot fits selection. Therefore, if you want to display the fit in the data plot, you might need to check the relevant check box in Plot fits.

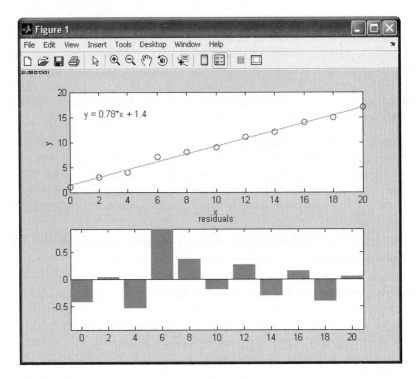

Figure 6.3–2 A figure produced by the Basic Fitting interface.
Source: MATLAB

- **Coefficients and norm of residuals.** Displays the numerical results for the equation selected in Fit. Note that when you first open the Numerical Results panel, the results of the last fit you selected in Plot fits are displayed.
- **Save to workspace.** Launches a dialog box that allows you to save the fit results to workspace variables.

The third pane of the Basic Fitting interface contains three items.

- **Find $Y = f(X)$.** Use this to interpolate or extrapolate the current fit. Enter a scalar or a vector of values corresponding to the independent variable (X). The current fit is evaluated after you click on the **Evaluate** button, and the results are displayed in the associated window. The current fit is displayed in the Fit window.
- **Save to workspace.** Launches a dialog box that allows you to save the evaluated results to workspace variables.
- **Plot evaluated results.** If checked, the evaluated results are displayed on the data plot.

Test Your Understanding

T6.3–1 The U.S. census data from 1790 to 1990 is stored in the file census. dat, which is supplied with MATLAB. Type load(census) to load this file. The first column, cdate, contains the years, and the second column, pop, contains the population in millions. Use the Basic Fitting interface to solve this problem. First try to fit a cubic polynomial to the data. If you get a warning message, center and scale the data by checking **Center and scale x data** in the Interface, and fit a cubic. Use the interface to interpolate to estimate the population in 1965.

(Answer:

$$y = 0.921 z^3 + 25.183 z^2 + 73.86z + 61.744$$

where $z = (cdate - 1890)/62.048$. Estimated population in 1965 is 189 million.)

6.4 Summary

In this chapter you learned an important application of plotting—function discovery—which is the technique for using data plots to obtain a mathematical function that describes the data. Regression can be used to develop a model for cases where there is considerable scatter in the data.

Many physical processes can be modeled with functions that produce a straight line when plotted using a suitable set of axes. In some cases, we can find a transformation that produces a straight line in the transformed variable.

When such a function or transformation cannot be found, we resort to polynomial regression, multiple linear regression, or linear-in-parameters regression to obtain an approximate functional description of the data. The MATLAB Basic Fitting interface is a powerful aid in obtaining regression models.

Key Terms

Coefficient of determination,   288
Linear-in-parameters regression,   294
Multiple linear regression,   293

Regression,   284
Residuals,   285

Problems

You can find the answers to problems marked with an asterisk at the end of the text.

Section 6.1

1. The distance a spring stretches from its "free length" is a function of how much tension force is applied to it. The following table gives the spring length y that the given applied force f produced in a particular spring. The spring's free length is 4.7 in. Find a functional relation between f and x, the extension from the free length ($x = y - 4.7$).

Force f (lb)	Spring length y (in.)
0	4.7
0.94	7.2
2.30	10.6
3.28	12.9

2.* In each of the following problems, determine the best function $y(x)$ (linear, exponential, or power function) to describe the data. Plot the function on the same plot with the data. Label and format the plots appropriately.

a.

x	25	30	35	40	45
y	5	260	480	745	1100

b.

x	2.5	3	3.5	4	4.5	5	5.5	6	7	8	9	10
y	1500	1220	1050	915	810	745	690	620	520	480	410	390

c.

x	550	600	650	700	750
y	41.2	18.62	8.62	3.92	1.86

3. The population data for a certain country are as follows:

Year	2012	2013	2014	2015	2016	2017
Population (millions)	10	10.9	11.7	12.6	13.8	14.9

Obtain a function that describes these data. Plot the function and the data on the same plot. Estimate when the population will be double its 2004 size.

4.* The *half-life* of a radioactive substance is the time it takes to decay by one-half. The half-life of carbon 14, which is used for dating previously living things, is 5500 years. When an organism dies, it stops accumulating carbon 14. The carbon 14 present at the time of death decays with time. Let $C(t)/C(0)$ be the fraction of carbon 14 remaining at time t. In radioactive carbon dating, scientists usually assume that the remaining fraction decays exponentially according to the following formula:

$$\frac{C(t)}{C(0)} = e^{-bt}$$

a. Use the half-life of carbon 14 to find the value of the parameter b, and plot the function.

b. If 90 percent of the original carbon 14 remains, estimate how long ago the organism died.

c. Suppose our estimate of b is off by ± 1 percent. How does this error affect the age estimate?

5. *Quenching* is the process of immersing a hot metal object in a bath for a specified time to obtain certain properties such as hardness. A copper sphere 25 mm in diameter, initially at 300°C, is immersed in a bath at 0°C. The following table gives measurements of the sphere's temperature versus time. Find a functional description of these data. Plot the function and the data on the same plot.

Time (s)	0	1	2	3	4	5	6
Temperature (°C)	300	150	75	35	12	5	2

6. The useful life of a machine bearing depends on its operating temperature, as the following data show. Obtain a functional description of these data. Plot the function and the data on the same plot. Estimate a bearing's life if it operates at 150°F.

Temperature (°F)	100	120	140	160	180	200	220
Bearing life (hours $\times 10^3$)	28	21	15	11	8	6	4

7. A certain electric circuit has a resistor and a capacitor. The capacitor is initially charged to 100 V. When the power supply is detached, the capacitor voltage decays with time, as the following data table shows. Find a functional description of the capacitor voltage v as a function of time t. Plot the function and the data on the same plot.

Time (s)	0	0.5	1	1.5	2	2.5	3	3.5	4
Voltage (V)	100	62	38	21	13	7	4	2	3

Sections 6.2 and 6.3

8.* The distance a spring stretches from its free length is a function of how much tension force is applied to it. The following table gives the spring length y that was produced in a particular spring by the given applied force f. The spring's free length is 4.7 in. Find a functional relation between f and x, the extension from the free length $(x = y - 4.7)$.

Force f (lb)	Spring length y (in.)
0	4.7
0.94	7.2
2.30	10.6
3.28	12.9

9. The following data give the drying time T of a certain paint as a function of the amount of a certain additive A.

a. Find the first-, second-, third-, and fourth-degree polynomials that fit the data, and plot each polynomial with the data. Determine the quality of the curve fit for each by computing J, S, and r^2.

b. Use the polynomial giving the best fit to estimate the amount of additive that minimizes the drying time.

A (oz)	0	1	2	3	4	5	6	7	8	9
T (min)	130	115	110	90	89	89	95	100	110	125

10.* The following data give the stopping distance d as a function of initial speed v, for a certain car model. Find a quadratic polynomial that fits the data. Determine the quality of the curve fit by computing J, S, and r^2.

v (mi/hr)	20	30	40	50	60	70
d (ft)	45	80	130	185	250	330

11.* The number of twists y required to break a certain rod is a function of the percentage x_1 and x_2 of each of two alloying elements present in the rod. The following table gives some pertinent data. Use linear multiple regression to obtain a model $y = a_0 + a_1 x_1 + a_2 x_2$ of the relationship between the number of twists and the alloy percentages. In addition, find the maximum percent error in the predictions.

Number of twists y	Percentage of element 1 x_1	Percentage of element 2 x_2
40	1	1
51	2	1
65	3	1
72	4	1
38	1	2
46	2	2
53	3	2
67	4	2

Continued

Number of twists y	Percentage of element 1 x_1	Percentage of element 2 x_2
31	1	3
39	2	3
48	3	3
56	4	3

12. Obtain a linear model $y = a_0 + a_1x_1 + a_2x_2$ for the following data to describe the relationship.

y	x_1	x_2
2.85	10	8
4.2	16	12
4.5	18	14
3.75	22	24
4.35	26	28
4.2	28	34

13. The following represents pressure samples, in pounds per square inch (psi), taken in a fuel line once every second for 10 sec.

Time (sec)	Pressure (psi)
1	26.1
2	27.0
3	28.2
4	29.0
5	29.8
6	30.6
7	31.1
8	31.3
9	31.0
10	30.5

a. Fit a first-degree polynomial, a second-degree polynomial, and a third-degree polynomial to these data. Plot the curve fits along with the data points.

b. Use the results from part *a* to predict the pressure at $t = 11$ sec. Explain which curve fit gives the most reliable prediction. Consider the coefficients of determination and the residuals for each fit in making your decision.

14. A liquid boils when its vapor pressure equals the external pressure acting on the surface of the liquid. This is why water boils at a lower temperature at higher altitudes. This information is important for people who must design processes utilizing boiling liquids. Data on the vapor pressure P of water as a function of temperature T are given in the following table. From theory we know that $\ln P$ is proportional to $1/T$. Obtain a curve fit for $P(T)$ from these data. Use the fit to estimate the vapor pressure at 285 and 300 K.

T (K)	P (torr)
273	4.579
278	6.543
283	9.209
288	12.788
293	17.535
298	23.756

15. The solubility of salt in water is a function of the water temperature. Let S represent the solubility of NaCl (sodium chloride) as grams of salt in 100 g of water. Let T be temperature in °C. Use the following data to obtain a curve fit for S as a function of T. Use the fit to estimate S when $T = 25$°C.

T (°C)	S (g NaCl/100 g H$_2$O)
10	35
20	35.6
30	36.25
40	36.9
50	37.5
60	38.1
70	38.8
80	39.4
90	40

16. The solubility of oxygen in water is a function of the water temperature. Let S represent the solubility of O$_2$ as millimoles of O$_2$ per liter of water. Let T be temperature in °C. Use the following data to obtain a curve fit for S as a function of T. Use the fit to estimate S when $T = 8$°C and $T = 50$°C.

T (°C)	S (mmol O$_2$/L H$_2$O)
5	1.95
10	1.7
15	1.55
20	1.40
25	1.30
30	1.15

Continued

T (°C)	S (mmol O_2/L H_2O)
35	1.05
40	1.00
45	0.95

17. The following function is linear in the parameters a_1 and a_2.

$$y(x) = a_1 + a_2 \ln(x)$$

Use least-squares regression with the following data to estimate the values of a_1 and a_2. Use the curve fit to estimate the values of y at $x = 2.5$ and at $x = 11$.

x	1	2	3	4	5	6	7	8	9	10
y	10	14	16	18	19	20	21	22	23	23

18. A mass attached to a spring and damper is displaced a distance x_0 (cm) while being given an initial velocity v_0 (cm/s). We know from physics and mathematics (see Chapter 8) that the displacement x as a function of time is given by

$$x(t) = \left(\frac{6x_0}{3} + \frac{v_0}{3}\right)e^{-3t} - \left(\frac{3x_0 + v_0}{3}\right)e^{-6t}$$

The displacement is measured every 0.2 s. The measured displacement versus time is given by

t (s)	0	0.2	0.4	0.6	0.8	1	1.2	1.4
x (cm)	1.3	1.2	0.8	0.5	0.3	0.2	0.1	0

Estimate the initial displacement and velocity.

19. Chemists and engineers must be able to predict the changes in chemical concentration in a reaction. A model used for many single-reactant processes is

$$\text{Rate of change of concentration} = -kC^n$$

where C is the chemical concentration and k is the rate constant. The order of the reaction is the value of the exponent n. Solution methods for differential equations (which are discussed in Chapter 9) can show that the solution for a first-order reaction ($n = 1$) is

$$C(t) = C(0)e^{-kt}$$

The following data describe the reaction:

$$(CH_3)_3CBr + H_2O \rightarrow (CH_3)_3COH + HBr$$

Use these data to obtain a least-squares fit to estimate the value of k.

Time t (h)	C (mol of $(CH3)_3$ CBr/L)
0	0.1039
3.15	0.0896
6.20	0.0776
10.0	0.0639
18.3	0.0353
30.8	0.0207
43.8	0.0101

20. Chemists and engineers must be able to predict the changes in chemical concentration in a reaction. A model used for many single-reactant processes is

$$\text{Rate of change of concentration} = -kC^n$$

where C is the chemical concentration and k is the rate constant. The order of the reaction is the value of the exponent n. Solution methods for differential equations (which are discussed in Chapter 9) can show that the solution for a first-order reaction ($n = 1$) is

$$C(t) = C(0)e^{-kt}$$

and the solution for a second-order reaction ($n = 2$) is

$$\frac{1}{C(t)} = \frac{1}{C(0)} + kt$$

The following data (from Brown, 1994) describes the gas-phase decomposition of nitrogen dioxide at 300°C.

$$2NO_2 \rightarrow 2NO + O_2$$

Time t (s)	C (mol NO_2/L)
0	0.0100
50	0.0079
100	0.0065
200	0.0048
300	0.0038

Determine whether this is a first-order or second-order reaction, and estimate the value of the rate constant k.

21. Chemists and engineers must be able to predict the changes in chemical concentration in a reaction. A model used for many single-reactant processes is

$$\text{Rate of change of concentration} = -kC^n$$

where C is the chemical concentration and k is the rate constant. The order of the reaction is the value of the exponent n. Solution methods for differential equations (which are discussed in Chapter 9) can show that the solution for a first-order reaction ($n = 1$) is

$$C(t) = C(0) e^{-kt}$$

The solution for a second-order reaction ($n = 2$) is

$$\frac{1}{C(t)} = \frac{1}{C(0)} + kt$$

and the solution for a third-order reaction ($n = 3$) is

$$\frac{1}{2C^2(t)} = \frac{1}{2C^2(0)} + kt$$

Time t (min)	C (mol of reactant/L)
5	0.3575
10	0.3010
15	0.2505
20	0.2095
25	0.1800
30	0.1500
35	0.1245
40	0.1070
45	0.0865

The preceding data describe a certain reaction. By examining the residuals, determine whether this is a first-order, second-order, or third-order reaction, and estimate the value of the rate constant k.

22. Consider the following data. Find the best-fit line that passes through the point $x_0 = 10$, $y_0 = 11$.

x	0	5	10
y	2	6	11

Engineering in the 21st Century. . .

Energy-Efficient Vehicles

Modern societies have become very dependent on transportation powered by carbon-based fuels. Climate change and dwindling resources mean that novel engineering developments in both personal and mass transportation will be needed to reduce our dependence on such fuels. These developments, as well as development of alternative energy sources, will be required in a number of areas such as engine design, electric motor and battery technology, lightweight materials, and aerodynamics.

A number of such initiatives are underway. Several projects have the goal of designing a six-passenger car that is one-third lighter and 40 percent more aerodynamic than today's sleekest cars. A hybrid gas-electric vehicle is the most promising at present. An internal combustion engine and an electric motor drive the wheels. A fuel cell or a battery is charged either by a generator driven by the engine or by energy recovered by regenerative braking.

The weight reduction can be achieved with all-aluminum unibody construction and by improved design of the engine, radiator, and brakes to make use of advanced materials such as composites and magnesium. Other manufacturers are investigating plastic bodies made from recycled materials.

A true energy analysis involves more than just engine operating efficiency and emissions, but must be based on a total life-cycle assessment including production and post-use considerations such as recyclability. In such a total analysis even all-electric vehicles may not be energy efficient. They contain lightweight materials such as carbon composites and aluminum that are energy intensive to produce. Batteries contain compounds such as lithium, copper, and nickel that require much energy to mine and process. In addition to energy efficiency, we must also consider the efficient use of materials that are scarce, such as the rare earth metals, and materials that are environmentally dangerous, such as lithium.

There is still much room for improved efficiency, and research and development engineers in this area will remain busy for some time. MATLAB is widely used to assist these efforts with modeling and analysis tools for designing new vehicle systems. ■

CHAPTER 7

Statistics, Probability, and Interpolation

OUTLINE

7.1   Statistics and Histograms

7.2   The Normal Distribution

7.3   Random Number Generation

7.4   Interpolation

7.5   Summary

Problems

This chapter begins with an introduction to basic statistics in Section 7.1. You will see how to obtain and interpret *histograms,* which are specialized plots for displaying statistical results. The *normal distribution,* commonly called the *bell-shaped curve,* forms the basis of much of probability theory and many statistical methods. It is covered in Section 7.2. In Section 7.3 you will see how to include random processes in your simulation programs. In Section 7.4 you will see how to use interpolation with data tables to estimate values that are not in the table.

When you have finished this chapter, you should be able to use MATLAB to do the following:

- Solve basic problems in statistics and probability.
- Create simulations incorporating random processes.
- Apply interpolation techniques.

7.1 Statistics and Histograms

MEAN

MODE

MEDIAN

BINS

With MATLAB you can compute the *mean* (the average), the *mode* (the most frequently occurring value), and the *median* (the middle value) of a set of data. MATLAB provides the mean(x), mode(x), and median(x) functions to compute the mean, mode, and median of the data values stored in x, if x is a vector. However, if x is a matrix, a row vector is returned containing the mean (or mode or median) value of each column of x. These functions do not require the elements in x to be sorted in ascending or descending order.

The way the data are spread around the mean can be described by a *histogram* plot. A *histogram* is a plot of the frequency of occurrence of data values versus the values themselves. It is a bar plot of the number of data values that occur within each range, with the bar centered in the middle of the range.

To plot a histogram, the data must be grouped into subranges, called *bins*. The choice of the bin width and bin center can drastically change the shape of the histogram. If the number of data values is relatively small, the bin width cannot be small because some of the bins will contain no data and the resulting histogram might not usefully illustrate the distribution of the data.

You can use the bar function to plot the number of values in each bin versus the bin centers as a bar chart. The function bar(x,y) creates a bar chart of y versus x. This syntax will give a plot with the bin rectangles shaded in the default color. To obtain unshaded rectangles like the plots shown in this section, use the syntax bar(x,y,'w'), where w stands for white fill.

MATLAB provides the histogram function to generate a histogram. This command has several forms. Its basic form is histogram(y), where y is a vector containing the data. This form aggregates the data into a number of bins between the minimum and maximum values in y, of uniform width automatically chosen to reveal the underlying distribution. The second form is histogram(y,n), where n is a user-specified scalar indicating the number of bins. Unshaded rectangles can be obtained by using the syntax histogram(y,'FaceColor','none'). There are several other forms of the function, which we will not need here. See the MATLAB documentation for details.

EXAMPLE 7.1–1	Breaking Strength of Thread

To ensure proper quality control, a thread manufacturer selects samples and tests them for breaking strength. Suppose that 20 thread samples are pulled until they break, and the breaking force is measured in newtons rounded off to integer values. The breaking force values recorded were 92, 94, 93, 96, 93, 94, 95, 96, 91, 93, 95, 95, 95, 92, 93, 94, 91, 94, 92, and 93. Plot the histogram of the data.

■ Solution

Store the data in the vector y, which is shown in the following script file. The following script file generates the histogram shown in Figure 7.1–1.

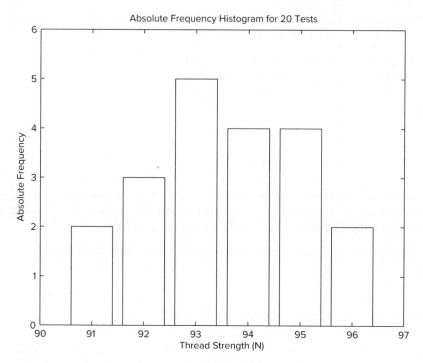

Figure 7.1–1 Histograms for 20 tests of thread strength.

```
% Thread breaking strength data for 20 tests.
y = [92,94,93,96,93,94,95,96,91,93,...
    95,95,95,92,93,94,91,94,92,93];
histogram(y,'FaceColor','none'),...
    axis([90 97 0 6]),...
    ylabel('Absolute Frequency'),...
    xlabel('Thread Strength (N)'),...
    title('Absolute Frequency Histogram for 20 Tests')
```

Because there are six outcomes, six bins are sufficient, and that is what the histogram function chose. We would have obtained the same plot if we had specify the number of bins to be six.

The *absolute frequency* is the number of times a particular outcome occurs. For example, in 20 tests these data show that a 95 occurred 4 times. The absolute frequency is 4, and its *relative frequency* is 4/20, or 20 percent of the time.

When there is a large amount of data, you can avoid typing in every data value by first aggregating the data. The following example shows how this is done using the ones function. The following data were generated by testing 100

ABSOLUTE FREQUENCY

RELATIVE FREQUENCY

thread samples. The number of times 91, 92, 93, 94, 95, or 96 N was measured is 13, 15, 22, 19, 17, and 14, respectively.

```
% Thread strength data for 100 tests.
y = [91*ones(1,13),92*ones(1,15),93*ones(1,22),...
    94*ones(1,19),95*ones(1,17),96*ones(1,14)];
histogram(y,'FaceColor','none'),ylabel('Absolute Frequency'),...
    xlabel('Thread Strength (N)'),...
    title('Absolute Frequency Histogram for 100 Tests');
```

The result appears in Figure 7.1–2.

Suppose you want to obtain a *relative* frequency histogram. In such cases you can use the bar function to generate the histogram. The following script file generates the relative frequency histogram for the 100 thread tests. Note that if you use the bar function, you must aggregate the data first.

```
% Relative frequency histogram using the bar function.
tests = 100;
y - [13,15,22,19,17,14]/tests;
x = 91:96;
bar(x,y,'w'),ylabel('Relative Frequency'),...
```

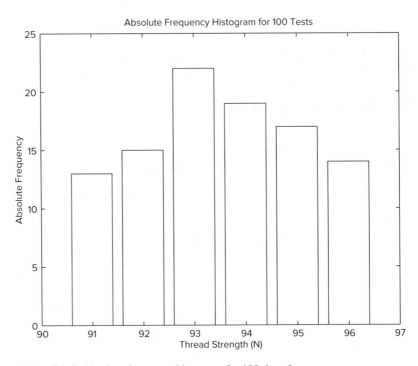

Figure 7.1–2 Absolute frequency histogram for 100 thread tests.

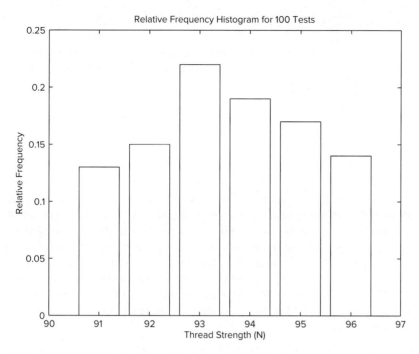

Figure 7.1–3 Relative frequency histogram for 100 thread tests.

```
xlabel('Thread Strength (N)'),...
title('Relative Frequency Histogram for 100 Tests')
```

The result appears in Figure 7.1–3.

These commands are summarized in Table 7.1–1.

Table 7.1–1 Histogram functions

Command	Description
`bar(x,y)`	Creates a bar chart of y versus x using the default color scheme.
`bar(x,y,'w')`	Creates a bar chart of y versus x using unshaded rectangles.
`histogram(y)`	Aggregates the data in the vector y into bins of uniform width between the minimum and maximum values in y, using the default color.
`histogram(y,n)`	Aggregates the data in the vector y into n bins of uniform width between the minimum and maximum values in y.
`histogram(y,'FaceColor','w')`	Aggregates the data in the vector y into bins of uniform width between the minimum and maximum values in y, using unshaded (white) rectangles.

Test Your Understanding

T7.1–1 In 50 tests of thread, the number of times 91, 92, 93, 94, 95, or 96 N was measured was 7, 8, 10, 6, 12, and 7, respectively. Obtain the absolute and relative frequency histograms.

The Data Statistics Tool

With the Data Statistics tool you can calculate statistics for data and add plots of the statistics to a graph of the data. The tool is accessed from the Figure window after you plot the data. Click on the **Tools** menu, then select **Data Statistics.** The menu appears as shown in Figure 7.1–4. To show the mean of the dependent variable (y) on the plot, click the box in the row labeled mean under the column labeled Y, as shown in the figure. A horizontal line is then placed on the plot at the mean. You can plot other statistics as well; these are shown in the figure. You can save the statistics to the workspace as a structure by clicking on the **Save to Workspace** button. This opens a dialog box that prompts you for a name for the structure containing the x data, and a name for the y data structure.

Figure 7.1–4 The Data Statistics tool.

7.2 The Normal Distribution

Rolling a die is an example of a process whose possible outcomes are a limited set of numbers, namely, the integers from 1 to 6. For such processes the probability is a function of a discrete-valued variable, that is, a variable having a limited number of values. For example, Table 7.2–1 gives the measured heights of 100 men 20 years of age. The heights were recorded to the nearest 1/2 in., so the height variable is discrete-valued.

Scaled Frequency Histogram

You can plot the data as a histogram using either the absolute or relative frequencies. However, another useful histogram uses data scaled so that the total area under the histogram's rectangles is 1. This *scaled frequency histogram* is the absolute frequency histogram divided by the total area of that histogram. The area of each rectangle on the absolute frequency histogram equals the bin width times the absolute frequency for that bin. Because all the rectangles have the same width, the total area is the bin width times the sum of the absolute frequencies. The following M-file produces the scaled histogram shown in Figure 7.2–1.

```
% Absolute frequency data.
y_abs=[1,0,0,0,2,4,5,4,8,11,12,10,9,8,7,5,4,4,3,1,1,0,1];
binwidth = 0.5;
% Compute scaled frequency data.
area = binwidth*sum(y_abs);
y_scaled = y_abs/area;
% Define the bins.
bins = 64:binwidth:75;
% Plot the scaled histogram.
bar(bins,y_scaled,'w'),...
   ylabel('Scaled Frequency'),xlabel('Height (in.)')
```

Table 7.2–1 Height data for men 20 years of age

Height (in.)	Frequency	Height (in.)	Frequency
64	1	70	9
64.5	0	70.5	8
65	0	71	7
65.5	0	71.5	5
66	2	72	4
66.5	4	72.5	4
67	5	73	3
67.5	4	73.5	1
68	8	74	1
68.5	11	74.5	0
69	12	75	1
69.5	10		

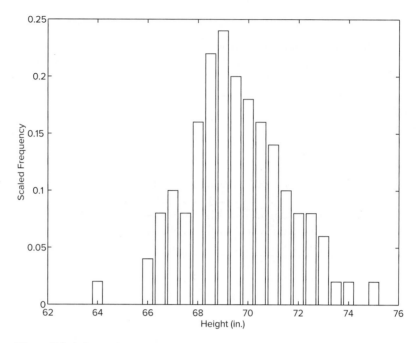

Figure 7.2–1 Scaled histogram of height data.

Because the total area under the scaled histogram is 1, the fractional area corresponding to a range of heights gives the probability that a randomly selected 20-year-old man will have a height in that range. For example, the heights of the scaled histogram rectangles corresponding to heights of 67 through 69 in. are 0.1, 0.08, 0.16, 0.22, and 0.24. Because the bin width is 0.5, the total area corresponding to these rectangles is $(0.1 + 0.08 + 0.16 + 0.22 + 0.24)(0.5) = 0.4$. Thus 40 percent of the heights lie between 67 and 69 in.

You can use the cumsum function to calculate areas under the scaled frequency histogram and therefore to calculate probabilities. If x is a vector, cumsum(x) returns a vector the same length as x, whose elements are the sum of the previous elements. For example, if x = [2, 5, 3, 8], cumsum(x) = [2, 7, 10, 18]. If A is a matrix, cumsum(A) computes the cumulative sum of each row. The result is a matrix the same size as A.

After running the previous script, the last element of cumsum(y_scaled)* binwidth is 1, which is the area under the scaled frequency histogram. To compute the probability of a height lying between 67 and 69 in. (that is, above the 6th value up to the 11th value), type

```
>>prob = cumsum(y_scaled) *binwidth;
>>prob67_69 = prob(11)-prob(6)
```

The result is prob67_69 = 0.4000, which agrees with our previous calculation of 40 percent.

Continuous Approximation to the Scaled Histogram

For processes having an infinite number of possible outcomes, the probability is a function of a *continuous* variable and is plotted as a curve rather than as rectangles. It is based on the same concept as the scaled histogram; that is, the total area under the curve is 1, and the fractional area gives the probability of occurrence of a specific range of outcomes. A probability function that describes many processes is the *normal* or *Gaussian function*, which is shown in Figure 7.2–2.

    This function is also known as the *bell-shaped curve*. Outcomes that can be described by this function are said to be *normally distributed*. The normal probability function is a two-parameter function; one parameter, μ, is the mean of the outcomes, and the other parameter, σ, is the *standard deviation*. The mean μ locates the peak of the curve and is the most likely value to occur. The width, or spread, of the curve is described by the parameter σ. Sometimes the term *variance* is used to describe the spread of the curve. The variance is the square of the standard deviation σ.

    The normal probability function is described by the following equation:

$$p(x) = \frac{1}{\sigma\sqrt{2\pi}} e^{-(x-\mu)^2/2\sigma^2} \tag{7.2–1}$$

    It can be shown that approximately 68 percent of the area lies between the limits of $\mu - \sigma \leq x \leq \mu + \sigma$. Consequently, if a variable is normally distributed, there is a 68 percent chance that a randomly selected sample will lie within one standard deviation of the mean. In addition, approximately 96 percent of the area lies between the limits of $\mu - 2\sigma \leq x \leq \mu + 2\sigma$, and 99.7 percent, or practically 100 percent, of the area lies between the limits of $\mu - 3\sigma \leq x \leq \mu + 3\sigma$.

    The functions `mean(x)`, `var(x)`, and `std(x)` compute the mean, variance, and standard deviation of the elements in the vector `x`.

NORMAL FUNCTION

GAUSSIAN FUNCTION

NORMALLY DISTRIBUTED

STANDARD DEVIATION

VARIANCE

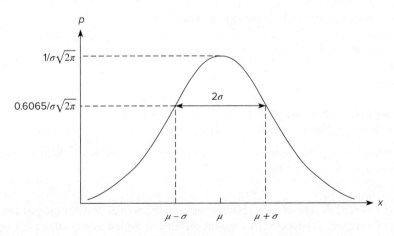

Figure 7.2–2 The basic shape of the normal distribution curve.

| EXAMPLE 7.2–1 | Mean and Standard Deviation of Heights |

Statistical analysis of data on human proportions is required in many engineering applications. For example, designers of submarine crew quarters need to know how small they can make bunk lengths without eliminating a large percentage of prospective crew members. Use MATLAB to estimate the mean and standard deviation for the height data given in Table 7.2–1.

■ **Solution**

The script file follows. The data given in Table 7.2–1 are the absolute frequency data and are stored in the vector y_abs. A bin width of 1/2 in. is used because the heights were measured to the nearest 1/2 in. The vector bins contains the heights in 1/2 in. increments.

To compute the mean and standard deviation, reconstruct the original (raw) height data from the absolute frequency data. Note that these data have some zero entries. For example, none of the 100 men had a height of 65 in. Thus to reconstruct the raw data, start with an empty vector y_raw and fill it with the height data obtained from the absolute frequencies. The for loop checks to see whether the absolute frequency for a particular bin is nonzero. If it is nonzero, append the appropriate number of data values to the vector y_raw. If the particular bin frequency is 0, y_raw is left unchanged.

```
% Absolute frequency data.
y_abs = [1,0,0,0,2,4,5,4,8,11,12,10,9,8,7,5,4,4,3,1,1,0,1];
binwidth = 0.5;
% Define the bins.
bins = [64:binwidth:75];
% Fill the vector y_raw with the raw data.
% Start with an empty vector.
y_raw = [];
for i = 1:length(y_abs)
   if y_abs(i)>0
      new = bins(i) *ones(1,y_abs(i));
   else
      new = [];
   end
y_raw = [y_raw,new];
end
% Compute the mean and standard deviation.
mu = mean(y_raw),sigma = std(y_raw)
```

When you run this program, you will find that the mean is $\mu = 69.6$ in. and the standard deviation is $\sigma = 1.96$ in.

If you need to compute probabilities based on the normal distribution, you can use the erf function. Typing erf(x) returns the area to the left of the value $t = x$ under the curve of the function $2e^{-t^2}/\sqrt{\pi}$. This area, which is a function of x, is known as the *error function* and is written as erf(x). The probability that the random variable *x* is less than or equal to *b* is written as $P(x \leq b)$ if the outcomes are normally distributed. This probability can be computed from the error function as follows:

$$P(x \leq b) = \frac{1}{2}\left[1 + \text{erf}\left(\frac{b - \mu}{\sigma\sqrt{2}}\right)\right] \tag{7.2–2}$$

The probability that the random variable *x* is no less than *a* and no greater than *b* is written as $P(a \leq x \leq b)$. It can be computed as follows:

$$P(a \leq x \leq b) = \frac{1}{2}\left[\text{erf}\left(\frac{b - \mu}{\sigma\sqrt{2}}\right) - \text{erf}\left(\frac{a - \mu}{\sigma\sqrt{2}}\right)\right] \tag{7.2–3}$$

Estimation of Height Distribution

EXAMPLE 7.2–2

Use the results of Example 7.2–1 to estimate how many 20-year-old men are no taller than 68 in. How many are within 3 in. of the mean?

■ Solution

In Example 7.2–1 the mean and standard deviation were found to be $\mu = 69.3$ in. and $\sigma = 1.96$ in. In Table 7.2–1, note that few data points are available for heights less than 68 in. However, if you assume that the heights are normally distributed, you can use Equation (7.2–2) to estimate how many men are shorter than 68 in. Use (7.2–2) with $b = 68$, that is,

$$P(x \leq 68) = \frac{1}{2}\left[1 + \text{erf}\left(\frac{68 - 69.3}{1.96\sqrt{2}}\right)\right]$$

To determine how many men are within 3 in. of the mean, use Equation (7.2–3) with $a = \mu - 3 = 66.3$ and $b = \mu + 3 = 72.3$, that is,

$$P(66.3 \leq x \leq 72.3) = \frac{1}{2}\left[\text{erf}\left(\frac{3}{1.96\sqrt{2}}\right) - \text{erf}\left(\frac{-3}{1.96\sqrt{2}}\right)\right]$$

In MATLAB these expressions are computed in a script file as follows:

```
mu = 69.3;
s = 1.96;
% How many are no taller than 68 inches?
b1 = 68;
P1 = (1+erf((b1-mu)/(s*sqrt(2))))/2
% How many are within 3 inches of the mean?
a2 = 66.3;
b2 = 72.3;
P2 = (erf((b2-mu)/(s*sqrt(2)))-erf((a2-mu)/(s*sqrt(2))))/2
```

When you run this program, you obtain the results P1 = 0.2536 and P2 = 0.8741. Thus 25 percent of 20-year-old men are estimated to be 68 in. or less in height, and 87 percent are estimated to be between 66.3 and 72.3 in. tall.

T7.2–1 Suppose that 10 more height measurements are obtained so that the following numbers must be *added* to Table 7.2–1.

Height (in.)	Additional data
64.5	1
65	2
66	1
67.5	2
70	2
73	1
74	1

(a) Plot the scaled frequency histogram. (b) Find the mean and standard deviation. (c) Use the mean and standard deviation to estimate how many 20-year-old men are no taller than 69 in. (d) Estimate how many are between 68 and 72 in. tall.

(Answers: (b) mean $= 69.4$ in., standard deviation $= 2.14$ in.; (c) 43 percent; (d) 63 percent.)

Sums and Differences of Random Variables

It can be proved that the mean of the sum (or difference) of two independent normally distributed random variables equals the sum (or difference) of their means, but the variance is always the sum of the two variances. That is, if x and y are normally distributed with means μ_x and μ_y and variances σ_x^2 and σ_y^2, and if $u = x + y$ and $v = x - y$, then

$$\mu_u = \mu_x + \mu_y \tag{7.2–4}$$

$$\mu_v = \mu_x - \mu_y \tag{7.2–5}$$

$$\sigma_u^2 = \sigma_v^2 = \sigma_x^2 + \sigma_y^2 \tag{7.2–6}$$

These properties are applied in some of the homework problems.

7.3 Random Number Generation

We often do not have a simple probability distribution to describe the distribution of outcomes in many engineering applications. For example, the probability that a circuit consisting of many components will fail is a function of the number and the age of the components, but we often cannot obtain a function to describe the failure probability. In such cases we often resort to simulation to make predictions. The simulation program is executed many times, using a random set of numbers to represent the failure of one or more components, and the results are used to estimate the desired probability.

Rolling a pair of "fair" dice generates numbers that are truly random, but "random" numbers created with software are not and are called *pseudorandom* because they result from a process within the computer that determines the next random number. However, MATLAB uses algorithms called *random number generators* that give results that pass certain tests for being random and independent. From now on we will ignore the distinction between random and pseudorandom and refer to these numbers as random, as is done in the MATLAB documentation.

One advantage of using random numbers generated in software is that you can repeat a random number calculation at any time. This is useful when comparing different simulations. However, you can accidentally repeat results if you are not careful. We will discuss how to avoid this.

Uniformly Distributed Numbers

In a sequence of *uniformly distributed* random numbers, all values within a given interval are equally likely to occur. The MATLAB function rand generates random numbers uniformly distributed over the *open* interval (0,1) using an algorithm called a *random number generator*, which requires a "*seed*" number to start. Type rand to obtain a single random number in the open interval (0,1). Typing rand again generates a different number. For example,

```
>>rand
ans =
    0.7502
>>rand
ans =
    0.5184
```

For example, the following script makes a random choice between two equally probable alternatives and computes the statistics for 100 simulated tosses of a fair coin.

```
% Simulates multiple tosses of a fair coin.
heads = 0;
tails = 0;
for k = 1:100
    if rand < 0.5
        heads = heads + 1;
    else
        tails = tails + 1;
    end
end
heads
tails
```

Every time MATLAB starts, the generator is reset to the same state. Therefore, a rand command gives an identical result every time it is executed immediately following startup, and you will see the same sequence you saw in a previous *startup*. In fact, any script or function that calls rand returns the same result whenever MATLAB restarts. To avoid getting the same random number when MATLAB restarts, use the command rng('shuffle') before calling rand. The function rng('shuffle') initializes the random number generator based on the current time given by the computer's CPU clock. To repeat a result obtained at startup without restarting, reset the generator to the startup state by using rng('default'). For example,

```
>>rand
ans =
   0.7502
>>rng('default')
>>rand
ans =
   0.7502
```

The rand function has an extended syntax. Type rand(n) to obtain an $n \times n$ matrix of uniformly distributed random numbers in the open interval (0, 1). Type rand(m,n) to obtain an $m \times n$ matrix of random numbers. For example, to create a 1 × 100 vector y having 100 random values in the open interval (0, 1), type y = rand(1,100). Using the rand function this way is equivalent to typing rand 100 times. Even though there is a single call to the rand function, the rand function's calculation has the effect of using a different state to obtain each of the 100 numbers so that they will be random.

Use Y = rand(m,n,p,...) to generate a multidimensional array Y having random elements. Typing rand(size(A)) produces an array of random entries that is the same size as A.

Tables 7.3–1 and 7.3–2 summarize these functions.

You can use the rand function to generate random numbers in an interval other than (0, 1). For example, to generate values in the interval (2, 10), first generate a random number between 0 and 1, multiply it by 8 (the difference between the upper and lower bounds), and add the lower bound (2). The result is a value that is uniformly distributed in the interval (2, 10). The general formula for generating a uniformly distributed random number y in the interval (a, b) is

$$y = (b - a)x + a \tag{7.3–1}$$

where x is a random number uniformly distributed in the interval (0,1). For example, to generate a vector y containing 1000 uniformly distributed random numbers in the interval (2, 10), you type y = 8*rand(1,1000) + 2. You can check the results with the mean, min, and max functions. You should obtain values close to 6, 2, and 10, respectively.

Normally Distributed Random Numbers

In a sequence of normally distributed random numbers, the values near the mean are more likely to occur. We have noted that the outcomes of many processes

Table 7.3–1 Random number functions

Command	Description
rand	Generates a single uniformly distributed random number between 0 and 1.
rand(n)	Generates an $n \times n$ matrix containing uniformly distributed random numbers between 0 and 1.
rand(m,n)	Generates an $m \times n$ matrix containing uniformly distributed random numbers between 0 and 1.
randi(b, [m,n])	Generates an $m \times n$ matrix containing random integer values between 1 and b.
randi([a,b], [m,n])	Generates an $m \times n$ matrix containing random integer values between a and b.
randi(imax)	Generates a single uniformly distributed random integer between 1 and imax.
randi(imax, size(A))	Same as randi(imax) but returns a matrix the size of A.
randn	Generates a single normally distributed random number having a mean of 0 and a standard deviation of 1.
randn(n)	Generates an $n \times n$ matrix containing normally distributed random numbers having a mean of 0 and a standard deviation of 1.
randn(m,n)	Generates an $m \times n$ matrix containing normally distributed random numbers having a mean of 0 and a standard deviation of 1.
randperm(n)	Generates a random unique permutation of the integers from 1 to n.
randperm(n,k)	Generates a row vector containing k unique integers selected randomly from 1 to n inclusive.

can be described by the normal distribution. Although a uniformly distributed random variable has definite upper and lower bounds, a normally distributed random variable does not.

The MATLAB function randn will generate a single number that is normally distributed with a mean equal to 0 and a standard deviation equal to 1. Type randn(n) to obtain an $n \times n$ matrix of such numbers. Type randn(m,n) to obtain an $m \times n$ matrix of random numbers.

The functions for retrieving and specifying the state of the normally distributed random number generator are identical to those for the uniformly distributed generator, except that randn(...) replaces rand(...) in the syntax. These functions are summarized in Table 7.3–1.

Table 7.3–2 Random number generator functions

Function	Description
s = rng	Saves the current generator settings in the structure s.
rng(s)	Restores the settings of the random number generator back to the values captured previously by s = rng.
rng(n)	Initializes the random number generator using the non-negative integer n.
rng('default')	Initializes the random number generator to the state it has at MATLAB startup.
rng('shuffle')	Initializes the random number generator based on the current time obtained from the CPU clock.
rng(n, 'twister')	Like rng(n) but specifies the random number generator to be the Mersenne Twister algorithm.

You can generate a sequence of normally distributed numbers having a mean μ and standard deviation σ from a normally distributed sequence having a mean of 0 and a standard deviation of 1. You do this by multiplying the values by σ and adding μ to each result. Thus if x is a random number with a mean of 0 and a standard deviation of 1, use the following equation to generate a new random number y having a standard deviation of σ and a mean of μ.

$$y = \sigma x + \mu \qquad (7.3\text{--}2)$$

For example, to generate a vector `y` containing 2000 random numbers normally distributed with a mean of 5 and a standard deviation of 3, you type `y = 3*randn(1,2000) + 5`. You can check the results with the `mean` and `std` functions. You should obtain values close to 5 and 3, respectively.

The `rng` function operates with `randn` exactly the same way as with `rand`.

Test Your Understanding

T7.3–1   Use MATLAB to generate a vector `y` containing 1800 random numbers normally distributed with a mean of 7 and a standard deviation of 10. Check your results with the `mean` and `std` functions. Why can't you use the `min` and `max` functions to check your results?

Functions of Random Variables   If y and x are linearly related as

$$y = bx + c \qquad (7.3\text{--}3)$$

and if x is normally distributed with a mean μ_x and standard deviation σ_x, it can be shown that the mean and standard deviation of y are given by

$$\mu_y = b\mu_x + c \qquad (7.3\text{--}4)$$

$$\sigma_y = |b|\sigma_x \qquad (7.3\text{--}5)$$

However, it is easy to see that the means and standard deviations do not combine in a straightforward fashion when the variables are related by a nonlinear function. For example, if x is normally distributed with a mean of 0, and if $y = x^2$, it is easy to see that the mean of y is not 0, but is positive. In addition, y is not normally distributed.

Some advanced methods are available for deriving a formula for the mean and variance of $y = f(x)$, but for our purposes, the simplest way is to use random number simulation.

It was noted in the previous section that the mean of the sum (or difference) of two independent normally distributed random variables equals the sum (or difference) of their means, but the variance is always the sum of the two variances. However, if z is a nonlinear function of x and y, then the mean and variance of z cannot be found with a simple formula. In fact, the distribution of z will not even be normal. This outcome is illustrated by the following example.

Statistical Analysis and Manufacturing Tolerances

EXAMPLE 7.3–1

Suppose you must cut a triangular piece off the corner of a square plate by measuring the distances x and y from the corner (see Figure 7.3–1). The desired value of x is 10 in., and the desired value of θ is 20°. This requires that $y = 3.64$ in. We are told that measurements of x and y are normally distributed with means of 10 and 3.64, respectively, with a standard deviation equal to 0.05 in. Determine the standard deviation of θ and plot the relative frequency histogram for θ.

■ **Solution**

From Figure 7.3–1, we see that the angle θ is determined by $\theta = \tan^{-1}(y/x)$. We can find the statistical distribution of θ by creating random variables x and y that have means of 10 and 3.64, respectively, with a standard deviation of 0.05. The random variable θ is then found by calculating $\theta = \tan^{-1}(y/x)$ for each random pair (x, y). The following script file shows this procedure.

```
s = 0.05; % standard deviation of x and y
n = 8000; % number of random simulations
x = 10 + s*randn(1,n);
y = 3.64 + s*randn(1,n);
theta = (180/pi) *atan(y./x);
mean_theta = mean(theta)
sigma_theta = std(theta)
xp = 19:0.1:21;
histogram(theta,xp,'Normalization','probability'),...
    xlabel('Theta (degrees)'),...
    ylabel('Relative Frequency')
```

The choice of 8000 simulations was a compromise between accuracy and the amount of time required to do the calculations. You should try different values of n and compare the results. The results gave a mean of 19.9993° for θ with a standard deviation

Figure 7.3–1 Dimensions of a triangular cut.

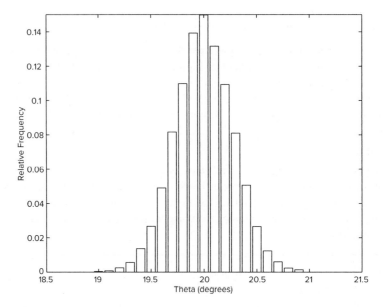

Figure 7.3–2 Scaled histogram of the angle θ.

of 0.2730°. The histogram is shown in Figure 7.3–2. We have used the probability normalization option of the histogram function to plot the relative probability of occurrence. Although the plot resembles the normal distribution, the values of θ are not distributed normally. From the histogram we can calculate that approximately 65 percent of the values of θ lie between 19.8 and 20.2. This range corresponds to a standard deviation of 0.2°, not 0.273° as calculated from the simulation data. Thus the curve is not a normal distribution.

This example shows that the interaction of two of more normally distributed variables does not produce a result that is normally distributed. In general, the result is normally distributed if and only if the result is a linear combination of the variables.

Generating Random Integers

If you want to generate random results for games involving dice, for example, but you must be able to generate integers. You can do this with the `randperm(n)` function, which generates a row vector containing a random permutation of the integers from 1 to n inclusive. For example, `randperm(6)` might generate the vector [3 2 6 4 1 5], or some other permutation of the numbers from 1 to 6. Note that `randperm` calls `rand` and therefore changes the state of the generator.

The function `randi(b,[m,n])` returns an m-by-n matrix containing random integer values between 1 and b. The function `randi([a,b],[m,n])` returns an m-by-n matrix containing random integer values between a and b. Typing `randi(imax)` returns a scalar between 1 and `imax`. Typing `randi(imax,size(A))` returns an array the same size as `A`. For example,

```
>> randi(20,[1,5])
ans =
    1    7    3    9   19   16
>>randi([5,20],[1,5])
ans =
    5   12   11   17   17
>> randi(6)
ans =
    3
```

Note that `randperm` returns *unique* integers, whereas the arrays returned by `randi` may contain *repeated* integer values. So to get unique integer values, use `randperm`. The sequence of numbers produced by `randi` is determined by the settings of the same uniform random number generator used by `rand`, `randn`, and `randperm`.

Random Walks   A *random walk* is a random process that describes a path produced by successive random steps. The "walk" may simply be motion back and forth on a straight line (a 1-dimensional walk), or it can take place on a plane (a 2-dimensional walk), or in 3-dimensional space, or mathematically in even higher dimensions. Random walk methods provide a basis for understanding *Brownian motion*, which describes the seemingly random motion of particles in a fluid caused by collisions with the fluid's molecules. Random walk theory has been applied to understanding a variety of processes including diffusion, stock prices, and games of chance.

A Random Walk with Drift   **EXAMPLE 7.3–2**

The `randi` function can be used to simulate a 1-dimensional random walk. Suppose that a particle starts at $x = 0$ and at each stage of the process it either stays still, moves back one space, or moves forward either one or two spaces, all with equal probability. We can obtain these moves with the `randi([-1,2],[1,99])` function, which will generate the four possible moves with equal probability. Because this will eventually produce increasing positive values for the location x, we say this is a random walk *with drift*. Create a MATLAB program to simulate this process for 100 steps. Generate the statistics for the particle's final location using 1000 trials and time the program.

■ **Solution**
We use two loops; an inner loop for the random walk itself, and an outer loop for the 1000 trials. We use the functions `tic` and `toc` to time the process.

```
% random_walk_1.m
clear
tic
for n = 1:1000
    clear x p
    x(1) = 0;
```

```
      p = randi([-1,2],[1,100]);
      for k = 1:100
            x(k+1) = x(k) + p(k);
      end
      y(n) = x(101);
end
toc
maximum = max(y)
minimum = min(y)
mean = mean(y)
st_dev = std(y)
histogram(y)
```

If you run this program several times, the resulting values of the minimum and maximum distance moved will be quite variable. The mean distance reached after 100 steps should be about 50 with a standard deviation of about 11. The histogram should resemble a bell-shaped curve. The run time depends heavily on the specific computer. Since the step lengths have an average value of 0.5, it is not surprising that the average distance covered in 100 steps is about 0.5(100) = 50. What may be unexpected is that the histogram resembles that of the normal distribution, even though the input is uniformly distributed. This is an example of how the output of a process can have a different distribution than the input.

A simple example of how a process can change the input distribution is given by the process $y = x^2$. Consider the following script.

```
x = rand(1,1000);
y = x.^2;
histogram(x)
histogram(x),hold on
histogram(y)
```

The histogram for x will be that of a uniform distribution, whereas that for y will resemble a decaying exponential with a peak near 0.

Test Your Understanding

T7.3-2 Suppose a particle conducts a one-dimensional random walk where the particle starts at $x = 0$ and moves either 0, 1, 2, 3, 4, 5, or 6 spaces forward at each stage, all with equal probability. Without writing a program, how far do you think the particle will move after 100 steps on average? Then write a MATLAB program to solve the problem.

T7.3–3 Suppose x consists of 1000 uniformly distributed numbers between 0 and 1. Plot the histogram of y where y is the square root of x. Compare the histogram with the case where y is the square of x.

Comparing the Results of Two or More Simulations   To compare the results of two or more simulations, sometimes you will need to generate the same sequence of random numbers each time the simulation runs. One way to do this is to use rng('default'), to repeat a result obtained at startup without restarting, as we have seen earlier. However, you need not start with the initial state to generate the same sequence. To initialize the generator differently, we can use the rng(seed) function, where seed is a positive integer. Every time you use rng(seed) to initialize the generator using the same seed, you always get the same result. Consider the following example. First, we initialize the random number generator to make the results in this example repeatable.

```
>>rng('default')
```

Now, we initialize the generator using an arbitrary seed number, say 4.

```
>>rng(4)
```

Then, create a vector of random numbers.

```
>> v1 = rand(1,5)
v1 =
    0.9670    0.5472    0.9727    0.7148    0.6977
```

Repeat the same command.

```
>> v2 = rand(1,5)
v2 =
    0.2161    0.9763    0.0062    0.2530    0.4348
```

The first use of rand changed the state of the generator, so the second result v2 is different.

   If we reinitialize the generator using the same seed as before, we can reproduce the first vector, v1, as follows:

```
>> rng(4)
>> v3 = rand(1,5)
v3 =
    0.9670    0.5472    0.9727    0.7148    0.6977
```

If you run your code in a different MATLAB release, or if you run your code after running someone else's random number code, setting the seed alone may not guarantee the same results. To ensure repeatability, you can specify the seed and the generator type together, using the function rng(n,'twister') where n is an integer seed number. The input 'twister' refers to the to *Mersenne Twister* random number generator, which is the preferred generator.

7.4 Interpolation

Paired data might represent a *cause and effect,* or *input-output relationship,* such as the current produced in a resistor as a result of an applied voltage, or a *time history,* such as the temperature of an object as a function of time. Another type of paired data represents a *profile*, such as a road profile (which shows the height of the road along its length). In some applications we want to estimate a variable's value between the data points. This process is called *interpolation*. In other cases we might need to estimate the variable's value outside the given data range. This process is called *extrapolation*. Interpolation and extrapolation are greatly aided by plotting the data. Such plots, some perhaps using logarithmic axes, often help to discover a functional description of the data.

Suppose we have the following temperature measurements, taken once an hour starting at 7:00 A.M. The measurements at 8 and 10 A.M. are missing for some reason, perhaps because of equipment malfunction.

Time	7 A.M.	9 A.M.	11 A.M.	12 noon
Temperature (°F)	49	57	71	75

A plot of these data is shown in Figure 7.4–1 with the data points connected by dashed lines. If we need to estimate the temperature at 10 A.M., we can read the value from the dashed line that connects the data points at 9 and 11 A.M. From the plot we thus estimate the temperature at 8 A.M. to be 53°F and at 10 A.M. to be 64°F. We have just performed *linear interpolation* on the data to obtain an *estimate* of the missing data. Linear interpolation is so named because it is equivalent to connecting the data points with a linear function (a straight line).

Figure 7.4–1 A plot of temperature data versus time.

Of course we have no reason to believe that the temperature follows the straight lines shown in the plot, and our estimate of 64°F will most likely be incorrect, but it might be close enough to be useful. Using straight lines to connect the data points is the simplest form of interpolation. Another function could be used if we have a good reason to do so. Later in this section we use polynomial functions to do the interpolation.

Linear interpolation in MATLAB is obtained with the `interp1` and `interp2` functions. Suppose that `x` is a vector containing the independent variable data and that `y` is a vector containing the dependent variable data. If `x_int` is a vector containing the value or values of the independent variable at which we wish to estimate the dependent variable, then typing `interp1(x,y,x_int)` produces a vector the same size as `x_int` containing the interpolated values of `y` that correspond to `x_int`. For example, the following session produces an estimate of the temperatures at 8 and 10 A.M. from the preceding data. The vectors `x` and `y` contain the times and temperatures, respectively.

```
>>x = [7, 9, 11, 12];
>>y = [49, 57, 71, 75];
>>x_int = [8, 10];
>>interp1(x,y,x_int)
ans =
      53
      64
```

You must keep in mind two restrictions when using the `interp1` function. The values of the independent variable in the vector `x` must be in ascending order, and the values in the interpolation vector `x_int` must lie within the range of the values in `x`. Thus we cannot use the `interp1` function to estimate the temperature at 6 A.M., for example.

The `interp1` function can be used to interpolate in a table of values by defining `y` to be a matrix instead of a vector. For example, suppose that we now have temperature measurements at three locations and the measurements at 8 and 10 A.M. are missing for all three locations. The data are as follows:

	Temperature (°F)		
Time	**Location 1**	**Location 2**	**Location 3**
7 A.M.	49	52	54
9 A.M.	57	60	61
11 A.M.	71	73	75
12 noon	75	79	81

We define `x` as before, but now we define `y` to be a matrix whose three columns contain the second, third, and fourth columns of the preceding table. The following session produces an estimate of the temperatures at 8 and 10 A.M. at each location.

```
>>x = [7, 9, 11, 12]';
>>y(:,1) = [49, 57, 71, 75]';
>>y(:,2) = [52, 60, 73, 79]';
>>y(:,3) = [54, 61, 75, 81]';
>>x_int = [8, 10]';
>>interp1(x,y,x_int);
ans =
        53.0000    56.0000    57.5000
        64.0000    65.5000    68.0000
```

Thus the estimated temperatures at 8 A.M. at each location are 53, 56, and 57.5°F, respectively. At 10 A.M. the estimated temperatures are 64, 65.5, and 68°F. From this example we see that if the first argument x in the interp1 (x,y,x_int) function is a *vector* and the second argument y is a *matrix,* then the function interpolates between the rows of y and computes a matrix having the same number of columns as y and the same number of rows as the number of values in x_int.

Note that we need not define two separate vectors x and y. Rather, we can define a single matrix that contains the entire table. For example, by defining the matrix temp to be the preceding table, the session will look like this:

```
>>temp(:,1) = [7, 9, 11, 12]';
>>temp(:,2) = [49, 57, 71, 75]';
>>temp(:,3) = [52, 60, 73, 79]';
>>temp(:,4) = [54, 61, 75, 81]';
>>x_int = [8, 10]';
>>interp1(temp(:,1),temp(:,2:4),x_int);
ans =
        53.0000    56.0000    57.5000
        64.0000    65.5000    68.0000
```

Two-Dimensional Interpolation

Now suppose that we have temperature measurements at four locations at 7 A.M. These locations are at the corners of a rectangle 1 mi wide and 2 mi long. Assigning a coordinate system origin (0, 0) to the first location, the coordinates of the other locations are (1, 0), (1, 2), and (0, 2); see Figure 7.4–2. The temperature measurements are shown in the figure. The temperature is a function of two variables, the coordinates x and y. MATLAB provides the interp2 function to interpolate functions of two variables. If the function is written as $z = f(x, y)$ and we wish to estimate the value of z for $x = x_i$ and $y = y_i$, the syntax is interp2(x,y,z,x_i,y_i).

Suppose we want to estimate the temperature at the point whose coordinates are (0.6, 1.5). Put the x coordinates in the vector x and the y coordinates in the vector y. Then put the temperature measurements in a matrix z such that going

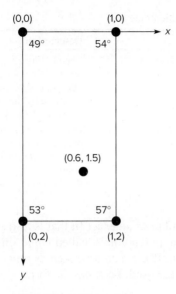

Figure 7.4–2 Temperature
measurements at four locations.

across a row represents an increase in x and going down a column represents an
increase in y. The session to do this is as follows:

```
>>x = [0,1];
>>y = [0,2];
>>z = [49,54;53,57]
z =
    49    54
    53    57
>>interp2(x,y,z,0.6,1.5)
ans =
      54.5500
```

Thus the estimated temperature is 54.55°F.

The syntax of the interp1 and interp2 functions is summarized in
Table 7.4–1. MATLAB also provides the interpn function for interpolating multidimensional arrays.

Cubic Spline Interpolation

High-order polynomials can exhibit undesired behavior between the data points,
and this can make them unsuitable for interpolation. A widely used alternative
procedure is to fit the data points using a lower-order polynomial between *each
pair* of adjacent data points. This method is called *spline* interpolation and is so
named for the splines used by illustrators to draw a smooth curve through a set
of points.

Table 7.4–1 Linear interpolation functions

Command	Description
`y_int=interp1(x,y,x_int)`	Used to linearly interpolate a function of one variable: $y = f(x)$. Returns a linearly interpolated vector `y_int` at the points specified by the vector `x_int`, using data stored in `x` and `y`.
`z_int=interp2(x,y,z,x_,y_int)`	Used to linearly interpolate a function of two variables: $y = f(x, y)$. Returns a linearly interpolated vector `z_int` at the points specified by the vectors `x_int` and `y_int`, using data stored in `x`, `y`, and `z`.

Spline interpolation obtains an exact fit that is also smooth. The most common procedure uses cubic polynomials, called *cubic splines*, and thus is called *cubic spline interpolation*. If the data are given as n pairs of (x, y) values, then $n - 1$ cubic polynomials are used. Each has the form

$$y_i(x) = a_i(x - x_i)^3 + b_i(x - x_i)^2 + c_i(x - x_i) + d_i$$

for $x_i \leq x \leq x_{i+1}$ and $i = 1, 2, \ldots, n - 1$. The coefficients a_i, b_i, c_i, and d_i for each polynomial are determined so that the following three conditions are satisfied for each polynomial:

1. The polynomial must pass through the data points at its endpoints at x_i and x_{i+1}.
2. The slopes of adjacent polynomials must be equal at their common data point.
3. The curvatures of adjacent polynomials must be equal at their common data point.

For example, a set of cubic splines for the temperature data given earlier follows (y represents the temperature values, and x represents the hourly values). The data are repeated here.

x	7	9	11	12
y	49	57	71	75

We will shortly see how to use MATLAB to obtain these polynomials. For $7 \leq x \leq 9$,

$$y_1(x) = -0.35(x - 7)^3 + 2.85(x - 7)^2 - 0.3(x - 7) + 49$$

For $9 \leq x \leq 11$,

$$y_2(x) = -0.35(x - 9)^3 + 0.75(x - 9)^2 + 6.9(x - 9) + 57$$

For $11 \leq x \leq 12$,

$$y_3(x) = -0.35(x - 11)^3 - 1.35(x - 11)^2 + 5.7(x - 11) + 71$$

MATLAB provides the `spline` command to obtain a cubic spline interpolation. Its syntax is `y_int = spline(x,y,x_int)`, where `x` and `y` are vectors containing the data and `x_int` is a vector containing the values of the independent variable `x` at which we wish to estimate the dependent variable `y`. The result `y_int` is a vector the same size as `x_int` containing the interpolated values of `y` that correspond to `x_int`. The spline fit can be plotted by plotting the vectors `x_int` and `y_int`. For example, the following session produces and plots a cubic spline fit to the preceding data, using an increment of 0.01 in the `x` values.

```
>>x = [7,9,11,12];
>>y = [49,57,71,75];
>>x_int = 7:0.01:12;
>>y_int = spline(x,y,x_int);
>>plot(x,y,'o',x,y,'- -',x_int,y_int),...
    xlabel('Time (hr)'),ylabel('Temperature (deg F)'),...
    title('Measurements at a Single Location'),...
    axis([7 12 45 80])
```

The plot is shown in Figure 7.4–3. The dashed lines represent linear interpolation, and the solid curve is the cubic spline. If we evaluate the spline polynomial at $x = 8$, we obtain $y(8) = 51.2°F$. This estimate is different from the 53°F estimate

Figure 7.4–3 Linear and cubic spline interpolation of temperature data.

obtained from linear interpolation. It is impossible to say which estimate is more accurate without having greater understanding of the temperature dynamics.

We can obtain an estimate more quickly by using the following variation of the `interp1` function.

```
y_est = interp1(x,y,x_est,'spline')
```

In this form the function returns a column vector `y_est` that contains the estimated values of y that correspond to the x values specified in the vector `x_est`, using cubic spline interpolation.

In some applications it is helpful to know the polynomial coefficients, but we cannot obtain the spline coefficients from the `interp1` function. However, we can use the form

```
[breaks, coeffs, m, n] = unmkpp(spline(x,y))
```

to obtain the coefficients of the cubic polynomials. The vector `breaks` contains the x values of the data, and the matrix `coeffs` is an $m \times n$ matrix containing the coefficients of the polynomials. The scalars `m` and `n` give the dimensions of the matrix `coeffs`; `m` is the number of polynomials, and `n` is the number of coefficients for each polynomial (MATLAB will fit a lower-order polynomial if possible, so there can be fewer than four coefficients). For example, using the same data, the following session produces the coefficients of the polynomials given earlier:

```
>>x = [7,9,11,12];
>>y = [49,57,71,75];
>> [breaks, coeffs, m, n] = unmkpp(spline(x,y))
breaks =
        7   9   11   12
coeffs =
        -0.3500   2.8500  -0.3000  49.0000
        -0.3500   0.7500   6.900   57.0000
        -0.3500  -1.3500   5.7000  71.0000
m =
    3
n =
    4
```

The first row of the matrix `coeffs` contains the coefficients of the first polynomial, and so on. The `spline`, `unmkpp`, and extended syntax of the `interp1` function are summarized in Table 7.4–2. Besides 'spline', other interpolation methods can be used with the `interp1` function by specifying the parameter `'method'`. These are listed in Table 7.4–2. See the MATLAB documentation for information on these methods. The Basic Fitting interface, which is available on the **Tools** menu of the Figure window, can also be used for cubic spline interpolation. See Section 6.3 for instructions for using the interface.

Table 7.4–2 Polynomial interpolation functions

Command	Description
`y_est = interp1(x,y,x_est, method)`	Returns a column vector `y_est` that contains the estimated values of y that correspond to the x values specified in the vector `x_est`, using interpolation specified by `method`. The choices for `method` are 'nearest', 'linear', 'next', previous','spline', and 'pchip'.
`y_int = spline(x,y,x_int)`	Computes a cubic spline interpolation where `x` and `y` are vectors containing the data and `x_int` is a vector containing the values of the independent variable x at which we wish to estimate the dependent variable y. The result `y_int` is a vector the same size as `x_int` containing the interpolated values of y that correspond to `x_int`.
`y_int = pchip(x,y,x_int)`	Similar to `spline` but uses piecewise cubic Hermite polynomials for interpolation to preserve shape and respect monotonicity.
`[breaks, coeffs, m, n] = unmkpp(spline(x,y))`	Computes the coefficients of the cubic spline polynomials for the data in `x` and `y`. The vector `breaks` contains the x values, and the matrix `coeffs` is an $m \times n$ matrix containing the polynomial coefficients. The scalars `m` and `n` give the dimensions of the matrix `coeffs`; `m` is the number of polynomials, and `n` is the number of coefficients for each polynomial.

As another example of interpolation, consider 10 evenly spaced data points generated by the function $y = 1/(3 - 3x + x^2)$ over the range $0 \leq x \leq 4$. The top graph in Figure 7.4–4 shows the results of fitting a cubic polynomial and an eighth-order polynomial to the data. Clearly the cubic is not suitable for interpolation. As we increase the order of the fitted polynomial, we find that the polynomial does not pass through all the data points if the order is less than 7. However, there are two problems with the eighth-order polynomial: we should not use it to interpolate over the interval $0 < x < 0.5$, and its coefficients must be stored with very high accuracy if we use the polynomial to interpolate. The bottom graph in Figure 7.4–4 shows the results of fitting a cubic spline, which is clearly a better choice here.

Interpolation with Hermite Polynomials

The `pchip` function, summarized in Table 7.4-2, uses piecewise continuous Hermite interpolation polynomials (*pchips*). Its syntax is identical to that of the `spline` function. With `pchip` the slopes at the data points are computed to preserve the "shape" of the data and to "respect" monotonicity. That is, the fitted function will be monotonic on intervals where the data are monotonic and will

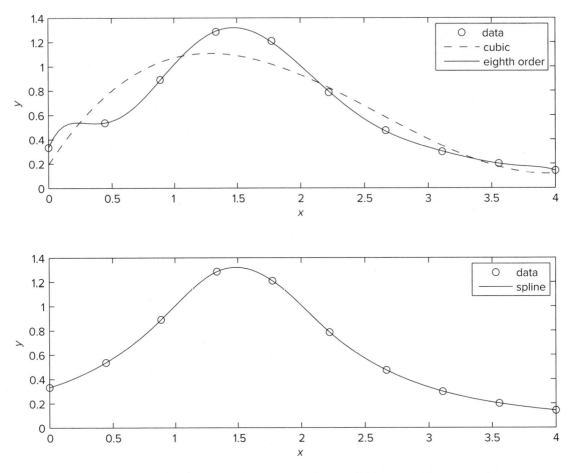

Figure 7.4–4 *Top graph:* Interpolation with a cubic polynomial and an eighth-order polynomial. *Bottom graph:* Interpolation with a cubic spline.

have a local extremum on intervals where the data have a local extremum. The differences between the two functions are that

■ The second derivatives are continuous with spline but may be discontinuous with pchip, so spline may give a smoother function.

■ Therefore, the spline function is more accurate if the data are "smoother."

■ There are no overshoots and less oscillation in the function produced by pchip, even if the data are not smooth.

Consider the data given by $x = [0, 1, 2, 3, 4, 5]$ and $y = [0, -10, 60, 40, 41, 47]$. The top graph in Figure 7.4–5 shows the results of fitting a fifth-order polynomial and a cubic spline to the data. Clearly the fifth-order polynomial is less suitable for interpolation because of the large excursions it makes, especially over the ranges $0 < x < 1$ and $4 < x < 5$. These excursions are often seen with

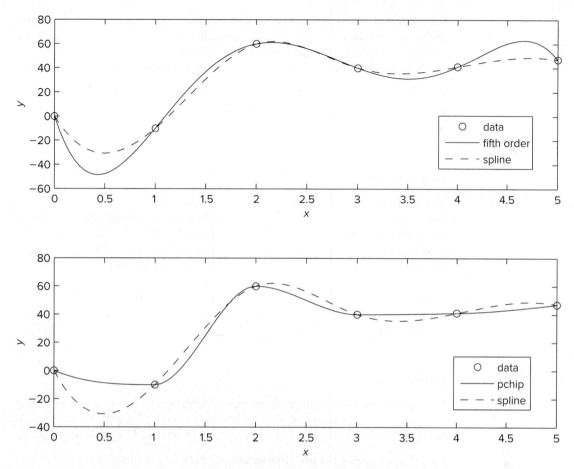

Figure 7.4–5 *Top graph:* Interpolation with a fifth-order polynomial and a cubic spline. *Bottom graph:* Interpolation with piecewise continuous Hermite polynomials (pchip) and a cubic spline.

high-order polynomials. Here, the cubic spline is more useful. The bottom graph in Figure 7.4–5 compares the results of a cubic spline fit with a piecewise continuous Hermite polynomial fit (using pchip), which is clearly a better choice here.

MATLAB provides a number of other functions to support interpolation for three-dimensional data. See griddata, interp3, and interpn in the MATLAB Help.

7.5 Summary

This chapter introduces MATLAB functions that have widespread and important uses in statistics and data analysis. Section 7.1 gives an introduction to basic statistics and probability, including histograms, which are specialized plots for displaying statistical results. The normal distribution that forms the basis of many statistical methods is covered in Section 7.2. Section 7.3 covers random number generators and their use in simulation programs. Section 7.4 covers interpolation methods, including linear and spline interpolation.

Now that you have finished this chapter, you should be able to use MATLAB to

■ Solve basic problems in statistics and probability.

■ Create simulations incorporating random processes.

■ Apply interpolation to data.

Key Terms

Absolute frequency, 313

Bins, 312

Cubic splines, 336

Error function, 321

Gaussian function, 319

Histogram, 312

Interpolation, 332

Mean, 312

Median, 312

Mode, 312

Normally distributed, 319

Normal, 319

Relative frequency, 313

Scaled frequency histogram, 317

Standard deviation, 319

Uniformly distributed, 323

Variance, 319

Problems

You can find the answers to problems marked with an asterisk at the end of the text.

Section 7.1

1. The following list gives the measured gas mileage in miles per gallon for 22 cars of the same model. Plot the absolute frequency histogram and the relative frequency histogram.

23	25	26	25	27	25	24	22	23	25	26
26	24	24	22	25	26	24	24	24	27	23

2. Thirty pieces of structural timber of the same dimensions were subjected to an increasing lateral force until they broke. The measured force in pounds required to break them is given in the following list. Plot the absolute frequency histogram. Try bin widths of 50, 100, and 200 lb. Which gives the most meaningful histogram? Try to find a better value for the bin width.

243	236	389	628	143	417	205
404	464	605	137	123	372	439
497	500	535	577	441	231	675
132	196	217	660	569	865	725
457	347					

3.   The following list gives the measured breaking force in newtons for a sample of 60 pieces of certain type of cord. Plot the absolute frequency histogram. Try bin widths of 10, 30, and 50 N. Which gives the most meaningful histogram? Try to find a better value for the bin width.

311	138	340	199	270	255	332	279	231	296	198	269
257	236	313	281	288	225	216	250	259	323	280	205
279	159	276	354	278	221	192	281	204	361	321	282
254	273	334	172	240	327	261	282	208	213	299	318
356	269	355	232	275	234	267	240	331	222	370	226

Section 7.2

4.   For the data given in Problem 1:
   a. Plot the scaled frequency histogram.
   b. Compute the mean and standard deviation and use them to estimate the lower and upper limits of gas mileage corresponding to 68 percent of cars of this model. Compare these limits with those of the data.

5.   For the data given in Problem 2:
   a. Plot the scaled frequency histogram.
   b. Compute the mean and standard deviation and use them to estimate the lower and upper limits of strength corresponding to 68 and 96 percent of such timber pieces. Compare these limits with those of the data.

6.   For the data given in Problem 3:
   a. Plot the scaled frequency histogram.
   b. Compute the mean and standard deviation, and use them to estimate the lower and upper limits of breaking force corresponding to 68 and 96 percent of cord pieces of this type. Compare these limits with those of the data.

7.*   Data analysis of the breaking strength of a certain fabric shows that it is normally distributed with a mean of 300 lb and a variance of 9.
   a. Estimate the percentage of fabric samples that will have a breaking strength no less than 294 lb.
   b. Estimate the percentage of fabric samples that will have a breaking strength no less than 297 lb and no greater than 303 lb.

8.   Data from service records show that the time to repair a certain machine is normally distributed with a mean of 65 min and a standard deviation of 5 min. Estimate how often it will take more than 75 min to repair a machine.

9.   Measurements of a number of fittings show that the pitch diameter of the thread is normally distributed with a mean of 8.007 mm and a standard deviation of 0.005 mm. The design specifications require that the pitch diameter be 8 ± 0.01 mm. Estimate the percentage of fittings that will be within tolerance.

10.   A certain product requires that a shaft be inserted into a bearing. Measurements show that the diameter d_1 of the cylindrical hole in the bearing is normally distributed with a mean of 3 cm and a variance of 0.0064. The diameter d_2 of the shaft is normally distributed with a mean of 2.96 cm and a variance of 0.0036.

   a. Compute the mean and the variance of the clearance $c = d_1 - d_2$.

   b. Find the probability that a given shaft will not fit into the bearing.
       (*Hint:* Find the probability that the clearance is negative.)

11.* A shipping pallet holds 10 boxes. Each box holds 300 parts of different types. The part weight is normally distributed with a mean of 1 lb and a standard deviation of 0.2 lb.

   a. Compute the mean and standard deviation of the pallet weight.

   b. Compute the probability that the pallet weight will exceed 3015 lb.

12.   A certain product is assembled by placing three components end to end. The components' lengths are L_1, L_2, and L_3. Each component is manufactured on a different machine, so the random variations in their lengths are independent of one another. The lengths are normally distributed with means of 1, 2, and 1.5 ft and variances of 0.00014, 0.0002, and 0.0003, respectively.

   a. Compute the mean and variance of the length of the assembled product.

   b. Estimate the percentage of assembled products that will be no less than 4.48 and no more than 4.52 ft in length.

13.   Use a random number generator to produce 1000 uniformly distributed numbers with a mean of 10, a minimum of 2, and a maximum of 18. Obtain the mean and the histogram of these numbers, and discuss whether they appear uniformly distributed with the desired mean.

14.   Use a random number generator to produce 1000 normally distributed numbers with a mean of 20 and a variance of 4. Obtain the mean, variance, and histogram of these numbers, and discuss whether they appear normally distributed with the desired mean and variance.

15.   The mean of the sum (or difference) of two independent random variables equals the sum (or difference) of their means, but the variance is always the sum of the two variances. Use random number generation to verify this statement for the case where $z = x + y$, where x and y are independent and normally distributed random variables. The mean and variance of x are $\mu_x = 8$ and $\sigma_x^2 = 2$. The mean and variance of y are $\mu_y = 15$ and $\sigma_y^2 = 4$. Find the mean and variance of z by simulation, and compare the results with the theoretical prediction. Do this for 100, 1000, and 5000 trials.

16.   Suppose that $z = xy$, where x and y are independent and normally distributed random variables. The mean and variance of x are $\mu_x = 10$ and $\sigma_x^2 = 2$.

The mean and variance of y are $\mu_y = 15$ and $\sigma_y^2 = 3$. Find the mean and variance of z by simulation. Does $\mu_z = \mu_x \mu_y$? Does $\sigma_z^2 = \sigma_x^2 \sigma_y^2$? Do this for 100, 1000, and 5000 trials.

17. Suppose that $y = x^2$, where x is a normally distributed random variable with a mean and variance of $\mu_x = 0$ and $\sigma_x^2 = 4$. Find the mean and variance of y by simulation. Does $\mu_y = \mu_x^2$? Does $\sigma_y = \sigma_y^2$? Do this for 100, 1000, and 5000 trials.

18.* Suppose you have analyzed the price behavior of a certain stock by plotting the scaled frequency histogram of the price over a number of months. Suppose that the histogram indicates that the price is normally distributed with a mean $100 and a standard deviation of $5. Write a MATLAB program to simulate the effects of buying 50 shares of this stock whenever the price is below the $100 mean, and selling all your shares whenever the price is above $105. Analyze the outcome of this strategy over 250 days (the approximate number of business days in a year). Define the profit as the yearly income from selling stock plus the value of the stocks you own at year's end, minus the yearly cost of buying stock. Compute the mean yearly profit you would expect to make, the minimum expected yearly profit, the maximum expected yearly profit, and the standard deviation of the yearly profit. The broker charges 6 cents per share bought or sold with a minimum fee of $40 per transaction. Assume you make only one transaction per day.

19. Suppose that data show that a certain stock price is normally distributed with a mean of $150 and a variance of 100. Create a simulation to compare the results of the following two strategies over 250 days. You start the year with 1000 shares. With the first strategy, every day the price is below $140 you buy 100 shares, and every day the price is above $160 you sell all the shares you own. With the second strategy, every day the price is below $150 you buy 100 shares, and every day the price is above $160 you sell all the shares you own. The broker charges 5 cents per share traded with a minimum of $35 per transaction.

20. Write a script file to simulate 100 plays of a game in which you flip two coins. You win the game if you get two heads, lose if you get two tails, and flip again if you get one head and one tail. Create three user-defined functions to use in the script. Function flip_coin simulates the flip of one coin, with the state s of the random number generator as the input argument, and the new state s and the result of the flip (0 for a tail and 1 for a head) as the outputs. Function flips simulates the flipping of two coins and calls flip_coin. The input of flips is the state s, and the outputs are the new state s and the result (0 for two tails, 1 for a head and a tail, and 2 for two heads). Function match simulates a turn at the game. Its input is the state s, and its outputs are the result (1 for win, 0 for lose) and the new state s. The script should reset the random number generator to its initial state, compute the state s, and pass this state to the user-defined functions.

21. Write a script file to play a simple number guessing game as follows. The script should generate a random integer in the range 1, 2, 3, ..., 14, 15. It should provide for the player to make repeated guesses of the number, and it should indicate if the player has won or give the player a hint after each wrong guess. The responses and hints are as follows:

- ■ "You won" and then stop the game.
- ■ "Very close" if the guess is within 1 of the correct number.
- ■ "Getting close" if the guess is within 2 or 3 of the correct number.
- ■ "Not close" if the guess is not within 3 of the correct number.

22. Suppose a particle conducts a one-dimensional random walk where the particle starts at $x = 0$ and moves forward according to a normal distribution of mean of one space with a standard deviation of two spaces at each stage. This motion resembles that of Brownian motion. Without writing a program, how far do you think the particle will move after 100 steps on average? Then write a MATLAB program to solve the problem. Compute the statistics and plot the histogram. Is the mean motion what you would expect?

23. Suppose x consists of 1000 uniformly distributed numbers between 0 and 1. Plot the histogram of y where (a) $y = e^{-x}$ and (b) $y = e^{-10x}$. Compare the histograms for each case. Interpret the results in terms of the time constant.

Section 7.4

24.* Interpolation is useful when one or more data points are missing. This situation often occurs with environmental measurements, such as temperature, because of the difficulty of making measurements around the clock. The following table of temperature versus time data is missing readings at 5 and 9 hours. Use linear interpolation with MATLAB to estimate the temperature at those times.

Time (hours, P.M.)	1	2	3	4	5	6	7	8	9	10	11	12
Temperature (°C)	10	9	18	24	?	21	20	18	?	15	13	11

25. The following table gives temperature data in °C as a function of time of day and day of the week at a specific location. Data are missing for the entries marked with a question mark (?). Use linear interpolation with MATLAB to estimate the temperature at the missing points.

Hour	Mon	Tues	Wed	Thurs	Fri
			Day		
1	17	15	12	16	16
2	13	?	8	11	12
3	14	14	9	?	15
4	17	15	14	15	19
5	23	18	17	20	24

26. Computer-controlled machines are used to cut and to form metal and other materials when manufacturing products. These machines often use cubic splines to specify the path to be cut or the contour of the part to be shaped. The following coordinates specify the shape of a certain car's front fender. Fit a series of cubic splines to the coordinates, and plot the splines along with the coordinate points.

x (ft)	0	0.25	0.75	1.25	1.5	1.75	1.875	2	2.125	2.25
y (ft)	1.2	1.18	1.1	1	0.92	0.8	0.7	0.55	0.35	0

27. The following data are the measured temperature T of water flowing from a hot water faucet after it is turned on at time $t = 0$.

t (sec)	$T(°F)$	t (sec)	$T(°F)$
0	72.5	6	109.3
1	78.1	7	110.2
2	86.4	8	110.5
3	92.3	9	109.9
4	110.6	10	110.2
5	111.5		

   a. Plot the data, connecting them first with straight lines and then with a cubic spline.

   b. Estimate the temperature values at the following times, using linear interpolation and then cubic spline interpolation: $t = 0.6, 2.5, 4.7, 8.9$.

   c. Use both the linear and cubic spline interpolations to estimate the time it will take for the temperature to equal the following values: $T = 75$, 85, 90, 105.

28. The U.S. census data from 1790 to 1990 is stored in the file census.dat, which is supplied with MATLAB. Type load census to load this file. The first column, cdate, contains the years, and the second column, pop, contains the population in millions. In Test Your Understanding Problem T6.2–2 in Chapter 6, we used a cubic polynomial to estimate the population in 1965 to be 189 million. Compare this prediction with that obtained using (a) linear interpolation, and (b) cubic spline interpolation.

Engineering in the 21st Century. . .

Additive Manufacturing

Three-dimensional (3D) printing builds a three-dimensional object by laying down successive layers of material. The process is controlled by a computer using solid modeling software. The original process used an inkjet printer to deposit a layer of liquid bonding agent onto a powder bed and is called binder jetting. Some newer methods can also be used to create the software in addition to using computer-aided-design (CAD) software. These include using a 3D scanner on an existing part, a small-scale model, or a sculpted model. Other sources use digital photos and photogrammetry software.

Later developments have led to a wider variety of techniques now described by the term *additive manufacturing* (AM). In addition to binder jetting, six other categories of AM are generally recognized. These are: directed energy deposition, material extrusion, material jetting, powder bed fusion, sheet lamination, and vat photopolymerization.

With directed energy deposition a high energy heating source such as a laser is used to fuse materials by melting. Material jetting deposits droplets of the building material. In powder bed fusion, thermal energy is used to fuse certain regions of a powder bed. The sheet lamination process bonds sheets of material to form an object. With vat photopolymerization, a liquid photopolymer in a vat is cured by light-activated cross-linking of adjoining polymer chains.

These technologies let manufacturers increase speed to market, eliminate costly tooling, molds, or dies, and produce small batches on order. Parts with more complex geometries and internal features can be produced. Shipping costs and shipping times are reduced because the low hardware cost enables many local and small manufacturing centers to be established.

MATLAB supports additive manufacturing in several ways. MATLAB files are used for converting 3D surface data into Standard Tessellation Language (STL) files, a format widely used in AM. MATLAB is used for topology optimization, a mathematical method for optimizing material layout within a given design space. It provides a new way to optimize load-bearing designs resulting in structures that are lighter yet stronger, whose interiors are cellular and look almost bone-like. These structures cannot be manufactured with traditional methods but can be made with AM. ∎

Linear Algebraic Equations

Linear algebraic equations such as

$$5x - 2y = 13$$
$$7x + 3y = 24$$

occur in many engineering applications. For example, electrical engineers use them to predict the power requirements for circuits; civil, mechanical, and aerospace engineers use them to design structures and machines; chemical engineers use them to compute material balances in chemical processes; and industrial engineers apply them to design schedules and operations. The examples and homework problems in this chapter explore some of these applications.

Linear algebraic equations can be solved "by hand" using pencil and paper, by calculator, or with software such as MATLAB. The choice depends on the circumstances. For equations with only two unknown variables, hand solution is easy and adequate. Some calculators can solve equation sets that have many variables. However, the greatest power and flexibility is obtained by using software.

For example, MATLAB can obtain and plot equation solutions as we vary one or more parameters.

Systematic solution methods have been developed for sets of linear equations. In Section 8.1 we introduce some matrix notation that is required for use with MATLAB and that is also useful for expressing solution methods in a compact way. The conditions for the existence and uniqueness of solutions are then introduced. Methods using MATLAB are treated in four sections: Section 8.2 covers the left-division method for solving equation sets that have unique solutions. Section 8.3 covers the case where the equation set does not contain enough information to determine all the unknown variables. This is the *underdetermined* case. The *overdetermined* case occurs when the equation set has more independent equations than unknowns (Section 8.4). A general solution program is given in Section 8.5.

8.1 Matrix Methods for Linear Equations

Sets of linear algebraic equations can be expressed as a single equation, using matrix notation. This standard and compact form is useful for expressing solutions and for developing software applications with an arbitrary number of variables. In this application, a vector is taken to be a column vector unless otherwise specified.

Matrix notation enables us to represent multiple equations as a single matrix equation. For example, consider the following set:

$$2x_1 + 9x_2 = 5$$
$$3x_1 - 4x_2 = 7$$

This set can be expressed in vector-matrix form as

$$\begin{bmatrix} 2 & 9 \\ 3 & -4 \end{bmatrix} \begin{bmatrix} x_1 \\ x_2 \end{bmatrix} = \begin{bmatrix} 5 \\ 7 \end{bmatrix}$$

which can be represented in the following compact form

$$\mathbf{Ax} = \mathbf{b} \tag{8.1-1}$$

where we have defined the following matrices and vectors:

$$\mathbf{A} = \begin{bmatrix} 2 & 9 \\ 3 & -4 \end{bmatrix} \qquad \mathbf{x} = \begin{bmatrix} x_1 \\ x_2 \end{bmatrix} \qquad \mathbf{b} = \begin{bmatrix} 5 \\ 7 \end{bmatrix}$$

In general, the set of m equations in n unknowns can be expressed in the form Equation (8.1-1), where $\mathbf{A}$ is $m \times n$, $\mathbf{x}$ is $n \times 1$, and $\mathbf{b}$ is $m \times 1$.

Matrix Inverse

The solution of the scalar equation $ax = b$ is $x = b/a$ if $a \neq 0$. The division operation of scalar algebra has an analogous operation in matrix algebra. For example, to solve the matrix Equation (8.1-1) for $\mathbf{x}$, we must somehow "divide" $\mathbf{b}$ by $\mathbf{A}$.

The procedure for doing this is developed from the concept of a *matrix inverse*. The inverse of a matrix $\mathbf{A}$ is denoted by $\mathbf{A}^{-1}$ and has the property that

$$\mathbf{A}^{-1}\mathbf{A} = \mathbf{A}\mathbf{A}^{-1} = \mathbf{I}$$

where $\mathbf{I}$ is the identity matrix. Using this property, we multiply both sides of Equation (8.1–1) from the left by $\mathbf{A}^{-1}$ to obtain $\mathbf{A}^{-1}\mathbf{A}\mathbf{x} = \mathbf{A}^{-1}\mathbf{b}$. Because $\mathbf{A}^{-1}\mathbf{A}\mathbf{x} = \mathbf{I}\mathbf{x} = \mathbf{x}$, we obtain the solution

$$\mathbf{x} = \mathbf{A}^{-1}\mathbf{b} \tag{8.1–2}$$

The inverse of a matrix $\mathbf{A}$ is defined only if $\mathbf{A}$ is square and nonsingular. A matrix is *singular* if its determinant $|\mathbf{A}|$ is zero. If $\mathbf{A}$ is singular, then a unique solution to Equation (8.1–1) does not exist. The MATLAB functions `inv(A)` and `det(A)` compute the inverse and the determinant of the matrix $\mathbf{A}$. If the `inv(A)` function is applied to a *singular matrix*, MATLAB will issue a warning to that effect.

SINGULAR MATRIX

An *ill-conditioned* set of equations is a set that is close to being singular. The ill-conditioned status depends on the accuracy with which the solution calculations are made. When internal numerical accuracy used by MATLAB is insufficient to obtain a solution, it prints the message warning that the matrix is close to singular and that the results might be inaccurate.

ILL-CONDITIONED

For a 2×2 matrix $\mathbf{A}$,

$$\mathbf{A} = \begin{bmatrix} a & b \\ c & d \end{bmatrix} \qquad \mathbf{A}^{-1} = \frac{1}{ad - bc} \begin{bmatrix} d & -b \\ -c & a \end{bmatrix}$$

where $\det(\mathbf{A}) = ad - bc$. Thus $\mathbf{A}$ is singular if $ad - bc = 0$.

The Matrix Inverse Method

EXAMPLE 8.1–1

Solve the following equations, using the matrix inverse.

$$2x_1 + 9x_2 = 5$$
$$3x_1 - 4x_2 = 7$$

■ Solution

The matrix $\mathbf{A}$ and the vector $\mathbf{b}$ are

$$\mathbf{A} = \begin{bmatrix} 2 & 9 \\ 3 & -4 \end{bmatrix} \qquad \mathbf{b} = \begin{bmatrix} 5 \\ 7 \end{bmatrix}$$

The session is

```
>>A = [2,9;3,-4]; b = [5;7];
>>x = inv(A)*b
x =
   2.3714
   0.0286
```

The solution is $x_1 = 2.3714$ and $x_2 = 0.0286$. MATLAB did not issue a warning, so the solution is unique.

The solution form $\mathbf{x} = \mathbf{A}^{-1}\mathbf{b}$ is rarely applied in practice to obtain numerical solutions to sets of many equations, because calculation of the matrix inverse is likely to introduce greater numerical inaccuracy than the left-division method to be introduced.

Test Your Understanding

T8.1–1   For what values of c will the following set (a) have a unique solution and (b) have an infinite number of solutions? Find the relation between x_1 and x_2 for these solutions.

$$6x_1 + cx_2 = 0$$
$$2x_1 + 4x_2 = 0$$

(Answers: (a) $c \neq 12$, $x_1 = x_2 = 0$; (b) $c = 12$, $x_1 = -2x_2$)

T8.1–2   Use the matrix inverse method to solve the following set.

$$3x_1 - 4x_2 = 5$$
$$6x_1 - 10x_2 = 2$$

(Answer: $x_1 = 7$, $x_2 = 4$)

T8.1–3   Use the matrix inverse method to solve the following set.

$$3x_1 - 4x_2 = 5$$
$$6x_1 - 8x_2 = 2$$

(Answer: No solution.)

Existence and Uniqueness of Solutions

The matrix inverse method will warn us if a *unique* solution does not exist, but it does not tell us whether there is no solution or an infinite number of solutions. In addition, the method is limited to cases where the matrix $\mathbf{A}$ is square, that is, cases where the number of equations equals the number of unknowns. For this reason we now introduce a method that allows us to determine easily whether an equation set has a solution and whether it is unique. The method requires the concept of the *rank of a matrix*.

RANK OF A MATRIX

Consider the 3×3 determinant

$$|\mathbf{A}| = \begin{vmatrix} 3 & -4 & 1 \\ 6 & 10 & 2 \\ 9 & -7 & 3 \end{vmatrix} = 0 \qquad (8.1\text{–}3)$$

If we eliminate one row and one column in the determinant, we are left with a 2×2 determinant. Depending on which row and column we choose to eliminate, there are nine possible 2×2 determinants we can obtain. These are called *subdeterminants*. For example, if we eliminate the second row and third column, we obtain

$$\begin{vmatrix} 3 & -4 \\ 9 & -7 \end{vmatrix} = 3(-7) - 9(-4) = 15$$

SUBDETER-
MINANTS

Subdeterminants are used to define the *rank* of a matrix. The definition of matrix rank is as follows.

> **Definition of Matrix Rank.** An $m \times n$ matrix **A** has a *rank* $r \geq 1$ if and only if $|A|$ contains a nonzero $r \times r$ determinant and every square subdeterminant with $r + 1$ or more rows is zero.

For example, the rank of **A** in Equation (8.1–3) is 2 because $|A| = 0$ while $|A|$ contains at least one nonzero 2×2 subdeterminant. To determine the rank of a matrix **A** in MATLAB, type `rank(A)`. If **A** is $n \times n$, its rank is n if $\det(A) \neq 0$.

We can use the following test to determine if a solution exists to $Ax = b$ and whether it is unique. The test requires that we first form the *augmented matrix* $[A\ b]$.

AUGMENTED
MATRIX

> **Existence and Uniqueness of Solutions.** The set $Ax = b$ with m equations and n unknowns has solutions if and only if (1) rank(A) = rank($[A\ b]$). Let $r =$ rank(A). If condition (1) is satisfied and if $r = n$, then the solution is unique. If condition (1) is satisfied but $r < n$, there are an infinite number of solutions, and r unknown variables can be expressed as linear combinations of the other $n - r$ unknown variables, whose values are arbitrary.

> **Homogeneous Case.** The homogeneous set $Ax = 0$ is a special case in which $b = 0$. For this case, rank(A) = rank($[A\ b]$) always, and thus the set always has the trivial solution $x = 0$. A nonzero solution, in which at least one unknown is nonzero, exists if and only if rank(A) $< n$. If $m < n$, the homogeneous set always has a nonzero solution.

This test implies that if **A** is square and of dimension $n \times n$, then rank($[A\ b]$) = rank(A), and a unique solution exists for any **b** if rank(A) = n.

8.2 The Left-Division Method

MATLAB provides the *left-division method* for solving the equation set $Ax = b$. This method is based on *Gauss elimination*. To use the left-division method to solve for **x**, you type `x = A\b`. If $|A| = 0$ or if the number of equations does not equal the number of unknowns, then you need to use the other methods to be presented later.

GAUSS
ELIMINATION

Left-Division Method with Three Unknowns

EXAMPLE 8.2–1

Use the left-division method to solve the following set.

$$\begin{aligned} 3x_1 + 2x_2 - 9x_3 &= -65 \\ -9x_1 - 5x_2 + 2x_3 &= 16 \\ 6x_1 + 7x_2 + 3x_3 &= 5 \end{aligned}$$

■ **Solution**

The matrices **A** and **b** are

$$\mathbf{A} = \begin{bmatrix} 3 & 2 & -9 \\ -9 & -5 & 2 \\ 6 & 7 & 3 \end{bmatrix} \qquad \mathbf{b} = \begin{bmatrix} -65 \\ 16 \\ 5 \end{bmatrix}$$

The session is

```
>>A = [3,2,-9;-9,-5,2;6,7,3];
>>rank(A)
ans =
    3
```

Because **A** is 3×3 and rank(**A**) = 3, which is the number of unknowns, a unique solution exists. It is obtained by continuing the session as follows:

```
>>b = [-65;16;5];
>>x = A\b
x =
    2.0000
   -4.0000
    7.0000
```

This answer gives the vector **x,** which corresponds to the solution $x_1 = 2$, $x_2 = -4$, $x_3 = 7$.

For the solution $\mathbf{x} = \mathbf{A}^{-1}\mathbf{b}$, vector **x** is proportional to the vector **b.** We can use this linearity property to obtain a more generally useful algebraic solution in cases where the right-hand sides are all multiplied by the same scalar. For example, suppose the matrix equation is $\mathbf{Ay} = \mathbf{b}c$, where c is a scalar. The solution is $\mathbf{y} = \mathbf{A}^{-1}\mathbf{b}c = \mathbf{x}c$. Thus if we obtain the solution to $\mathbf{Ax} = \mathbf{b}$, the solution to $\mathbf{Ay} = \mathbf{b}c$ is given by $\mathbf{y} = \mathbf{x}c$.

| EXAMPLE 8.2–2 | Calculation of Cable Tension |

A mass m is suspended by three cables attached at three points B, C, and D, as shown in Figure 8.2–1. Let T_1, T_2, and T_3 be the tensions in the three cables AB, AC, and AD, respectively. If the mass m is stationary, the sum of the tension components in the x, in the y, and in the z directions must each be zero. This gives the following three equations:

$$\frac{T_1}{\sqrt{35}} - \frac{3T_2}{\sqrt{34}} + \frac{T_3}{\sqrt{42}} = 0$$

$$\frac{3T_1}{\sqrt{35}} - \frac{4T_3}{\sqrt{42}} = 0$$

$$\frac{5T_1}{\sqrt{35}} + \frac{5T_2}{\sqrt{34}} + \frac{5T_3}{\sqrt{42}} - mg = 0$$

Determine T_1, T_2, and T_3 in terms of an unspecified value of the weight mg.

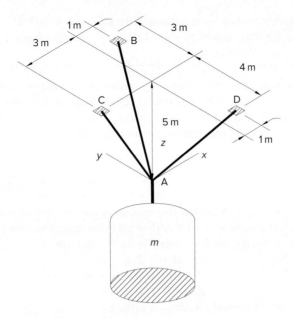

Figure 8.2–1 A mass suspended by three cables.

■ Solution

If we set $mg = 1$, the equations have the form $\mathbf{AT} = \mathbf{b}$ where

$$
\mathbf{A} = \begin{bmatrix} \dfrac{1}{\sqrt{35}} & -\dfrac{3}{\sqrt{34}} & \dfrac{1}{\sqrt{42}} \\[2mm] \dfrac{3}{\sqrt{35}} & 0 & -\dfrac{4}{\sqrt{42}} \\[2mm] \dfrac{5}{\sqrt{35}} & \dfrac{5}{\sqrt{34}} & \dfrac{5}{\sqrt{42}} \end{bmatrix} \qquad \mathbf{T} = \begin{bmatrix} T_1 \\ T_2 \\ T_3 \end{bmatrix} \qquad \mathbf{b} = \begin{bmatrix} 0 \\ 0 \\ 1 \end{bmatrix}
$$

The script file to solve this system is

```
% File cable.m
s34 = sqrt(34); s35 = sqrt(35); s42 = sqrt(42);
A1 = [1/s35, -3/s34, 1/s42];
A2 = [3/s35, 0, -4/s42];
A3 = [5/s35, 5/s34, 5/s42];
A = [A1; A2; A3];
b = [0; 0; 1];
rank(A)
rank([A, b])
T = A\b
```

When this file is executed by typing `cable`, we find that rank($\mathbf{A}$) = rank ([$\mathbf{A}$ $\mathbf{b}$]) = 3 and obtain the values $T_1 = 0.5071$, $T_2 = 0.2915$, and $T_3 = 0.4166$. Because $\mathbf{A}$ is 3×3 and rank($\mathbf{A}$) = 3, which is the number of unknowns, the solution is unique. Using the linearity property, we multiply these results by mg and obtain the general solution $T_1 = 0.5071 \ mg$, $T_2 = 0.2915 \ mg$, and $T_3 = 0.4166 \ mg$.

Linear equations are useful in many engineering fields. Electric circuits are a common source of linear equation models. The circuit designer must be able to solve them to predict the currents that will exist in the circuit. This information is often needed to determine the power supply requirements among other things.

EXAMPLE 8.2–3   An Electric Resistance Network

The circuit shown in Figure 8.2–2 has five resistances and two applied voltages. Assuming that the positive directions of current flow are in the directions shown in the figure, Kirchhoff's voltage law applied to each loop in the circuit gives

$$-v_1 + R_1 i_1 + R_4 i_4 \quad = 0$$
$$-R_4 i_4 + R_2 i_2 + R_5 i_5 = 0$$
$$-R_5 i_5 + R_3 i_3 + v_2 \quad = 0$$

Conservation of charge applied at each node in the circuit gives

$$i_1 = i_2 + i_4$$
$$i_2 = i_3 + i_5$$

You can use these two equations to eliminate i_4 and i_5 from the first three equations. The result is

$$(R_1 + R_4)i_1 - R_4 i_2 \quad = v_1$$
$$-R_4 i_1 + (R_2 + R_4 + R_5)i_2 - R_5 i_3 = 0$$
$$R_5 i_2 - (R_3 + R_5)i_3 = v_2$$

Thus we have three equations in the three unknowns i_1, i_2, and i_3.

Write a MATLAB script file that uses given values of the applied voltages v_1 and v_2 and given values of the five resistances to solve for the currents i_1, i_2, and i_3. Use the

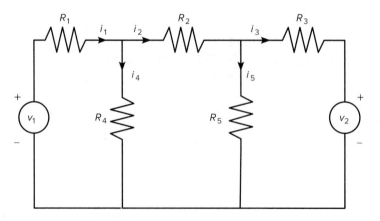

Figure 8.2–2 An electric resistance network.

program to find the currents for the case $R_1 = 5$, $R_2 = 100$, $R_3 = 200$, $R_4 = 150$, and $R_5 = 250$ kΩ and $v_1 = 100$ and $v_2 = 50$ V. (Note that 1 kΩ = 1000 Ω.)

■ **Solution**

Because there are as many unknowns as equations, there will be a unique solution if $|\mathbf{A}| \neq 0$; in addition, the left-division method will generate an error message if $|\mathbf{A}| = 0$. The following script file, named `resist.m`, uses the left-division method to solve the three equations for i_1, i_2, and i_3.

```
% File resist.m
% Solves for the currents i_1, i_2, i_3
R = [5,100,200,150,250]*1000;
v1 = 100; v2 = 50;
A1 = [R(1) + R(4), -R(4), 0];
A2 = [-R(4), R(2) + R(4) + R(5), -R(5)];
A3 = [0, R(5), -(R(3) + R(5))];
A = [A1; A2; A3];
b=[v1; 0; v2];
current = A\b;
disp('The currents are:')
disp(current)
```

The row vectors `A1`, `A2`, and `A3` were defined to avoid typing the lengthy expression for `A` in one line. This script is executed from the command prompt as follows:

```
>>resist
The currents are:
   1.0e-003*
   0.9544
   0.3195
   0.0664
```

Because MATLAB did not generate an error message, the solution is unique. The currents are $i_1 = 0.9544$, $i_2 = 0.3195$, and $i_3 = 0.0664$ mA, where 1 mA = 1 milliampere = 0.001 A.

Ethanol Production

EXAMPLE 8.2–4

Engineers in the food and chemical industries use fermentation in many processes. The following equation describes Baker's yeast fermentation.

$$a(C_6H_{12}O_6) + b(O_2) + c(NH_3)$$
$$\rightarrow C_6H_{10}NO_3 + d(H_2O) + e(CO_2) + f(C_2H_6O)$$

The variables a, b, . . . , f represent the masses of the products involved in the reaction. In this formula $C_6H_{12}O_6$ represents glucose, $C_6H_{10}NO_3$ represents yeast, and C_2H_6O represents ethanol. This reaction produces ethanol in addition to water and carbon dioxide.

We want to determine the amount of ethanol f produced. The number of C, O, N, and H atoms on the left must balance those on the right side of the equation. This gives four equations:

$$6a = 6 + e + 2f$$
$$6a + 2b = 3 + d + 2e + f$$
$$c = 1$$
$$12a + 3c = 10 + 2d + 6f$$

The fermenter is equipped with an oxygen sensor and a carbon dioxide sensor. These enable us to compute the respiratory quotient R:

$$R = \frac{CO_2}{O_2} = \frac{e}{b}$$

Thus the fifth equation is $Rb - e = 0$. The yeast yield Y (grams of yeast produced per gram of glucose consumed) is related to a as follows:

$$Y = \frac{144}{180a}$$

where 144 is the molecular weight of yeast and 180 is the molecular weight of glucose. By measuring the yeast yield Y we can compute a as follows: $a = 144/180Y$. This is the sixth equation.

Write a user-defined function that computes f, the amount of ethanol produced, with R and Y as the function's arguments. Test your function for two cases where Y is measured to be 0.5: (a) $R = 1.1$ and (b) $R = 1.05$.

■ Solution

First note that there are only four unknowns because the third equation directly gives $c = 1$, and the sixth equation directly gives $a = 144/180Y$. To write these equations in matrix form, let $x_1 = b$, $x_2 = d$, $x_3 = e$, and $x_4 = f$. Then the equations can be written as

$$-x_3 - 2x_4 = 6 - 6(144/180Y)$$
$$2x_1 - x_2 - 2x_3 - x_4 = 3 - 6(144/180Y)$$
$$-2x_2 - 6x_4 = 7 - 12(144/180Y)$$
$$Rx_1 - x_3 = 0$$

In matrix form these become

$$\begin{bmatrix} 0 & 0 & -1 & -2 \\ 2 & -1 & -2 & -1 \\ 0 & -2 & 0 & -6 \\ R & 0 & -1 & 0 \end{bmatrix} \begin{bmatrix} x_1 \\ x_2 \\ x_3 \\ x_4 \end{bmatrix} = \begin{bmatrix} 6 - 6(144/180Y) \\ 3 - 6(144/180Y) \\ 7 - 12(144/180Y) \\ 0 \end{bmatrix}$$

The function file is shown below.

```
function E = ethanol(R,Y)
% Computes ethanol produced from yeast reaction.
A = [0,0,-1,-2;2,-1,-2,-1;...
     0,-2,0,-6;R,0,-1,0];
b = [6-6*(144./(180*Y));3-6*(144./(180*Y));...
```

```
        7-12*(144./(180*Y));0];
x = A\b;
E = x(4);
```

The session is as follows:

```
>>ethanol(1.1,0.5)
ans =
        0.0654
>>ethanol(1.05,0.5)
ans =
       -0.0717
```

The negative value for **E** in the second case indicates that ethanol is being consumed rather than produced.

Test Your Understanding

T8.2–1 Use the left-division method to solve the following set:

$$5x_1 - 3x_2 = 21$$
$$7x_1 - 2x_2 = 36$$

(Answers: $x_1 = 6$, $x_2 = 3$)

T8.2–2 Use MATLAB to solve the following equations:

$$6x - 4y + 3z = 5$$
$$4x + 3y - 2z = 23$$
$$2x + 6y + 3z = 63$$

(Answers: $x = 3$, $y = 7$, $z = 5$)

8.3 Underdetermined Systems

An *underdetermined system* does not contain enough information to determine all the unknown variables, usually but not always because it has fewer equations than unknowns. Thus an infinite number of solutions can exist, with one or more of the unknowns dependent on the remaining unknowns. The left-division method works for square and nonsquare **A** matrices. However, if **A** is not square, the left-division method can give answers that might be misinterpreted. We will show how to interpret MATLAB results correctly.

When there are fewer equations than unknowns, the left-division method might give a solution with some of the unknowns set equal to zero, but this is not the general solution. An infinite number of solutions might exist even when the number of equations equals the number of unknowns. This can occur when $|A| = 0$. For such systems the left-division method generates an error message

**PSEUDOINVERSE
METHOD**

**MINIMUM-NORM
SOLUTION**

warning us that the matrix **A** is singular. In such cases the *pseudoinverse method* x = pinv(A)*b gives one solution, the *minimum-norm solution*. In cases where there are an infinite number of solutions, the rref function can be used to express some of the unknowns in terms of the remaining unknowns, whose values are arbitrary.

An equation set can be underdetermined even though it has as many equations as unknowns. This can happen if some of the equations are not independent. Determining by hand whether all the equations are independent might not be easy, especially if the set has many equations, but it is easily done in MATLAB.

EXAMPLE 8.3–1   **An Underdetermined Set with Three Equations and Three Unknowns**

Show that the following set does not have a unique solution. How many of the unknowns will be undetermined? Interpret the results given by the left-division method.

$$2x_1 - 4x_2 + 5x_3 = -4$$
$$-4x_1 - 2x_2 + 3x_3 = 4$$
$$2x_1 + 6x_2 - 8x_3 = 0$$

■ Solution

A MATLAB session to check the ranks is

```
>>A = [2,-4,5;-4,-2,3;2,6,-8];
>>b = [-4;4;0];
>>rank(A)
ans =
   2
>>rank([A, b])
ans =
   2
>>x = A\b
Warning: Matrix is singular to working precision.
ans =
    NaN
    NaN
    NaN
```

Because the ranks of **A** and [**A b**] are equal, a solution exists. However, because the number of unknowns is 3 and is 1 greater than the rank of **A,** one of the unknowns will be undetermined. An infinite number of solutions exist, and we can solve for only two of the unknowns in terms of the third unknown. The set is underdetermined because there are fewer than three independent equations; the third equation can be obtained from the first two. To see this, add the first and second equations, to obtain $-2x_1 - 6x_2 + 8x_3 = 0$, which is equivalent to the third equation.

Note that we could also tell that the matrix **A** is singular because its rank is less than 3. If we use the left-division method, MATLAB returns a message warning that the problem is singular, and it does not produce an answer.

The pinv Function and the Euclidean Norm

The pinv function (which stands for "pseudoinverse") can be used to obtain a solution of an underdetermined set. To solve the equation set $\mathbf{Ax} = \mathbf{b}$ using the pinv function, you type x = pinv(A)*b. The pinv function gives a solution that gives the minimum value of the *Euclidean norm*, which is the magnitude of the solution vector $\mathbf{x}$. The magnitude of a vector $\mathbf{v}$ in three-dimensional space, having components x, y, z, is $\sqrt{x^2 + y^2 + z^2}$. It can be computed using matrix multiplication and the transpose as follows:

$$\sqrt{\mathbf{v}^T \mathbf{v}} = \sqrt{[x\ y\ z]^T \begin{bmatrix} x \\ y \\ z \end{bmatrix}} = \sqrt{x^2 + y^2 + z^2}$$

The generalization of this formula to an n-dimensional vector $\mathbf{v}$ gives the magnitude of the vector and is the Euclidean norm N. Thus

$$N = \sqrt{\mathbf{v}^T \mathbf{v}} \tag{8.3–1}$$

The MATLAB function norm(v) computes the Euclidean norm.

A Statically Indeterminate Problem

EXAMPLE 8.3–2

Determine the forces in the three equally spaced supports that hold up a light fixture. The supports are 5 ft apart. The fixture weighs 400 lb, and its mass center is 4 ft from the right end. Obtain the solution using the MATLAB left-division method and the pseudoinverse method.

■ Solution

Figure 8.3–1 shows the fixture and the free-body diagram, where T_1, T_2, and T_3 are the tension forces in the supports. For the fixture to be in equilibrium, the vertical forces must cancel, and the total moments about an arbitrary fixed point—say, the right endpoint—must be zero. These conditions give the two equations

$$T_1 + T_2 + T_3 - 400 = 0$$
$$400(4) - 10T_1 - 5T_2 = 0$$

or

$$T_1 + T_2 + T_3 = 400 \tag{8.3–2}$$
$$10T_1 + 5T_2 + 0T_3 = 1600 \tag{8.3–3}$$

Because there are more unknowns than equations, the set is underdetermined. Thus we cannot determine a unique set of values for the forces. Such a problem, when the equations of statics do not give enough equations, is called *statically indeterminate*. These equations can be written in the matrix form $\mathbf{AT} = \mathbf{b}$ as follows:

STATICALLY INDETERMINATE

$$\begin{bmatrix} 1 & 1 & 1 \\ 10 & 5 & 0 \end{bmatrix} \begin{bmatrix} T_1 \\ T_2 \\ T_3 \end{bmatrix} = \begin{bmatrix} 400 \\ 1600 \end{bmatrix}$$

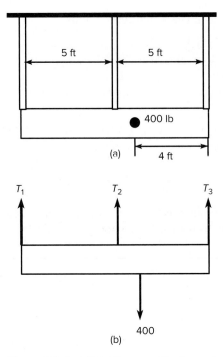

Figure 8.3–1 A light fixture and its free-body diagram.

The MATLAB session is

```
>>A = [1,1,1;10,5,0];
>>b = [400;1600];
>>rank(A)
ans =
   2
>>rank([A, b])
ans =
   2
>>T = A\b
T =
   160.0000
   0
   240.0000
>>T = pinv(A)*b
T =
   93.3333
   133.3333
   173.3333
```

The left-division answer corresponds to $T_1 = 160$, $T_2 = 0$, and $T_3 = 240$. This illustrates how the MATLAB left-division operator produces a solution with one or more variables set to zero, for underdetermined sets having more unknowns than equations.

Because the ranks of $\mathbf{A}$ and $[\mathbf{A} \; \mathbf{b}]$ are both 2, a solution exists, but it is not unique. Because the number of unknowns is 3, and is 1 greater than the rank of $\mathbf{A}$, an infinite number of solutions exist, and we can solve for only two of the unknowns in terms of the third.

The pseudoinverse solution gives $T_1 = 93.3333$, $T_2 = 133.3333$, and $T_3 = 173.3333$. This is the minimum-norm solution for real values of the variables. The minimum-norm solution consists of the real values of T_1, T_2, and T_3 that minimize

$$N = \sqrt{T_1^2 + T_2^2 + T_3^2}$$

To understand what MATLAB is doing, note that we can solve Equations (8.3–2) and (8.3–3) to obtain T_1 and T_2 in terms of T_3 as $T_1 = T_3 - 80$ and $T_2 = 480 - 2T_3$. Then the Euclidean norm can be expressed as

$$N = \sqrt{(T_3 - 80)^2 + (480 - 2T_3)^2 + T_3^2} = \sqrt{6T_3^2 - 2080T_3 + 236,800}$$

The real value of T_3 that minimizes N can be found by plotting N versus T_3, or by using calculus. The answer is $T_3 = 173.3333$, the same as the minimum-norm solution given by the pseudoinverse method.

Where there are an infinite number of solutions, we must decide whether the solutions given by the left-division and the pseudoinverse methods are useful for applications. This must be done in the context of the specific application.

Test Your Understanding

T8.3–1 Find two solutions to the following set:
$$x_1 + 3x_2 + 2x_3 = 2$$
$$x_1 + x_2 + x_3 = 4$$

(Answer: Minimum-norm solution: $x_1 = 4.33$, $x_2 = -1.67$, $x_3 = 1.34$. Left-division solution: $x_1 = 5$, $x_2 = -1$, $x_3 = 0$.)

The Reduced Row-Echelon Form

We can express some of the unknowns in an underdetermined set as functions of the remaining unknowns. In Example 8.3–2, we wrote the solutions for two of the unknowns in terms of the third: $T_1 = T_3 - 80$ and $T_2 = 480 - 2T_3$. These two equations are equivalent to

$$T_1 - T_3 = -80 \qquad T_2 + 2T_3 = 480$$

In matrix form these are

$$\begin{bmatrix} 1 & 0 & -1 \\ 0 & 1 & 2 \end{bmatrix} \begin{bmatrix} T_1 \\ T_2 \\ T_3 \end{bmatrix} = \begin{bmatrix} -80 \\ 480 \end{bmatrix}$$

The augmented matrix [**A b**] for the above set is

$$\begin{bmatrix} 1 & 0 & -1 & -80 \\ 0 & 1 & 2 & 480 \end{bmatrix}$$

Note that the first two columns form a 2×2 identity matrix. This indicates that the corresponding equations can be solved directly for T_1 and T_2 in terms of T_3.

We can always reduce an underdetermined set to such a form by multiplying the set's equations by suitable factors and adding the resulting equations to eliminate an unknown variable. The MATLAB rref function provides a procedure for reducing an equation set to this form, which is called the *reduced row-echelon form*. Its syntax is rref([A b]). Its output is the augmented matrix [**C d**] that corresponds to the equation set **Cx = d**. This set is in reduced row-echelon form.

EXAMPLE 8.3-3	Three Equations in Three Unknowns, *Continued*

The following underdetermined equation set was analyzed in Example 8.3–1. There it was shown that an infinite number of solutions exist. Use the rref function to obtain the solutions.

$$\begin{aligned} 2x_1 - 4x_2 + 5x_3 &= -4 \\ -4x_1 - 2x_2 + 3x_3 &= 4 \\ 2x_1 + 6x_2 - 8x_3 &= 0 \end{aligned}$$

■ **Solution**

The MATLAB session is

```
>>A = [2,-4,5;-4,-2,3;2,6,-8];
>>b = [-4;4;0];
>>rref([A, b])
ans =
    1     0    -0.1    -1.2000
    0     1    -1.3     0.4000
    0     0       0        0
```

The answer corresponds to the augmented matrix [**C d**], where

$$[\mathbf{C} \quad \mathbf{d}] = \begin{bmatrix} 1 & 0 & -0.1 & -1.2 \\ 0 & 1 & -1.3 & 0.4 \\ 0 & 0 & 0 & 0 \end{bmatrix}$$

This matrix corresponds to the matrix equation $\mathbf{Cx} = \mathbf{d}$, or

$$
\begin{aligned}
x_1 + 0x_2 - 0.1x_3 &= -1.2 \\
0x_1 + x_2 - 1.3x_3 &= 0.4 \\
0x_1 + 0x_2 - 0x_3 &= 0
\end{aligned}
$$

These can be easily solved for x_1 and x_2 in terms of x_3 as follows: $x_1 = 0.1x_3 - 1.2$, $x_2 = 1.3x_3 + 0.4$. This is the general solution to the problem, where x_3 is taken to be the arbitrary variable.

Supplementing Underdetermined Systems

Often the linear equations describing the application are underdetermined because not enough information has been specified to determine unique values of the unknowns. In such cases we might be able to include additional information, objectives, or constraints to find a unique solution. We can use the `rref` command to reduce the number of unknown variables in the problem, as illustrated in the next two examples.

Production Planning

EXAMPLE 8.3–4

The following table shows how many hours reactors A and B need to produce 1 ton each of the chemical products 1, 2, and 3. The two reactors are available for 40 and 30 hr per week, respectively. Determine how many tons of each product can be produced each week.

Hours	Product 1	Product 2	Product 3
Reactor A	5	3	3
Reactor B	3	3	4

■ **Solution**

Let x, y, and z be the number of tons each of products 1, 2, and 3 that can be produced in one week. Using the data for reactor A, the equation for its usage in one week is

$$5x + 3y + 3z = 40$$

The data for reactor B gives

$$3x + 3y + 4z = 30$$

This system is underdetermined. The matrices for the equation $\mathbf{Ax} = \mathbf{b}$ are

$$
\mathbf{A} = \begin{bmatrix} 5 & 3 & 3 \\ 3 & 3 & 4 \end{bmatrix} \qquad \mathbf{b} = \begin{bmatrix} 40 \\ 30 \end{bmatrix} \qquad \mathbf{x} = \begin{bmatrix} x \\ y \\ z \end{bmatrix}
$$

Here the rank($\mathbf{A}$) = rank($[\mathbf{A} \ \mathbf{b}]$) = 2, which is less than the number of unknowns. Thus an infinite number of solutions exist, and we can determine two of the variables in terms of the third.

Using the `rref` command `rref([A b])`, where `A = [5,3,3;3,3,4]` and `b = [40;30]`, we obtain the following augmented matrix:

$$\begin{bmatrix} 1 & 0 & -0.5 & 5 \\ 0 & 1 & 1.8333 & 5 \end{bmatrix}$$

This matrix gives the reduced system

$$\begin{aligned} x - 0.5z &= 5 \\ y + 1.8333z &= 5 \end{aligned}$$

which can be easily solved as follows:

$$x = 5 + 0.5z \tag{8.3-4}$$
$$y = 5 - 1.8333z \tag{8.3-5}$$

where z is arbitrary. However, z cannot be completely arbitrary if the solution is to be meaningful. For example, negative values of the variables have no meaning here; thus we require that $x \geq 0$, $y \geq 0$, and $z \leq 0$. Equation (8.3–4) shows that $x \geq 0$ if $z \geq -10$. From Equation (8.3–5), $y \geq 0$ implies that $z \leq 5/1.8333 = 2.727$. Thus valid solutions are those given by Equations (8.3–4) and (8.3–5), where $0 \leq z \leq 2.737$ tons. The choice of z within this range must be made on some other basis, such as profit.

For example, suppose we make a profit of \$400, \$600, and \$100 per ton for products 1, 2, and 3, respectively. Then our total profit P is

$$\begin{aligned} P &= 400x + 600y + 100z \\ &= 400(5 + 0.5z) + 600(5 - 1.8333z) + 100z \\ &= 5000 - 800z \end{aligned}$$

Thus to maximize profit, we should choose z to be the smallest possible value, namely, $z = 0$. This choice gives $x = y = 5$ tons.

However, if the profits for each product were \$3000, \$600, and \$100, the total profit would be $P = 18{,}000 + 500z$. Thus we should choose z to be its maximum, namely, $z = 2.727$ tons. From Equations (8.3–4) and (8.3–5), we obtain $x = 6.36$ and $y = 0$ tons.

EXAMPLE 8.3–5   Traffic Engineering

A traffic engineer wants to know if measurements of traffic flow entering and leaving a road network are sufficient to predict the traffic flow on each street in the network. For example, consider the network of one-way streets shown in Figure 8.3–2. The numbers shown are the measured traffic flows in vehicles per hour. Assume that no vehicles park anywhere within the network. If possible, calculate the traffic flows f_1, f_2, f_3, and f_4. If this is not possible, suggest how to obtain the necessary information.

■ **Solution**

The flow *into* intersection 1 must equal the flow *out* of the intersection. This gives

$$100 + 200 = f_1 + f_4$$

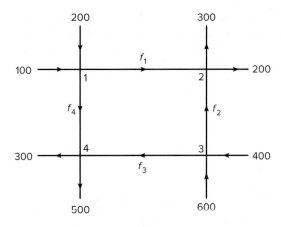

Figure 8.3–2 A network of one-way streets.

Similarly, for the other three intersections, we have

$$f_1 + f_2 = 300 + 200$$
$$600 + 400 = f_2 + f_3$$
$$f_3 + f_4 = 300 + 500$$

Putting these in the matrix form $\mathbf{Ax} = \mathbf{b}$, we obtain

$$\mathbf{A} = \begin{bmatrix} 1 & 0 & 0 & 1 \\ 1 & 1 & 0 & 0 \\ 0 & 1 & 1 & 0 \\ 0 & 0 & 1 & 1 \end{bmatrix} \quad \mathbf{b} = \begin{bmatrix} 300 \\ 500 \\ 1000 \\ 800 \end{bmatrix} \quad \mathbf{x} = \begin{bmatrix} f_1 \\ f_2 \\ f_3 \\ f_4 \end{bmatrix}$$

First, check the ranks of $\mathbf{A}$ and $[\mathbf{A}\ \mathbf{b}]$, using the MATLAB `rank` function. Both have a rank of 3, which is 1 less than the number of unknowns, so we can determine three of the unknowns in terms of the fourth. Thus we cannot determine all the traffic flows based on the given measurements.

Using the `rref([A b])` function produces the augmented matrix

$$\begin{bmatrix} 1 & 0 & 0 & 1 & 300 \\ 0 & 1 & 0 & -1 & 200 \\ 0 & 0 & 1 & 1 & 800 \\ 0 & 0 & 0 & 0 & 0 \end{bmatrix}$$

which corresponds to the reduced system

$$f_1 + f_4 = 300$$
$$f_2 - f_4 = 200$$
$$f_3 + f_4 = 800$$

These can be solved easily as follows: $f_1 = 300 - f_4$, $f_2 = 200 + f_4$, and $f_3 = 800 - f_4$. If we could measure the flow on one of the internal roads, say f_4, then we could compute the other flows. So we recommend that the engineer arrange to have this additional measurement made.

Test Your Understanding

T8.3–2 Use the `rref`, `pinv`, and the left-division methods to solve the following set.

$$3x_1 + 5x_2 + 6x_3 = 6$$
$$8x_1 - x_2 + 2x_3 = 1$$
$$5x_1 - 6x_2 - 4x_3 = -5$$

(Answer: There are an infinite number of solutions. The result obtained with the `rref` function is $x_1 = 0.2558 - 0.3721x_3$, $x_2 = 1.0465 - 0.9767x_3$, x_3 arbitrary. The `pinv` function gives $x_1 = 0.0571$, $x_2 = 0.5249$, $x_3 = 0.5340$. The left-division method generates an error message.)

T8.3–3 Use the `rref`, `pinv`, and left-division methods to solve the following set.

$$3x_1 + 5x_2 + 6x_3 = 4$$
$$x_1 - 2x_2 - 3x_3 = 10$$

(Answer: There are an infinite number of solutions. The result obtained with the `rref` function is $x_1 = 0.2727x_3 + 5.2727$, $x_2 = -1.3636x_3 - 2.2626$, x_3 arbitrary. The solution obtained with left division is $x_1 = 4.8000$, $x_2 = 0$, $x_3 = -1.7333$. The one obtained with the pseudoinverse method is $x_1 = 4.8394$, $x_2 = -0.1972$, $x_3 = -1.5887$.)

8.4 Overdetermined Systems

An *overdetermined system* is a set of equations that has more independent equations than unknowns. Some overdetermined systems have exact solutions, and they can be obtained with the left-division method `x = A\b`. For other overdetermined systems, no exact solution exists; in some of these cases, the left-division method does not yield an answer, while in other cases the left-division method gives an answer that satisfies the equation set only in a "least-squares" sense. We will show what this means in the next example. When MATLAB gives an answer to an overdetermined set, it does not tell us whether the answer is the exact solution. We must determine this information ourselves, and we will now show how to do this.

The Least-Squares Method

EXAMPLE 8.4–1

Suppose we have the following three data points, and we want to find the straight line $y = c_1 x + c_2$ that best fits the data in some sense.

x	y
0	2
5	6
10	11

(a) Find the coefficients c_1 and c_2 using the least-squares criterion. (b) Find the coefficients by using the left-division method to solve the three equations (one for each data point) for the two unknowns c_1 and c_2. Compare with the answer from part (a).

■ **Solution**
(a) Because two points define a straight line, unless we are extremely lucky, our three data points will not lie on the same straight line. A common criterion for obtaining the straight line that best fits the data is the *least-squares* criterion. According to this criterion, the line that minimizes J, the sum of the squares of the vertical differences between the line and the data points, is the "best" fit. Here J is

LEAST-SQUARES METHOD

$$J = \sum_{i=1}^{i=3}(c_1 x_i + c_2 - y_i)^2 = (0c_1 + c_2 - 2)^2 + (5c_1 + c_2 - 6)^2 + (10c_1 + c_2 - 11)^2$$

If you are familiar with calculus, you know that the values of c_1 and c_2 that minimize J are found by setting the partial derivatives $\partial J/\partial c_1$ and $\partial J/\partial c_2$ equal to zero.

$$\frac{\partial J}{\partial c_1} = 250c_1 + 30c_2 - 280 = 0$$

$$\frac{\partial J}{\partial c_2} = 30c_1 + 6c_2 - 38 = 0$$

The solution is $c_1 = 0.9$ and $c_2 = 11/6$. The best straight line in the least-squares sense is $y = 0.9x + 11/6$.

(b) Evaluating the equation $y = c_1 x + c_2$ at each data point gives the following three equations, which are overdetermined because there are more equations than unknowns.

$$
\begin{align}
0c_1 + c_2 &= 2 \tag{8.4-1}\\
5c_1 + c_2 &= 6 \tag{8.4-2}\\
10c_1 + c_2 &= 11 \tag{8.4-3}
\end{align}
$$

These equations can be written in the matrix form $\mathbf{Ax} = \mathbf{b}$ as follows:

$$\mathbf{Ax} = \begin{bmatrix} 0 & 1 \\ 5 & 0 \\ 10 & 1 \end{bmatrix} \begin{bmatrix} c_1 \\ c_2 \end{bmatrix} = \begin{bmatrix} 2 \\ 6 \\ 11 \end{bmatrix} = \mathbf{b}$$

where

$$[\mathbf{A}\ \mathbf{b}] = \begin{bmatrix} 0 & 1 & 2 \\ 5 & 1 & 6 \\ 10 & 1 & 11 \end{bmatrix}$$

To use left division, the MATLAB session is

```
>>A = [0,1;5,1;10,1];
>>b = [2;6;11];
>>rank(A)
ans =
   2
>>rank([A, b])
ans =
   3
>>x = A\b
x =
    0.9000
    1.8333
>>A*x
ans =
    1.833
    6.333
    10.8333
```

This result for $\mathbf{x}$ agrees with the least-squares solution obtained previously: $c_1 = 0.9$, $c_2 = 11/6 = 1.8333$. The rank of $\mathbf{A}$ is 2, but the rank of $[\mathbf{A}\ \mathbf{b}]$ is 3, so no exact solution exists for c_1 and c_2. Note that $\mathbf{A*x}$ gives the y values generated by the line $y = 0.9x + 1.8333$ at the x data values $x = 0, 5, 10$. These are different from the right-hand sides of the original three Equations (8.4–1) through (8.4–3). This is not unexpected, because the least-squares solution is not an exact solution of the equations.

Some overdetermined systems have an exact solution. The left-division method sometimes gives an answer for overdetermined systems, but it does not indicate whether the answer is the exact solution. We need to check the ranks of $\mathbf{A}$ and $[\mathbf{A}\ \mathbf{b}]$ to know if the answer is the exact solution. The next example illustrates this situation.

An Overdetermined Set

EXAMPLE 8.4–2

Solve the following equations and discuss the solution for two cases: $c = 9$ and $c = 10$.

$$x_1 + x_2 = 1$$
$$x_1 + 2x_2 = 3$$
$$x_1 + 5x_2 = c$$

■ **Solution**

The coefficient matrix and the augmented matrix for this problem are

$$\mathbf{A} = \begin{bmatrix} 1 & 1 \\ 1 & 2 \\ 1 & 5 \end{bmatrix} \qquad [\mathbf{A} \ \mathbf{b}] = \begin{bmatrix} 1 & 1 & 1 \\ 1 & 2 & 3 \\ 1 & 5 & c \end{bmatrix}$$

Making the computations in MATLAB, we find that for $c = 9$, rank $(\mathbf{A}) =$ rank$([\mathbf{A} \ \mathbf{b}]) = 2$. Thus the system has a solution, and because the number of unknowns (2) equals the rank of $\mathbf{A}$, there is a unique solution. The left-division method $\mathtt{A \backslash b}$ gives this solution, which is $x_1 = -1$ and $x_2 = 2$.

For $c = 10$ we find that rank$(\mathbf{A}) = 2$, but rank$([\mathbf{A} \ \mathbf{b}]) = 3$. Because rank$(\mathbf{A}) \neq$ rank $([\mathbf{A} \ \mathbf{b}])$, there is no solution. However, the left-division method $\mathtt{A \backslash b}$ gives $x_1 = -1.3846$ and $x_2 = 2.2692$, which is *not* an exact solution! This can be verified by substituting these values into the original equation set. This answer is the solution to the equation set in a least-squares sense. That is, these values are the values of x_1 and x_2 that minimize J, the sum of the squares of the differences between the equations' left- and right-hand sides.

$$J = (x_1 + x_2 - 1)^2 + (x_1 + 2x_2 - 3)^2 + (x_1 + 5x_2 - 10)^2$$

To interpret MATLAB answers correctly for an overdetermined system, first check the ranks of $\mathbf{A}$ and $[\mathbf{A} \ \mathbf{b}]$ to see if an exact solution exists; if one does not exist, then we know that the left-division answer is a least-squares solution. In Section 8.5 we develop a general-purpose program that checks the ranks and solves a general set of linear equations.

Test Your Understanding

T8.4–1 Solve the following set.

$$x_1 - 3x_2 = 2$$
$$3x_1 + 5x_2 = 7$$
$$70x_1 - 28x_2 = 153$$

(Answer: There is a unique solution: $x_1 = 2.2143$, $x_2 = 0.0714$, which is given by the left-division method.)

T8.4–2 Show why there is no solution to the following set.

$$x_1 - 3x_2 = \ \ \ 2$$
$$3x_1 + 5x_2 = \ \ \ 7$$
$$5x_1 - 2x_2 = -4$$

8.5 A General Solution Program

In this chapter you saw that the set of linear algebraic equations $\mathbf{Ax} = \mathbf{b}$ with m equations and n unknowns has solutions if and only if (1) rank[$\mathbf{A}$] = rank[$\mathbf{A}\ \mathbf{b}$]. Let r = rank[$\mathbf{A}$]. If condition (1) is satisfied and if $r = n$, then the solution is unique. If condition (1) is satisfied but $r < n$, an infinite number of solutions exist; in addition, r unknown variables can be expressed as linear combinations of the other $n - r$ unknown variables, whose values are arbitrary. In this case we can use the rref command to find the relations between the variables. The pseudocode in Table 8.5–1 can be used to outline an equation solver program before writing it.

A condensed flowchart is shown in Figure 8.5–1. From this chart or the pseudocode, we can develop the script file shown in Table 8.5–2. The program uses the given arrays A and b to check the rank conditions; the left-division method to obtain the solution, if one exists; and the rref method if there are an infinite number of solutions. Note that the number of unknowns equals the number of columns in A, which is given by size_A(2), the second element in size_A. Note also that the rank of $\mathbf{A}$ cannot exceed the number of columns in $\mathbf{A}$.

Test Your Understanding

T8.5–1 Type in the script file lineq.m given in Table 8.5–2 and run it for the following cases. Hand-check the answers.

  a. A = [1,-1;1,1], b = [3;5]
  b. A = [1,-1;2,-2], b = [3;6]
  c. A = [1,-1;2,-2], b = [3;5]

Table 8.5–1 Pseudocode for the linear equation solver

If the rank of $\mathbf{A}$ equals the rank of [$\mathbf{A}\ \mathbf{b}$], then
    determine whether the rank of $\mathbf{A}$ equals the number of unknowns. If so, there is a
    unique solution, which can be computed using left division. Display the results
    and stop.
    Otherwise, there are an infinite number of solutions, which can be found from the
    augmented matrix. Display the results and stop.
Otherwise (if the rank of $\mathbf{A}$ does not equal the rank of [$\mathbf{A}\ \mathbf{b}$]), there are no solutions.
Display this message and stop.

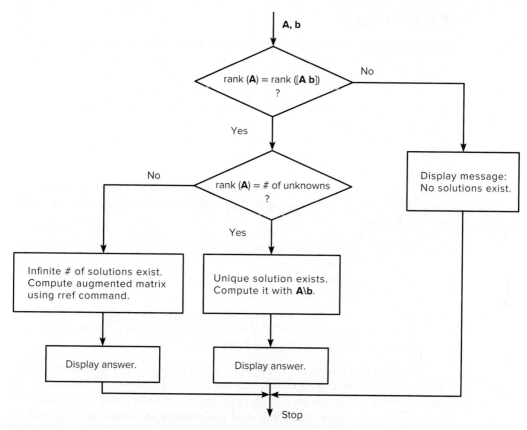

Figure 8.5–1 Flowchart illustrating a program to solve linear equations.

Table 8.5–2 MATLAB program to solve linear equations

```
% Script file lineq.m
% Solves the set Ax = b, given A and b.
% Check the ranks of A and [A b].
if rank(A) == rank([A b])
    % The ranks are equal.
    size_A = size(A);
    % Does the rank of A equal the number of unknowns?
    if rank(A) == size_A(2)
        % Yes. Rank of A equals the number of unknowns.
        disp('There is a unique solution, which is:')
        x = A\b % Solve using left division.
    else
        % Rank of A does not equal the number of unknowns.
        disp('There is an infinite number of solutions.')
        disp('The augmented matrix of the reduced system is:')
        rref([A b]) % Compute the augmented matrix.
    end
else
    % The ranks of A and [A b] are not equal.
    disp('There are no solutions.')
end
```

8.6 Summary

If the number of equations in the set *equals* the number of unknown variables, MATLAB provides two ways of solving the equation set $\mathbf{Ax} = \mathbf{b}$: the matrix inverse method, x = inv(A)*b, and the matrix left-division method, x = A\b. If MATLAB does not generate an error message when you use one of these methods, then the set has a unique solution. You can always check the solution for x by typing Ax to see if the result is the same as b. If you receive an error message, the set is underdetermined (even though it may have an equal number of equations and unknowns), and either it does not have a solution, or it has more than one solution.

For underdetermined sets, MATLAB provides three ways of dealing with the equation set $\mathbf{Ax} = \mathbf{b}$ (note that the matrix inverse method will never work with such sets):

1. The matrix left-division method (which gives one specific solution, but not the general solution).

2. The pseudoinverse method. Solve for x by typing x = pinv(A)*b. This gives the minimum-norm solution.

3. The reduced row-echelon form (RREF) method. This method uses the MATLAB command rref to obtain a general solution for some of the unknowns in terms of the other unknowns.

The four methods are summarized in Table 8.6–1. You should be able to determine whether a unique solution, an infinite number of solutions, or no solution exists. You can do this by applying the existence and uniqueness test given in the subsection **Existence and Uniqueness of Solutions** at the end of Section 8.1.

Some overdetermined systems have exact solutions, and they can be obtained with the left-division method, but the method does not indicate that the solution is exact. To determine this, first check the ranks of $\mathbf{A}$ and $[\mathbf{A} \ \mathbf{b}]$ to see if a solution exists; if one does not exist, then we know that the left-division solution is a least-squares answer.

Table 8.6–1 Matrix functions and commands for solving linear equations

Function	Description
det(A)	Computes the determinant of the array **A**.
inv(A)	Computes the inverse of the matrix **A**.
pinv(A)	Computes the pseudoinverse of the matrix **A**.
rank(A)	Computes the rank of the matrix **A**.
rref([A b])	Computes the reduced row-echelon form corresponding to the augmented matrix [**A b**].
x = inv(A)*b	Solves the matrix equation $\mathbf{Ax} = \mathbf{b}$ using the matrix inverse.
x = A\b	Solves the matrix equation $\mathbf{Ax} = \mathbf{b}$ using left division.

Key Terms

Problems

You can find the answers to problems marked with an asterisk at the end of the text.

Section 8.1

1. Solve the following problems using matrix inversion. Check your solutions by computing $\mathbf{A}^{-1}\mathbf{A}$.

a. $\quad 2x + y = 5$
$\quad\quad 3x - 9y = 7$

b. $-8x - 5y = 4$
$\quad\quad -2x + 7y = 10$

c. $\quad\quad 12x - 5y = 11$
$\quad\quad -3x + 4y + 7x_3 = -3$
$\quad\quad 6x + 2y + 3x_3 = 22$

d. $\quad 6x - 3y + 4x_3 = 41$
$\quad\quad 12x + 5y - 7x_3 = -26$
$\quad\quad -5x + 2y + 6x_3 = 16$

2.* *a.* Solve the following matrix equation for the matrix $\mathbf{C}$.

$$\mathbf{A}(\mathbf{BC} + \mathbf{A}) = \mathbf{B}$$

b. Evaluate the solution obtained in part *a* for the case

$$\mathbf{A} = \begin{bmatrix} 7 & 9 \\ -2 & 4 \end{bmatrix} \quad\quad \mathbf{B} = \begin{bmatrix} 4 & -3 \\ 7 & 6 \end{bmatrix}$$

3.   Use MATLAB to solve the following problems.

a. $-2x + y = -5$
$-2x + y = 3$

b. $-2x + y = 3$
$-8x + 4y = 12$

c. $-2x + y = -5$
$-2x + y = -5.00001$

d. $x_1 + 5x_2 - x_3 + 6x_4 = 19$
$2x_1 - x_2 + x_3 - 2x_4 = 7$
$-x_1 + 4x_2 - x_3 + 3x_4 = 30$
$3x_1 - 7x_2 - 2x_3 + x_4 = -75$

Section 8.2

4.   The circuit shown in Figure P4 has five resistances and one applied voltage. Kirchhoff's voltage law applied to each loop in the circuit shown gives

$$v - R_2 i_2 - R_4 i_4 = 0$$
$$-R_2 i_2 + R_1 i_1 + R_3 i_3 = 0$$
$$-R_4 i_4 - R_3 i_3 + R_5 i_5 = 0$$

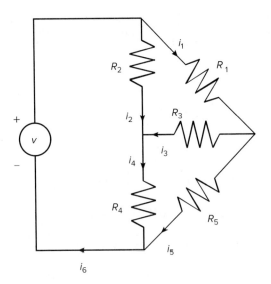

Figure P4

Conservation of charge applied at each node in the circuit gives

$$i_6 = i_1 + i_2$$
$$i_2 + i_3 = i_4$$
$$i_1 = i_3 + i_5$$
$$i_4 + i_5 = i_6$$

a. Write a MATLAB script file that uses given values of the applied voltage v and the values of the five resistances and solves for the six currents.

b. Use the program developed in part a to find the currents for the case where $R_1 = 1$, $R_2 = 5$, $R_3 = 2$, $R_4 = 10$, $R_5 = 5$ kΩ, and $v = 100$ V. (1 k$\Omega = 1000$ Ω.)

5.* a. Use MATLAB to solve the following equations for x, y, and z as functions of the parameter c.

$$x - 5y - 2z = 11c$$
$$6x + 3y + z = 13c$$
$$7x + 3y - 5z = 10c$$

b. Plot the solutions for x, y, and z versus c on the same plot, for $-10 \le c \le 10$.

6. Fluid flows in pipe networks can be analyzed in a manner similar to that used for electric resistance networks. Figure P6 shows a network with three pipes. The volume flow rates in the pipes are q_1, q_2, and q_3. The pressures at the pipe ends are p_a, p_b, and p_c. The pressure at the junction is p_1. Under certain conditions, the pressure–flow rate relation in a pipe has the same form as the voltage-current relation in a resistor. Thus, for the three pipes, we have

$$q_1 = \frac{1}{R_1}(p_a - p_1)$$

$$q_2 = \frac{1}{R_2}(p_1 - p_b)$$

$$q_3 = \frac{1}{R_3}(p_1 - p_c)$$

where the R_i are the pipe resistances. From conservation of mass, $q_1 = q_2 + q_3$.

a. Set up these equations in a matrix form $\mathbf{Ax} = \mathbf{b}$ suitable for solving for the three flow rates q_1, q_2, and q_3 and the pressure p_1, given the values of pressures p_a, p_b, and p_c and the values of resistances R_1, R_2, and R_3. Find the expressions for $\mathbf{A}$ and $\mathbf{b}$.

(a)

(b)

Figure P6

b. Use MATLAB to solve the matrix equations obtained in part *a* for the case where $p_a = 4320$ lb/ft^2, $p_b = 3600$ lb/ft^2, and $p_c = 2880$ lb/ft^2. These correspond to 30, 25, and 20 psi, respectively (1 psi = 1 lb/in^2, and atmospheric pressure is 14.7 psi). Use the resistance values $R_1 = 10,000$, $R_2 = 14,000$ lb sec/ft^5. These values correspond to fuel oil flowing through pipes 2 ft long, with 2- and 1.4-in. diameters, respectively. The units of the answers are ft^3/sec for the flow rates and lb/ft^2 for pressure.

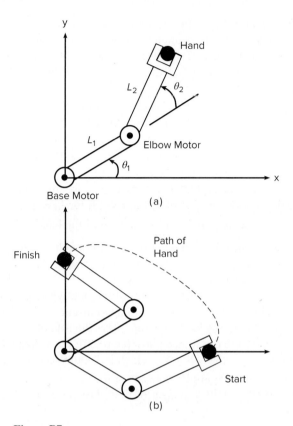

y

Hand

L_2

θ_2

L_1

Elbow Motor

θ_1

x

Base Motor

(a)

Path of
Hand

Finish

Start

(b)

Figure P7

7. Figure P7 illustrates a robot arm that has two "links" connected by two "joints"—a shoulder or base joint and an elbow joint. There is a motor at each joint. The joint angles are θ_1 and θ_2. The (x, y) coordinates of the hand at the end of the arm are given by

$$x = L_1 \cos \theta_1 + L_2 \cos(\theta_1 + \theta_2)$$
$$y = L_1 \sin \theta_1 + L_2 \sin(\theta_1 + \theta_2)$$

where L_1 and L_2 are the lengths of the links.

Polynomials are used for controlling the motion of robots. If we start the arm from rest with zero velocity and acceleration, the following polynomials are used to generate commands to be sent to the joint motor controllers

$$\theta_1(t) = \theta_1(0) + a_1 t^3 + a_2 t^4 + a_3 t^5$$
$$\theta_2(t) = \theta_2(0) + b_1 t^3 + b_2 t^4 + b_3 t^5$$

where $\theta_1(0)$ and $\theta_2(0)$ are the starting values at time $t = 0$. The angles $\theta_1(t_f)$ and $\theta_2(t_f)$ are the joint angles corresponding to the desired destination of the arm at time t_f. The values of $\theta_1(0)$, $\theta_2(0)$, $\theta_1(t_f)$, and $\theta_2(t_f)$ can be found from trigonometry, if the starting and ending (x, y) coordinates of the hand are specified.

a. Set up a matrix equation to be solved for the coefficients a_1, a_2, and a_3, given values for $\theta_1(0)$, $\theta_1(t_f)$, and t_f. Obtain a similar equation for the coefficients b_1, b_2, and b_3.

b. Use MATLAB to solve for the polynomial coefficients given the values $t_f = 2$ sec, $\theta_1(0) = -19°$, $\theta_2(0) = 44°$, $\theta_1(t_f) = 43°$, and $\theta_2(t_f) = 151°$. (These values correspond to a starting hand location of $x = 6.5$, $y = 0$ ft and a destination location of $x = 0$, $y = 2$ ft for $L_1 = 4$ and $L_2 = 3$ ft.)

c. Use the results of part b to plot the path of the hand.

8.* Engineers must be able to predict the rate of heat loss through a building wall to determine the heating system requirements. They do this by using the concept of *thermal resistance R*, which relates the heat flow rate q through a material to the temperature difference ΔT across the material: $q = \Delta T/R$. This relation is like the voltage-current relation for an electric resistor: $i = v/R$. So the heat flow rate plays the role of electric current, and the temperature difference plays the role of the voltage difference. The SI unit for q is the *watt* (W), which is 1 joule/second (J/s).

The wall shown in Figure P8 consists of four layers: an inner layer of plaster/lathe 10 mm thick, a layer of fiberglass insulation 125 mm thick,

(a)

(b)

Figure P8

a layer of wood 60 mm thick, and an outer layer of brick 50 mm thick. If we assume that the inner and outer temperatures T_i and T_o have remained constant for some time, then the heat energy stored in the layers is constant, and thus the heat flow rate through each layer is the same. Applying conservation of energy gives the following equations.

$$q = \frac{1}{R_1}(T_i - T_1) = \frac{1}{R_2}(T_1 - T_2) = \frac{1}{R_3}(T_2 - T_3) = \frac{1}{R_4}(T_3 - T_o)$$

The thermal resistance of a solid material is given by $R = D/k$, where D is the material thickness and k is the material's *thermal conductivity*. For the given materials, the resistances for a wall area of 1 m² are $R_1 = 0.036$, $R_2 = 4.01$, $R_3 = 0.408$, and $R_4 = 0.038$ K/W.

Suppose that $T_i = 20°C$ and $T_o = -10°C$. Find the other three temperatures and the heat loss rate q, in watts. Compute the heat loss rate if the wall's area is 10 m².

9. The concept of thermal resistance described in Problem 8 can be used to find the temperature distribution in the flat square plate shown in Figure P9(a).

(a)

(b)

Figure P9

The plate's edges are insulated so that no heat can escape, except at two points where the edge temperature is heated to T_a and T_b, respectively. The temperature varies through the plate, so no single point can describe the plate's temperature. One way to estimate the temperature distribution is to imagine that the plate consists of four subsquares and to compute the temperature in each subsquare. Let R be the thermal resistance of the material between the centers of adjacent subsquares. Then we can think of the problem as a network of electric resistors, as shown in part (b) of the figure. Let q_{ij} be the heat flow rate between the points whose temperatures are T_i and T_j. If T_a and T_b remain constant for some time, then the heat energy stored in each subsquare is constant also, and the heat flow rate between each subsquare is constant. Under these conditions, conservation of energy says that the heat flow into a subsquare equals the heat flow out. Applying this principle to each subsquare gives the following equations.

$$q_{a1} = q_{12} + q_{13}$$

$$q_{12} = q_{24}$$

$$q_{13} = q_{34}$$

$$q_{34} + q_{24} = q_{4b}$$

Substituting $q = (T_i - T_j)/R$, we find that R can be canceled out of every equation, and they can be rearranged as follows:

$$T_1 = \frac{1}{3}(T_a + T_2 + T_3)$$

$$T_2 = \frac{1}{2}(T_1 + T_4)$$

$$T_3 = \frac{1}{2}(T_1 + T_4)$$

$$T_4 = \frac{1}{3}(T_2 + T_3 + T_5)$$

These equations tell us that the temperature of each subsquare is the average of the temperatures in the adjacent subsquares!

Solve these equations for the case where $T_a = 150°C$ and $T_b = 20°C$.

10. Use the averaging principle developed in Problem 9 to find the temperature distribution of the plate shown in Figure P10, using the 3×3 grid and the given values $T_a = 150°C$ and $T_b = 20°C$.

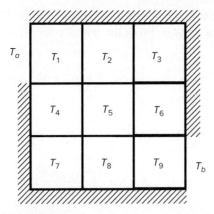

Figure P10

11. Consider Example 8.2–3 (a) in Section 8.2, except that now the voltage v_2 is unspecified. Suppose that each resistance is rated to carry a current of no more than 1 mA ($= 0.001$ A). Determine the allowable range of positive values for the voltage v_2.

12. A weight W is supported by two cables anchored a distance D apart (see Figure P12). The cable length L_{AB} is given, but the length L_{AC} is to be selected. Each cable can support a maximum tension force equal to W. For the weight to remain stationary, the total horizontal force and total vertical force must each be zero. This principle gives the equations

$$-T_{AB} \cos \theta + T_{AC} \cos \phi = 0$$
$$T_{AB} \sin \theta + T_{AC} \sin \phi = W$$

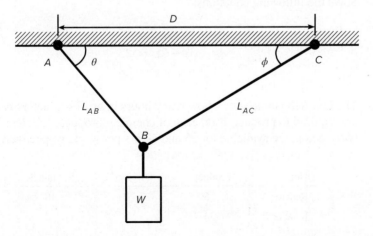

Figure P12

We can solve these equations for the tension forces T_{AB} and T_{AC} if we know the angles θ and ϕ. From the law of cosines

$$\theta = \cos^{-1}\left(\frac{D^2 + L_{AB}^2 - L_{AC}^2}{2DL_{AB}}\right)$$

From the law of sines

$$\phi = \sin^{-1}\left(\frac{L_{AB}\sin\theta}{L_{AC}}\right)$$

For the given values $D = 6$ ft, $L_{AB} = 3$ ft, and $W = 2000$ lb, use a loop in MATLAB to find $L_{AC\text{min}}$, the shortest length L_{AC} we can use without T_{AB} or T_{AC} exceeding 2000 lb. Note that the largest L_{AC} can be is 6.7 ft (which corresponds to $\theta = 90°$). Plot the tension forces T_{AB} and T_{AC} on the same graph versus L_{AC} for $L_{AC\text{min}} \le L_{AC} \le 6.7$.

Section 8.3

13.* Solve the following equations:

$$
\begin{aligned}
7x + 9y - 9z &= 22 \\
3x + 2y - 4z &= 12 \\
x + 5y - z &= -2
\end{aligned}
$$

14. Solve the following equations:

$$
\begin{aligned}
6x - 4y + 3z &= 5 \\
4x + 3y - 2z &= 23 \\
10x - y + z &= 28
\end{aligned}
$$

15. The following table shows how many hours in process reactors A and B are required to produce 1 ton each of chemical products 1, 2, and 3. The two reactors are available for 35 and 40 hrs per week, respectively.

Hours	Product 1	Product 2	Product 3
Reactor A	6	2	10
Reactor B	3	5	2

Let x, y, and z be the number of tons each of products 1, 2, and 3 that can be produced in one week.

a. Use the data in the table to write two equations in terms of x, y, and z. Determine whether a unique solution exists. If not, use MATLAB to find the relations between x, y, and z.

b. Note that negative values x, y, and z have no meaning here. Find the allowable ranges for x, y, and z.

c. Suppose the profits for each product are $200, $300, and $100 for products 1, 2, and 3, respectively. Find the values of x, y, and z to maximize the profit.

d. Suppose the profits for each product are $200, $500, and $100 for products 1, 2, and 3, respectively. Find the values of x, y, and z to maximize the profit.

16. See Figure P16. Assume that no vehicles stop within the network. A traffic engineer wants to know if the traffic flows $f_1, f_2, \ldots, f_7$ (in vehicles per hour) can be computed given the measured flows shown in the figure. If not, then determine how many more traffic sensors need to be installed, and obtain the expressions for the other traffic flows in terms of the measured quantities.

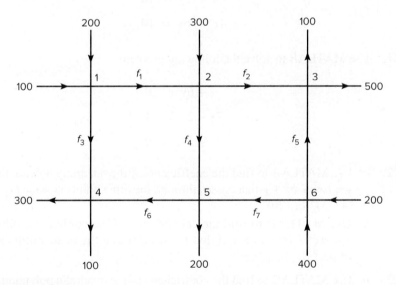

Figure P16

17. Use MATLAB to find the coefficients of the cubic polynomial $ax^3 + bx^2 + cx + d$ that passes through the four points $(x, y) = (1, 6), (2, 38), (4, 310), (5, 580)$.

Section 8.4

18.* Use MATLAB to solve the following problem:

$$x - 3y = 2$$
$$x + 5y = 18$$
$$4x - 6y = 20$$

19. Use MATLAB to solve the following problem:

$$x + 6y = 32$$
$$7x - 2y = 4$$
$$2x + 3y = 19$$

20.* Use MATLAB to solve the following problem:

$$x - 3y = 2$$
$$x + 5y = 18$$
$$4x - 6y = 10$$

21. Use MATLAB to solve the following problem:

$$x + 6y = 20$$
$$7x - 2y = 4$$
$$2x + 3y = 19$$

22. *a.* Use MATLAB to find the coefficients of the quadratic polynomial
$y = ax^2 + bx + c$ that passes through the three points $(x, y) = (1, 4)$,
$(4, 73)$, $(5, 120)$.

  b. Use MATLAB to find the coefficients of the cubic polynomial
$y = ax^3 + bx^2 + cx + d$ that passes through the three points given
in part *a*.

23. *a.* Use MATLAB to find the coefficients of the quadratic polynomial
$y = ax^2 + bx + c$ that passes through the three points $(x, y) = (1, 10)$,
$(3, 30)$, $(5, 74)$.

  b. Use MATLAB to find the coefficients of the cubic polynomial
$ax^3 + bx^2 + cx + d$ that passes through the three points given in
part *a*.

24. Use the MATLAB program given in Table 8.5–2 to solve the following problems:

   a. Problem 3d

   b. Problem 13

   c. Problem 18

   d. Problem 20

Engineering in the 21st Century...

Rebuilding the Infrastructure

During the Great Depression, many public works projects that improved the nation's infrastructure were undertaken to stimulate the economy and provide employment. These projects included highways, bridges, water supply systems, sewer systems, and electrical power distribution networks. Following World War II another burst of such activity culminated in the construction of the interstate highway system. Much of the infrastructure is 40 to 80 years old and is literally crumbling or not up to date. One survey showed that more than 25 percent of the nation's bridges are substandard. These need to be repaired or replaced. A 2013 study estimated that the needed improvements for all types of infrastructure would cost about $3.3 trillion, about $1.4 trillion more than the current funding level.

Rebuilding the infrastructure requires engineering methods different from those in the past because labor and material costs are now higher and environmental and social issues have greater importance than before. Infrastructure engineers must take advantage of new materials, inspection technology, construction techniques, and labor-saving machines.

Also, some infrastructure components, such as communications networks, need to be replaced because they are outdated and do not have sufficient capacity or ability to take advantage of new technology. An example is the *information infrastructure*, which includes physical facilities to transmit, store, process, and display voice, data, and images. Better communications and computer networking technology will be needed for such improvements.

Obviously every engineering discipline will be engaged in such work and many of the MATLAB toolboxes will provide advanced support for these disciplines, including the Financial, Communications, Image Processing, Signal Processing, PDE, and Wavelet toolboxes. ∎

Numerical Methods for Calculus and Differential Equations

This chapter covers numerical methods for computing integrals and derivatives and for solving ordinary differential equations. Some integrals cannot be evaluated analytically, and we need to compute them numerically with an approximate method (Section 9.1). In addition, it is often necessary to use data to estimate rates of change, and this requires a numerical estimate of the derivative (Section 9.2). Finally, many differential equations cannot be solved analytically, and so we must be able to solve them by using appropriate numerical techniques. Section 9.3 covers first-order differential equations, and Section 9.4 extends the methods to higher-order equations. More powerful methods are available for linear equations. Section 9.5 treats these methods.

When you have finished this chapter, you should be able to

■   Use MATLAB to numerically evaluate integrals.

■   Use numerical methods with MATLAB to estimate derivatives.

■   Use the MATLAB numerical differential equation solvers to obtain solutions.

9.1 Numerical Integration

The integral of a function $f(x)$ for $a \leq x \leq b$ can be interpreted as the area between the $f(x)$ curve and the x axis, bounded by the limits $x = a$ and $x = b$. If we denote this area by A, then we can write A as

$$A = \int_a^b f(x)\, dx \tag{9.1–1}$$

DEFINITE INTEGRAL

INDEFINITE INTEGRAL

IMPROPER INTEGRAL

SINGULARITY

An integral is called a *definite integral* if it has specified limits of integration. *Indefinite integrals* have no specified limits. *Improper integrals* can have infinite values, depending on their integration limits. For example, the following integral can be found in most integral tables:

$$\int \frac{1}{x-1}\, dx = \ln |x-1|$$

However, it is an improper integral if the integration limits include the point $x = 1$. So, even though an integral can be found in an integral table, you should examine the integrand to check for *singularities*, which are points at which the integrand is undefined. The same warning applies when you are using numerical methods to evaluate integrals.

Integration of Discrete Points

The simplest way to find the area under a curve is to split the area into rectangles (Figure 9.1–1a). If the widths of the rectangles are small enough, the sum of their areas gives the approximate value of the integral. A more sophisticated method is to use trapezoidal elements (Figure 9.1–1b). Each trapezoid is called a *panel*. It is not necessary to use panels of the same width; to increase the method's accuracy,

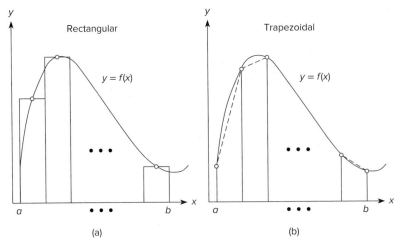

Figure 9.1–1 Illustration of (a) rectangular and (b) trapezoidal numerical integration.

you can use narrow panels where the function is changing rapidly. When the widths are adjusted according to the function's behavior, the method is said to be *adaptive*. MATLAB implements *trapezoidal integration* with the `trapz` function. Its syntax is `trapz(x, y)`, where the array `x` contains the function values at the points contained in the array `x`. If you want the integral of a single function, then `y` is a vector. To integrate more than one function, place their values in a matrix `y`; `trapz(x, y)` will compute the integral of each column of `y`.

You cannot directly specify a function to integrate with the `trapz` function; you must first compute and store the function's values ahead of time in an array. Later we will discuss another integration function, the `integral` function, that can accept functions directly. However, it cannot handle arrays of values. So the functions complement each other. The `trapz` function is summarized in Table 9.1–1.

As a simple example of the use of the `trapz` function, let us compute the integral

$$A = \int_0^\pi \sin x \, dx \qquad (9.1\text{–}2)$$

whose exact answer is $A = 2$. To investigate the effect of panel width, let us first use 10 panels with equal widths of $\pi/10$. The script file is

```
x = linspace(0,pi,10);
y = sin(x);
A = trapz(x,y)
```

Table 9.1–1 Basic syntax of numerical integration functions

Command	Description
`integral(fun,a,b)`	Uses an adaptive Simpson rule to compute the integral of the function `fun` between the limits a and b. The input `fun`, which represents the integrand $f(x)$, is a function handle for the integrand function. It must accept a vector argument x and return the vector result y.
`integral2` `(fun,a,b,c,d)`	Computes the double integral of the function $f(x, y)$ between the limits $a \le x \le b$ and $c \le y \le d$. The input `fun` specifies the function that computes the integrand. It must accept a vector argument x and scalar y, and it must return a vector result.
`integral3` `(fun,a,b,c,d,e,f)`	Computes the triple integral of the function $f(x, y, z)$ between the limits $a \le x \le b$, $c \le y \le d$, and $e \le y \le f$. The input `fun` specifies the function that computes the integrand. It must accept a vector argument x, scalar y, and scalar z, and it must return a vector result.
`polyint(p,C)`	Computes the integral of the polynomial `p` using an optional user-specified constant of integration `C`.
`trapz(x,y)`	Uses trapezoidal integration to compute the integral of `y` with respect to `x`, where the array `y` contains the function values at the points contained in the array `x`.

The answer is $A = 1.9797$, which gives a relative error of $100(2 - 1.9797)/2 = 1$ percent. Now try 100 panels of equal width; replace the array x with x = linspace(0,pi,100). The answer is $A = 1.9998$ for a relative error of $100(2 - 1.9998)/2 = 0.01$ percent. If we examine the plot of the integrand sin x, we see that the function is changing faster near $x = 0$ and $x = \pi$ than near $x = \pi/2$. Thus we could achieve the same accuracy by using fewer panels if narrower panels are used near $x = 0$ and $x = \pi$.

We normally use the trapz function when the integrand is given as a table of values. Otherwise, if the integrand is given as a function, use the integral function, to be introduced shortly.

| EXAMPLE 9.1–1 | Velocity from an Accelerometer

An *accelerometer* is used in aircraft, rockets, and other vehicles to estimate the vehicle's velocity and displacement. The accelerometer integrates the acceleration signal to produce an estimate of the velocity, and it integrates the velocity estimate to produce an estimate of displacement. Suppose the vehicle starts from rest at time $t = 0$, and its measured acceleration is given in the following table.

Time (s)	0	1	2	3	4	5	6	7	8	9	10
Acceleration (m/s^2)	0	2	4	7	11	17	24	32	41	48	51

(a) Estimate the velocity v after 10 s.
(b) Estimate the velocity at times $t = 1, 2, \ldots , 10$ s.

■ **Solution**
(a) The initial velocity is zero, so $v(0) = 0$. The relation between the velocity and acceleration $a(t)$ is

$$v(10) = \int_0^{10} a(t)\, dt + v(0) = \int_0^{10} a(t)\, dt$$

The script file is shown below.

```
t = 0:10;
a = [0,2,4,7,11,17,24,32,41,48,51];
v10 = trapz(t,a)
```

The answer for the velocity after 10 s is v10, and it is 211.5 m/s.
(b) The following script file uses the fact that the velocity can be expressed as

$$v(t_{k+1}) = \int_{t_k}^{t_{k+1}} a(t)\, dt + v(t_k) \qquad k = 1, 2, \ldots , 10$$

where $v(t_1) = 0$.

```
t = 0:10;
a = [0,2,4,7,11,17,24,32,41,48,51];
v(1) = 0;
for k = 1:10
   v(k+1) = trapz(t(k:k+1), a(k:k+1))+v(k);
end
disp([t',v'])
```

The answers are given in the following table.

Time (s)	0	1	2	3	4	5	6	7	8	9	10
Velocity (m/s)	0	1	4	9.5	18.5	32.5	53	81	117	162	211.5

Test Your Understanding

T9.1–1 Modify the script file given in part (b) of Example 9.1–1 to estimate the displacement at times $t = 1, 2, \ldots, 10$ s. (Partial answer: The displacement after 10 s is 584.25 m.)

Integration of Functions

Another approach to numerical integration is *Simpson's rule*, which divides the integration range $b - a$ into an even number of sections and uses a different quadratic polynomial to represent the integrand for each panel. A quadratic polynomial has three parameters, and Simpson's rule computes these parameters by requiring that the quadratic pass through the function's three points corresponding to the two adjacent panels. To obtain greater accuracy, we can use polynomials of degree higher than 2.

The MATLAB function `integral` implements an adaptive version of Simpson's rule. The function `integral(fun,a,b)` computes the integral of the function `fun` between the limits a and b. The input `fun`, which represents the integrand $f(x)$, is either a function handle of the integrand function or the name of an anonymous function. The function $y = f(x)$ must accept a vector argument x and must return the vector result y. The basic syntax is summarized in Table 9.1–1.

To illustrate, let us compute the integral given in Equation (9.1–2). The session consists of one command: `A = integral(@sin, 0, pi)`. The answer given by MATLAB is $A = 2.0000$, which is correct to four decimal places.

Because the `integral` function calls the integrand function using vector arguments, you must always use array operations when defining the function. The following example shows how this is done.

EXAMPLE 9.1–2   Evaluation of Fresnel's Cosine Integral

Some simple-looking integrals cannot be evaluated in closed form. An example is Fresnel's cosine integral

$$A = \int_0^b \cos x^2 \, dx \qquad (9.1\text{–}3)$$

(a) Demonstrate two ways to compute the integral when the upper limit is $b = \sqrt{2\pi}$.

(b) Demonstrate the use of a nested function to compute the more general integral

$$A = \int_0^b \cos x^n \, dx \qquad (9.1\text{–}4)$$

for $n = 2$ and for $n = 3$.

■ **Solution**

(a) The integrand $\cos x^2$ obviously does not contain any singularities that might cause problems for the integration function. We demonstrate two ways to use the `integral` function.

1. With a function file: Define the integrand with a user-defined function as shown by the following function file.

```
function c2 = cossq(x)
c2 = cos(x.^2);
```

The `integral` function is called as follows:

```
>>A = integral(@cossq,0,sqrt(2*pi))
```

The result is $A = 0.6119$.

2. With an anonymous function (anonymous functions are discussed in Section 3.3): The session is

```
>>cossq = @(x)cos(x.^2);
>>A = integral(cossq,0,sqrt(2*pi))
  A =
       0.6119
```

The two lines can be combined into one as follows:

```
A = integral(@(x)cos(x.^2),0,sqrt(2*pi))
```

The advantage of using an anonymous function is that you need not create and save a function file. However, for complicated integrand functions, using a function file is preferable.

(b) Because `integral` requires that the integrand function have only one argument, the following code will not work.

```
>>cossq = @(x)cos(x.^n);
>>n = 2;
>>A = integral(cossq,0,sqrt(2*pi))
??? Undefined function or variable 'n'.
```

Instead we will use parameter passing with a nested function (nested functions are discussed in Section 3.3). First create and save the following function.

```
function A = integral_n(n)
A = integral(@cossq_n,0,sqrt(2*pi));

% Nested function
   function integrand = cossq_n(x)
       integrand = cos(x.^n);
   end
end
```

The session for $n = 2$ and $n = 3$ is as follows.

```
>>A = integral_n(2)
   A =
       0.6119
>>A = integral_n(3)
   A =
       0.7734
```

The `integral` function has some optional arguments for analyzing and adjusting the algorithm's efficiency and accuracy. Type `help integral` for details.

Test Your Understanding

T9.1–2 Use the `integral` function to compute the integral

$$A = \int_2^5 \frac{1}{x} dx$$

and compare the answer with that obtained from the closed-form solution, which is $A = 0.9163$.

Polynomial Integration

MATLAB provides the `polyint` function to compute the integral of a polynomial. The syntax `q = polyint(p, C)` returns a polynomial `q` representing the integral of polynomial `p` with a user-specified scalar constant of integration `C`. The elements of the vector `p` are the coefficients of the polynomial, arranged in descending powers. The syntax `polyint(p)` assumes the constant of integration `C` is zero.

For example, the integral of $12x^3 + 9x^2 + 8x + 5$ is obtained from `q = polyint([12,9,8,5], 10)`. The answer is `q = [3, 3, 4, 5, 10]`, which corresponds to $3x^4 + 3x^3 + 4x^2 + 5x + 10$. Because polynomial integrals can be obtained from a symbolic formula, the `polyint` function is not a numerical integration operation.

Double Integrals

The function `integral2` computes double integrals. Consider the integral

$$A = \int_c^d \int_a^b f(x,y)dx\,dy$$

The basic syntax is

```
A = integral2(fun, a, b, c, d)
```

where `fun` is the handle to a user-defined function that defines the integrand $f(x, y)$. The function must accept a vector x and a scalar y, and it must return a vector result, so the appropriate array operations must be used. The extended syntax enables the user to adjust the accuracy. See the MATLAB Help for details.

For example, using an anonymous function to compute the integral

$$A = \int_0^1 \int_1^3 xy^2 dx\,dy$$

you type

```
>>fun = @(x,y)x.*y.^2;
>>A = integral2(fun, 1, 3, 0, 1)
```

The answer is $A = 1.3333$.

The preceding integral is carried out over the rectangular region specified by $1 \le x \le 3, 0 \le y \le 1$. Some double integrals are specified over a nonrectangular region. These problems can be handled by a transformation of variables. You can also use a rectangular region that encloses the nonrectangular region and force the integrand to be zero outside of the nonrectangular region, by using the MATLAB relational operators, for example. See Problem 16. The following example illustrates the former approach.

EXAMPLE 9.1–3   Double Integral over a Nonrectangular Region

Compute the integral

$$A = \iint_R (x - y)^4 (2x + y)^2 dx\,dy$$

over the region R bounded by the lines

$$x - y = \pm 1 \qquad 2x + y = \pm 2$$

■ **Solution**

We must convert the integral into one that is specified over a rectangular region. To do this, let $u = x - y$ and $v = 2x + y$. Thus, using the Jacobian, we obtain

$$dxdy = \begin{vmatrix} \partial x/\partial u & \partial x/\partial v \\ \partial y/\partial u & \partial y/\partial v \end{vmatrix} du\,dv = \begin{vmatrix} 1/3 & 1/3 \\ -2/3 & 1/3 \end{vmatrix} du\,dv = \frac{1}{3} du\,dv$$

Then the region R is specified as a rectangular region in terms of u and v. Its boundaries are given by $u = \pm 1$ and $v = \pm 2$, and the integral becomes

$$A = \frac{1}{3} \int_{-2}^{2} \int_{-1}^{1} u^4 v^2 \, du \, dv$$

and the MATLAB session is

```
>>fun = @(u,v)u.^4*v^2;
>>A = (1/3)*integral2(fun, -1, 1, -2, 2)
```

The answer is $A = 0.7111$.

Triple Integrals

The function `integral3` computes triple integrals. Consider the integral

$$A = \int_{e}^{f} \int_{c}^{d} \int_{a}^{b} f(x, y, z) dx \, dy \, dz$$

The basic syntax is

```
A = integral3(fun, a, b, c, d, e, f)
```

where `fun` is the handle to a user-defined function that defines the integrand $f(x, y, z)$. The function must accept a vector x, a scalar y, and a scalar z, and it must return a vector result, so the appropriate array operations must be used. The extended syntax enables the user to adjust the accuracy. See the MATLAB Help for details. For example, to compute the integral

$$A = \int_{1}^{2} \int_{0}^{2} \int_{1}^{3} \left(\frac{xy - y^2}{z} \right) dx \, dy \, dz$$

You type

```
>>fun = @(x,y,z)(x.*y-y.^2)./z;
>>A = integral3(fun, 1, 3, 0, 2, 1, 2)
```

The answer is $A = 1.8484$.

Test Your Understanding

T9.1–3 Use MATLAB to evaluate the following double integral:

$$\int_{1}^{2} \int_{0}^{1} (x^2 + xy^3) dx \, dy$$

(Answer: 2.2083)

T9.1–4 Use MATLAB to evaluate the following triple integral:

$$\int_{0}^{1} \int_{1}^{2} \int_{2}^{3} xyz \, dx \, dy \, dz$$

(Answer: 1.875)

9.2 Numerical Differentiation

The derivative of a function can be interpreted graphically as the slope of the function. This interpretation leads to various methods for computing the derivative of a set of data. Figure 9.2–1 shows three data points that represent a function $y(x)$. Recall that the definition of the derivative is

$$\frac{dy}{dx} = \lim_{\Delta x \to 0} \frac{\Delta y}{\Delta x} \tag{9.2–1}$$

The success of numerical differentiation depends heavily on two factors: the spacing of the data points and the scatter present in the data due to measurement error. The greater the spacing, the more difficult it is to estimate the derivative. We assume here that the spacing between the measurements is regular; that is, $x_3 - x_2 = x_2 - x_1 = \Delta x$. Suppose we want to estimate the derivative dy/dx at the point x_2. The correct answer is the slope of the straight line passing through the point (x_2, y_2); but we do not have a second point on that line, so we cannot find its slope. Therefore, we must estimate the slope by using nearby data points. One estimate can be obtained from the straight line labeled A in the figure. Its slope is

$$m_A = \frac{y_2 - y_1}{x_2 - x_1} = \frac{y_2 - y_1}{\Delta x} \tag{9.2–2}$$

BACKWARD DIFFERENCE

This estimate of the derivative is called the *backward difference* estimate, and it is actually a better estimate of the derivative at $x = x_1 + (\Delta x)/2$ than at $x = x_2$. Another estimate can be obtained from the straight line labeled B. Its slope is

$$m_B = \frac{y_3 - y_2}{x_3 - x_2} = \frac{y_3 - y_2}{\Delta x} \tag{9.2–3}$$

FORWARD DIFFERENCE

This estimate is called the *forward difference* estimate, and it is a better estimate of the derivative at $x = x_2 + (\Delta x)/2$ than at $x = x_2$. Examining the plot, you might

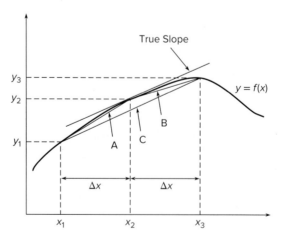

Figure 9.2–1 Illustration of methods for estimating the derivative dy/dx.

think that the average of these two slopes would provide a better estimate of the derivative at $x = x_2$, because the average tends to cancel out the effects of measurement error. The average of m_A and m_B is

$$m_C = \frac{m_A + m_B}{2} = \frac{1}{2}\left(\frac{y_2 - y_1}{\Delta x} + \frac{y_3 - y_2}{\Delta x}\right) = \frac{y_3 - y_1}{2\Delta x} \qquad (9.2\text{–}4)$$

This is the slope of the line labeled C, which connects the first and third data points. This estimate of the derivative is called the *central difference* estimate.

CENTRAL DIFFERENCE

The diff Function

MATLAB provides the `diff` function to use for computing derivative estimates. Its syntax is d = `diff(x)`, where x is a vector of values, and the result is a vector d containing the differences between adjacent elements in x. That is, if x has n elements, d will have $n - 1$ elements, where $d = [x(2) - x(1), x(3) - x(2), \ldots, x(n) - x(n - 1)]$. For example, if x = [5, 7, 12, -20], then `diff(x)` returns the vector [2, 5, -32]. The derivative dy/dx can be estimated from `diff(y)./diff(x)`.

The following script file implements the backward difference and central difference methods for artificial data generated from a sinusoidal signal that is measured 51 times during one half-period. The measurement error is uniformly distributed between −0.025 and 0.025.

```
x = 0:pi/50:pi;
n = length(x);
% Data-generation function with +/-0.025 random error.
y = sin(x)+.05*(rand(1,51)-0.5);
% Backward difference estimate of dy/dx.
d1 = diff(y)./diff(x);
subplot(2,1,1)
plot(x(2:n),d1,x(2:n),d1,'o')
% Central difference estimate of dy/dx.
d2 = (y(3:n)-y(1:n-2))./(x(3:n)-x(1:n-2));
subplot(2,1,2)
plot(x(2:n-1),d2,x(2:n-1),d2,'o')
```

Test Your Understanding

T9.2–1 Modify the previous program to use the forward difference method to estimate the derivative. Plot the results, and compare with the results from the backward and central difference methods.

Polynomial Derivatives

MATLAB provides the `polyder` function to compute the derivative of a polynomial. Its syntax has several forms. The basic form is d = `polyder(p)`, where p is a vector whose elements are the coefficients of the polynomial, arranged in descending powers. The output d is a vector containing the coefficients of the derivative polynomial.

The second syntax form is d = polyder(p1,p2). This form computes the derivative of the *product* of the two polynomials p1 and p2. The third form is [num, den] = polyder(p2,p1). This form computes the derivative of the *quotient* p2/p1. The vector of coefficients of the numerator of the derivative is given by num. The denominator is given by den.

Here are some examples of the use of polyder. Let $p_1 = 5x + 2$ and $p_2 = 10x^2 + 4x - 3$. Then

$$\frac{dp_2}{dx} = 20x + 4$$

$$p_1 p_2 = 50x^3 + 40x^2 - 7x - 6$$

$$\frac{d(p_1 p_2)}{dx} = 150x^2 + 80x - 7$$

$$\frac{d(p_2/p_1)}{dx} = \frac{50x^2 + 40x + 23}{25x^2 + 20x + 4}$$

These results can be obtained with the following program.

```
p1 = [5, 2];p2 = [10, 4, -3];
% Derivative of p2.
der2 = polyder(p2)
% Derivative of p1*p2.
prod = polyder(p1,p2)
% Derivative of p2/p1.
[num, den] = polyder(p2,p1)
```

The results are der2 = [20, 4], prod = [150, 80, -7], num = [50, 40, 23], and den = [25, 20, 4].

Because polynomial derivatives can be obtained from a symbolic formula, the polyder function is not a numerical differentiation operation.

Gradients

The *gradient* ∇f of a function $f(x, y)$ is a vector pointing in the direction of increasing values of $f(x, y)$. It is defined by

$$\nabla f = \frac{\partial f}{\partial x}\mathbf{i} + \frac{\partial f}{\partial y}\mathbf{j}$$

where $\mathbf{i}$ and $\mathbf{j}$ are the unit vectors in the x and y directions, respectively. The concept can be extended to functions of three or more variables.

In MATLAB the gradient of a set of data representing a two-dimensional function $f(x, y)$ can be computed with the gradient function. Its syntax

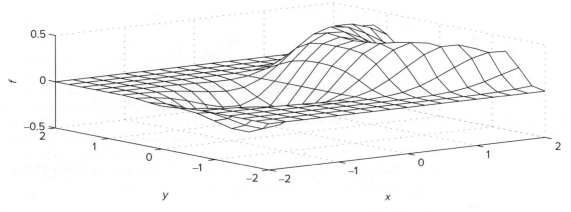

Figure 9.2–2 Gradient, contour, and surface plots of the function $f(x, y) = x e^{-(x^2+y^2)^2} + y^2$.

is [df_dx, df_dy] = gradient(f, dx, dy), where df_dx and df_dy represent $\partial f/\partial x$ and $\partial f/\partial y$ and dx and dy are the spacing in the x and y values associated with the numerical values of f. The syntax can be extended to include functions of three or more variables.

The following program plots the contour plot and the gradient (shown by arrows) for the function

$$f(x, y) = x e^{-(x^2+y^2)^2} + y^2$$

The plots are shown in Figure 9.2–2. The arrows point in the direction of increasing f.

```
[x,y] = meshgrid(-2:0.25:2);
f = x.*exp(-((x-y.^2).^2+y.^2));
dx = x(1,2) - x(1,1); dy = y(2,1) - y(1,1);
```

```
[df_dx, df_dy] = gradient(f, dx, dy);
subplot(2,1,1)
contour(x,y,f), xlabel('x'), ylabel('y'),...
    hold on, quiver(x,y,df_dx, df_dy), hold off
subplot(2,1,2)
mesh(x,y,f),xlabel('x'),ylabel('y'),zlabel('f')
```

LAPLACIAN

The curvature is given by the second-order derivative expression called the *Laplacian*.

$$\nabla^2 f(x, \ y) = \frac{\partial^2 f}{\partial x^2} + \frac{\partial^2 f}{\partial y^2}$$

It can be computed with the del2 function. See the MATLAB Help for details.

The MATLAB differentiation functions discussed here are summarized in Table 9.2–1.

9.3 First-Order Differential Equations

ORDINARY DIFFERENTIAL EQUATION

INITIAL-VALUE PROBLEM

In this section, we introduce numerical methods for solving first-order differential equations. In Section 9.4 we show how to extend the techniques to higher-order equations.

An *ordinary differential equation* (ODE) is an equation containing ordinary derivatives of the dependent variable. An equation containing partial derivatives with respect to two or more independent variables is a *partial differential equation* (PDE). Solution methods for PDEs are an advanced topic, and we will not treat them in this text. In this chapter we limit ourselves to *initial-value problems* (IVPs). These are problems where the ODE must be solved for a given set of values specified at some initial time, which is usually taken to be $t = 0$. Other types of ODE problems are discussed at the end of Section 9.6.

Table 9.2–1 Numerical differentiation functions

Command	Description
d = diff(x)	Returns a vector d containing the differences between adjacent elements in the vector x.
[df_dx,df_dy] = gradient(f,dx,dy)	Computes the gradient of the function $f(x, y)$, where df_dx and df_dy represent $\partial f/\partial x$ and $\partial f/\partial y$, and dx and dy are the spacing in the x and y values associated with the numerical values of f.
d = polyder(p)	Returns a vector d containing the coefficients of the derivative of the polynomial represented by the vector p.
d = polyder(p1,p2)	Returns a vector d containing the coefficients of the polynomial that is the derivative of the product of the polynomials represented by p1 and p2.
[num, den] = polyder(p2,p1)	Returns the vectors num and den containing the coefficients of the numerator and denominator polynomials of the derivative of the quotient p2/p1, where p1 and p2 are polynomials.

It will be convenient to use the following abbreviated "dot" notation for derivatives.

$$\dot{y}(t) = \frac{dy}{dt} \qquad \ddot{y}(t) = \frac{d^2y}{dt^2}$$

The *free response* of a differential equation, sometimes called the homogeneous solution or the initial response, is the solution for the case where there is no forcing function. The free response depends on the initial conditions. The *forced response* is the solution due to the forcing function when the initial conditions are zero. For linear differential equations, the complete or total response is the sum of the free and the forced responses. Nonlinear ODEs can be recognized by the fact that the dependent variable or its derivatives appear raised to a power or in a transcendental function. For example, the equations $\dot{y} = y^2$ and $\dot{y} = \cos y$ are nonlinear.

FREE RESPONSE

FORCED RESPONSE

The essence of a numerical method is to convert the differential equation into a difference equation that can be programmed. Numerical algorithms differ partly as a result of the specific procedure used to obtain the difference equations. It is important to understand the concept of "step size" and its effects on solution accuracy. To provide a simple introduction to these issues, we consider the simplest numerical methods, the *Euler method* and the *predictor-corrector method*.

The Euler Method

The *Euler method* is the simplest algorithm for numerical solution of a differential equation. Consider the equations

$$\frac{dy}{dt} = f(t, y) \qquad y(0) = y_0 \tag{9.3-1}$$

where $f(t, y)$ is a known function and y_0 is the initial condition, which is the given value of $y(t)$ at $t = 0$. From the definition of the derivative,

$$\frac{dy}{dt} = \lim_{\Delta t \to 0} \frac{y(t + \Delta t) - y(t)}{\Delta t}$$

If the time increment Δt is chosen small enough, the derivative can be replaced by the approximate expression

$$\frac{dy}{dt} \approx \frac{y(t + \Delta t) - y(t)}{\Delta t} \tag{9.3-2}$$

Assume that the function $f(t, y)$ in Equation (9.3–1) remains constant over the time interval $(t, t + \Delta t)$, and replace Equation (9.3–1) by the following approximation:

$$\frac{y(t + \Delta t) - y(t)}{\Delta t} = f(t, y)$$

or

$$y(t + \Delta t) = y(t) + f(t, y)\,\Delta t \tag{9.3-3}$$

The smaller Δt is, the more accurate are our two assumptions leading to Equation (9.3–3). This technique for replacing a differential equation with a difference equation is the Euler method. The increment Δt is called the *step size*.

STEP SIZE

Equation (9.3–3) can be written in a more convenient form as

$$y(t_{k+1}) = y(t_k) + \Delta t f[t_k, y(t_k)] \qquad (9.3\text{–}4)$$

where $t_{k+1} = t_k + \Delta t$. This equation can be applied successively at the times t_k by putting it in a for loop. The accuracy of the Euler method can be improved sometimes by using a smaller step size. However, very small step sizes require longer run times and can result in a large accumulated error due to roundoff effects.

The Predictor-Corrector Method

The Euler method can have a serious deficiency in problems where the variables are rapidly changing, because the method assumes the variables are constant over the time interval Δt. One way of improving the method is to use a better approximation to the right-hand side of Equation (9.3–1). Suppose instead of the Euler approximation (9.3–4) we use the average of the right-hand side of Equation (9.3–1) on the interval (t_k, t_{k+1}). This gives

$$y(t_{k+1}) = y(t_k) + \frac{\Delta t}{2}(f_k + f_{k+1}) \qquad (9.3\text{–}5)$$

where

$$f_k = f[t_k, y(t_k)] \qquad (9.3\text{–}6)$$

with a similar definition for f_{k+1}. Equation (9.3–5) is equivalent to integrating Equation (9.3–1) with the trapezoidal rule.

The difficulty with Equation (9.3–5) is that f_{k+1} cannot be evaluated until $y(t_{k+1})$ is known, but this is precisely the quantity being sought. A way out of this difficulty is to use the Euler formula (9.3–4) to obtain a preliminary estimate of $y(t_{k+1})$. This estimate is then used to compute f_{k+1} for use in Equation (9.3–5) to obtain the required value of $y(t_{k+1})$.

The notation can be changed to clarify the method. Let $h = \Delta t$ and $y_k = y(t_k)$, and let x_{k+1} be the estimate of $y(t_{k+1})$ obtained from the Euler formula (9.3–4). Then, by omitting the t_k notation from the other equations, we obtain the following description of the predictor-corrector process.

Euler predictor

$$x_{k+1} = y_k + hf(t_k, y_k) \qquad (9.3\text{–}7)$$

Trapezoidal corrector

$$y_{k+1} = y_k + \frac{h}{2}[f(t_k, y_k) + f(t_{k+1}, x_{k+1})] \qquad (9.3\text{–}8)$$

MODIFIED EULER METHOD

This algorithm is sometimes called the *modified Euler method*. However, note that any algorithm can be tried as a predictor or a corrector. Thus many other methods can be classified as predictor-corrector.

Runge-Kutta Methods

The Taylor series representation forms the basis of several methods of solving differential equations, including the Runge-Kutta methods. The Taylor series may be used to represent the solution $y(t + h)$ in terms of $y(t)$ and its derivatives as follows:

$$y(t + h) = y(t) + h\dot{y}(t) + \frac{1}{2}h^2\ddot{y}(t) + \cdots \qquad (9.3\text{--}9)$$

The number of terms kept in the series determines its accuracy. The required derivatives are calculated from the differential equation. If these derivatives can be found, Equation (9.3–9) can be used to march forward in time. In practice, the high-order derivatives can be difficult to calculate, and the series (9.3–9) is truncated at some term. The Runge-Kutta methods were developed because of the difficulty in computing the derivatives. These methods use several evaluations of the function $f(t, y)$ in a way that approximates the Taylor series. The number of terms in the series that is duplicated determines the order of the Runge-Kutta method. Thus, a fourth-order Runge-Kutta algorithm duplicates the Taylor series through the term involving h^4.

MATLAB ODE Solvers

In addition to the many variations of the predictor-corrector and Runge-Kutta algorithms that have been developed, there are more-advanced algorithms that use a variable step size. These "adaptive" algorithms use larger step sizes when the solution is changing more slowly. MATLAB provides several functions, called *solvers*, that implement the Runge-Kutta and other methods with variable step size. Two of these are the ode45 and ode15s functions. The ode45 function uses a combination of fourth- and fifth-order Runge-Kutta methods. It is a general-purpose solver, whereas ode15s is suitable for more-difficult equations called "stiff" equations. These solvers are more than sufficient to solve the problems in this text. It is recommended that you try ode45 first. If the equation proves difficult to solve (as indicated by a lengthy solution time or by a warning or error message), then use ode15s.

In this section we limit our coverage to first-order equations. Solution of higher-order equations is covered in Section 9.4. When used to solve the equation $\dot{y} = f(t, y)$, the basic syntax is (using ode45 as the example)

```
[t,y] = ode45(@ydot, tspan, y0)
```

where @ydot is the handle of the function file whose inputs must be t and y, and whose output must be a column vector representing dy/dt, that is, $f(t, y)$. The number of rows in this column vector must equal the order of the equation. The syntax for ode15s is identical. The function file ydot may also be specified by a character string (i.e., its name placed in single quotes), but use of the function handle is now the preferred approach.

The vector tspan contains the starting and ending values of the independent variable t, and optionally any intermediate values of t where the solution is desired. For example, if no intermediate values are specified, tspan is [t0 tfinal], where t0 and tfinal are the desired starting and ending values of the independent parameter t. As another example, using tspan = [0, 5, 10] tells MATLAB to find the solution at $t = 5$ and at $t = 10$. You can solve equation backward in time by specifying t0 to be greater than tfinal.

The parameter y0 is the initial value $y(0)$. The function file must have its first two input arguments as t and y in that order, even for equations where $f(t, y)$ is not a function of *t*. You need not use array operations in the function file because the ODE solvers call the file with scalar values for the arguments.

First consider an equation whose solution is known in closed form, so that we can make sure we are using the method correctly.

EXAMPLE 9.3–1   Response of an *RC* Circuit

The model of the *RC* circuit shown in Figure 9.3–1 can be found from Kirchhoff's voltage law and conservation of charge. It is $RC\dot{y} + y = v(t)$. Suppose the value of *RC* is 0.1 s. Use a numerical method to find the free response for the case where the applied voltage y is zero and the initial capacitor voltage is $y(0) = 2$ V. Compare the results with the analytical solution, which is $y(t) = 2e^{-10t}$.

■ Solution

The equation for the circuit becomes $0.1\dot{y} + y = 0$. First solve this for y: $\dot{y} = -10y$. Next define and save the following function file. Note that the order of the input arguments must be *t* and *y* even though *t* does not appear on the right hand side of the equation.

```
function ydot = RC_circuit(t,y)
% Model of an RC circuit with no applied voltage.
ydot = -10*y;
```

The initial time is $t = 0$, so set t0 to be 0. Here we know from the analytical solution that $y(t)$ will be close to 0 for $t \geq 0.5$ s, so we choose tfinal to be 0.5 s. In other problems we generally do not have a good guess for tfinal, so we must try several increasing values of tfinal until we see enough of the response on the plot.

The function ode45 is called as follows, and the solution plotted along with the analytical solution y_true.

```
[t, y] = ode45(@RC_circuit, [0 0.5], 2);
y_true = 2*exp(-10*t);
plot(t,y,'o',t,y_true), xlabel('Time(s)'),...
    ylabel('Capacitor Voltage')
```

Note that we need not generate the array t to evaluate y_true because t is generated by the ode45 function. The plot is shown in Figure 9.3–2. The numerical solution is marked by the circles, and the analytical solution is indicated by the solid line. Clearly

Figure 9.3–1 An *RC* circuit.

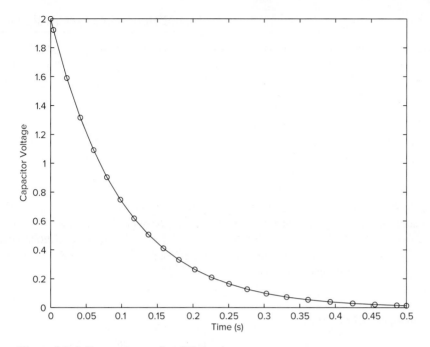

Figure 9.3–2 Free response of an *RC* circuit.

the numerical solution gives an accurate answer. Note that the step size has been automatically selected by the ode45 function.

Earlier versions of MATLAB required that the function name, here RC_circuit, be enclosed within single quotes, but this might not be allowed in future versions. The use of function handles is now preferred, such as @RC_circuit. As we will see, additional capabilities are available with function handles.

Test Your Understanding

T9.3–1 Use MATLAB to compute and plot the solution of the following equation.

$$10\frac{dy}{dt} + y = 20 + 7 \ \sin 2t \qquad y(0) = 15$$

When the differential equation is nonlinear, we often have no analytical solution to use for checking our numerical results. In such cases we can use our physical insight to guard against grossly incorrect results. We can also check the equation for singularities that might affect the numerical procedure. Finally, we can sometimes use an approximation to replace the nonlinear equation with a linear one that can be solved analytically. Although the linear approximation does not give the exact answer, it can be used to see if our numerical answer is "in the ballpark." The following example illustrates this approach.

EXAMPLE 9.3–2

Liquid Height in a Spherical Tank

Figure 9.3–3 shows a spherical tank for storing water. The tank is filled through a hole in the top and drained through a hole in the bottom. If the tank's radius is r, you can use integration to show that the volume of water in the tank as a function of its height h is given by

$$V(h) = \pi r h^2 - \pi \frac{h^3}{3} \tag{9.3–10}$$

Torricelli's principle states that the liquid flow rate through the hole is proportional to the square root of the height h. Further studies in fluid mechanics have identified the relation more precisely, and the result is that the volume flow rate through the hole is given by

$$q = C_d A \sqrt{2gh} \tag{9.3–11}$$

where A is the area of the hole, g is the acceleration due to gravity, and C_d is an experimentally determined value that depends partly on the type of liquid. For water, $C_d = 0.6$ is a common value. We can use the principle of conservation of mass to obtain a differential equation for the height h. Applied to this tank, the principle says that the rate of change of liquid volume in the tank must equal the flow rate out of the tank; that is,

$$\frac{dV}{dt} = -q \tag{9.3–12}$$

From Equation (9.3–10),

$$\frac{dV}{dt} = 2\pi r h \frac{dh}{dt} - \pi h^2 \frac{dh}{dt} = \pi h (2r - h) \frac{dh}{dt}$$

Substituting this and Equation (9.3–11) into Equation (9.3–12) gives the required equation for h.

$$\pi(2rh - h^2)\frac{dh}{dt} = -C_d A \sqrt{2gh} \tag{9.3–13}$$

Use MATLAB to solve this equation to determine how long it will take for the tank to empty if the initial height is 9 ft. The tank has a radius of $r = 5$ ft and has a 1-in.-diameter hole in the bottom. Use $g = 32.2$ ft/sec^2. Discuss how to check the solution.

■ **Solution**

With $C_d = 0.6$, $r = 5$, $g = 32.2$, and $A = \pi(1/24)^2$, Equation (9.3–13) becomes

$$\frac{dh}{dt} = -\frac{0.0334\sqrt{h}}{10h - h^2} \tag{9.3–14}$$

Figure 9.3–3 Draining of a spherical tank.

We can first check the above expression for *dh/dt* for singularities. The denominator does not become zero unless $h = 0$ or $h = 10$, which correspond to a completely empty and a completely full tank. So we will avoid singularities if $0 < h < 10$.

Finally, we can use the following approximation to estimate the time to empty. Replace *h* on the right side of Equation (9.3–14) with its average value, namely, $(9 - 0)/2 = 4.5$ ft. This gives *dh/dt* $= -0.00286$, whose solution is $h(t) = h(0) - 0.00286t = 9 - 0.00286t$. According to this equation, the tank will be empty at $t = 9/0.00286 = 3147$ sec, or 52 min. We will use this value as a "reality check" on our answer.

The function file based on Equation (9.3–14) is

```
function hdot = height(t,h)
hdot = -(0.0334*sqrt(h))/(10*h-h^2);
```

The file is called as follows, using the `ode45` solver:

```
[t, h] = ode45(@height, [0 2475], 9);
plot(t,h),xlabel('Time (sec)'),ylabel('Height (ft)')
```

The resulting plot is shown in Figure 9.3–4. Note how the height changes more rapidly when the tank is nearly full or nearly empty. This is to be expected because of the effects of the tank's curvature. The tank empties in 2475 sec, or 41 min. This value is not grossly different from our rough estimate of 52 min, so we should feel comfortable accepting the numerical results. The value of the final time of 2475 sec was found by increasing the final time until the plot showed that the height became 0.

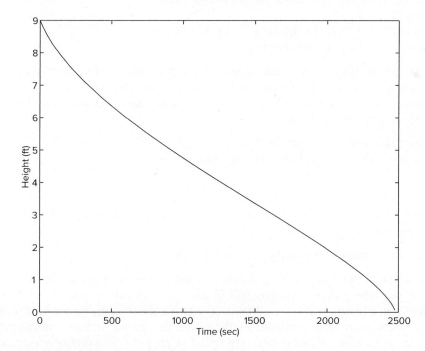

Figure 9.3–4 Plot of water height in a spherical tank.

9.4 Higher-Order Differential Equations

To use the ODE solvers to solve an equation higher than order 1, you must first write the equation as a set of first-order equations. This is easily done. Consider the second-order equation

$$5\ddot{y} + 7\dot{y} + 4y = f(t) \tag{9.4–1}$$

Solve it for the highest derivative:

$$\ddot{y} = \frac{1}{5}f(t) - \frac{4}{5}y - \frac{7}{5}\dot{y} \tag{9.4–2}$$

Define two new variables x_1 and x_2 to be y and its derivative $\dot{y}$. That is, define $x_1 = y$ and $x_2 = \dot{y}$. This implies that

$$\dot{x}_1 = x_2$$

$$\dot{x}_2 = \frac{1}{5}f(t) - \frac{4}{5}x_1 - \frac{7}{5}x_2$$

CAUCHY FORM

STATE-VARIABLE FORM

This form is sometimes called the *Cauchy form* or the *state-variable form*.

Now write a function file that computes the values of x_1 and $\dot{x}_2$ and stores them in a *column* vector. To do this, we must first have a function specified for $f(t)$. Suppose that $f(t) = \sin t$. Then the required file is

```
function xdot = example_1(t,x)
% Computes derivatives of two equations
xdot(1) = x(2);
xdot(2) = (1/5)*(sin(t)-4*x(1)-7*x(2));
xdot = [xdot(1); xdot(2)];
```

Note that xdot(1) represents $\dot{x}_1$, xdot(2) represents $\dot{x}_2$, x(1) represents x_1, and x(2) represents x_2. Once you become familiar with the notation for the state-variable form, you will see that the previous code could be replaced with the following shorter form.

```
function xdot = example_1(t,x)
% Computes derivatives of two equations
xdot = [x(2); (1/5)*(sin(t)-4*x(1)-7*x(2))];
```

Suppose we want to solve Equation (9.4–1) for $0 \le t \le 6$ with the initial conditions $x(0) = 3$, $\dot{x}(0) = 9$. Then the initial condition for the vector **x** is [3, 9]. To use ode45, you type

```
[t, x] = ode45(@example_1, [0 6], [3 9]);
```

Each row in the vector **x** corresponds to a time returned in the column vector t. If you type plot(t,x), you will obtain a plot of both x_1 and x_2 versus t. Note **x** is a matrix with two columns. The first column contains the values of x_1 at the various times generated by the solver; the second column contains the values of x_2. Thus, to plot only x_1, type plot(t,x(:,1)). To plot only x_2, type plot(t,x(:,2)).

When we are solving nonlinear equations, sometimes it is possible to check the numerical results by using an approximation that reduces the equation to a linear one. The following example illustrates such an approach with a second-order equation.

A Nonlinear Pendulum Model

EXAMPLE 9.4–1

The pendulum shown in Figure 9.4–1 consists of a concentrated mass m attached to a rod whose mass is small compared to m. The rod's length is L. The equation of motion for this pendulum is

$$\ddot{\theta} + \frac{g}{L}\sin\theta = 0 \qquad (9.4\text{–}3)$$

Suppose that $L = 1$ m and $g = 9.81$ m/s². Use MATLAB to solve this equation for $\theta(t)$ for two cases: $\theta(0) = 0.5$ rad and $\theta(0) = 0.8\pi$ rad. In both cases $\dot{\theta}(0) = 0$. Discuss how to check the accuracy of the results.

■ **Solution**

If we use the small-angle approximation $\sin\theta \approx \theta$, the equation becomes

$$\ddot{\theta} + \frac{g}{L}\theta = 0 \qquad (9.4\text{–}4)$$

which is linear and has the solution

$$\theta(t) = \theta(0)\cos\sqrt{\frac{g}{L}}\,t \qquad (9.4\text{–}5)$$

if $\dot{\theta}(0) = 0$. Thus the amplitude of oscillation is $\theta(0)$, and the period is $P = 2\pi\sqrt{L/g} = 2.006$ s. We can use this information to select a final time and to check our numerical results.

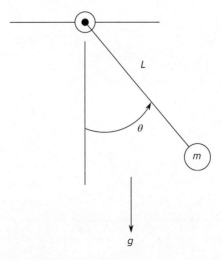

Figure 9.4–1 A pendulum.

First rewrite the pendulum Equation (9.4–3) as two first-order equations. To do this, let $x_1 = \theta$ and $x_2 = \dot\theta$. Thus

$$\dot{x}_1 = \dot\theta = x_2$$
$$\dot{x}_2 = \ddot\theta = -\frac{g}{L}\sin x_1$$

The following function file is based on the last two equations. Remember that the output xdot must be a *column* vector.

```
function xdot = pendulum(t,x)
g = 9.81; L = 1;
xdot = [x(2); -(g/L)*sin(x(1))];
```

This file is called as follows. The vectors ta and xa contain the results for the case where $\theta(0) = 0.5$. In both cases, $\dot\theta(0) = 0$. The vectors tb and xb contain the results for $\theta(0) = 0.8\pi$.

```
[ta, xa] = ode45(@pendulum, [0 5], [0.5 0]);
[tb, xb] = ode45(@pendulum, [0 5], [0.8*pi 0]);
plot(ta, xa(:,1), tb,xb(:,1)),xlabel('Time (s)'), . . .
    ylabel('Angle (rad)'),gtext('Case 1'),gtext('Case 2')
```

The results are shown in Figure 9.4–2. The amplitude remains constant, as predicted by the small-angle analysis, and the period for the case where $\theta(0) = 0.5$ is a little larger

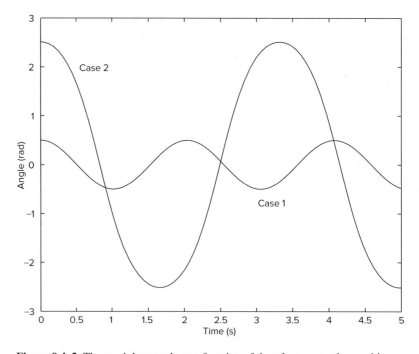

Figure 9.4–2 The pendulum angle as a function of time for two starting positions.

than 2 s, the value predicted by the small-angle analysis. So we can place some confidence in the numerical procedure. For the case where $\theta(0) = 0.8\pi$, the period of the numerical solution is about 3.3 s. This illustrates an important property of nonlinear differential equations. The free response of a linear equation has the same period for any initial conditions; however, the form and therefore the period of the free response of a nonlinear equation often depend on the particular values of the initial conditions.

In this example, the values of g and L were encoded in the function `pendulum(t, x)`. Now suppose you want to obtain the pendulum response for different lengths L or different gravitational accelerations g. You could use the `global` command to declare g and L as global variables, or you could pass parameter values through an argument list in the `ode45` function; but the preferred method is to use a nested function. Nested functions are discussed in Section 3.3. The following program shows how this is done.

```
function pendula
g = 9.81; L = 0.75; % First case.
tF = 6*pi*sqrt(L/g); % Approximately 3 periods.
[t1, x1] = ode45(@pendulum, [0,tF], [0.4, 0];
%
g = 1.63; L = 2.5; % Second case.
tF = 6*pi*sqrt(L/g); % Approximately 3 periods.
[t2, x2] = ode45(@pendulum, [0 tF], [0.2 0];
plot(t1, x1(:,1), t2, x2(:,1)), . . .
   xlabel ('time (s)'), ylabel ('\theta (rad)')
   % Nested function.
      function xdot = pendulum(t,x)
      xdot = [x(2);-(g/L)*sin(x(1))];
   end
end
```

Table 9.4–1 summarizes the syntax of the ODE solvers using `ode45` as an example.

Table 9.4–1 Syntax of the ODE solver `ode45`

Command	Description
`[t, y] = ode45(@ydot,tspan, y0, options)`	Solves the vector differential equation $\dot{\mathbf{y}} = \mathbf{f}(t, \mathbf{y})$ specified by the function file whose handle is `@ydot` and whose inputs must be t and $\mathbf{y}$, and whose output must be a *column* vector representing $d\mathbf{y}/dt$; that is, $\mathbf{f}(t, \mathbf{y})$. The number of rows in this column vector must equal the order of the equation. The vector `tspan` contains the starting and ending values of the independent variable t, and optionally any intermediate values of t where the solution is desired. The vector `y0` contains the initial values. The function file must have two input arguments, t and $\mathbf{y}$, even for equations where $f(t, \mathbf{y})$ is not a function of t. The `options` argument is created with the `odeset` function. The syntax is identical for the solver `ode15s`.

9.5 Special Methods for Linear Equations

MATLAB provides some convenient tools to use if the differential equation model is linear. Even though there are general methods available for finding the analytical solutions of linear differential equations, it is sometimes more convenient to use a numerical method to find the solution. Examples of such situations occur when the forcing function is a complicated function or when the order of the differential equation is higher than 2. In such cases the algebra involved in obtaining the analytical solution might not be worth the effort, especially if the main objective is to obtain a plot of the solution.

Matrix Methods

We can use matrix operations to reduce the number of lines to be typed in the derivative function file. For example, the following equation describes the motion of a mass connected to a spring, with viscous friction acting between the mass and the surface. Another force $u(t)$ also acts on the mass.

$$m\ddot{y} + c\dot{y} + ky = u(t) \tag{9.5--1}$$

This can be put into Cauchy form by letting $x_1 = y$ and $x_2 = \dot{y}$. This gives

$$\dot{x}_1 = x_2$$

$$\dot{x}_2 = \frac{1}{m}u(t) - \frac{k}{m}x_1 - \frac{c}{m}x_2$$

This can be written as one matrix equation as follows.

$$\begin{bmatrix} \dot{x}_1 \\ \dot{x}_2 \end{bmatrix} = \begin{bmatrix} 0 & 1 \\ -\dfrac{k}{m} & -\dfrac{c}{m} \end{bmatrix} \begin{bmatrix} x_1 \\ x_2 \end{bmatrix} + \begin{bmatrix} 0 \\ \dfrac{1}{m} \end{bmatrix} u(t)$$

In compact form this is

$$\dot{\mathbf{x}} = \mathbf{A}\mathbf{x} + \mathbf{B}u(t) \tag{9.5--2}$$

where

$$\mathbf{A} = \begin{bmatrix} 0 & 1 \\ -\dfrac{k}{m} & -\dfrac{c}{m} \end{bmatrix} \qquad \mathbf{B} = \begin{bmatrix} 0 \\ \dfrac{1}{m} \end{bmatrix} \qquad \mathbf{x} = \begin{bmatrix} x_1 \\ x_2 \end{bmatrix}$$

The following function file shows how to use matrix operations. In this example, $m = 1$, $c = 2$, $k = 5$, and the applied force is $u(t) = 10$.

```
function xdot = msd(t,x)
% Function file for mass with spring and damping.
% Position is first variable, velocity is second variable.
u = 10;
m = 1;c = 2;k = 5;
A = [0, 1;-k/m, -c/m];
B = [0; 1/m];
xdot = A*x+B*u;
```

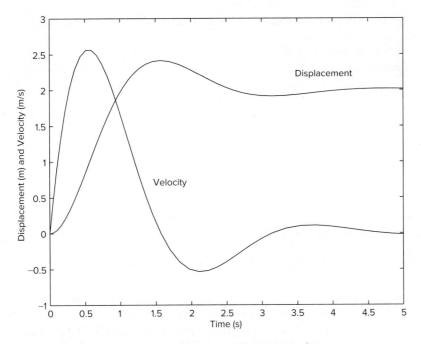

Figure 9.5–1 Displacement and velocity of the mass as a function of time.

Note that the output `xdot` will be a column vector because of the definition of matrix-vector multiplication. We try different values of the final time until we see the entire response. Using a final time of 5 and the initial conditions $x_1(0) = 0$ and $x_2(0) = 0$, we call the solver and plot the solution as follows:

```
[t, x] = ode45(@msd, [0,5], [0,0]);
plot(t,x(:,1),t,x(:,2))
```

Figure 9.5–1 shows the edited plot. Note that we could have avoided embedding the values of the parameters m, c, k, and u by making `msd` a nested function as was done with the functions `pendulum` and `pendula` in Section 9.4.

Test Your Understanding

T9.5–1 Plot the position and velocity of a mass with a spring and damping, having the parameter values $m = 2$, $c = 3$, and $k = 7$. The applied force is $u = 35$, the initial position is $y(0) = 2$, and the initial velocity is $\dot{y}(0) = -3$.

Characteristic Roots from the eig Function

The characteristic roots of a linear differential equation give information about the speed of response and the oscillation frequency, if any.

MATLAB provides the `eig` function to compute the characteristic roots when the model is given in the state-variable form (9.5–2). Its syntax is `eig(A)`, where `A` is the matrix that appears in Equation (9.5–2). (The function's name is an abbreviation of *eigenvalue*, which is another name for characteristic root.) For example, consider the equations

$$\dot{x}_1 = -3x_1 + x_2 \tag{9.5–3}$$
$$\dot{x}_2 = -x_1 - 7x_2 \tag{9.5–4}$$

The matrix **A** for these equations is

$$\mathbf{A} = \begin{bmatrix} -3 & 1 \\ -1 & -7 \end{bmatrix}$$

To find the characteristic roots, type

```
>>A = [-3, 1;-1, -7];
>>r = eig(A)
```

The answer so obtained is `r = [-6.7321, -3.2679]`. To find the time constants, which are the negative reciprocals of the real parts of the roots, you type `tau = -1./real(r)`. The time constants are 0.1485 and 0.3060. Four times the dominant time constant, or $4(0.3060) = 1.224$, gives the time it takes for the free response to become approximately zero.

EIGENVALUE

ODE Solvers in the Control System Toolbox

Many of the functions from the Control System toolbox are available in the Student Edition of MATLAB. Some of these can be used to solve linear, time-invariant (constant-coefficient) differential equations. They are sometimes more convenient to use and more powerful than the ODE solvers discussed thus far, because general solutions can be found for linear, time-invariant equations. Here we discuss several of these functions. These are summarized in Table 9.5–1. The other features of the Control System toolbox require advanced methods, and will not be covered here. See [Palm, 2014] for coverage of these methods.

An *LTI object* describes a linear, time-invariant equation, or sets of equations, here referred to as the *system*. An LTI object can be created from different descriptions of the system, it can be analyzed with several functions, and it can be accessed to provide alternate descriptions of the system. For example, the equation

$$2\ddot{x} + 3\dot{x} + 5x = u(t) \tag{9.5–5}$$

is one description of a particular system. This description is called the *reduced form*. The following is a state-model description of the same system

$$\dot{\mathbf{x}} = \mathbf{A}\mathbf{x} + \mathbf{B}u \tag{9.5–6}$$

LTI OBJECT

Table 9.5–1 LTI object functions

Command	Description
sys = ss(A, B, C, D)	Creates an LTI object in state-space form, where the matrices A, B, C, and D correspond to those in the model $\dot{x} = Ax + Bu$, $y = Cx + Du$.
[A, B, C, D] = ssdata(sys)	Extracts the matrices A, B, C, and D corresponding to those in the model $\dot{x} = Ax + Bu$, $y = Cx + Du$.
sys = tf(right,left)	Creates an LTI object in transfer function form, where the vector right is the vector of coefficients of the right-hand side of the equation, arranged in descending derivative order, and left is the vector of coefficients of the left-hand side of the equation, also arranged in descending derivative order.
sys2 = tf(sys1)	Creates the transfer function model sys2 from the state model sys1.
sys1 = ss(sys2)	Creates the state model sys1 from the transfer function model sys2.
[right, left] = tfdata(sys,'v')	Extracts the coefficients on the right- and left-hand sides of the reduced-form model specified in the transfer function model sys. When the optional parameter 'v' is used, the coefficients are returned as vectors rather than as cell arrays.

where $x_1 = x$, $x_2 = \dot{x}$, and

$$A = \begin{bmatrix} 0 & 1 \\ -\frac{5}{2} & -\frac{3}{2} \end{bmatrix} \qquad B = \begin{bmatrix} 0 \\ \frac{1}{2} \end{bmatrix} \qquad x = \begin{bmatrix} x_1 \\ x_2 \end{bmatrix} \qquad (9.5–7)$$

Both model forms contain the same information. However, each form has its own advantages, depending on the purpose of the analysis.

Because there are two or more state variables in a state model, we need to be able to specify which state variable, or combination of variables, constitutes the output of the simulation. For example, models (9.5–6) and (9.5–7) can represent the motion of a mass, with x_1 the position and x_2 the velocity of the mass. We need to be able to specify whether we want to see a plot of the position, or the velocity, or both. This specification of the output, denoted by the vector **y**, is done in general with the matrices **C** and **D**, which must be compatible with the equation

$$y = Cx + Du(t) \qquad (9.5–8)$$

where the vector **u**(t) allows for multiple inputs. To continue the previous example, if we want the output to be the position $x = x_1$, then $y = x_1$, and we would select **C** = [1, 0] and **D** = 0. Thus, in this case, Equation (9.5–8) reduces to $y = x_1$.

To create an LTI object from the reduced form (9.5–5), use the tf(right, left) function, and type

```
>>sys1 = tf(1, [2, 3, 5]);
```

where the vector right is the vector of coefficients of the right-hand side of the equation, arranged in descending derivative order, and left is the vector of coefficients of the left-hand side of the equation, also arranged in descending derivative order. The result, sys1, is the LTI object that describes the system in reduced form, also called the *transfer function form*. (The name of the function, tf, stands for *transfer function*, which is an equivalent way of describing the coefficients on the left- and right-hand sides of the equation.)

The LTI object sys2 in transfer function form for the equation

$$6\frac{d^3x}{dt^3} - 4\frac{d^2x}{dt^2} + 7\frac{dx}{dt} + 5x = 3\frac{d^2u}{dt^2} + 9\frac{du}{dt} + 2u \tag{9.5-9}$$

is created by typing

```
>>sys2 = tf([3, 9, 2], [6, -4, 7, 5]);
```

To create an LTI object from a state model, you use the ss(A, B, C, D) function, where ss stands for *state space*. For example, to create an LTI object in state-model form for the system described by Equations (9.5–6) through (9.5–8), you type

```
>>A - [0, 1; -5/2, -3/2]; B = [0; 1/2];
>>C = [1, 0]; D = 0;
>>sys3 = ss(A,B,C,D);
```

An LTI object defined using the tf function can be used to obtain an equivalent state-model description of the system. To create a state model for the system described by the LTI object sys1 created previously in transfer function form, you type ss(sys1). You will then see the resulting **A, B, C**, and **D** matrices on the screen. To extract and save the matrices, use the ssdata function as follows.

```
>>[A1, B1, C1, D1] = ssdata(sys1);
```

The results are

$$\mathbf{A1} = \begin{bmatrix} -1.5 & -1.25 \\ 2 & 0 \end{bmatrix} \quad \mathbf{B1} = \begin{bmatrix} 0.5 \\ 0 \end{bmatrix} \quad \mathbf{C1} = [0 \quad 0.5] \quad \mathbf{D1} = [0]$$

When using ssdata to convert a transfer function form to a state model, note that the output y will be a scalar that is identical to the solution variable of the reduced form; in this case the solution variable of Equation (9.5–1) is the variable y. To interpret the state model, we need to relate its state variables x_1 and x_2 to y. The values of the matrices **C1** and **D1** tell us that the output variable $y = 0.5x_2$. Thus we see that $x_2 = 2y$. The other state variable x_1 is related to x_2 by $\dot{x}_2 = 2x_1$. Thus $x_1 = \dot{y}$.

To create a transfer function description of the system sys3, previously created from the state model, you type tfsys3 = tf(sys3). To extract and save the coefficients of the reduced form, use the tfdata function as follows:

```
[right, left] = tfdata(sys3, 'v')
```

For this example, the vectors returned are right = 1 and left = [1, 1.5, 2.5]. The optional parameter 'v' tells MATLAB to return the coefficients as vectors;

otherwise, they are returned as cell arrays. These functions are summarized in Table 9.5–1.

T9.5–2 Obtain the state model for the reduced-form model

$$5\ddot{x} + 7\dot{x} + 4x = u(t)$$

Then convert the state model back to reduced form, and see if you get the original reduced-form model.

Linear ODE Solvers

The Control System toolbox provides several solvers for linear models. These solvers are categorized by the type of input function they can accept: zero input, impulse input, step input, and a general input function. These are summarized in Table 9.5–2.

The initial Function   The `initial` function computes and plots the free response of a state model. This is sometimes called the *initial-condition response* or the *undriven response* in the MATLAB documentation. The basic syntax is `initial(sys,x0)`, where `sys` is the LTI object in state-model form and `x0` is the initial-condition vector. The time span and number of solution points are chosen automatically. For example, to find the free response of the state model (9.5–5) through (9.5–8), for $x_1(0) = 5$ and $x_2(0) = -2$, first define it in state-model form. This was done previously to obtain the system `sys3`. Then use the `initial` function as follows.

```
>>initial(sys3, [5, -2])
```

The plot shown in Figure 9.5–2 will be displayed on the screen. Note that MATLAB automatically labels the plot, computes the steady-state response, and displays it with a dotted line.

Table 9.5–2 Basic syntax of the LTI ODE solvers

Command	Description
`impulse(sys)`	Computes and plots the impulse response of the LTI object `sys`.
`initial(sys,x0)`	Computes and plots the free response of the LTI object `sys` given in state-model form, for the initial conditions specified in the vector `x0`.
`lsim(sys,u,t)`	Computes and plots the response of the LTI object `sys` to the input specified by the vector `u`, at the times specified by the vector `t`.
`step(sys)`	Computes and plots the step response of the LTI object `sys`.

See the text for description of extended syntax.

Initial Condition Results

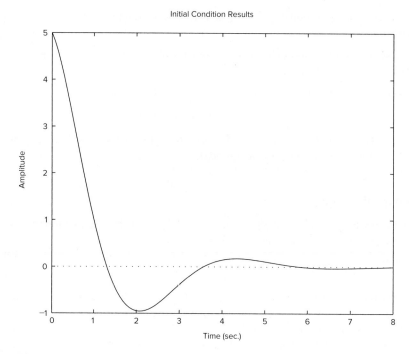

Figure 9.5–2 Free response of the model given by Equations (9.5–5) through (9.5–8) for $x_1(0) = 5$ and $x_2(0) = -2$.

To specify the final time tF, use the syntax $initial(sys,x0,tF)$. To specify a vector of times of the form $t = 0:dt:tF$, at which to obtain the solution, use the syntax $initial(sys,x0,t)$.

When called with left-hand arguments, as $[y,t,x] = initial(sys, x0,\ .\ .\ .\ )$, the function returns the output response y, the time vector t used for the simulation, and the state vector x evaluated at those times. The columns of the matrices y and x are the outputs and the states, respectively. The number of rows in y and x equals $length(t)$. No plot is drawn. The syntax $initial(sys1, sys2,\ .\ .\ .\ ,x0,t)$ plots the free response of multiple LTI systems on a single plot. The time vector t is optional. You can specify line color, line style, and marker for each system, for example, $initial(sys1,'r',\ sys2,'y-\ -', sys3,'gx',x0)$.

The impulse Function The $impulse$ function plots the unit-impulse response for each input-output pair of the system, assuming that the initial conditions are zero. (The unit impulse is also called the Dirac delta function.) The basic syntax is $impulse(sys)$, where sys is the LTI object. Unlike the $initial$ function, the $impulse$ function can be used with either a state model or a transfer function model. The time span and number of solution points are chosen automatically. For example, the impulse response of Equation (9.5–5) is found as follows:

```
>>sys1 = tf(1, [2, 3, 5]);
>>impulse(sys1)
```

The extended syntax of the `impulse` function is similar to that of the `initial` function.

The step Function The `step` function plots the unit-step response for each input-output pair of the system, assuming that the initial conditions are zero. [The unit step function $u(t)$ is 0 for $t < 0$ and 1 for $t > 0$.] The basic syntax is `step(sys)`, where `sys` is the LTI object. The `step` function can be used with either a state model or a transfer function model. The time span and number of solution points are chosen automatically. The extended syntax of the `step` function is similar to that of the `initial` and the `impulse` functions.

To find the unit-step response, for zero initial conditions, of the state model (9.5–6) through (9.5–8), and the reduced-form model

$$5\ddot{x} + 7\dot{x} + 5x = 5\dot{f} + f \qquad (9.5\text{–}10)$$

the session is (assuming `sys3` is still available in the workspace)

```
>>sys4 = tf([5, 1], [5, 7, 5]);
>>step(sys3,'b',sys4,'- -')
```

The result is shown in Figure 9.5–3. The steady-state response is indicated by the horizontal dotted line. Note how the steady-state response and the time to reach that state are automatically determined.

Step response can be characterized by the following parameters.

- *Steady-state value*: The limit of the response as $t \to \infty$.
- *Settling time*: The time for the response to reach and stay within a certain percentage (usually 2 percent) of its steady-state value.
- *Rise time*: The time required for the response to rise from 10 to 90 percent of its steady-state value.
- *Peak response*: The largest value of the response.
- *Peak time*: The time at which the peak response occurs.

When the `step(sys)` function puts a plot on the screen, you may use the plot to calculate these parameters by right-clicking anywhere within the plot area. This brings up a menu. Choose **Characteristics** to obtain a submenu that contains the response characteristics. When you select a specific characteristic, for example, "peak response," MATLAB puts a large dot on the peak and displays dashed lines indicating the value of the peak response and the peak time. Move the cursor over this dot to see a display of the values. You can use the other solvers in the same way, although the menu choices may be different. For example, peak response

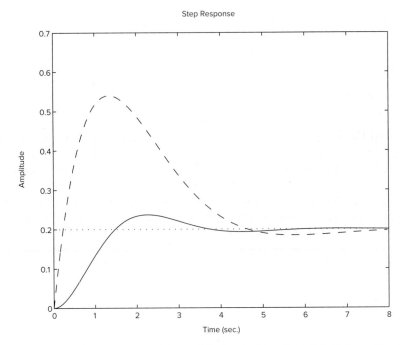

Figure 9.5–3 Step response of the model given by Equations (9.5–6) through (9.5–8) and the model (9.5–10), for zero initial conditions.

and settling time are available when you use the `impulse(sys)` function, but not the rise time. If instead of choosing **Characteristics** you choose **Properties** and select the **Options** tab, you can change the defaults for the settling time and rise time, which are 2 percent and 10 to 90 percent.

Using this method, we find that the solid curve in Figure 9.5–3 has the following characteristics:

■ Steady-state value: 0.2
■ 2 percent settling time: 5.22
■ 10 to 90 percent rise time: 1.01
■ Peak response: 0.237
■ Peak time: 2.26

You can also read values off any part of the curve by placing the cursor on the curve at the desired point. You can move the cursor along the curve and read the values as they change. Using this method, we find that the solid curve in Figure 9.5–3 crosses the steady-state value of 0.2 for the second time at $t = 3.74$.

You can suppress the plot generated by `step` and create your own plot as follows, assuming `sys3` is still available in the workspace.

```
[x,t] = step(sys3);
plot(t,x)
```

You can then use the Plot Editor tools to edit the plot. However, with this approach, right-clicking on the plot will no longer give you information about the step response characteristics.

Suppose the step input is not a *unit* step but instead is 0 for $t < 0$ and 10 for $t > 0$. There are two ways to obtain the solution with the factor 10. Using sys3 as the example, these are step(10*sys3) and

```
[x,t] = step(sys3);
plot(t,10*x)
```

The lsim Function   The lsim function plots the response of the system to an arbitrary input. The basic syntax for zero initial conditions is lsim(sys,u,t), where sys is the LTI object, t is a time vector having regular spacing, as t = 0:dt:tF, and u is a matrix with as many columns as inputs, and whose *i*th row specifies the value of the input at time t(i). To specify nonzero initial conditions for a state-space model, use the syntax lsim(sys,u,t,x0). This computes and plots the total response (the free plus forced response). Right-clicking on the plot brings up the menu containing the **Characteristics** choice, although the only characteristic available is the peak response.

When called with left-hand arguments, as [y, t] = lsim(sys, u,...), the function returns the output response y and the time vector t used for the simulation. The columns of the matrix y are the outputs, and the number of its rows equals length(t). No plot is drawn. To obtain the state vector solution for state-space models, use the syntax [y, t, x] = lsim(sys,u,...). The syntax lsim(sys1,sys2,...,u,t,x0) plots the responses of multiple LTI systems on a single plot. The initial-condition vector xo is needed only if the initial conditions are nonzero. You can specify line color, line style, and marker for each system, for example, lsim(sys1,'r',sys2, 'y--',sys3,'gx',u,t).

We will see an example of the lsim function shortly.

Programming Detailed Forcing Functions

As a final example of higher-order equations, we now show how to program a detailed forcing function for use with the lsim function. We use a dc motor as the application. The equations for an armature-controlled dc motor (such as a permanent-magnet motor) shown in Figure 9.5–4 are the following. They result

Figure 9.5–4 An armature-controlled dc motor.

from Kirchhoff's voltage law and Newton's law applied to a rotating inertia. The motor's current is i and its rotational velocity is ω.

$$L\frac{di}{dt} = -Ri - K_e\omega + v(t) \tag{9.5–11}$$

$$I\frac{d\omega}{dt} = K_T i - c\omega \tag{9.5–12}$$

where L, R, and I are the motor's inductance, resistance, and inertia; K_T and K_e are the torque constant and back emf constant; c is a viscous damping constant; and $v(t)$ is the applied voltage. These equations can be put into matrix form as follows, where $x_1 = i$ and $x_2 = \omega$.

$$\begin{bmatrix} \dot{x}_1 \\ \dot{x}_2 \end{bmatrix} = \begin{bmatrix} -\dfrac{R}{L} & -\dfrac{K_e}{L} \\ \dfrac{K_T}{I} & -\dfrac{c}{I} \end{bmatrix} \begin{bmatrix} x_1 \\ x_2 \end{bmatrix} + \begin{bmatrix} \dfrac{1}{L} \\ 0 \end{bmatrix} v(t)$$

EXAMPLE 9.5–1	Trapezoidal Profile for a DC Motor

In many applications we want to accelerate the motor to a desired speed and allow it to run at that speed for some time before decelerating to a stop. Investigate whether an applied voltage having a trapezoidal profile will accomplish this. Use the values $R = 0.6\ \Omega$, $L = 0.002$ H, $K_T = 0.04$ N $\cdot$ m/A, $K_e = 0.04$ V $\cdot$ s/rad, $c = 0$, and $I = 6 \times 10^{-5}$ kg $\cdot$ m^2. The applied voltage in volts is given by

$$y(t) = \begin{cases} 100t & 0 \le t < 0.1 \\ 10 & 0.1 \le t \le 0.4 \\ -100(t - 0.4) + 10 & 0.4 < t \le 0.5 \\ 0 & t > 0.5 \end{cases}$$

This is shown in the top graph in Figure 9.5–5.

■ **Solution**
The following program first creates the model `sys` from the matrices **A, B, C,** and **D.** We choose **C** and **D** to obtain the speed x_2 as the only output. (To obtain both the speed and the current as outputs, we would choose C = [1, 0; 0, 1] and D = [0; 0].) The program computes the time constants using the `eig` function and then creates `time`, the array of time values to be used by `lsim`. We choose the time increment 0.0001 to be a very small fraction of the total time, 0.6 s.

The trapezoidal voltage function is then created with a `for` loop. This is perhaps the easiest way because the `if-elseif-else` structure mimics the equations that define $v(t)$. The initial conditions $x_1(0)$ and $x_2(0)$ are assumed to be zero, and so they need not be specified in the `lsim` function.

```
% File dcmotor.m
R = 0.6; L = 0.002; c = 0;
K_T = 0.04; K_e = 0.04; I = 6e-5;
```

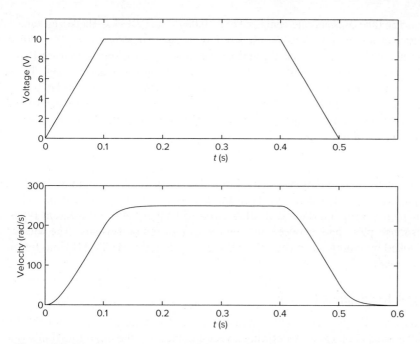

Figure 9.5–5 Voltage input and resulting velocity response of a dc motor.

```
A = [-R/L, -K_e/L; K_T/I, -c/I];
B = [1/L; 0]; C = [0,1]; D = [0];
sys = ss(A,B,C,D);
Time_constants = -1./real(eig(A))
time = 0:0.0001:0.6;
k = 0;
for t = 0:0.0001:0.6
    k = k + 1;
    if t < 0.1
        v(k) = 100*t;
    elseif t < = 0.4
        v(k) = 10;
    elseif t < = 0.5
        v(k) = -100*(t-0.4) + 10;
    else
        v(k) = 0;
    end
end
[y,t] = lsim(sys, v, time);
subplot(2,1,1), plot(time,v)
subplot(2,1,2), plot(time,y)
```

The time constants are computed to be 0.0041 and 0.0184 s. The largest time constant indicates that the motor's response time is approximately 4(0.0184) = 0.0736 s. Because this time is less than the time needed for the applied voltage to reach 10 V, the motor should be able to follow the desired trapezoidal profile reasonably well. To know for certain, we must solve the motor's differential equations. The results are plotted in the bottom graph of Figure 9.5–5. The motor's velocity follows a trapezoidal profile as expected, although there is some slight deviation because of its electric resistance and mechanical inertia.

Linear System Analyzer   The Control System toolbox contains the Linear System Analyzer, which assists in the analysis of LTI systems. It provides an interactive user interface that allows you to switch between different types of response plots and between the analysis of different systems. The viewer is invoked by typing linearSystemAnalyzer. See the MATLAB Help for more information.

Predefined Input Functions

You can always create any complicated input function to use with the ODE solver ode45 or lsim by defining a vector containing the input function's values at specified times, as was done in Example 9.5–1 for the trapezoidal profile. However, MATLAB provides the gensig function that makes it easy to construct periodic input functions.

The syntax [u, t] = gensig(type, period) generates a periodic input of a specified type type, having a period period. The following types are available: sine wave (type = 'sin'), square wave (type = 'square'), and narrow-width periodic pulse (type = 'pulse'). The vector t contains the times, and the vector u contains the input values at those times. All generated inputs have unit amplitudes. The syntax [u,t] = gensig(type, period, tF,dt) specifies the time duration tF of the input and the spacing dt between the time instants.

For example, suppose a square wave with period 5 is applied to the following reduced-form model.

$$\ddot{x} + 2\dot{x} + 4x = 4f \tag{9.5–13}$$

To find the response for zero initial conditions, over the interval $0 \le t \le 10$, using a step size of 0.01, the session is

```
>>sys5 = tf(4,[1,2,4]);
>>[u, t] = gensig('square',5,10,0.01);
>>[y, t] = lsim (sys5,u,t);plot(t,y,u), . . .
   axis([0 10 -0.5 1.5]), . . .
   xlabel('Time'),ylabel('Response')
```

The result is shown in Figure 9.5–6.

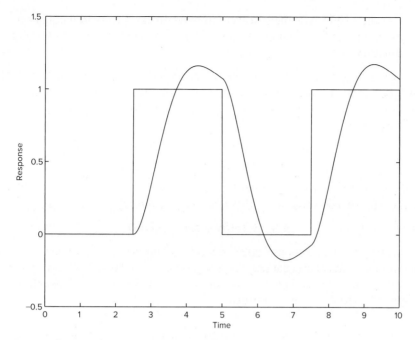

Figure 9.5–6 Square wave response of the model $\ddot{x} + 2\dot{x} + 4x = 4f$.

9.6 Summary

This chapter covered numerical methods for computing integrals and derivatives, and for solving ordinary differential equations. Now that you have finished this chapter, you should be able to do the following.

■ Numerically evaluate single, double, and triple integrals whose integrands are given functions.
■ Numerically evaluate single integrals whose integrands are given as numerical values.
■ Numerically estimate the derivative of a set of data.
■ Compute the gradient and Laplacian of a given function.
■ Obtain in closed form the integral and derivative of a polynomial function.
■ Use the MATLAB ODE solvers to solve single first-order ordinary differential equations whose initial conditions are specified.
■ Convert higher-order ordinary differential equations into a set of first-order equations.
■ Use the MATLAB ODE solvers to solve sets of higher-order ordinary differential equations whose initial conditions are specified.
■ Use MATLAB to convert a model from transfer function form to state-variable form, and vice versa.

■ Use the MATLAB linear solvers to solve linear differential equations to obtain the free response and the step response for arbitrary forcing functions.

We have not covered all the differential equation solvers provided in MATLAB, but limited our coverage to ordinary differential equations whose initial conditions are specified. MATLAB provides algorithms for solving *boundary-value problems* (BVPs) such as

$$\ddot{x} + 7\dot{x} + 10x = 0 \qquad x(0) = 2 \qquad x(5) = 8 \qquad 0 \leq t \leq 5$$

See the Help for the function `bvp4c`. Some differential equations are specified *implicitly* as $f(t, y, \dot{y}) = 0$. The solver `ode15i` can be used for such problems. MATLAB can also solve *delay-differential equations* (DDEs) such as

$$\ddot{x} + 7\dot{x} + 10x + 5x(t - 3) = 0$$

See the help for the functions `dde23`, `ddesd`, and `deval`. The function `pdepe` can solve *partial* differential equations. See also `pdeval`. In addition, MATLAB provides support for analyzing and plotting the solver's output. See the functions `odeplot`, `odephas2`, `odephas3`, and `odeprint`.

Key Terms

Backward difference,   398

Cauchy form,   410

Central difference,   399

Definite integral,   390

Eigenvalue,   416

Euler method,   403

Forced response,   403

Forward difference,   398

Free response,   403

Improper integral,   390

Indefinite integral,   390

Initial-value problem,   402

Laplacian,   402

LTI object,   416

Modified Euler method,   404

Ordinary differential equation,   402

Predictor-corrector method,   404

Singularity,   390

State-variable form,   410

Step size,   404

Problems

You can find answers to problems marked with an asterisk at the end of the text.

Section 9.1

1.* An object moves at a velocity $v(t) = 5 + 7t^2$ m/s starting from the position $x(2) = 5$ m at $t = 2$ s. Determine its position at $t = 10$ s.

2. The total distance traveled by an object moving at velocity $v(t)$ from the time $t = a$ to the time $t = b$ is

$$x(b) = \int_a^b |v(t)|\, dt + x(a)$$

The absolute value $|v(t)|$ is used to account for the possibility that $v(t)$ might be negative. Suppose an object starts at time $t = 0$ and moves with a velocity of $v(t) = \cos(\pi t)$ m. Find the object's location at $t = 1$ s if $x(0) = 2$ m.

3. An object starts with an initial velocity of 3 m/s at $t = 0$, and it accelerates with an acceleration of $a(t) = 7t$ m/s^2. Find the total distance the object travels in 4 s.

4. The equation for the voltage $v(t)$ across a capacitor as a function of time is

$$v(t) = \frac{1}{C}\left[\int_0^t i(t)\, dt + Q_0\right]$$

where $i(t)$ is the applied current and Q_0 is the initial charge. A certain capacitor initially holds no charge. Its capacitance is $C = 10^{-7}$ F. If a current $i(t) = 0.2[1 + \sin(0.2t)]$ A is applied to the capacitor, compute the voltage $v(t)$ at $t = 1.2$ s if its initial velocity is zero.

5. A certain object's acceleration is given by $a(t) = 7t \sin 5t$ m/s^2. Compute its velocity at $t = 10$ s if its initial velocity is zero.

6. A certain object moves with the velocity $v(t)$ given in the table below. Determine the object's position $x(t)$ at $t = 10$ s if $x(0) = 3$.

Time (s)	0	1	2	3	4	5	6	7	8	9	10
Velocity (m/s)	0	2	5	7	9	12	15	18	22	20	17

7.* A tank having vertical sides and a bottom area of 100 ft^2 is used to store water. The tank is initially empty. To fill the tank, water is pumped into the top at the rate given in the following table. Determine the water height $h(t)$ at $t = 10$ min.

Time (min)	0	1	2	3	4	5	6	7	8	9	10
Flow rate (ft^3/min)	0	80	130	150	150	160	165	170	160	140	120

8. A cone-shaped paper drinking cup (like the kind supplied at water fountains) has a radius R and a height H. If the water height in the cup is h, the water volume is given by

$$V = \frac{1}{3}\pi\left(\frac{R}{H}\right)^2 h^3$$

Suppose that the cup's dimensions are $R = 1.5$ in. and $H = 4$ in.

a. If the flow rate from the fountain into the cup is 2 in.^{3}/s, how long will it take to fill the cup to the brim?

  b. If the flow rate from the fountain into the cup is given by $2(1 - e^{-2t})$ in.^{3}/s, how long will it take to fill the cup to the brim?

9. A certain object has a mass of 100 kg and is acted on by a force $f(t) = 500[2 - e^{-t}\sin(5\pi t)]$ N. The mass is at rest at $t = 0$. Determine the object's velocity at $t = 5$ s.

10.* A rocket's mass decreases as it burns fuel. The equation of motion for a rocket in vertical flight can be obtained from Newton's law, and it is

$$m(t)\frac{dv}{dt} = T - m(t)g$$

where T is the rocket's thrust and its mass as a function of time is given by $m(t) = m_0(1 - rt/b)$. The rocket's initial mass is m_0, the burn time is b, and r is the fraction of the total mass accounted for by the fuel.

Use the values $T = 48{,}000$ N, $m_0 = 2200$ kg, $r = 0.8$, $g = 9.81$ m/s^2, and $b = 40$ s. Determine the rocket's velocity at burnout.

11. The equation for the voltage $v(t)$ across a capacitor as a function of time is

$$y(t) = \frac{1}{C}\left[\int_0^t i(t)\,dt + Q_0\right]$$

where $i(t)$ is the applied current and Q_0 is the initial charge. Suppose that $C = 10^{-7}$ F and that $Q_0 = 0$. Suppose the applied current is $i(t) = 0.3 + 0.1e^{-5t}\sin(25\pi t)$ A. Plot the voltage $v(t)$ for $0 \le t \le 7$ s.

12. Compute the indefinite integral of $p(x) = 5x^2 - 9x + 8$.

13. Compute the double integral

$$A = \int_0^3 \int_1^3 (x^2 + 3xy)\,dx\,dy$$

14. Compute the double integral

$$A = \int_0^4 \int_0^\pi x^2 \sin y\,dx\,dy$$

15. Use MATLAB to evaluate the following double integral:

$$\int_1^2 \int_0^3 (1 + 10xy)\,dx\,dy$$

16. Compute the double integral

$$A = \int_0^1 \int_y^3 x^2(x + y)\,dx\,dy$$

Note that the region of integration lies to the right of the line $y = x$. Use this fact and a MATLAB relational operator to eliminate values for which $y > x$.

17. Compute the triple integral

$$A = \int_1^2 \int_0^1 \int_1^3 xe^{yz}\, dx\, dy\, dz$$

18. Use MATLAB to evaluate the following triple integral:

$$\int_0^3 \int_0^2 \int_0^1 xyz^2\, dx\, dy\, dz$$

Section 9.2

19. Plot the estimate of the derivative dy/dx from the following data. Do this by using forward, backward, and central differences. Compare the results.

x	0	1	2	3	4	5	6	7	8	9	10
y	0	2	5	7	9	12	15	18	22	20	17

20. At a relative maximum of a curve $y(x)$, the slope dy/dx is zero. Use the following data to estimate the values of x and y that correspond to a maximum point.

x	0	1	2	3	4	5	6	7	8	9	10
y	0	2	5	7	9	10	8	7	6	4	5

21. Compare the performance of the forward, backward, and central difference methods for estimating the derivative of $y(x) = e^{x} \sin(3x)$. Use 101 points from $x = 0$ to $x = 4$. Use a random additive error of ± 0.01.

22. Compute the expressions for dp_2/dx, $d(p_1p_2)/dx$, and $d(p_2/p_1)/dx$ for $p_1 = 5x^2 + 7$ and $p_2 = 5x^2 - 6x + 7$.

23. Plot the contour plot and the gradient (shown by arrows) for the function

$$f(x,\, y) = -x^2 + 2xy + 3y^2$$

Section 9.3

24. Plot the solution of the equation

$$6\dot{y} + y = f(t)$$

if $f(t) = 0$ for $t < 0$ and $f(t) = 15$ for $t \geq 0$. The initial condition is $y(0) = 7$.

25. The equation for the voltage y across the capacitor of an RC circuit is

$$RC\frac{dy}{dt} + y = v(t)$$

where $v(t)$ is the applied voltage. Suppose that $RC = 0.2$ s and that the capacitor voltage is initially 2 V. Suppose also that the applied voltage goes from 0 to 10 V at $t = 0$. Plot the voltage $y(t)$ for $0 \leq t \leq 1$ s.

26. The following equation describes the temperature $T(t)$ of a certain object immersed in a liquid bath of constant temperature T_b.

$$10\frac{dT}{dt} + T = T_b$$

Suppose the object's temperature is initially $T(0) = 70°F$ and the bath temperature is $T_b = 170°F$.

a. How long will it take for the object's temperature T to reach the bath temperature?

b. How long will it take for the object's temperature T to reach 168°F?

c. Plot the object's temperature $T(t)$ as a function of time.

27.* The equation of motion of a rocket-propelled sled is, from Newton's law,

$$m\dot{v} = f - cv$$

where m is the sled mass, f is the rocket thrust, and c is an air resistance coefficient. Suppose that $m = 1000$ kg and $c = 500$ N · s/m. Suppose also that $v(0) = 0$ and $f = 75,000$ N for $t \geq 0$. Determine the speed of the sled at $t = 10$ s.

28. The following equation describes the motion of a mass connected to a spring, with viscous friction on the surface.

$$m\ddot{y} + c\dot{y} + ky = 0$$

Plot $y(t)$ for $y(0) = 10$, $\dot{y}(0) = 5$ if

a. $m = 3$, $c = 18$, and $k = 102$

b. $m = 3$, $c = 39$, and $k = 120$

29. The equation for the voltage y across the capacitor of an RC circuit is

$$RC\frac{dy}{dt} + y = v(t)$$

where $v(t)$ is the applied voltage. Suppose that $RC = 0.2$ s and that the capacitor voltage is initially 2 V. Suppose also that the applied voltage is $v(t) = 10[2 - e^{-t}\sin(5\pi t)]$ V. Plot the voltage $y(t)$ for $0 \leq t \leq 5$ s.

30. The equation describing the water height h in a spherical tank with a drain at the bottom is

$$\pi(2rh - h^2)\frac{dh}{dt} = -C_d A\sqrt{2gh}$$

Suppose the tank's radius is $r = 3$ m and the circular drain hole has a radius of 2 cm. Assume that $C_d = 0.5$ and that the initial water height is $h(0) = 5$ m. Use $g = 9.81$ m/s^2.

a. Use an approximation to estimate how long it takes for the tank to empty.

b. Plot the water height as a function of time until $h(t) = 0$.

31. The following equation describes a certain dilution process, where $y(t)$ is the concentration of salt in a tank of freshwater to which salt brine is being added.

$$\frac{dy}{dt} + \frac{5}{10 + 2t}y = 4$$

Suppose that $y(0) = 0$. Plot $y(t)$ for $0 \leq t \leq 10$.

Section 9.4

32. The following equation describes the motion of a certain mass connected to a spring, with viscous friction on the surface

$$3\ddot{y} + 18\dot{y} + 102y = f(t)$$

where $f(t)$ is an applied force. Suppose that $f(t) = 0$ for $t < 0$ and $f(t) = 10$ for $t \geq 0$.

a. Plot $y(t)$ for $y(0) = \dot{y}(0) = 0$.

b. Plot $y(t)$ for $y(0) = 0$ and $\dot{y}(0) = 10$. Discuss the effect of the nonzero initial velocity.

33. The following equation describes the motion of a certain mass connected to a spring, with viscous friction on the surface

$$3\dot{y} + 39\dot{y} + 120y = f(t)$$

where $f(t)$ is an applied force. Suppose that $f(t) = 0$ for $t < 0$ and $f(t) = 10$ for $t \geq 0$.

a. Plot $y(t)$ for $y(0) = \dot{y}(0) = 0$.

b. Plot $y(t)$ for $y(0) = 0$ and $\dot{y}(0) = 10$. Discuss the effect of the nonzero initial velocity.

34. The following equation describes the motion of a certain mass connected to a spring, with no friction

$$3\ddot{y} + 75y = f(t)$$

where $f(t)$ is an applied force. Suppose the applied force is sinusoidal with a frequency of ω rad/s and an amplitude of 10 N: $f(t) = 10 \sin(\omega t)$.

Suppose that the initial conditions are $y(0) = \dot{y}(0) = 0$. Plot $y(t)$ for $0 \leq t \leq 20$ s. Do this for the following three cases. Compare the results of each case.

a. $\omega = 1$ rad/s

b. $\omega = 5$ rad/s

c. $\omega = 10$ rad/s

35. Van der Pol's equation has been used to describe many oscillatory processes. It is

$$\ddot{y} - \mu(1 - y^2)\dot{y} + y = 0$$

Plot $y(t)$ for $\mu = 1$ and $0 \le t \le 20$, using the initial conditions $y(0) = 5$, $\dot{y}(0) = 0$.

36. The equation of motion for a pendulum whose base is accelerating horizontally with an acceleration $a(t)$ is

$$L\ddot{\theta} + g \sin \theta = a(t) \cos \theta$$

Suppose that $g = 9.81$ m/s^2, $L = 1$ m, and $\dot{\theta}(0) = 0$. Plot $\theta(t)$ for $0 \le t \le 10$ s for the following three cases.

a. The acceleration is constant: $a = 5$ m/s^2, and $\theta(0) = 0.5$ rad.

b. The acceleration is constant: $a = 5$ m/s^2, and $\theta(0) = 3$ rad.

c. The acceleration is linear with time: $a = 0.5t$ m/s^2, and $\theta(0) = 3$ rad.

37. Van der Pol's equation is

$$\ddot{y} - \mu(1 - y^2)\,\dot{y} + y = 0$$

This equation is stiff for large values of the parameter μ. Compare the performance of ode45 and ode15s for this equation. Use $\mu = 1000$ and $0 \le t \le 3000$, with the initial conditions $y(0) = 2$, $\dot{y}(0) = 0$. Plot $y(t)$ versus t.

38. Consider a mass-spring-damper system in which the spring element gets weaker with time due to metal fatigue. Suppose the spring constant varies with time as follows.

$$k = 20\left(1 + e^{-t/10}\right)$$

The equation of motion is

$$m\ddot{x} + c\dot{x} + 20\left(1 + e^{-t/10}\right)x = f(t)$$

Use the values $m = 1$, $c = 2$, and $f = 10$, and solve the equation and plot $x(t)$ for zero initial conditions over the interval $0 \le t \le 4$.

39. Two similar mechanical systems are shown in Figure P39. In both cases the input is the displacement $y(t)$ of the base and the spring constant is nonlinear, so the differential equations are nonlinear. Their equations of motion are, for part (a),

$$m\ddot{x} = c(\dot{y} - \dot{x}) + k_1(y - x) + k_2(y - x)^3$$

and for part (b)

$$m\ddot{x} = -c\dot{x} + k_1(y - x) + k_2(y - x)^3$$

Figure P39

The only difference between these systems is that the system in Figure P39a has an equation of motion containing the derivative of the input function $y(t)$. A step function is difficult to use with a numerical solution method, especially when the input derivative $\dot{y}$ is present, due to the discontinuity at $t = 0$. So we will model the unit-step input with the function $y(t) = 1 - e^{-t/\tau}$.

The parameter τ should be chosen to be small compared to the oscillation period and the time constant, both of which we do not know. We can make an estimate by using the characteristic roots of the linear model obtained by setting $k_2 = 0$.

Use the values $m = 100$, $c = 600$, $k_1 = 8000$, and $k_2 = 24{,}000$. Choose the parameter τ to be small compared to the period and time constant of the linear model with $k_2 = 0$. Plot the solution $x(t)$ for both systems on the same graph. Use zero initial conditions.

Section 9.5

40. The equations for an armature-controlled dc motor are the following. The motor's current is i and its rotational velocity is ω.

$$L\frac{di}{dt} = -Ri - K_e\omega + v(t) \tag{9.6-1}$$

$$I\frac{d\omega}{dt} = K_T i - c\omega \tag{9.6-2}$$

where L, R, and I are the motor's inductance, resistance, and inertia; K_T and K_e are the torque constant and back emf constant; c is a viscous damping constant; and $v(t)$ is the applied voltage.

Use the values $R = 0.8\ \Omega$, $L = 0.003$ H, $K_T = 0.05$ N $\cdot$ m/A, $K_e = 0.05$ V $\cdot$ s/rad, $c = 0$, and $I = 8 \times 10^{-5}$ kg $\cdot$ m^2.

a. Suppose the applied voltage is 20 V. Plot the motor's speed and current versus time. Choose a final time large enough to show the motor's speed becoming constant.

b. Suppose the applied voltage is trapezoidal as given below.

$$v(t) = \begin{cases} 400t & 0 \le t \le 0.05 \\ 20 & 0.05 \le t \le 0.2 \\ -400(t-0.2) + 20 & 0.2 < t \le 0.25 \\ 0 & t > 0.25 \end{cases}$$

Plot the motor's speed versus time for $0 \le t \le 0.3$ s. Also plot the applied voltage versus time. How well does the motor speed follow a trapezoidal profile?

41. Compute and plot the unit-impulse response of the following model.

$$10\ddot{y} + 3\dot{y} + 7y = f(t)$$

42. Compute and plot the unit-step response of the following model.

$$10\ddot{y} + 6\dot{y} + 2y = f + 7\dot{f}$$

43. Find the reduced form of the following state model.

$$\begin{bmatrix} \dot{x}_1 \\ \dot{x}_2 \end{bmatrix} = \begin{bmatrix} -4 & -1 \\ 2 & -3 \end{bmatrix} \begin{bmatrix} x_1 \\ x_2 \end{bmatrix} + \begin{bmatrix} 2 \\ 5 \end{bmatrix} u(t)$$

44. The following state model describes the motion of a certain mass connected to a spring, with viscous friction on the surface, where $m = 1$, $c = 2$, and $k = 5$.

$$\begin{bmatrix} \dot{x}_1 \\ \dot{x}_2 \end{bmatrix} = \begin{bmatrix} 0 & 1 \\ -5 & -2 \end{bmatrix} \begin{bmatrix} x_1 \\ x_2 \end{bmatrix} + \begin{bmatrix} 0 \\ 1 \end{bmatrix} f(t)$$

a. Use the `initial` function to plot the position x_1 of the mass, if the initial position is 5 and the initial velocity is 3.

b. Use the `step` function to plot the step response of the position and velocity for zero initial conditions, where the magnitude of the step input is 10. Compare your plot with that shown in Figure 9.5–1.

45. Consider the following equation.

$$5\ddot{y} + 2\dot{y} + 10y = f(t)$$

a. Plot the free response for the initial conditions $y(0) = 10$, $\dot{y}(0) = -5$.

b. Plot the unit-step response (for zero initial conditions).

c. The *total response* to a step input is the sum of the free response and the step response. Demonstrate this fact for this equation by plotting the sum of the solutions found in parts *a* and *b* and comparing the plot with that generated by solving for the total response with $y(0) = 10$, $\dot{y}(0) = -5$.

46. The model for the RC circuit shown in Figure P46 is

$$RC\frac{dv_o}{dt} + v_o = v_i$$

For $RC = 0.2$ s, plot the voltage response $v_o(t)$ for the case where the applied voltage is a single square pulse of height 10 V and duration 0.4 s, starting at $t = 0$. The initial capacitor voltage is zero.

Figure P46

Engineering in the 21st Century. . .

Embedded Control Systems

An embedded control system is a microprocessor and sensor suite designed to be an integral part of a product. The aerospace and automotive industries have used embedded controllers for some time, but the decreasing cost of components now makes embedded controllers feasible for more consumer and biomedical applications.

For example, embedded controllers can greatly increase the performance of orthopedic devices. One model of an artificial leg now uses sensors to measure in real time the walking speed, the knee joint angle, and the loading due to the foot and ankle. These measurements are used by the controller to adjust the hydraulic resistance of a piston to produce a more stable, natural, and efficient gait. The controller algorithms are adaptive in that they can be tuned to an individual's characteristics and their settings changed to accommodate different physical activities.

Engines incorporate embedded controllers to improve efficiency. Embedded controllers in new active suspensions use actuators to improve on the performance of traditional passive systems consisting only of springs and dampers. One design phase of such systems is *hardware-in-the-loop testing*, in which the controlled object (the engine or vehicle suspension) is replaced with a real-time simulation of its behavior. This enables the embedded system hardware and software to be tested faster and less expensively than with the physical prototype, and perhaps even before the prototype is available.

Simulink is often used to create the simulation model for hardware-in-the-loop testing. The Control Systems and the Signal Processing toolboxes, and the DSP and Fixed Point block sets, are also useful for such applications. ■

Simulink

Simulink is built on top of MATLAB, so you must have MATLAB to use Simulink. It is included in the Student Edition of MATLAB and is also available separately from The MathWorks, Inc. Simulink is widely used in industry to model complex systems and processes that are difficult to model with a simple set of differential equations.

Simulink provides a graphical user interface that uses various types of elements called *blocks* to create a simulation of a dynamic system, that is, a system that can be modeled with differential or difference equations whose independent variable is time. For example, one block type is a multiplier, another performs a sum, and still another is an integrator. The Simulink graphical interface enables you to position the blocks, resize them, label them, specify block parameters, and interconnect the blocks to describe complicated systems for simulation.

This chapter starts with simulations of simple systems that require few blocks. Gradually, through a series of examples, more block types are introduced. The chosen applications require only a basic knowledge of physics and thus can be appreciated by readers from any engineering or scientific discipline. By the time you have finished this chapter, you will have seen the block types needed to simulate a large variety of common applications.

10.1 Simulation Diagrams

You develop Simulink models by constructing a diagram that shows the elements of the problem to be solved. Such diagrams are called *simulation diagrams* or *block diagrams*. Consider the equation $\dot{y} = 10f(t)$. Its solution can be represented symbolically as

BLOCK DIAGRAMS

$$y(t) = \int 10f(t)dt$$

which can be thought of as two steps, using an intermediate variable x:

$$x(t) = 10f(t) \quad \text{and} \quad y(t) = \int x(t)dt$$

This solution can be represented graphically by the simulation diagram shown in Figure 10.1–1a. The arrows represent the variables y, x, and f. The blocks represent cause-and-effect processes. Thus, the block containing the number 10 represents the process $x(t) = 10f(t)$, where $f(t)$ is the cause (the *input*) and $x(t)$ represents the effect (the *output*). This type of block is called a *multiplier* or *gain block*.

The block containing the integral sign $\int$ represents the integration process $y(t) = \int x(t)dt$, where $x(t)$ is the cause (the *input*) and $y(t)$ represents the effect (the *output*). This type of block is called an *integrator block*.

INTEGRATOR BLOCK

There is some variation in the notation and symbols used in simulation diagrams. Figure 10.1–1b shows one variation. Instead of being represented by a box, the multiplication process is now represented by a triangle like that used to represent an electrical amplifier, hence the name *gain block*.

GAIN BLOCK

In addition, the integration symbol in the integrator block has been replaced by the operator symbol $1/s$, which derives from the notation used for the Laplace transform (see Section 11.7 for a discussion of this transform). Thus the equation $\dot{y} = 10f(t)$ is represented by $sy = 10f$, and the solution is represented as

$$y = \frac{10f}{s}$$

or as the two equations

$$x = 10f \qquad \text{and} \qquad y = \frac{1}{s}x$$

(a)　　　　　　　　　　　　(b)

Figure 10.1–1 Simulation diagrams for $\dot{y} = 10f(t)$.

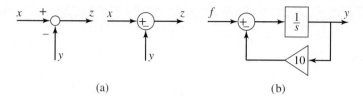

Figure 10.1–2 (a) The summer element. (b) Simulation diagram for
$\dot{y} = f(t) - 10y$.

Another element used in simulation diagrams is the *summer* that, despite its **SUMMER**
name, is used to subtract as well as to sum variables. Two versions of its symbol
are shown in Figure 10.1–2a. In each case the symbol represents the equation $z = x - y$. Note that a plus or minus sign is required for each input arrow.

The summer symbol can be used to represent the equation $\dot{y} = f(t) - 10y$,
which can be expressed as

$$y(t) = \int [f(t) - 10y]dt$$

or as

$$y = \frac{1}{s}(f - 10y)$$

You should study the simulation diagram shown in Figure 10.1–2b to confirm
that it represents this equation. This figure forms the basis for developing a
Simulink model to solve the equation.

10.2 Introduction to Simulink

Type `simulink` in the MATLAB Command window to start Simulink, or click
on the icon under the HOME tab. The Start window opens. For now, click on
Blank Model. An untitled model window will open. Then click on the Simulink
Library Browser window icon under the View menu. See Figure 10.2–1. The **LIBRARY**
Simulink blocks are located in "libraries." These libraries are displayed under **BROWSER**
the Simulink heading in Figure 10.2–1. Depending on what other MathWorks
products are installed, you might see additional items in this window, such as
the Control System Toolbox and Stateflow. These provide additional Simulink
blocks, which can be displayed by clicking on the plus sign to the left of the
item. As Simulink evolves through new versions, some libraries are renamed and
some blocks are moved to different libraries, so the library we specify here might
change in later releases. The best way to locate a block, given its name, is to type
its name in the search pane at the top of the Simulink Library Browser. When you
press Enter, Simulink will take you to the block location.

To select a block from the Library Browser, double-click on the appropriate
library, and a list of blocks within that library then appears.

Click on the block name or icon, hold the mouse button down, drag the block to
the new model window, and release the button. You can access help for that block
by right-clicking on its name or icon and selecting **Help** from the drop-down menu.

Figure 10.2–1 The Simulink Library Browser.
Source: MATLAB

Simulink model files have the extensions `.slx,` and `.mdl` for older files. Use the **File** menu in the model window to open, close, and save model files. To print the block diagram of the model, select **Print** on the **File** menu. Use the **Edit** menu to copy, cut, and paste blocks. You can also use the mouse for these operations. For example, to delete a block, click on it and press the **Delete** key.

Getting started with Simulink is best done through examples, which we now present.

<table>
<tr><td>**EXAMPLE 10.2–1**</td><td align="right">Simulink Solution of $\dot{y} = 10 \sin t$</td></tr>
</table>

Use Simulink to solve the following problem for $0 \le t \le 13$.

$$\frac{dy}{dt} = 10 \sin t \qquad y(0) = 0$$

The exact solution is $y(t) = 10(1 - \cos t)$.

■ **Solution**

To construct the simulation, do the following steps. Refer to Figure 10.2–2. Figure 10.2–3 shows the Model window after completing the following steps.

1. Start Simulink and open a new model window as described previously.

2. Select and place in the new window the Sine Wave block from the Sources library. Double-click on it to open the Block Parameters window, and make sure the Amplitude is set to 1, the Bias to 0, the Frequency to 1, the Phase to 0, and the Sample time to 0. Then click **OK.**

3. Select and place the Gain block from the Math Operations library, double-click on it, and set the Gain value to 10 in the Block Parameters window. Then click **OK.** Note that the value 10 then appears in the triangle. To make the number more visible, click on the block, and drag one of the corners to expand the block so that all the text is visible.

4. Select and place the Integrator block from the Continuous library, double-click on it to obtain the Block Parameters window, and set the Initial condition to 0 [because $y(0) = 0$]. Then click **OK.**

5. Select and place the Scope block from the Sinks library.

6. Once the blocks have been placed as shown in Figure 10.2–2, connect the input port on each block to the outport port on the preceding block. To do this, move the cursor to an input port or an output port; the cursor will change to a cross. Hold the mouse button down, and drag the cursor to a port on another block. When you

Sine Wave   Gain   Integrator   Scope

Figure 10.2–2 Simulink model for $\dot{y} = 10 \sin t$.

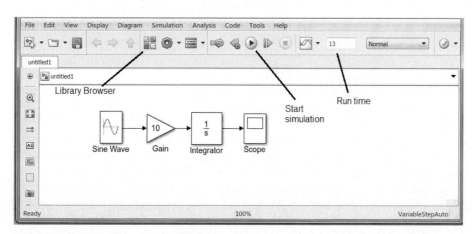

Figure 10.2–3 The Simulink Model window showing the model created in Example 10.2–1.
Source: MATLAB

Figure 10.2–4 The Scope window after running the model in Example 10.2–1.
Source: MATLAB

release the mouse button, Simulink will connect them with an arrow pointing at the input port. Your model should now look like that shown in Figure 10.2–2.

7. Enter 13 for the Stop time in the window to the right of the **Start Simulation** icon (the black triangle). See Figure 10.2–3. The default value is 10, which can be deleted and replaced with 13.

8. Run the simulation by clicking on the **Start Simulation** icon on the toolbar.

9. You will hear a bell sound when the simulation is finished. Then double-click on the Scope block to view the response. You should see an oscillating curve with an amplitude of 10 and a period of 2π (Figure 10.2–4). The independent variable in the Scope block is time t; the input to the block is the dependent variable y. This completes the simulation.

To have Simulink automatically connect two blocks, select the Source block, hold down the **Ctrl** key, and left-click on the Destination block. Simulink also provides easy ways to connect multiple blocks and lines; click on the Help for information.

Note that blocks have a Block Parameters window that opens when you double-click on the block. This window contains several items, the number and nature of which depend on the specific type of block. In general, you can use the default values of these parameters, except where we have explicitly indicated that they should be changed. You can always click on **Help** within the Block Parameters window to obtain more information.

When you click on **Apply,** any changes immediately take effect and the window remains open. If you click on **OK,** the changes take effect but the window closes.

Note that most blocks have default labels. You can edit text associated with a block by clicking on the text and making the changes. You can save the Simulink model by selecting **Save** from the **File** menu. The model file can then be reloaded at a later time. You can also print the diagram by selecting **Print** on the **File** menu.

The Scope block is useful for examining the solution, but if you want to obtain a labeled and printed plot, you can use the To Workspace block, which is described in the next example.

Exporting to the MATLAB Workspace

EXAMPLE 10.2–2

We now demonstrate how to export the results of the simulation to the MATLAB workspace, where they can be plotted or analyzed with any of the MATLAB functions.

■ Solution
Modify the Simulink model constructed in Example 10.2–1 as follows. Refer to Figure 10.2–5.

1. Delete the arrow connecting the Scope block by clicking on it and pressing the **Delete** key. Delete the Scope block in the same way.

2. Select and place the To Workspace block from the Sinks library.

3. Select and place the Mux block from the Signal Routing library, double-click on it, and set the Number of inputs to 2. Click **OK.** (The name *Mux* is an abbreviation for multiplexer, which is an electrical device for combining several signals.)

4. Connect the top input port of the Mux block to the output port of the Gain block. Then use the same technique to connect the bottom input port of the Mux block to the outport port of the Integrator block. Your model should now look somewhat like that shown in Figure 10.2–5.

5. Double-click on the To Workspace block. You can specify any variable name you want as the output; the default is `simout`. Change its name to `y`. The output variable `y` will have as many rows as there are simulation time steps, and as many columns as there are inputs to the block. The first column contains the output of the Gain block. The second column in `y` will be the output of the integrator block. In the To Workspace block, specify the Save format as Array. Use the default values for the other parameters. Click on **OK.**

Figure 10.2–5 Simulink model using a Mux and the To Workspace blocks.

6. After running the simulation, you can use the MATLAB plotting commands from the Command window to plot the columns of y (or `simout` in general). Simulink automatically puts the time variable `tout` into the MATLAB workspace when you are using the To Workspace block. To plot both outputs, in the MATLAB Command window type

```
>>plot(tout,y(:,1),tout,y(:,2)),xlabel('t'),ylabel('y')
```

EXAMPLE 10.2–3   Simulink Model for $\dot{y} = -10y + f(t)$

Construct a Simulink model to solve

$$\dot{y} = -10y + f(t) \qquad y(0) = 1$$

where $f(t) = 2 \sin 4t$, for $0 \leq t \leq 3$.

■ **Solution**

To construct the simulation, do the following steps.

1. You can use the model shown in Figure 10.2–2 by rearranging the blocks as shown in Figure 10.2–6. You will need to add a Sum block.

2. Select the Sum block from the Math Operations library and place it as shown in the simulation diagram. Its default setting adds two input signals. To change this, double-click on the block, and in the List of Signs window, type |+−. The signs are ordered counterclockwise from the top. The symbol | is a spacer indicating here that the top port is to be empty.

3. To reverse the direction of the Gain block, right-click on the block, select **Format** from the pop-up menu, and select **Flip Block.**

4. When you connect the negative input port of the Sum block to the output port of the Gain block, Simulink will attempt to draw the shortest line. To obtain the more standard appearance shown in Figure 10.2–6, first extend the line vertically down from the Sum input port. Release the mouse button, and then click on the end of the line and attach it to the Gain block. The result will be a line with a right angle. Do the same to connect the input of the Gain block to the arrow connecting the Integrator and the Scope blocks. A small dot appears to indicate that the lines have been successfully connected. This point is called a *takeoff point* because it takes the value of the variable represented by the arrow (here, the variable y) and makes that value available to another block.

Figure 10.2–6 Simulink model for $\dot{y} = -10y + f(t)$.

5. Set the Stop time to 3.

6. Run the simulation as before and observe the results in the Scope.

10.3 Linear State-Variable Models

State-variable models, unlike transfer-function models, can have more than one input and more than one output. Simulink has the State-Space block that represents the *linear* state-variable model $\dot{\mathbf{x}} = \mathbf{Ax} + \mathbf{Bu}$, $\mathbf{y} = \mathbf{Cx} + \mathbf{Du}$. (See Section 9.5 for discussion of this model form.) The vector $\mathbf{u}$ represents the inputs, and the vector $\mathbf{y}$ represents the outputs. Thus, when you are connecting inputs to the State-Space block, care must be taken to connect them in the proper order. Similar care must be taken when connecting the block's outputs to another block. The following example illustrates how this is done.

Simulink Model of a Two-Mass Suspension System

EXAMPLE 10.3–1

The following are the equations of motion of the two-mass suspension model shown in Figure 10.3–1.

$$m_1\ddot{x}_1 = k_1(x_2 - x_1) + c_1(\dot{x}_2 - \dot{x}_1)$$
$$m_2\ddot{x}_2 = -k_1(x_2 - x_1) - c_1(\dot{x}_2 - \dot{x}_1) + k_2(y - x_2)$$

Develop a Simulink model of this system to obtain the plots of x_1 and x_2. The input $y(t)$ is a unit step function, and the initial conditions are zero. Use the following values: $m_1 = 250$ kg, $m_2 = 40$ kg, $k_1 = 1.5 \times 10^4$ N/m, $k_2 = 1.5 \times 10^5$ N/m, and $c_1 = 1917$ N·s/m.

■ **Solution**

The equations of motion can be expressed in state-variable form by letting $z_1 = x_1$, $z_2 = \dot{x}_1$, $z_3 = x_2$, $z_4 = \dot{x}_2$. The equations of motion become

$$\dot{z}_1 = z_2 \quad \dot{z}_2 = \frac{1}{m_1}(-k_1 z_1 - c_1 z_2 + k_1 z_3 + c_1 z_4)$$
$$\dot{z}_3 = z_4 \quad \dot{z}_4 = \frac{1}{m_2}[k_1 z_1 + c_1 z_2 - (k_1 + k_2)z_3 - c_1 z_4 + k_2 y]$$

Figure 10.3–1 Two-mass suspension model.

These equations are expressed in vector-matrix form as

$$\dot{z} = Az + By(t)$$

where

$$A = \begin{bmatrix} 0 & 1 & 0 & 0 \\ -\dfrac{k_1}{m_1} & -\dfrac{c_1}{m_1} & \dfrac{k_1}{m_1} & \dfrac{c_1}{m_1} \\ 0 & 0 & 0 & 1 \\ \dfrac{k_1}{m_2} & \dfrac{c_1}{m_2} & -\dfrac{k_1+k_2}{m_2} & -\dfrac{c_1}{m_2} \end{bmatrix} \qquad B = \begin{bmatrix} 0 \\ 0 \\ 0 \\ \dfrac{k_2}{m_2} \end{bmatrix}$$

and

$$z = \begin{bmatrix} z_1 \\ z_2 \\ z_3 \\ z_4 \end{bmatrix} = \begin{bmatrix} x_1 \\ \dot{x}_1 \\ x_3 \\ \dot{x}_2 \end{bmatrix}$$

To simplify the notation, let $a_1 = k_1/m_1$, $a_2 = c_1/m_1$, $a_3 = k_1/m_2$, $a_4 = c_1/m_2$, $a_5 = k_2/m_2$, and $a_6 = a_3 + a_5$. The matrices A and B become

$$A = \begin{bmatrix} 0 & 1 & 0 & 0 \\ -a_1 & -a_2 & a_1 & a_2 \\ 0 & 0 & 0 & 1 \\ a_3 & a_4 & -a_6 & -a_4 \end{bmatrix} \qquad B = \begin{bmatrix} 0 \\ 0 \\ 0 \\ a_5 \end{bmatrix}$$

Next, select appropriate values for the matrices in the output equation $y = Cz + By(t)$. Because we want to plot x_1 and x_2, which are z_1 and z_3, we must use the following matrices for C and D.

$$C = \begin{bmatrix} 1 & 0 & 0 & 0 \\ 0 & 0 & 1 & 0 \end{bmatrix} \qquad D = \begin{bmatrix} 0 \\ 0 \end{bmatrix}$$

Note that the dimensions of B tell Simulink that there is one input. The dimensions of C and D tell Simulink that there are two outputs.

Open a new model window, and then do the following to create the model shown in Figure 10.3–2.

1. Select and place the Step block from the Sources library. Double-click on it to open the Block Parameters window, and set the Step Time to 0, the Initial Value to 0, and the Final Value to 1. Do not change the default value of any other parameters in this window. Click **OK**. The Step Time is the time at which the step input begins.

2. Select and place the State-Space block from the Continuous library. Open its Block Parameters window and enter the following values for the matrices **A, B, C,** and **D**. For **A** enter

    `[0, 1, 0, 0; -a1, -a2, a1, a2; 0, 0, 0, 1; a3, a4, -a6, -a4]`

    For **B** enter `[0; 0; 0; a5]`. For **C** enter `[1, 0, 0, 0; 0, 0, 1, 0]`, and for **D** enter `[0; 0]`. Then enter `[0; 0; 0; 0]` for the initial conditions. Click **OK**.

Figure 10.3–2 Simulink model containing the State-Space block and Step block.

3. Select and place the Scope block from the Sinks library.
4. Connect the input and output ports as shown in Figure 10.3–2, and save the model.
5. In the Workspace window enter the parameter values and compute the a_i constants as shown in the following session.

```
>>m1 = 250; m2 = 40; k1 = 1.5e+4;
>>k2 = 1.5e+5; c1 = 1917;
>>a1 = k1/m1; a2 = c1/m1; a3 = k1/m2;
>>a4 = c1/m2; a5 = k2/m2; a6 = a3 + a5;
```

6. Experiment with different values of Stop Time until the Scope shows that steady state has been reached. Using this method a Stop Time of 1 s was found to be satisfactory. The plots of both x_1 and x_2 will appear in the Scope. A To Workspace block can be added to obtain the plot in MATLAB. Figure 10.3–3 was created in this way.

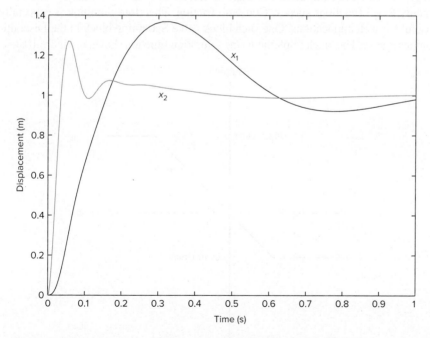

Figure 10.3–3 Unit-step response of the two-mass suspension model.

10.4 Piecewise-Linear Models

Unlike linear models, closed-form solutions are not available for most nonlinear differential equations, and we must therefore solve such equations numerically. A nonlinear ordinary differential equation can be recognized by the fact that the dependent variable or its derivatives appear raised to a power or in a transcendental function. For example, the following equations are nonlinear.

$$y\ddot{y} + 5\dot{y} + y = 0 \qquad \dot{y} + \sin y = 0 \qquad \dot{y} + \sqrt{y} = 0$$

Piecewise-linear models are actually nonlinear, although they may appear to be linear. They are composed of linear models that take effect when certain conditions are satisfied. The effect of switching back and forth between these linear models makes the overall model nonlinear. An example of such a model is a mass attached to a spring and sliding on a horizontal surface with Coulomb friction. The model is

$$m\ddot{x} + kx = \begin{cases} f(t) - \mu mg & \text{if } \dot{x} \geq 0 \\ f(t) + \mu mg & \text{if } \dot{x} < 0 \end{cases}$$

These two linear equations can be expressed as the single, nonlinear equation

$$m\ddot{x} + kx = f(t) - \mu mg \, \text{sign}(\dot{x}) \qquad \text{where} \qquad \text{sign}(\dot{x}) = \begin{cases} +1 & \text{if } \dot{x} \geq 0 \\ -1 & \text{if } \dot{x} < 0 \end{cases}$$

Solutions of models that contain piecewise-linear functions are very tedious to program. However, Simulink has built-in blocks that represent many of the commonly found functions such as Coulomb friction. Therefore Simulink is especially useful for such applications. One such block is the Saturation block in the Discontinuities library. The block implements the saturation function shown in Figure 10.4–1.

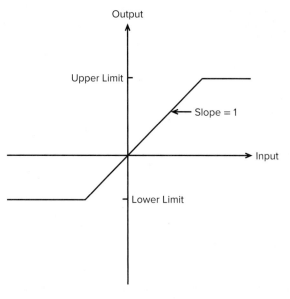

Figure 10.4–1 The saturation nonlinearity.

Simulink Model of a Rocket-Propelled Sled

EXAMPLE 10.4–1

A rocket-propelled sled on a track is represented in Figure 10.4–2 as a mass m with an applied force f that represents the rocket thrust. The rocket thrust initially is horizontal, but the engine accidentally pivots during firing and rotates with an angular acceleration of $\ddot{\theta} = \pi/50$ rad/s. Compute the sled's velocity v for $0 \le t \le 6$ if $v(0) = 0$. The rocket thrust is 4000 N and the sled mass is 450 kg.

The sled's equation of motion is

$$450\dot{v} = 4000 \cos\theta(t)$$

To obtain $\theta(t)$, note that

$$\dot{\theta} = \int_0^t \ddot{\theta}\, dt = \frac{\pi}{50}t$$

and

$$\theta = \int_0^t \dot{\theta}\, dt = \int_0^t \frac{\pi}{50}t\, dt = \frac{\pi}{100}t^2$$

Thus the equation of motion becomes

$$450\dot{v} = 4000 \cos\left(\frac{\pi}{100}t^2\right)$$

or

$$\dot{v} = \frac{80}{9}\cos\left(\frac{\pi}{100}t^2\right)$$

The solution is formally given by

$$v(t) = \frac{80}{9}\int_0^t \cos\left(\frac{\pi}{100}t^2\right)dt$$

Unfortunately, no closed-form solution is available for the integral, which is called *Fresnel's cosine integral*. The value of the integral has been tabulated numerically, but we will use Simulink to obtain the solution.

(a) Create a Simulink model to solve this problem for $0 \le t \le 10$ s.

(b) Now suppose that the engine angle is limited by a mechanical stop to 60°, which is $\pi/3$ rad. Create a Simulink model to solve the problem.

■ **Solution**

(a) There are several ways to create the input function $\theta = (\pi/100)t^2$. Here we note that $\ddot{\theta} = \pi/50$ rad/s and that

$$\dot{\theta} = \int_0^t \ddot{\theta}\, dt$$

Figure 10.4–2 A rocket-propelled sled.

Figure 10.4–3 Simulation diagram for $v = (80/9)\cos(\pi t^2/100)$.

Figure 10.4–4 Simulink model for $v = (80/9)\cos(\pi t^2/100)$.

and

$$\theta = \int_0^t \dot\theta \, dt = \frac{\pi}{100} t^2$$

Thus we can create $\theta(t)$ by integrating the constant $\ddot\theta = \pi/50$ twice. The simulation diagram is shown in Figure 10.4–3. This diagram is used to create the corresponding Simulink model shown in Figure 10.4–4.

There are two new blocks in this model. The Constant block is in the Sources library. After placing it, double-click on it and type `pi/50` in its Constant Value window.

The Trigonometric block is in the Math Operations library. After placing it, double-click on it and select `cos` in its Function window.

Set the Stop Time to 10, run the simulation, and examine the results in the Scope.

(b) Modify the model in Figure 10.4–4 as follows to obtain the model shown in Figure 10.4–5. We use the Saturation block in the Discontinuities library to limit the range of θ to $\pi/3$ rad. After placing the block as shown in Figure 10.4–5, double-click on it and type `pi/3` in its Upper Limit window. Then type `0` in its Lower Limit window.

Enter and connect the remaining elements as shown, and run the simulation. The upper Constant block and Integrator block are used to generate the solution when the engine angle is $\theta = 0$, as a check on our results. [The equation of motion for $\theta = 0$ is $\dot v = 80/9$, which gives $v(t) = 80t/9$.]

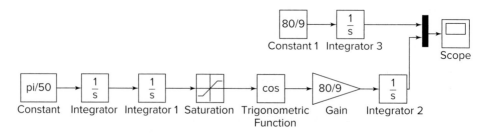

Figure 10.4–5 Simulink model for $v = (80/9)\cos(\pi t^2/100)$ with a Saturation block.

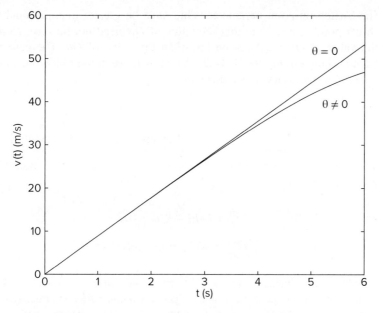

Figure 10.4–6 Speed response of the sled for $\theta = 0$ and $\theta \neq 0$.

If you prefer, you can substitute a To Workspace block for the Scope. Then you can plot the results in MATLAB. The resulting plot is shown in Figure 10.4–6.

The Relay Block

The Simulink *Relay block* is an example of something that is tedious to program in MATLAB but is easy to implement in Simulink. Figure 10.4–7a is a graph of the logic of a relay. The relay switches the output between two specified values, named *On* and *Off* in the figure. Simulink calls these values "Output when on" and "Output when off." When the relay output is *On,* it remains *On* until the input drops below the value of the Switch-off point parameter, named *SwOff* in the figure. When the relay output is *Off,* it remains *Off* until the input exceeds the value of the Switch-on point parameter, named *SwOn* in the figure.

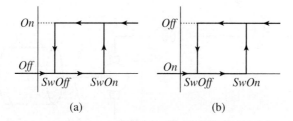

Figure 10.4–7 The relay function. (a) The case where *On > Off.* (b) The case where *On < Off.*

The Switch-on point parameter value must be greater than or equal to the Switch-off point value. Note that the value of *Off* need not be zero. Note also that the value of *Off* need not be less than the value of *On*. The case where *Off* > *On* is shown in Figure 10.4–7b. As we will see in the following example, it is sometimes necessary to use this case.

EXAMPLE 10.4–2

Model of a Relay-Controlled Motor

The model of an armature-controlled dc motor was discussed in Section 9.5. See Figure 10.4–8. The model is

$$L\frac{di}{dt} = -Ri - K_e\omega + v(t)$$

$$I\frac{d\omega}{dt} = K_T i - c\omega - T_d(t)$$

where the model now includes a torque $T_d(t)$ acting on the motor shaft due, for example, to some unwanted and unpredictable source such as Coulomb friction or external forces on a robot arm. Control system engineers call this a *disturbance*. These equations can be put into matrix form as follows, where $x_1 = i$ and $x_2 = \omega$.

$$\begin{bmatrix} \dot{x}_1 \\ \dot{x}_2 \end{bmatrix} = \begin{bmatrix} -\dfrac{R}{L} & -\dfrac{K_e}{L} \\ \dfrac{K_T}{I} & -\dfrac{c}{I} \end{bmatrix} \begin{bmatrix} x_1 \\ x_2 \end{bmatrix} + \begin{bmatrix} \dfrac{1}{L} & 0 \\ 0 & -\dfrac{1}{I} \end{bmatrix} \begin{bmatrix} v(t) \\ T_d(t) \end{bmatrix}$$

Use the values $R = 0.6\ \Omega$, $L = 0.002$ H, $K_T = 0.04$ N · m/A, $K_e = 0.04$ V · s/rad, $c = 0.01$ N · m · s/rad, and $I = 6 \times 10^{-5}$ kg · m^2.

Suppose we have a sensor that measures the motor speed, and we use the sensor's signal to activate a relay to switch the applied voltage $v(t)$ between 0 and 100 V to keep the speed between 250 and 350 rad/s. This corresponds to the relay logic shown in Figure 10.4–7b, with *SwOff* = 250, *SwOn* = 350, *Off* = 100, and *On* = 0. Investigate how well this scheme will work if the disturbance torque is a step function that increases from

Figure 10.4–8 An armature-controlled dc motor.

0 to 3 N · m, starting at $t = 0.05$ s. Assume that the system starts from rest with $\omega(0) = 0$ and $i(0) = 0$.

■ **Solution**

For the given parameter values,

$$\mathbf{A} = \begin{bmatrix} -300 & -20 \\ 666.7 & -167.7 \end{bmatrix} \qquad \mathbf{B} = \begin{bmatrix} 500 & 0 \\ 0 & -16{,}667 \end{bmatrix}$$

To examine the speed ω as output, we choose $\mathbf{C} = [0, 1]$ and $\mathbf{D} = [0, 0]$. To create this simulation, first obtain a new model window. Then do the following.

1. Select and place in the new window the Step block from the Sources library. Label it Disturbance Step as shown in Figure 10.4–9. Double-click on it to obtain the Block Parameters window, and set the Step Time to 0.05, the Initial and Final values to 0 and 3, and the Sample time to 0. Click **OK.**

2. Select and place the Relay block from the Discontinuities library. Double-click on it, and set the Switch-on and Switch-off points to 350 and 250, and set the Output when on and Output when off to 0 and 100. Click **OK.**

3. Select and place the Mux block from the Signal Routing library. The Mux block combines two or more signals into a vector signal. Double-click on it, and set the Display option to signals. Click **OK.** Then click on the Mux icon in the model window, and drag one of the corners to expand the box so that all the text is visible.

4. Select and place the State-Space block from the Continuous library. Double-click on it, and enter [-300, -20; 666.7, -166.7] for **A,** [500, 0; 0, -16667] for **B,** [0, 1] for **C,** and [0, 0] for **D.** Then enter [0; 0] for the initial conditions. Click **OK.** Note that the dimension of the matrix **B** tells Simulink that there are two inputs. The dimensions of the matrices **C** and **D** tell Simulink that there is one output.

5. Select and place the Scope block from the Sinks library.

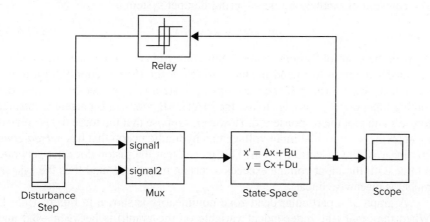

Figure 10.4–9 Simulink model of a relay-controlled motor.

6. Once the blocks have been placed, connect the input port on each block to the out-port port on the preceding block as shown in the figure. It is important to connect the top port of the Mux block [which corresponds to the first input, $v(t)$] to the output of the Relay block, and to connect the bottom port of the Mux block [(which corresponds to the second input, $T_d(t)$] to the output of the Disturbance Step block.

7. Set the Stop time to 0.1 (which is simply an estimate of how long is needed to see the complete response), run the simulation, and examine the plot of $\omega(t)$ in the Scope. If you want to examine the current $i(t)$, change the matrix **C** to [1, 0] and run the simulation again.

The results show that the relay logic control scheme keeps the speed within the desired limits of 250 and 350 before the disturbance torque starts to act. The speed oscillates because when the applied voltage is zero, the speed decreases as a result of the back-emf and the viscous damping. The speed drops below 250 when the disturbance torque starts to act, because the applied voltage is 0 at that time. As soon as the speed drops below 250, the relay controller switches the voltage to 100, but it now takes longer for the speed to increase because the motor torque must now work against the disturbance.

Note that the speed becomes constant, instead of oscillating. This is so because with $v = 100$, the system achieves a steady-state condition in which the motor torque equals the sum of the disturbance torque and the viscous damping torque. Thus the acceleration is zero.

One practical use of this simulation is to determine how long the speed is below the limit of 250. The simulation shows that this time is approximately 0.013 s. Other uses of the simulation include finding the period of the speed's oscillation (about 0.013 s) and the maximum value of the disturbance torque that can be tolerated by the relay controller (it is about 3.7 N $\cdot$ m).

10.5 Transfer-Function Models

The equation of motion of a mass-spring-damper system is

$$m\ddot{y} + c\dot{y} + ky = f(t) \qquad (10.5\text{–}1)$$

As with the Control System toolbox, Simulink can accept a system description in transfer-function form and in state-variable form. (See Section 9.5 for a discussion of these forms.) If the mass-spring system is subjected to a sinusoidal forcing function $f(t)$, it is easy to use the MATLAB commands presented thus far to solve and plot the response $y(t)$. However, suppose that the force $f(t)$ is created by applying a sinusoidal input voltage to a hydraulic piston that has a *dead-zone* nonlinearity due to static friction. This means that the piston does not generate a force until the input voltage exceeds a certain magnitude, and thus the system model is piecewise linear.

DEAD ZONE

A graph of a particular dead-zone nonlinearity is shown in Figure 10.5–1. When the input (the independent variable on the graph) is between −0.5 and 0.5, the output is zero. When the input is greater than or equal to the upper limit

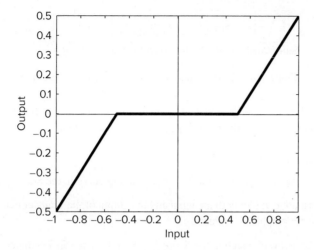

Figure 10.5–1 A dead-zone nonlinearity.

of 0.5, the output is the input minus the upper limit. When the input is less than or equal to the lower limit of −0.5, the output is the input minus the lower limit. In this example, the dead zone is symmetric about 0, but it need not be in general.

Simulations with dead-zone nonlinearities are somewhat tedious to program in MATLAB, but are easily done in Simulink. The following example illustrates how it is done.

Response with a Dead Zone

EXAMPLE 10.5–1

Create and run a Simulink simulation of a mass-spring-damper model (Equation 10.5–1) using the parameter values $m = 1$, $c = 2$, and $k = 4$. The forcing function is the function $f(t) = \sin 1.4t$. The system has the dead-zone nonlinearity shown in Figure 10.5–1.

■ Solution

To construct the simulation, do the following steps.

1. Start Simulink and open a new Model window as described previously.
2. Select and place in the new window the Sine Wave block from the Sources library. Double-click on it, and set the Amplitude to 1, the Frequency to 1.4, the Phase to 0, and the Sample time to 0. Click **OK.**
3. Select and place the Dead Zone block from the Discontinuities library, double-click on it, and set the Start of dead zone to −0.5 and the End of dead zone to 0.5. Click **OK.**
4. Select and place the Transfer Fcn block from the Continuous library, double-click on it, and set the Numerator to [1] and the Denominator to [1, 2, 4]. Click **OK.**
5. Select and place the Scope block from the Sinks library.

Figure 10.5–2 The Simulink model of dead-zone response.

6. Once the blocks have been placed, connect the input port on each block to the outport port on the preceding block. Your model should now look like Figure 10.5–2.

7. Set the Stop time to 10.

8. Run the simulation. You should see an oscillating curve in the Scope display.

It is informative to plot both the input and the output of the Transfer Fcn block versus time on the same graph. To do this:

1. Delete the arrow connecting the Scope block to the Transfer Fcn block. Do this by clicking on the arrow line and then pressing the **Delete** key.

2. Select and place the Mux block from the Signal Routing library, double-click on it, and set the Number of inputs to 2. Click **OK.**

3. Connect the top input port of the Mux block to the output port of the Transfer Fcn block. Then use the same technique to connect the bottom input port of the Mux block to the arrow from the output port of the Dead Zone block. Just remember to start with the input port. Simulink will sense the arrow automatically and make the connection. Your model should now look like Figure 10.5–3.

4. Set the Stop time to 10, run the simulation as before, and bring up the Scope display. You should see what is shown in Figure 10.5–4. This plot shows the effect of the dead zone on the sine wave.

You can bring the simulation results into the MATLAB workspace by using the To Workspace block. For example, suppose we want to examine the effects of the dead zone by comparing the response of the system with and without a dead zone. We can do this with the model shown in Figure 10.5–5. To create this model:

1. Copy the Transfer Fcn block by right-clicking on it, holding down the mouse button, and dragging the block copy to a new location. Then release the button. Copy the Mux block in the same way.

2. Double-click on the first Mux block and change the number of its inputs to 3.

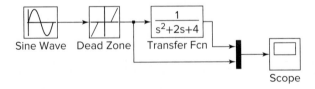

Figure 10.5–3 Modification of the dead-zone model to include a Mux block.

Figure 10.5–4 The response of the dead-zone model.
Source: MATLAB

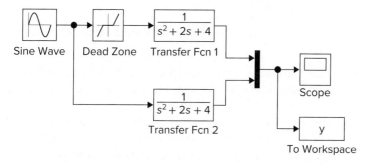

Figure 10.5–5 Modification of the dead-zone model to export variables to the MATLAB workspace.

3. Finish constructing the model in the usual way and run the simulation.
4. To plot the response of the two systems and the output of the Dead Zone block versus time, type

```
>>plot(tout,y(:,1),tout,y(:,2))
```

10.6 Nonlinear State-Variable Models

Nonlinear models cannot be put into transfer-function form or the state-variable form $\dot{x} = Ax + Bu.$ However, they can be simulated in Simulink. The following example shows how this can be done.

EXAMPLE 10.6–1

Model of a Nonlinear Pendulum

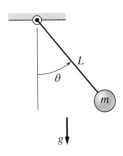

Figure 10.6–1
A pendulum.

The pendulum shown in Figure 10.6–1 has the following nonlinear equation of motion, if there is viscous friction in the pivot and if there is an applied moment $M(t)$ about the pivot

$$I\ddot{\theta} + c\dot{\theta} + mgL \sin \theta = M(t)$$

where I is the mass moment of inertia about the pivot. Create a Simulink model for this system for the case where $I = 4$, $mgL = 10$, $c = 0.8$, and $M(t)$ is a square wave with an amplitude of 3 and a frequency of 0.5 Hz. Assume that the initial conditions are $\theta(0) = \pi/4$ rad and $\dot{\theta}(0) = 0$.

■ Solution

To simulate this model in Simulink, define a set of variables that lets you rewrite the equation as two first-order equations. Thus let $\omega = \dot{\theta}$. Then the model can be written as

$$\dot{\theta} = \omega$$
$$\dot{\omega} = \frac{1}{I}[-c\omega - mgL \sin \theta + M(t)] = 0.25[-0.8\omega - 10 \sin \theta + M(t)]$$

Integrate both sides of each equation over time to obtain

$$\theta = \int \omega \, dt$$
$$\omega = 0.25 \int [-0.8\omega - 10 \sin \theta + M(t)] \, dt$$

We will introduce four new blocks to create this simulation. Obtain a new Model window and do the following.

1. Select and place in the new window the Integrator block from the Continuous library, and change its label to Integrator 1 as shown in Figure 10.6–2. You can edit text associated with a block by clicking on the text and making the changes. Double-click on the block to obtain the Block Parameters window, and set the Initial condition to 0 [this is the initial condition $\dot{\theta}(0) = 0$]. Click **OK.**

2. Copy the Integrator block to the location shown and change its label to Integrator 2. Set its initial condition to $\pi/4$ by typing `pi/4` in the Block Parameters window. This is the initial condition $\theta(0) = \pi/4$.

3. Select and place a Gain block from the Math Operations library, double-click on it, and set the Gain value to 0.25. Click **OK.** Change its label to `1/I`. Then click on the block, and drag one of the corners to expand the box so that all the text is visible.

4. Copy the Gain box, change its label to `c`, and place it as shown in Figure 10.6–2. Double-click on it, and set the Gain value to 0.8. Click **OK.** To flip the box left to right, right-click on it, select **Format,** and select **Flip.**

5. Select and place the Scope block from the Sinks library.

6. For the term $10 \sin \theta$, we cannot use the Trigonometric Function block in the Math Operations library because we need to multiply the $\sin \theta$ by 10. So we use the Fcn block under the User-Defined Functions library. Select and place this block as shown. Double-click on it, and type `10*sin(u)` in the expression window. This

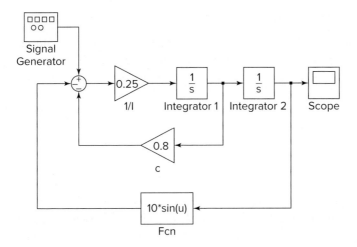

Figure 10.6–2 Simulink model of nonlinear pendulum dynamics.

block uses the variable u to represent the input to the block. Click **OK.** Then flip the block.

7. Select and place the Sum block from the Math Operations library. Double-click on it, and select round for the Icon shape. In the List of Signs window, type +-- . Click **OK.**

8. Select and place the Signal Generator block from the Sources library. Double-click on it, select square wave for the Wave form, 3 for the Amplitude, 0.5 for the Frequency, and Hertz for the Units. Click **OK.**

9. Once the blocks have been placed, connect arrows as shown in the figure.

10. Set the Stop time to 10, run the simulation, and examine the plot of $\theta(t)$ in the Scope. This completes the simulation.

10.7 Subsystems

One potential disadvantage of a graphical interface such as Simulink is that, to simulate a complex system, the diagram can become rather large and therefore somewhat cumbersome. Simulink, however, provides for the creation of *subsystem blocks*, which play a role analogous to that of subprograms in a programming language. A subsystem block is actually a Simulink program represented by a single block. A subsystem block, once created, can be used in other Simulink programs. We also introduce some other blocks in this section.

To illustrate subsystem blocks, we will use a simple hydraulic system whose model is based on the conservation of mass principle familiar to engineers. Because the governing equations are similar to other engineering applications, such as electric circuits and devices, the lessons learned from this example will enable you to use Simulink for other applications.

A Hydraulic System

The working fluid in a *hydraulic* system is an incompressible fluid such as water or a silicon-based oil. (*Pneumatic* systems operate with compressible fluids, such as air.) Consider a hydraulic system composed of a tank of liquid of mass density ρ (Figure 10.7–1). The tank shown in cross section in the figure is cylindrical with a bottom area A. A flow source dumps liquid into the tank at the mass flow rate $q_{mi}(t)$. The total mass in the tank is $m = \rho A h$, and from conservation of mass we have

$$\frac{dm}{dt} = \rho A \frac{dh}{dt} = q_{mi} - q_{mo} \qquad (10.7\text{–}1)$$

since ρ and A are constants.

If the outlet is a pipe that discharges to atmospheric pressure p_a and provides a resistance to flow that is proportional to the pressure difference across its ends, then the outlet flow rate is

$$q_{mo} = \frac{1}{R}[(\rho g h + p_a) - p_a] = \frac{\rho g h}{R}$$

where R is called the *fluid resistance*. Substituting this expression into Equation (10.7–1) gives the model

$$\rho A \frac{dh}{dt} = q_{mi}(t) - \frac{\rho g}{R} h \qquad (10.7\text{–}2)$$

The transfer function is

$$\frac{H(s)}{Q_{mi}(s)} = \frac{1}{\rho A s + \rho g / R}$$

On the other hand, the outlet may be a valve or other restriction that provides nonlinear resistance to the flow. In such cases, a common model is the signed-square-root (SSR) relation

$$q_{mo} = \frac{1}{R}\text{SSR}(\Delta p)$$

Figure 10.7–1 A hydraulic system with a flow source.

Figure 10.7–2 A hydraulic system with a flow source and two pumps.

where q_{mo} is the outlet mass flow rate, R is the resistance, Δp is the pressure difference across the resistance, and

$$\text{SSR}(\Delta p) = \begin{cases} \sqrt{\Delta p} & \text{if } \Delta p \geq 0 \\ -\sqrt{|\Delta p|} & \text{if } \Delta p < 0 \end{cases}$$

Note that we may express the $\text{SSR}(u)$ function in MATLAB as follows: `sgn(u)*sqrt(abs(u))`.

Consider the slightly different system shown in Figure 10.7–2, which has a flow source q and two pumps that supply liquid at the pressures p_l and p_r. Suppose the resistances are nonlinear and obey the signed-square-root relation. Then the model of the system is

$$\rho A \frac{dh}{dt} = q + \frac{1}{R_l} \text{SSR}(p_l - p) - \frac{1}{R_r} \text{SSR}(p - p_r)$$

where A is the bottom area and $p = \rho g h$. The pressures p_l and p_r are the *gauge* pressures at the left- and right-hand sides. Gauge pressure is the difference between the absolute pressure and atmospheric pressure. Note that the atmospheric pressure p_a cancels out of the model because of the use of gauge pressure.

We will use this application to introduce the following Simulink elements:

■ Subsystem blocks
■ Input and output ports

You can create a subsystem block in one of two ways: by dragging the Subsystem block from the library to the Model window or by first creating a Simulink model and then "encapsulating" it within a bounding box. We will illustrate the latter method.

We will create a subsystem block for the liquid-level system shown in Figure 10.7–2. First construct the Simulink model shown in Figure 10.7–3. The oval blocks are Input and Output Ports (In 1 and Out 1), which are available in the Ports and Subsystems library. Note that you can use MATLAB variables and expressions when entering the gains in each of the four Gain blocks.

Figure 10.7–3 Simulink model of the system shown in Figure 10.7–2.

Before running the program, we will assign values to these variables in the MATLAB Command window. Enter the gains for the four Gain blocks using the expressions shown in the block. You may also use a variable as the Initial condition of the Integrator block. Name this variable h0.

The SSR blocks are examples of the Fcn block. Double-click on the block and enter the MATLAB expression sgn(u)*sqrt(abs(u)). Note that this block requires you to use the variable u. The output of the block must be a scalar, as is the case here, and you cannot perform matrix operations in this block, but these are not needed here. (An alternative to this block is the MATLAB Function block to be discussed in Section 10.9.) Save the model and give it a name, such as Tank.

Now create a "bounding box" surrounding the diagram. Do this by placing the mouse cursor in the upper left, holding the mouse button down, and dragging the expanding box to the lower right to enclose the entire diagram. Then choose **Create Subsystem** from the **Edit** menu. Simulink will then replace the diagram with a single block having as many input and output ports as required and will assign default names. You can resize the block to make the labels readable. You may view or edit the subsystem by double-clicking on it. The result is shown in Figure 10.7–4.

Figure 10.7–4 The Subsystem block.

Figure 10.7–5 A hydraulic system with two tanks.

Connecting Subsystem Blocks

We now create a simulation of the system shown in Figure 10.7–5, where the mass inflow rate q is a step function. To do this, create the Simulink model shown in Figure 10.7–6. The square blocks are Constant blocks from the Sources library. These give constant inputs (which are not the same as step function inputs).

   The larger rectangular blocks are two subsystem blocks of the type just created. To insert them into the model, open the Tank subsystem model, select **Copy** from the **Edit** menu, then paste it twice into the new model window. Connect the input and output ports and edit the labels as shown. Then double-click on the Tank 1 subsystem block, set the left-side gain `1/R_l` equal to 0, the right-side gain `1/R_r` equal to `1/R_1`, and the gain `1/rho*A`

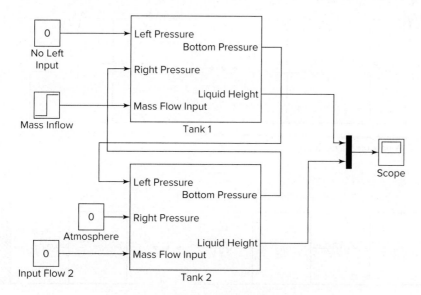

Figure 10.7–6 Simulink model of the system shown in Figure 10.7–5.

equal to 1/rho*A_1. Set the Initial condition of the integrator to h10. Note that setting the gain 1/R_l equal to 0 is equivalent to $R\_l = \infty$, which indicates no inlet on the left-hand side.

Then double-click on the Tank 2 subsystem block, set the left-side gain 1/R_l equal to 1/R_1, the right-side gain 1/R_r equal to 1/R_2, and the gain 1/rho*A equal to 1/rho*A_2. Set the Initial condition of the integrator to h20. For the Step block, set the Step time to 0, the Initial value to 0, the Final value to the variable q_1, and the Sample time to 0. Save the model using a name other than Tank.

Before you run the model, in the Command window assign numerical values to the variables. As an example, you may type the following values for water, in U.S. Customary units, in the Command window.

```
>>A_1 = 2;A_2 = 5;rho = 1.94;g = 32.2;
>>R_1 = 20;R_2 = 50;q_1 = 0.3;h10 = 1;h20 = 10;
```

After selecting a simulation Stop time, you may run the simulation. The Scope will display the plots of the heights h_1 and h_2 versus time.

Figures 10.7–7, 10.7–8, and 10.7–9 illustrate some electrical and mechanical systems that are likely candidates for application of subsystem blocks. In Figure 10.7–7, the basic element for the subsystem block is an *RC* circuit. In Figure 10.7–8, the basic element for the subsystem block is a mass connected to two elastic elements.

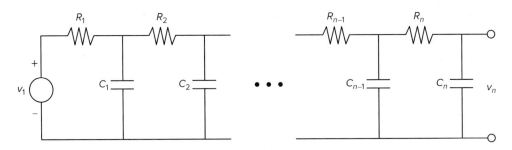

Figure 10.7–7 A network of *RC* loops.

Figure 10.7–8 A vibrating system.

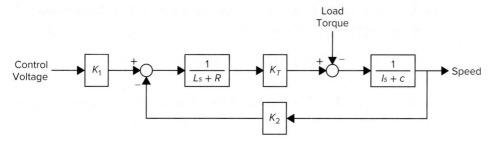

Figure 10.7–9 An armature-controlled dc motor.

Figure 10.7–9 is the block diagram of an armature-controlled dc motor, which may be converted to a subsystem block. The inputs for the block would be the voltage from a controller and a load torque, and the output would be the motor speed. Such a block would be useful in simulating systems containing several motors, such as a robot arm.

10.8 Dead Time in Models

TRANSPORT
DELAY

Dead time, also called *transport delay*, is a time delay between an action and its effect. It occurs, for example, when a fluid flows through a conduit. If the fluid velocity v is constant and the conduit length is L, it takes a time $T = L/v$ for the fluid to move from one end to the other. The time T is the dead time.

Let $\theta_1(t)$ denote the incoming fluid temperature and $\theta_2(t)$ the temperature of the fluid leaving the conduit. If no heat energy is lost, then $\theta_2(t) = \theta_1(t - T)$. From the shifting property of the Laplace transform,

$$\Theta_2(s) = e^{-Ts}\Theta_1(s)$$

So the transfer function for a dead-time process is e^{-Ts}.

Dead time may be described as a "pure" time delay, in which no response at all occurs for a time T, as opposed to the time lag associated with the time constant of a response, for which $\theta_2(t) = (1 - e^{-t/\tau})\theta_1(t)$.

Some systems have an unavoidable time delay in the interaction between components. The delay often results from the physical separation of the components and typically occurs as a delay between a change in the actuator signal and its effect on the system being controlled, or as a delay in the measurement of the output.

Another, perhaps unexpected, source of dead time is the computation time required for a digital control computer to calculate the control algorithm. This can be a significant dead time in systems using inexpensive and slower microprocessors.

The presence of dead time means the system does not have a characteristic equation of finite order. In fact, there are an infinite number of characteristic roots for a system with dead time. This can be seen by noting that the term e^{-Ts} can be expanded in an infinite series as

$$e^{-Ts} = \frac{1}{e^{Ts}} = \frac{1}{1 + Ts + T^2 s^2/2 + \cdots}$$

The fact that there are an infinite number of characteristic roots means that the analysis of dead-time processes is difficult, and often simulation is the only practical way to study such processes.

Systems having dead-time elements are easily simulated in Simulink. The block implementing the dead-time transfer function e^{-Ts} is called the *Transport Delay* block, which is in the Continuous library.

Consider the model of the height h of liquid in a tank, such as that shown in Figure 10.7–1, whose input is a mass flow rate q_i. Suppose that it takes a time T for the change in input flow to reach the tank following a change in the valve opening. Thus, T is a dead time. For specific parameter values, the transfer function has the form

$$\frac{H(s)}{Q_i(s)} = e^{-Ts} \frac{2}{5s + 1}$$

Figure 10.8–1 shows a Simulink model for this system. After placing the Transport Delay block, set the delay to 1.25. Set the Step time to 0 and the Final Value to 1 in the Step Function block. We will now discuss the other blocks in the model.

Initial Conditions and Transfer Functions

Some of the usefulness of transfer functions is due to the fact that a complicated transfer function can be decomposed into a series of simpler transfer functions by multiplication or division operations. However, these operations assume the initial conditions associated with each transfer function are zero. For this reason, Simulink assumes the initial conditions associated with the Transfer Fcn block to be zero. To specify initial conditions for a given transfer function, convert the transfer function to its equivalent state-space realization using the MATLAB function tf2ss. Then use the State-Space block instead of the Transfer Fcn block.

The Saturation and Rate Limiter Blocks

Suppose that the minimum and maximum flow rates available from the input flow valve are 0 and 2. These limits can be simulated with the *Saturation block*, discussed in Section 10.4. After placing the block as shown in Figure 10.8–1, double-click on it and type 2 in its Upper limit window and 0 in the Lower limit window.

In addition to being limited by saturation, some actuators have limits on how fast they can react. This limitation might be due to deliberate restrictions placed on the unit by its manufacturer to avoid damage to the unit. An example is a flow

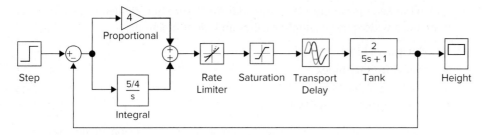

Figure 10.8–1 Simulink model of a hydraulic system with dead time.

control valve whose rate of opening and closing is controlled by a rate limiter. Simulink has such a block, and it can be used in series with the Saturation block to model the valve behavior. Place the *Rate Limiter block* as shown in Figure 10.8–1. Set the Rising slew rate to 1 and the Falling slew rate to −1.

A Control System

The Simulink model shown in Figure 10.8–1 is for a specific type of control system called a *PI controller*, whose response $f(t)$ to the error signal $e(t)$ is the sum of a term proportional to the error signal and a term proportional to the integral of the error signal. That is,

PI CONTROLLER

$$f(t) = K_p e(t) + K_I \int_0^t e(t)dt$$

where K_P and K_I are called the proportional and integral gains. Here the error signal $e(t)$ is the difference between the unit-step command representing the desired height and the actual height. In transform notation this expression becomes

$$F(s) = K_p E(s) + \frac{K_I}{s} E(s) = \left(K_p + \frac{K_I}{s} \right) E(s)$$

In Figure 10.8–1, we used the values $K_P = 4$ and $K_I = 5/4$. These values are computed using the methods of control theory. (For a discussion of control systems, see, for example, [Palm, 2014].) The simulation is now ready to be run. Set the Stop time to 50 and observe the behavior of the liquid height $h(t)$ in the Scope. Does it reach the desired height of 1?

10.9 Simulation of a Nonlinear Vehicle Suspension Model

Linear or linearized models are useful for predicting the behavior of dynamic systems because powerful analytical techniques are available for such models, especially when the inputs are relatively simple functions such as the impulse, step, ramp, and sine. Often in the design of an engineering system, however, we must eventually deal with nonlinearities in the system and with more complicated inputs such as trapezoidal functions, and this must often be done with simulation.

In this section we introduce four additional Simulink elements that enable us to model a wide range of nonlinearities and input functions:

- Derivative block
- Signal Builder block
- Look-Up Table block
- MATLAB Function block

As our example, we will use the single-mass suspension model shown in Figure 10.9–1, where the spring and damper forces f_s and f_d have the nonlinear models shown in Figures 10.9–2 and 10.9–3. The damper model is unsymmetric and represents a damper whose force during rebound is higher than during jounce (in order to minimize the force transmitted to the passenger compartment when the vehicle strikes a bump). The bump is represented by the trapezoidal function $y(t)$ shown in Figure 10.9–4. This function corresponds approximately to a vehicle traveling at 30 mi/h over a road surface elevation 0.2 m high and 48 m long.

The system model from Newton's law is

$$m\ddot{x} = f_s(y - x) + f_d(\dot{y} - \dot{x})$$

where $m = 400$ kg, $f_s(y - x)$ is the nonlinear spring function shown in Figure 10.9–2, and $f_d(\dot{y} - \dot{x})$ is the nonlinear damper function shown in Figure 10.9–3. The corresponding simulation diagram is shown in Figure 10.9–5.

The Derivative and Signal Builder Blocks

The simulation diagram shows that we need to compute $\dot{y}$. Because Simulink uses numerical and not analytical methods, it computes derivatives only approximately, using the *Derivative block*. We must keep this in mind when using rapidly changing or discontinuous inputs. The Derivative block has no settings, so merely place it in the Simulink diagram as shown in Figure 10.9–6.

Next, place the *Signal Builder block*, then double-click on it. A plot window appears that enables you to place points to define the input function. Follow the directions in the window to create the function shown in Figure 10.9–4.

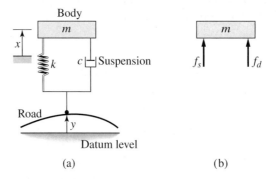

(a)   (b)

Figure 10.9–1 Single-mass model of a vehicle suspension.

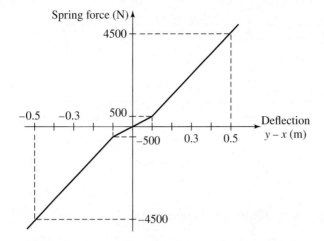

Figure 10.9–2 Nonlinear spring function.

Figure 10.9–3 Nonlinear damping function.

Figure 10.9–4 Road surface profile.

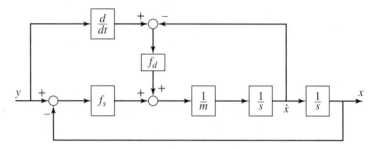

Figure 10.9–5 Simulation diagram of a vehicle suspension model.

Figure 10.9–6 Simulink model of a vehicle suspension system.

The Look-Up Table Block

The spring function f_s is created with the *Look-Up Table block*. After placing it as shown, double-click on it and enter [-0.5, -0.1, 0, 0.1, 0.5] for the Vector of input values and [-4500, -500, 0, 500, 4500] for the Vector of output values. Use the default settings for the remaining parameters.

   Place the two integrators as shown, and make sure the initial values are set to 0. Then place the Gain block and set its gain to 1/400. The To Workspace block will enable you to plot $x(t)$ and $y(t)-x(t)$ versus t in the MATLAB Command window.

The MATLAB Function Block

FCN BLOCK

In Section 10.7 we used the *Fcn block* to implement the signed-square-root function. We cannot use this block for the damper function shown in Figure 10.9–3 because we must write a user-defined function to describe it. This function is as follows.

```
function f = damper(v)
if v <= 0
    f = -800*(abs(v)).^(0.6);
else
    f = 200*v.^(0.6);
end
```

Create and save this function file. After placing the MATLAB Function block, double-click on it and the MATLAB Function editor opens. Type in the code for `damper`. When you close this editor the function is saved.

The Simulink model when completed should look like Figure 10.9–6. You can plot the response $x(t)$ in the Command window as follows:

```
>>x = simout(:,1);
>>t = simout(:,3);
>>plot(t,x),grid,xlabel('t (s)'),ylabel('x (m)')
```

The result is shown in Figure 10.9–7. The maximum overshoot is seen to be $0.26 - 0.2 = 0.06$ m, but the maximum *undershoot* is seen to be much greater, -0.168 m.

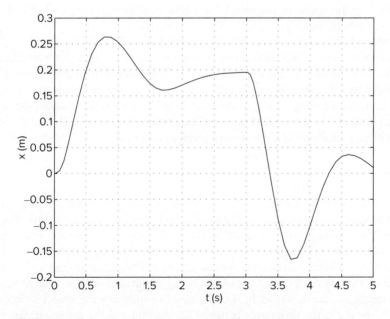

Figure 10.9–7 Output of the Simulink model shown in Figure 10.9–6.

10.10 Control Systems and Hardware-in-the-Loop Testing

As discussed on the facing page of this chapter, industry is using embedded controllers, and one design phase for such systems often involves hardware-in-the-loop testing, in which the physical controller, and sometimes the controlled object (say an engine), is replaced with a real-time simulation of its behavior. This enables the embedded system hardware and software to be tested faster and less expensively than with the physical prototype, and perhaps even before the prototype is available. Simulink is often used to create the simulation model for such testing.

The MathWorks provides Simulink support packages for hardware such as LEGO© MINDSTORMS©, Arduino©, and Raspberry Pi© that are popular with hobbyists and researchers. These packages let you develop and simulate algorithms that run standalone on the supported hardware. They include a library of Simulink blocks for configuring and accessing the hardware's sensors, actuators, and communication interfaces. You can also tune parameters live from your Simulink model while your algorithm runs on the hardware. The MathWorks supports an active user community online where you can see applications and download files.

Some of these applications involve only data collection but many are examples of control systems. Some control systems have the objective of regulating some variable such as temperature, but many of the projects are examples of controlling the speed or position of a mechanical device like a robot arm or a wheeled robotic vehicle.

Many in the user community show a need to understand the basics of feedback control theory, and this section is designed to help with that understanding. A feedback control system uses real-time measurements from a sensor to adjust the input to a device generically called an *actuator*, such as a heater or a motor. An algorithm running on the control computer decides how to adjust the actuator input to obtain the desired value of the controlled variable. One such simple algorithm, called on-off control, was given in Example 10.4–2 (Figure 10.4–9).

PID Control

A common control algorithm is the PID algorithm. The structure of a typical control system is illustrated in Figure 10.10–1. This is not a Simulink diagram but is a so-called *block diagram* that shows the physical structure. For a *speed* control system the *command input r* represents the requested speed and the *controlled variable c* represents the actual speed. The *actuator* is a motor and the *plant* is the generic term for the object being controlled (e.g., a vehicle wheel). The feedback sensor would be a tachometer to measure the wheel speed. The *error signal e* is the difference between the desired and the measured values of the speed; namely, $e = r - b$. The PID controller implements an algorithm that operates on the error signal e. The term "error signal" is an unfortunate choice

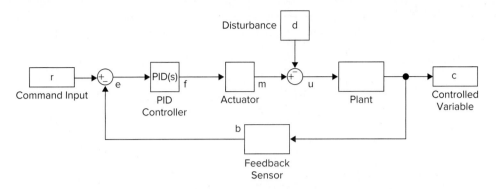

Figure 10.10–1 Structure of a feedback control system.

because it implies a mistake, but nevertheless the term remains in use; it just represents the difference between the desired and the actual value of the controlled variable. If the sensor is "perfect" then $b = c$ and $e = r - c$.

The mathematical expression for the PID algorithm in the *parallel form*, using the Simulink notation, is

$$f(t) = Pe(t) + I \int_0^t e(x)dx + D\frac{de}{dt} \qquad (10.10\text{–}1)$$

$$e(t) = r(t) - b(t) \qquad (10.10\text{–}2)$$

The transfer function form is

$$\frac{F(s)}{E(s)} = P + \frac{I}{s} + Ds \qquad (10.10\text{–}3)$$

Thus we see that PID stands for Proportional-Integral-Derivative, and the constants, P, I, and D are called the proportional, integral, and derivative gains. The parallel form is the default form in the Simulink *PID Controller block*, which is in the Continuous library. In the Simulink *ideal form*, the gain P is factored out and the algorithm is written as

PID CONTROLLER BLOCK

$$f(t) = P\left(e(t) + I \int_0^t e(x)dx + D\frac{de}{dt}\right) \qquad (10.10\text{–}4)$$

The PID controller block lets you choose which form to use. Some designers prefer to use the ideal form and set $P = 1$ initially, and then adjust P once the desired shape of the response curve is obtained by adjusting I and D.

The proportional term is the simplest to understand and is almost always used. The larger the error, the larger will be the actuator signal. For example, if the wheel speed is too slow, we want to increase the motor torque. The integral term "never gives up"; it keeps changing the actuator output as long as the error is nonzero. But this effort sometimes causes the controlled variable to overshoot the desired value and to oscillate. If so, we include the derivative term.

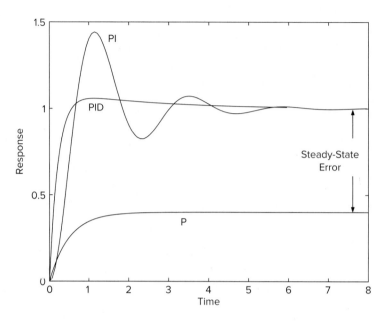

Figure 10.10–2 Typical response of P, PI, and PID control systems when subjected to a unit-step command input.

The proportional and integral terms are often used to counteract the effects of a disturbance. For example, if the vehicle encounters an incline, the wheel torque must be increased to counteract the effects of gravity. The effects of each term are shown in Figure 10.10–2, where the command input is assumed to be a unit-step function. With P control only, there is often a steady-state error. If so, then PI control is used, and this often eliminates any steady-state error. If overshoot or oscillation occur, adding the D term often reduces or eliminates overshoot and oscillations.

Choosing Gain Values

Choosing effective values for the gains is not always easy for several reasons. Mathematical methods based on transfer functions and differential equations are available but these require numerical values of motor-amplifier parameters and of masses/inertias (see Chapter 10 in [Palm, 2014]). For small mechanical devices, friction forces often dominate over inertial forces, and friction is very difficult to calculate. If the device is available for testing (and small devices usually are), we can test various algorithms and gain values, keeping in mind the contribution of each of the three PID terms (try P control first, etc.). This is what hardware-in-the-loop testing is all about.

In the following examples we will assume that the parameter values are known well enough to compute approximate values for the gains. The closed-loop transfer function can be found with methods covered in texts dealing with system dynamics and control systems. The denominator of the transfer function is the *characteristic polynomial*. Its roots determine the stability, response time, and oscillation frequency (if any) of the closed-loop response. The system is stable if all the roots are negative or have negative real parts. If the system is stable, its *time constants* are the negative reciprocals of the real roots and the negative reciprocals of the real parts of any complex roots. The *response time* can be estimated to be four times the *dominant* time constant (the largest time constant). For example, the polynomial $s^2 + 60s + 500$ has the roots $s = -10, -50$. The time constants are $0.1, 0.02$, and its response time is $4(0.1) = 0.4$. As another example, the polynomial $s^2 + 10s + 41$ has the roots $s = -5 \pm 4j$. The time constant is 0.2 and the response time is $4(0.2) = 0.8$. The response will oscillate with a radian frequency of 4.

If we know the desired response time, we can pick the gains to achieve the desired response. For example, suppose we want to achieve a response time of 0.4 for a system that has the following characteristic polynomial: $s^2 + Ps + I$. So the dominant time constant must be 0.1, and at least one root must have a real part equal to -10. The second root must have a real part equal to or more negative than -10. So we somewhat arbitrarily choose the second root to be $s = -50$. This means that the factored form of the polynomial must be $(s + 10)(s + 50)$, which expands to $s^2 + 60s + 500$. Comparing coefficients shows that $P = 60$ and $I = 500$.

Something often overlooked when choosing gains is the fact that the actuator has limitations; for example, an amplifier can produce only so much voltage or current and a motor can produce only so much torque. It is easy to get carried away when selecting the gains by looking only at the simulated response of the controlled variable. In simulation mode, you should also put a Scope on the actuator variable m, or put a Saturation block after the actuator block. This of course requires you to have some idea of the maximum actuator values.

Although Simulink has the PID controller block, it not always be possible to code it for use on certain hardware. In such cases it may be necessary to program the PID algorithm in hardware-specific code. The following discrete-time version can be used for such cases. It was derived using the rectangular integration formula for the integral term and the simplest difference formula for the derivative.

$$ f(t_k) = Pe(t_k) + IT \sum_{i=0}^{k} e(t_i) + \frac{D}{T} [e(t_k) - e(t_{k-1})] \qquad (10.10\text{--}5) $$

where $t_k = kT$ and T is the sampling period.

Speed Control

We now use speed and position control as our examples. Temperature control applications are very similar in form to our speed control examples, except the actuator is a heater instead of a motor. Permanent magnet electric motors are commonly used for speed control, and the speed sensor is either a tachometer (which is built like a motor) that outputs an analog voltage, or an encoder, which consists of slotted disks and outputs a digital signal.

Consider the simplest example. Suppose we apply a force f to a mass m, pushing it along a straight line starting from rest. Suppose the mass is acted on by a disturbance force d, which acts against f. Then the equation of motion is

$$m\frac{dv}{dt} = f - d \tag{10.10–6}$$

where v is the velocity. In transfer function form this is

$$V(s) = \frac{1}{ms}[F(s) - D(s)] \tag{10.10–7}$$

We note that a rotational system, such as a wheel driven by a motor, has the same form where v represents angular velocity, m represents mass moment of inertia, f represents motor torque, and d represents a disturbance torque. Thus the following analysis applies exactly to such a system.

Using PID control and assuming a perfect speed sensor, we obtain the Simulink diagram shown in Figure 10.10–3. Using advanced methods found for example, in Chapter 10 of [Palm, 1014] and other references on system dynamics and control systems, we can find the characteristic equation of the total system to be

$$(m + D)s^2 + Ps + I = 0 \tag{10.10–8}$$

This shows that if $D = 0$ we can place the two roots anywhere by properly choosing P and I. Thus D is unnecessary. Suppose that $m = 1$, the desired speed is 1, and that the disturbance force is $d - 10$ and begins acting at $t = 0.4$. To obtain a response time of 0.2, we choose $s = -20, -20$, which give the polynomial $s^2 + 40s + 400$. This gives $D = 0$, $P = 40$, and $I = 400$ with the parallel form. After setting $m = 1$ in the Command window, the simulation shows that the speed reaches the desired value of 1 in about 0.3 seconds with some overshoot, and that the maximum controller output is 40. The response time is longer than expected because of the overshoot. The effect of the disturbance is to temporarily reduce the speed to about 0.8 before recovering.

Put Scope blocks on v and the output of the PID block, and experiment with different values of P and I. Can you reduce the overshoot in v without the PID output exceeding 40?

The electric motor model given in Example 10.4–2 requires numerical values for several electrical and mechanical parameters. Obtaining these values can

Figure 10.10–3 Simulink model of the simplest speed control system.

Figure 10.10–4 Simulink model of a speed control system using an aggregated actuator-mass response time.

be the most difficult part of a robotics project. However, experience has shown that often a simple test consisting of applying a step voltage to the motor-and-mass and plotting its speed will yield a useful value for the time constant T (the time for the motor speed to reach steady state is $4T$). The value of T will include the effects of damping and of all the masses (inertias) in the system. The Simulink model of this system is shown in Figure 10.10–4, and its characteristic equation is

$$(T + D)s^2 + (P + 1)s + I = 0 \qquad (10.10–9)$$

Again we see that if $D = 0$ we can place the two roots anywhere by properly choosing P and I. Assuming that experiments determine that $T = 0.1$ (second), we set $s = -20, -20$ to obtain a total response time of 0.2 (seconds). This requires that $P = 3, I = 40, D = 0$.

The model shown in Figure 10.10–4 does not let us investigate the effects of a disturbance acting on the mass or on the output of the actuator (caused, for example, by the vehicle going up a slope). The model shown in Figure 10.10–5 shows this effect. Note that we must now test the motor separately from the load mass to obtain its value of T. The model's characteristic equation is

$$mTs^3 + (m + D)s^2 + Ps + I = 0 \qquad (10.10–10)$$

Note that in general we now must use all three gains to achieve any desired values of the roots. For example, if $m = 1$ and $T = 0.1$, to obtain a response time of 0.8, we choose $s = -5, -5, -5$. This gives $P = 7.5, I = 12.5$, and $D = 0.5$.

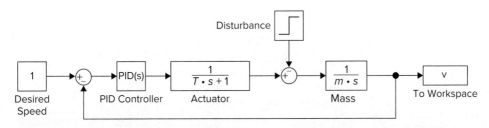

Figure 10.10–5 Simulink model of a speed control system using one block each for actuator and mass.

Position Control

Noting that velocity is the time derivative of displacement, we see that $dx/dt = v$. Substituting this into Equation (10.10–1), we obtain

$$m\frac{d^2x}{dt^2} = f - d \tag{10.10–11}$$

In transfer function form this is

$$X(s) = \frac{1}{ms^2}[F(s) - D(s)] \tag{10.10–12}$$

If we replace the integrator in Figure 10.10–3 with a double integrator, we obtain a simple model of position control, where m and x may represent either mass and rectilinear displacement or inertia and angular displacement (in radians). This gives the Simulink diagram shown in Figure 10.10–6. Its characteristic equation is

$$ms^3 + Ds^2 + Ps + I = 0 \tag{10.10–13}$$

Note that we need positive values for all three gains to achieve a stable system and to place the three roots where we want. For example, choosing the three roots to be $s = -1, -1, -1$ to obtain a system response time of 4, requires that $P = 3$, $I = 1, D = 3$.

We can obtain a more detailed model of position control by replacing the transfer function $1/ms$ with $1/ms^2$ in Figure 10.10–5.

Servo Motors   Our examples thus far assumed that we are controlling a device by adjusting its voltage or current input. There are motors that can be controlled with a digital input that specifies the desired position and have a built-in angular position sensor. These devices are often called servo motors (which stands for "servomechanism"), and they usually employ P control with a gain that is not adjustable by the user. They are often used in the remote control (RC) world, usually to control the steering of RC vehicles or the flaps on a RC plane. They are not useful for speed control. So when modeling such devices in Simulink, we assume that the controlled position equals the desired position.

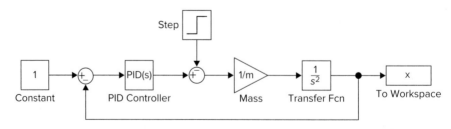

Figure 10.10–6 Simulink model of a simple position control system.

Simplified PID

Some computer hardware will not support the sophisticated PID algorithm used by the Simulink PID Controller block. In such cases a much simpler form of the algorithm can be tried and programmed in hardware-specific code. Using Equation 10.10–5 as a guide, the following MATLAB code implements a rudimentary PID algorithm where $tk = kT$ and T is the sampling period.

```
% Simplified PID algorithm
der(k) = e(k) + e(k-1);
sum(k) = e(k) + sum(k-1);
PID(k) = P*e(k) + I*T*sum(k) + (D/T)*der(k);
```

A similar approach can be implemented in Simulink, as shown in Figure 10.10–7. The model has been used in some applications on the MathWorks web site.

Trajectory Control of a Two-Wheeled Robot

The speed and position control systems require a command input that describes either the desired speed or the desired position. To take a specific example, consider the two-wheeled robot vehicle shown in Figure 10.10–8. The third wheel, at the front, is simply a free-swinging castor that is not driven. Suppose each of the two rear wheels is driven by its own motor and associated control system. The distance between the wheels is L. We take the axle midpoint to be our reference point and establish the coordinate system (x_1, y_1) there. We can choose to control the vehicle either by controlling the rotational speed or the rotational displacement of each wheel.

If we want the vehicle to move to a desired point specified by its (x,y) coordinates, we need to compute the rotational displacement required for each wheel.

Figure 10.10–7 Simulink diagram of a simplified PID algorithm.

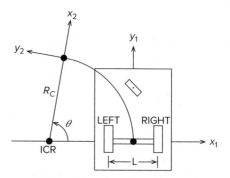

Figure 10.10–8 Turning geometry of a two-wheeled vehicle

Denote these angles by φ_L and φ_R for the left and right wheels. If we want the move to be completed in time T, then we divide the displacements by T to obtain the required wheel rotational speeds, $S_L = \varphi_L/T$ and $S_R = \varphi_R/T$. The distance traveled by each wheel is its radius multiplied by its rotational displacement. Denote these distances by D_L and D_R for the left and right wheels. If the wheel radius is R, then $D_L = R\varphi_L$ and $D_R = R\varphi_R$. So we must first develop a way to compute D_L and D_R. We then use these values to compute φ_L, φ_R, S_L, and S_R.

Consider the geometry of the circular turn shown in Figure 10.10–8. The point ICR is the instantaneous center of rotation, and R_C is the radius of the turn. Figure 10.10–9 shows the paths of the two wheels and the path of the center point. From the geometry of a circular arc we can see that

$$\frac{D_L}{R_C - L/2} = \frac{D_R}{R_C + L/2}$$

This can be solved for R_C as follows.

$$R_C = \frac{L}{2}\frac{D_L + D_R}{D_R - D_L} \tag{10.10–14}$$

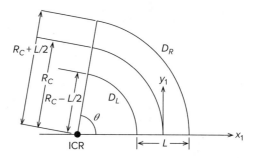

Figure 10.10–9 Wheel paths in a circular turn.

Note also from the figure that the turn angle is given by

$$\theta = \frac{D_R}{R_C + L/2} \tag{10.10-15}$$

The location of the center point after the turn is given by

$$x_C = R_C(\cos \theta - 1) \qquad y_C = R_C \sin \theta \tag{10.10-16}$$

These equations constitute the *forward solution*. Now we must obtain the backward or *inverse solution*. This will compute the wheel displacements required to place the vehicle at a desired location specified by coordinates (x_C, y_C). Equations (10.10–16) can be combined as follows.

$$\frac{1 - \cos \theta}{\sin \theta} = -\frac{x_C}{y_C} = A \tag{10.10-17}$$

This cannot be solved for θ analytically. However, for small angles, $\sin \theta \approx \theta$ and $\cos \theta \approx 1 - \theta^2/2$. If so, Equation (10.10–17) becomes

$$\theta \approx -2\frac{x_C}{y_C} = 2A \tag{10.10-18}$$

For large turns, θ will be large, and Equation (10.10–18) will not be useful. In such a case, we must solve Equation (10.10–17) numerically. The function file `turn_angle(A)` accomplishes this. Finally, from Equations (10.10–14) through (10.10–16) we obtain

$$R_C = \frac{y_C}{\sin \theta}$$

$$D_L = (R_C - L/2)\,\theta \qquad D_R = (R_C + L/2)\,\theta$$

These equations are implemented in the function `wheel_inverse`, which calls the function `turn_angle`.

```
function [theta,RC,DL,DR] = wheel_inverse(L,xC,yC)
% Two Wheel Drive Inverse Solution
A = -xC/yC;
theta = turn_angle(A);
RC = yC/sin(theta);
DL = (RC - L/2)*theta;
DR = (RC + L/2)*theta;
end


function theta = turn_angle(A)
% Computes turn angle for two wheel vehicle
theta_guess = 2*A;
myfun = @(th,A) (1-cos(th))/sin(th) - A;
theta = fzero(@(th) myfun(th,A),theta_guess);
end
```

These equations and two functions can be used either to plan a trajectory for the vehicle or to generate command inputs to either position or speed control systems for each wheel. Example 11.2–2 in Chapter 11 shows how a trajectory for a robot hand can be planned to generate position commands for the arm motors. A similar approach can be used with a robot vehicle. A Simulink model can be developed by adding these algorithms to one of the speed or position control models treated earlier in this section.

The kinematic equations derived from Figure 10.10–8 assume that the vehicle starts at the origin of the (x_1, y_1) coordinate system, with the axle aligned with the x_1 axis. To plan a continuation of the trajectory, the following coordinate transformation must be used.

$$\begin{bmatrix} x_2 \\ y_2 \end{bmatrix} = \begin{bmatrix} \cos\theta & \sin\theta \\ -\sin\theta & \cos\theta \end{bmatrix} \begin{bmatrix} x_1 - x_c \\ y_1 - y_c \end{bmatrix}$$

where (x_c, y_c) are the coordinates of the new location of the center point and are given by Equation (10.10–16).

10.11 Summary

The Simulink model window contains menu items we have not discussed. However, the ones we have discussed are the most important ones for getting started. We have introduced just a few of the blocks available within Simulink. Some of the blocks not discussed deal with discrete-time systems (ones modeled with difference, rather than differential, equations), digital logic systems, and other types of mathematical operations. In addition, some blocks have additional properties that we have not mentioned. However, the examples given here will help you get started in exploring the other features of Simulink. Consult the online help for information about these items.

Key Terms

Block diagrams,   440

Dead time,   467

Dead zone,   456

Derivative block,   470

Fcn block,   472

Gain block,   440

Integrator block,   440

Library Browser,   441

Look-Up Table block,   472

PI controller,   469

PID Controller block,   475

Piecewise-linear models,   450

Rate Limiter block,   469

Relay block,   453

Saturation block,   468

Signal Builder block,   470

Simulation diagrams,   440

State-variable models,   447

Subsystem block,   461

Summer,   441

Transfer-function models,   456

Transport delay,   467

Problems

Section 10.1

1. Draw a simulation diagram for the following equation.
$$\dot{y} = 5f(t) - 7y$$

2. Draw a simulation diagram for the following equation.
$$5\ddot{y} = 3\dot{y} + 7y = f(t)$$

3. Draw a simulation diagram for the following equation.
$$3\dot{y} + 7\sin y = f(t)$$

Section 10.2

4. Create a Simulink model to plot the solution of the following equation for $0 \le t \le 6$.
$$10\ddot{y} = 7\sin 4t + 5\cos 3t \quad y(0) = 3 \quad \dot{y}(0) = 2$$

5. A projectile is launched with a velocity of 100 m/s at an angle of 30° above the horizontal. Create a Simulink model to solve the projectile's equations of motion where x and y are the horizontal and vertical displacements of the projectile.

$$\ddot{x} = 0 \quad x(0) = 0 \quad \dot{x}(0) = 100\cos 30°$$
$$\ddot{y} = -g \quad y(0) = 0 \quad \dot{y}(0) = 100\sin 30°$$

    Use the model to plot the projectile's trajectory y versus x for $0 \le t \le 10$ s.

6. The following equation has no analytical solution even though it is linear.
$$\dot{x} + x = \tan t \quad x(0) = 0$$
    The approximate solution, which is less accurate for larger values of t, is
$$x(t) = \frac{1}{3}t^3 - t^2 + 3t - 3 + 3e^{-t}$$
    Create a Simulink model to solve this problem, and compare its solution with the approximate solution over the range $0 \le t \le 1$.

7. Construct a Simulink model to plot the solution of the following equation for $0 \le t \le 10$
$$15\dot{x} + 5x = 4u_s(t) - 4u_s(t - 2) \quad x(0) = 5$$
    where $u_s(t)$ is a unit-step function (in the Block Parameters window of the Step block, set the Step time to 0, the Initial value to 0, and the Final value to 1).

8. A tank having vertical sides and a bottom area of 100 ft² is used to store water. To fill the tank, water is pumped into the top at the rate given in the

following table. Use Simulink to solve for and plot the water height $h(t)$ for $0 \le t \le 10$ min.

Time (min)	0	1	2	3	4	5	6	7	8	9	10
Flow Rate (ft³/min)	0	80	130	150	150	160	165	170	160	140	120

Section 10.3

9. Construct a Simulink model to plot the solution of the following equations for $0 \le t \le 2$

$$\dot{x}_1 = -6x_1 + 4x_2$$
$$\dot{x}_2 = 5x_1 - 7x_2 + f(t)$$

where $f(t) = 3t$. Use the Ramp block in the Sources library.

10. Construct a Simulink model to plot the solution of the following equations for $0 \le t \le 3$

$$\dot{x}_1 = -6x_1 + 4x_2 + f_1(t)$$
$$\dot{x}_2 = 5x_1 - 7x_2 + f_2(t)$$

where $f_1(t)$ is a step function of height 3 starting at $t = 0$ and $f_2(t)$ is a step function of height -3 starting at $t = 1$.

Section 10.4

11. Use the Saturation block to create a Simulink model to plot the solution of the following equation for $0 \le t \le 6$.

$$3\dot{y} + y = f(t) \quad y(0) = 3$$

where

$$f(t) = \begin{cases} 8 & \text{if } 10 \sin 3t > 8 \\ -8 & \text{if } 10 \sin 3t < -8 \\ 10 \sin 3t & \text{otherwise} \end{cases}$$

12. Construct a Simulink model of the following problem.

$$5\dot{x} + \sin x = f(t) \quad x(0) = 0$$

The forcing function is

$$f(t) = \begin{cases} -5 & \text{if } g(t) \le -5 \\ g(t) & \text{if } -5 < g(t) < 5 \\ 5 & \text{if } g(t) \ge 5 \end{cases}$$

where $g(t) = 10 \sin 4t$.

13. If a mass-spring system has Coulomb friction on the surface rather than viscous friction, its equation of motion is

$$m\ddot{y} = \begin{cases} -ky + f(t) - \mu mg & \text{if } \dot{y} \ge 0 \\ -ky + f(t) + \mu mg & \text{if } \dot{y} < 0 \end{cases}$$

where μ is the coefficient of friction. Develop a Simulink model for the case where $m = 1$ kg, $k = 5$ N/m, $\mu = 0.4$, and $g = 9.8$ m/s^2. Run the simulation for two cases: (a) the applied force $f(t)$ is a step function with a magnitude of 10 N and (b) the applied force is sinusoidal: $f(t) = 10 \sin 2.5t$. Either the Sign block in the Math Operations library or the Coulomb and Viscous Friction block in the Discontinuities library can be used, but since there is no viscous friction in this problem, the Sign block is easier to use.

14. A certain mass, $m = 2$ kg, moves on a surface inclined at an angle $\phi = 30°$ above the horizontal. Its initial velocity is $v(0) = 3$ m/s up the incline. An external force of $f_1 = 5$ N acts on it parallel to and up the incline. The coefficient of Coulomb friction is $\mu = 0.5$. Use the Sign block and create a Simulink model to solve for the velocity of the mass until the mass comes to rest. Use the model to determine the time at which the mass comes to rest.

15. *a.* Develop a Simulink model of a thermostatic control system in which the temperature model is

$$RC\frac{dT}{dt} + T = Rq + T_a (t)$$

where T is the room air temperature in °F, T_a is the ambient (outside) air temperature in °F, time t is measured in hours, q is the input from the heating system in lb-ft/hr, R is the thermal resistance, and C is the thermal capacitance. The thermostat switches q on at the value q_{max} whenever the temperature drops below 69°F and switches q to $q = 0$ whenever the temperature is above 71°F. The value of q_{max} indicates the heat output of the heating system.

Run the simulation for the case where $T(0) = 70$°F, and $T_a(t) = 50 + 10 \sin (\pi t/12)$. Use the values $R = 5 \times 10^{-5}$ °F-hr/lb-ft and $C = 4 \times 10^4$ lb-ft/°F. Plot the temperatures T and T_a versus t on the same graph, for $0 \le t \le 24$ hr. Do this for two cases: $q_{max} = 4 \times 10^5$ and $q_{max} = 8 \times 10^5$ lb-ft/hr. Investigate the effectiveness of each case.

  b. The integral of q over time is the energy used. Plot $\int q \, dt$ versus t and determine how much energy is used in 24 hr for the case where $q_{max} = 8 \times 10^5$.

16. Refer to Problem 15. Use the simulation with $q_{max} = 8 \times 10^5$ to compare the energy consumption and the thermostat cycling frequency for the two temperature bands (69°, 71°) and (68°, 72°).

17. Consider the liquid-level system shown in Figure 10.7–1. The governing equation based on conservation of mass is Equation (10.7–2). Suppose that the height h is controlled by using a relay to switch the input flow rate between the values 0 and 50 kg/s. The flow rate is switched on when the height is less than 4.5 m and is switched off when the height reaches 5.5 m. Create a Simulink model for this application using the values $A = 2$ m^2, $R = 400$ m$^{-1} \cdot$ s^{-1}, $\rho = 1000$ kg/m^3, and $h(0) = 1$ m. Obtain a plot of $h(t)$.

Section 10.5

18. Use the Transfer Function block to construct a Simulink model to plot the solution of the following equation for $0 \leq t \leq 4$.

$$2\ddot{x} + 12\dot{x} + 10x = 5u_s(t) - 5u_s(t-2) \quad x(0) = \dot{x}(0) = 0$$

19. Use Transfer Function blocks to construct a Simulink model to plot the solution of the following equations for $0 \leq t \leq 2$.

$$3\ddot{x} + 15\dot{x} + 18x = f(t) \quad x(0) = \dot{x}(0) = 0$$
$$2\ddot{y} + 16\dot{y} + 50y = x(t) \quad y(0) = \dot{y}(0) = 0$$

where $f(t) = 75u_s(t)$.

20. Use Transfer Function blocks to construct a Simulink model to plot the solution of the following equations for $0 \leq t \leq 2$.

$$3\ddot{x} + 15\dot{x} + 18x = f(t) \quad x(0) = \dot{x}(0) = 0$$
$$2\ddot{y} + 16\dot{y} + 50y = x(t) \quad y(0) = \dot{y}(0) = 0$$

where $f(t) = 50u_s(t)$. At the output of the first block there is a dead zone for $-1 \leq x \leq 1$. This limits the input to the second block.

21. Use Transfer Function blocks to construct a Simulink model to plot the solution of the following equations for $0 \leq t \leq 2$.

$$3\ddot{x} + 15\dot{x} + 18x = f(t) \quad x(0) = \dot{x}(0) = 0$$
$$2\ddot{y} + 16\dot{y} + 50y = x(t) \quad y(0) = \dot{y}(0) = 0$$

where $f(t) = 50u_s(t)$. At the output of the first block there is a saturation that limits x to be $|x| \leq 1$. This limits the input to the second block.

Section 10.6

22. Construct a Simulink model to plot the solution of the following equation for $0 \leq t \leq 4$.

$$2\ddot{x} + 12\dot{x} + 10x^2 = 8 \sin 0.8t \quad x(0) = \dot{x}(0) = 0$$

23. Create a Simulink model to plot the solution of the following equation for $0 \leq t \leq 3$.

$$\dot{x} + 10x^2 = 5 \sin 3t \quad x(0) = 1$$

24. Construct a Simulink model of the following problem.

$$10\dot{x} + \sin x = f(t) \quad x(0) = 0$$

The forcing function is $f(t) = \sin 2t$. The system has the dead-zone nonlinearity shown in Figure 10.5–1.

25. The following model describes a mass supported by a nonlinear, hardening spring. The units are SI. Use $g = 9.81$ m/s^2.

$$5\ddot{y} = 5g - \left(900y + 1700y^3\right) \quad y(0) = 0.5 \quad \dot{y}(0) = 0$$

Create a Simulink model to plot the solution for $0 \leq t \leq 2$.

26. Consider the system for lifting a mast shown in Figure P26. The 70-ft-long mast weighs 500 lb. The winch applies a force $f = 380$ lb to the cable. The mast is supported initially at an angle of 30°, and the cable at A is initially horizontal. The equation of motion of the mast is

$$25,400\ddot{\theta} = -17,500\cos\theta + \frac{626,000}{Q}\sin(1.33 + \theta)$$

where

$$Q = \sqrt{2020 + 1650\cos(1.33 + \theta)}$$

Create and run a Simulink model to solve for and plot $\theta(t)$ for $\theta(t) \le \pi/2$ rad.

Figure P26

27. The equation describing the water height h in a spherical tank with a drain at the bottom is

$$\pi(2rh - h^2)\frac{dh}{dt} = -C_d A \sqrt{2gh}$$

Suppose that the tank's radius is $r = 3$ m and the circular drain hole of area A has a radius of 2 cm. Assume that $C_d = 0.5$ and that the initial water height is $h(0) = 5$ m. Use $g = 9.81$ m/s². Use Simulink to solve the nonlinear equation, and plot the water height as a function of time until $h(t) = 0$.

28. A cone-shaped paper drinking cup (like the kind used at water fountains) has a radius R and a height H. If the water height in the cup is h, the water volume is given by

$$V = \frac{1}{3}\pi\left(\frac{R}{H}\right)^2 h^3$$

Suppose that the cup's dimensions are $R = 1.5$ in. and $H = 4$ in.

a. If the flow rate from the fountain into the cup is 2 in.³/sec, use Simulink to determine how long will it take to fill the cup to the brim.

b. If the flow rate from the fountain into the cup is given by $2(1 - e^{-2t})$ in.³/sec, use Simulink to determine how long will it take to fill the cup to the brim.

Section 10.7

29. Refer to Figure 10.7–2. Assume that the resistances obey the linear relation, so that the mass flow q_l through the left-hand resistance is $q_l = (p_l - p)/R_l$, with a similar linear relation for the right-hand resistance.

a. Create a Simulink subsystem block for this element.

b. Use the subsystem block to create a Simulink model of the system shown in Figure 10.7–5. Assume that the mass inflow rate is a step function.

c. Use the Simulink model to obtain plots of $h_1(t)$ and $h_2(t)$ for the following parameter values: $A_1 = 2$ m^2, $A_2 = 5$ m^2, $R_1 = 400$ m$^{-1} \cdot$ s^{-1} $R_2 = 600$ m$^{-1} \cdot$ s^{-1}, $\rho = 1000$ kg/m^3, $q_{mi} = 50$ kg/s, $h_1(0) = 1.5$ m, and $h_2(0) = 0.5$ m.

30. *a.* Use the subsystem block developed in Section 10.7 to construct a Simulink model of the system shown in Figure P30. The mass inflow rate is a step function.

b. Use the Simulink model to obtain plots of $h_1(t)$ and $h_2(t)$ for the following parameter values: $A_1 = 3$ ft^2, $A_2 = 5$ ft^2, $R_1 = 30$ ft^{-1}–sec^{-1}, $R_2 = 40$ ft^{-1}–sec^{-1}, $\rho = 1.94$ slug/ft^3, $q_{mi} = 0.5$ slug/sec, $h_1(0) = 2$ ft, and $h_2(0) = 5$ ft.

Figure P30

31. Consider Figure 10.7–7 for the case where there are three RC loops with the values $R_1 = R_3 = 10^4 \,\Omega$, $R_2 = 5 \times 10^4 \,\Omega$, $C_1 = C_3 = 10^{-6}$ F, and $C_2 = 4 \times 10^{-6}$ F.

a. Develop a subsystem block for one RC loop.

b. Use the subsystem block to construct a Simulink model of the entire system of three loops. Plot $v_3(t)$ over $0 \leq t \leq 3$ for $v_1(t) = 12 \sin 10t$ V.

32. Consider Figure 10.7–8 for the case where there are three masses. Use the values $m_1 = m_3 = 10$ kg, $m_2 = 30$ kg, $k_1 = k_4 = 10^4$ N/m, and $k_2 = k_3 = 2 \times 10^4$ N/m.

    a. Develop a subsystem block for one mass.

    b. Use the subsystem block to construct a Simulink model of the entire system of three masses. Plot the displacements of the masses over $0 \leq t \leq 2$ s for if the initial displacement of m_1 is 0.1 m.

Section 10.8

33. Refer to Figure P30. Suppose there is a dead time of 10 sec between the outflow of the top tank and the lower tank. Use the subsystem block developed in Section 10.7 to create a Simulink model of this system. Using the parameters given in Problem 30, plot the heights h_1 and h_2 versus time.

Section 10.9

34. Redo the Simulink suspension model developed in Section 10.9, using the spring relation and input function shown in Figure P34 and the following damper relation.

$$f_d(v) = \begin{cases} -500|v|^{1.2} & v \leq 0 \\ 50v^{1.2} & v > 0 \end{cases}$$

Use the simulation to plot the response. Evaluate the overshoot and undershoot.

(a)

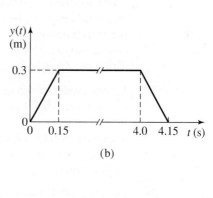

(b)

Figure P34

35. Consider the system shown in Figure P35. The equations of motion are

$$m_1\ddot{x}_1 + (c_1 + c_2)\dot{x}_1 + (k_1 + k_2)x_1 - c_2\dot{x}_2 - k_2x_2 = 0$$
$$m_2\ddot{x}_2 + c_2\dot{x}_2 + k_2x_2 - c_2\dot{x}_1 - k_2x_1 = f(t)$$

Figure P35

Suppose that $m_1 = m_2 = 1$, $c_1 = 3$, $c_2 = 1$, $k_1 = 1$, and $k_2 = 4$.

a. Develop a Simulink model of this system. In doing this, consider whether to use a state-variable representation or a transfer-function representation of the model.

b. Use the Simulink model to plot the response $x_1(t)$ for the following input. The initial conditions are zero.

$$f(t) = \begin{cases} t & 0 \le t \le 1 \\ 2 - t & 1 < t < 2 \\ 0 & t \ge 2 \end{cases}$$

Section 10.10

36. For the model shown in Figure 10.10–3 with a unit-step command input, let $m = 1$. Compute the PI gains required to obtain the roots $s = -50$, -100. a) Run the simulation and plot the speed. Is the response time what you would expect given the specified roots? b) Suppose the disturbance is a step function with a magnitude of 100 and a step time of 0.1. Run the simulation and plot the speed. How effective is the controller in counteracting the disturbance?

37. For the model shown in Figure 10.10–3 with a unit-step command input, let $m = 1$. Given the PI gains $P = 150$ and $I = 5000$, put a Scope block on the PID block output. a) Run the simulation until steady-state is reached and plot the speed. What is the maximum output of the PID block? b) Insert a Saturation block after the PID block and use the limits −50 and 50. Run the simulation again and plot the speed. How is the response affected by the limits?

38. For the model shown in Figure 10.10–3 with a unit-step command input, let $m = 1$. Compute the PI gains required to obtain a response time of 2. Run the simulation and check the speed response. Is the response time what you would expect?

39. For the model shown in Figure 10.10–3 with a unit-step command input, let $m = 1$. Compute the PI gains required to obtain the roots $s = -50 \pm 50j$.

Run the simulation and plot the speed. Is the response time what you would expect given the specified roots?

40. For the model shown in Figure 10.10–4 with a unit-step command input, let $T = 0.3$. Compute the PI gains required to obtain a response time of 2 without oscillations. Run the simulation and check the speed response. Is the response time what you would expect?

41. For the model shown in Figure 10.10–4 with a unit-step command input, let $m = 1$ and $T = 0.3$. Compute the PI gains required to obtain the roots $s = -50, -100$. Run the simulation and check the speed response. Is the response time what you would expect?

42. For the model shown in Figure 10.10–5 with a unit-step command input, let $m = 1$ and $T = 0.2$. Compute the PID gains required to obtain the roots $s = -10, -20, -20$. Run the simulation and check the speed response. Is the response time what you would expect?

43. For the model shown in Figure 10.10–5 with a unit-step command input, let $m = 1$ and $T = 0.2$. Compute the PID gains required to obtain the roots $s = -10, -20 \pm 20j$. Run the simulation and check the speed response. Is the response time what you would expect?

44. For the model shown in Figure 10.10–6 with a unit-step command input, let $m = 1$. Compute the PID gains required to obtain the roots $s = -10, -20, -20$. Run the simulation and check the position response. Is the response time what you would expect?

45. For the model shown in Figure 10.10–6 with a unit-step command input, let $m = 1$. Compute the PID gains required to obtain the roots $s = -10, -20 \pm 20j$. Run the simulation and check the position response. Is the response time what you would expect?

46. For a certain two-wheeled vehicle, the wheelbase is $L = 2$ and the wheel radii are $R = 0.5$. The wheel rotation angles are $\varphi_L = 2$ rad and $\varphi_R = 4$ rad. Compute the resulting turn radius R_C, the turn angle θ, and the coordinates of the new location of the vehicle reference point.

47. For a certain two-wheeled vehicle, the wheelbase is $L = 2$ and the wheel radii are $R = 0.5$. It is desired to place the vehicle reference point at $x = -0.4$, $y = 1.4$. Compute the required turn radius RC, the turn angle θ, and the wheel rotation angles φ_L and φ_R.

Engineering in the 21st Century. . .

Developing Alternative Sources of Energy

It now appears that the United States and much of the rest of the world have recognized the need to reduce their dependence on nonrenewable energy sources such as natural gas, oil, coal, and perhaps even uranium. Supplies of these fuels will eventually be exhausted, they have harmful environmental effects, and when imported, they cause huge trade imbalances that hurt the economy. One of the major engineering challenges of the 21st century will be to develop renewable energy sources.

Renewable energy sources include solar energy (both solar-thermal and solar-electric), geothermal power, tidal and wave power, wind power, as well as crops that can be converted to alcohol. In solar-thermal applications, energy from the sun is used to heat a fluid, which can be used to heat a building or power an electrical generator such as a steam turbine. In solar-electric applications, sunlight is directly converted to electricity.

Geothermal power is obtained from ground heat or steam vents. With tidal power the tidal currents are used to drive a turbine to generate electricity. Wave power uses the change in water surface level due to waves to drive water through a turbine or other device. Wind power uses a wind turbine to drive a generator.

The difficulty with most renewable energy sources is that they are diffuse, so the energy must be concentrated somehow, and they are intermittent, which requires a storage method. At present most renewable energy systems are not very efficient, and so the engineering challenge of the future is to improve their efficiency.

MATLAB supports engineering design of renewable energy systems. Examples include software to study the performance of a 9 MW wind farm connected to a distribution system, and the performance of a unified power flow controller to relieve congestion in a power transmission system. Numerous studies of photovoltaic systems have been done in MATLAB. ■

Symbolic Processing with MATLAB

Up to now we have used MATLAB to perform numerical operations only; that is, our answers have been numbers, not expressions. In this chapter, we show how to use MATLAB to perform *symbolic processing* to obtain answers in the form of expressions. Symbolic processing is the term used to describe how a computer performs operations on mathematical expressions in the way, for example that humans do algebra with pencil and paper. Whenever possible, we wish to obtain solutions in closed form, because they give us greater insight into the problem. For example, we often can see how to improve an engineering design by modeling it with mathematical expressions that do not have specific parameter values. Then we can analyze the expressions and decide which parameter values will optimize the design.

Symbolic expression   In this chapter we will show how to define a symbolic expression such as $y = \sin x/\cos x$ in MATLAB, and how to use MATLAB to simplify expressions wherever possible. For example, the previous function simplifies to $y = \sin x / \cos x = \tan x$. MATLAB can perform operations such as addition and multiplication on mathematical expressions, and we can use MATLAB to obtain symbolic solutions to algebraic equations such as $x^2 + 2x + a = 0$ (the solution for x is $x = -1 \pm \sqrt{1 - a}$). MATLAB can also perform symbolic differentiation and integration, and can solve ordinary differential equations in closed form.

To use the methods of this chapter, you must have either the Symbolic Math Toolbox or the Student Edition of MATLAB. This chapter is based on Version 8.9 (R2017b) of the toolbox.

Symbolic processing is a relatively new computer application, and such software is undergoing rapid development. As a result, software upgrades might produce changes in performance and in the syntax as capabilities are improved and bugs removed. For this reason, you should check the MathWorks web site if your software does not perform as expected or gives erroneous results. The MathWorks web site is http//www.mathworks.com. Look for answers to frequently asked questions (FAQ) and technical notes. These are arranged by category (for example, one category is the Symbolic Math Toolbox). You can also search for information using key words.

In this chapter we cover a subset of the capabilities of the Symbolic Math Toolbox. Specifically we treat

- symbolic algebra,
- symbolic methods for solving algebraic and transcendental equations,
- symbolic methods for solving ordinary differential equations,
- symbolic calculus, including integration, differentiation, limits, and series,
- Laplace transforms, and
- selected topics in linear algebra, including symbolic methods for obtaining determinants, matrix inverses, and eigenvalues.

The topic of Laplace transforms is included because Laplace transforms are one way of solving differential equations, and are often covered along with differential equations.

We do not discuss the following features of the Symbolic Math Toolbox: canonical forms of symbolic matrices; variable precision arithmetic that allows you to evaluate expressions to a specified numerical accuracy; and more advanced mathematical functions such as Fourier transforms. Details on these capabilities can be found in the online help.

When you have finished this chapter, you should be able to use MATLAB to

- create symbolic expressions and manipulate them algebraically,
- obtain symbolic solutions to algebraic and transcendental equations,
- perform symbolic differentiation and integration,
- evaluate limits and series symbolically,
- obtain symbolic solutions to ordinary differential equations,

■ obtain Laplace transforms, and

■ perform symbolic linear algebra operations, including obtaining expressions for determinants, matrix inverses, and eigenvalues.

11.1 Symbolic Expressions and Algebra

The `sym` function can be used to create "symbolic objects" in MATLAB. If the input argument to `sym` is a string, the result is a symbolic number or variable. If the input argument is a numeric scalar or matrix, the result is a symbolic representation of the given numeric values. For example, typing `x = sym('x')` creates the symbolic variable with name `x`, and typing `y = sym('y')` creates a symbolic variable named `y`. Typing `x = sym('x','real')` tells MATLAB to assume that `x` is real.

The `syms` command enables you to combine more than one such statement into a single statement. For example, typing `syms x` is equivalent to typing `x = sym('x')`, and typing `syms x y u v` creates the four symbolic variables `x`, `y`, `u`, and `v`. When used without arguments, `syms` lists the symbolic objects in the workspace. The `syms` command, however, cannot be used to create symbolic constants; you must use `sym` for this purpose.

The `syms` command enables you to specify that certain variables are real. For example,

```
>>syms x y real
```

or positive:

```
>>syms x y positive
```

To clear `x` and `y`, type

```
>>syms x y clear
```

You can use the `syms` function to create symbolic functions. For example,

```
>>syms x(t)
>>x(t) = t^2;
>>x(3)
    9
```

or

```
>>syms f(x,y)
>>f(x,y) = x + 4*y;
>>f(2,5)
    22
```

Symbolic constants The `sym` function can be used to create symbolic constants by using a numerical value for the argument. For example, typing `pi = sym('pi')`, or `syms pi`, and `fraction = sym('1/3')` create symbolic constants that avoid the floating point approximations inherent in the values of π

and 1/3. If you create the *symbolic constant* `pi` this way, it temporarily replaces the built-in numeric constant, and you no longer obtain a numerical value when you type its name. For example,

```
>>syms pi
>>b = 4*pi     % This gives a symbolic result.
b =
    4*pi
>>fraction = syms('1/3');
>>c = 5*fraction     % This gives a symbolic result.
c =
    5/3
```

The advantages of using symbolic constants is that they need not be evaluated (with the accompanying round-off error) until a numeric answer is required.

Symbolic constants can look like numbers, but are actually symbolic expressions. Symbolic expressions can look like character strings, but are a different sort of quantity. You can use the `class` function to determine whether or not a quantity is symbolic, numeric, or a character string. We will give examples of the `class` function later.

The effects of round-off error needs to be considered when converting MAT-LAB floating point values to symbolic constants in this way. You can use an optional second argument with the `sym` function to specify the technique for converting floating point numbers. Refer to the online help for more information.

Symbolic Expressions

You can use symbolic variables in expressions and as arguments of functions. You use the operators `+ - * / ^` and the built-in functions just as you use them with numerical calculations. For example, typing

```
>>syms x y
>>s = x + y;
>>r = sqrt(x^2 + y^2);
```

creates the symbolic variables `s` and `r`. The terms `s = x + y` and `r = sqrt(x^2 + y^2)` are examples of symbolic *expressions*. The variables `s` and `r` created this way are *not* the same as user-defined function files. That is, if you later assign `x` and `y` numeric values, typing `r` will not cause MATLAB to evaluate the equation $r = \sqrt{x^2 + y^2}$. We will see later how to evaluate symbolic expressions numerically.

The `syms` command enables you to specify that expressions have certain characteristics. For example, in the following session, MATLAB will treat the expression `w` as a non-negative number.

```
>>syms x y real
>>w = x^2 + y^2;
```

To clear `x` of its real property, type `syms x clear`.

The vector and matrix notation used in MATLAB also applies to symbolic variables. For example, you can create a symbolic matrix A as follows.

```
>>n = 3;
>>syms x
>>A = x.^((0:n)' *(0:n))
A =
    [ 1, 1, 1, 1]
    [ 1, x, x^2, x^3]
    [ 1, x^2, x^4, x^6]
    [ 1, x^3, x ^6, x^9]
```

Note that it was not necessary to use sym or syms to declare A to be a symbolic variable beforehand. It is recognized as a symbolic variable because it is created with a *symbolic expression*. Note also that we need the period following x to enable element-by-element exponentiation.

Default variable In MATLAB the variable x is the default independent variable, but other variables can be specified to be the independent variable. It is important to know which variable is the independent variable in an expression. The function findsym(E) can be used to determine the symbolic variable used by MATLAB in a particular expression E.

The function findsym(E) finds the symbolic variables in a symbolic expression or matrix, where E is a scalar or matrix symbolic expression, and returns a string containing all of the symbolic variables appearing in E. The variables are returned in alphabetical order and are separated by commas. If no symbolic variables are found, findsym returns the empty string.

By contrast, the function findsym(E,n) returns the n symbolic variables in E closest to x, with the tie breaker going to the variable closer to z.

The following session shows some examples of its use.

```
>>syms b x1 y
>>findsym(6*b+y)
ans =
    b,y
>>findsym(6*b+y+x) % Note:x has not been declared symbolic.
??? Undefined function or variable 'x'.
>>findsym(6*b+y,1) % Find the one variable closest to x.
ans =
    y
>>findsym(6*b+y+x1,1)
ans =
    x1
>>findsym(6*b+y*i) % Note: i is not symbolic
ans =
    b, y
```

Manipulating Expressions

The following functions can be used to manipulate expressions by collecting coefficients of like powers, expanding powers, and factoring expressions, for example.

The function `collect(E)` collects coefficients of like powers in the expression E. If there is more than one variable, you can use the optional form `collect(E,v)`, which collects all the coefficients with the same power of v.

```
>>syms x y
>>E = (x-5)^2+(y-3)^2;
>>collect(E)
ans =
    x^2-10*x+25+(y-3)^2
>>collect(E,y)
ans =
    y^2-6*y+(x-5)^2+9
```

The function `expand(E)` expands the expression E by carrying out powers. For example,

```
>>syms x y
>>expand((x+y)^2) % Applies algebra rules.
ans =
    x^2+2*x*y+y^2
>>expand(sin(x+y)) % Applies trig identities.
ans =
    sin(x)*cos(y)+cos(x)*sin(y)
ans =
    6
```

The function `factor(n)` returns the prime factors of the *number n*, whereas if the argument is a *symbolic expression* E, the function `factor(E)` factors the expression E. For example,

```
>>syms x y
>>factor(x^2-1)
ans =
    (x-1)*(x+1)
```

The function `simplify(E)` attempts to simplify the expression E. For example,

```
>>syms a x y
>>simplify(exp(a*log(sqrt(x))))
ans =
    x^(a/2)
>>simplify(6*((sin(x))^2+(cos(x))^2))
ans =
    6
```

```
>>simplify(sqrt(x^2)) % Does not assume that x is non-negative.
ans =
    sqrt(x^2)
>>simplify(sqrt(x^2),'IgnoreAnalyticConstraints',true)
ans =
    x % Assumes that x is non-negative.
```

The function `simplify(E,IgnoreAnalyticConstraints',value)` controls the level of mathematical rigor to use on the analytical constraints while simplifying (non-negativity, division by zero, etc). The options for `value` are `true` or `false`. Specify `true` to relax the level of mathematical rigor in the simplification process. The default is `false`.

You can use the operators `+` `-` `*` `/` and `^` with symbolic expressions to obtain new expressions. The following session illustrates how this is done.

```
>>syms x y
>>E1 = x^2 + 5      % Define two expressions.
>>E2 = y^3 - 2
>>S1 = E1 + E2      % Add the expressions.
S1 =
    x^2+3+y^3
>>S2 = E1*E2        % Multiply the expressions.
S2 =
    (x^2+5)*(y^3-2)
>>expand(S2)        % Expand the product.
ans =
    x^2*y^3-2*x^2+5*y^3-10
>>E3 =x^3+2*x^2+5*x+10      % Define a third expression.
>>S3 = E3/E1        % Divide two expressions.
    S3 = (x^3+2*x^2+5*x+10)/(x^2+5)
>>simplify(S3)      % See if some terms cancel.
ans =
    x+2
```

The function `[num den]` = `numden(E)` returns two symbolic expressions that represent the numerator num and denominator den for the rational representation of the expression E.

```
>>syms x
>>E1 = x^2+5;
>>E4 = 1/(x+6);
>>[num, den] = numden(E1+E4)
num =
    x^3+6*x^2+5*x+31
den =
    x+6
```

The function double(E) converts the expression E to numeric form. The term "double" stands for floating point, *double* precision. For example,

```
>>sym_num = sym([pi, 1/3]);
>>double(sym_num)
ans =
    3.1416        0.3333
```

The function poly2sym(p) converts a coefficient vector p to a symbolic polynomial. The *default variable* is x. The form poly2sym(p,'v') generates the polynomial in terms of the variable v. For example,

```
>>poly2sym([2,6,4])
ans =
    2*x^2+6*x+4
>>poly2sym([5,-3,7],'y')
ans =
    5*y^2-3*y+7
```

The function sym2poly(E) converts the expression E to a polynomial coefficient vector.

```
>>syms x
>>sym2poly(9*x^2+4*x+6)
ans =
    [9 4 6]
```

The function subs(E,old,new) substitutes new for old in the expression E, where old can be a symbolic variable or expression, new can be a symbolic variable, expression, or matrix, or a numeric value or matrix. For example,

```
>>syms x y
>>E = x^2+6*x+7
>>F = subs(E,x,y)
F =
  y^2+6*y+7
```

If old and new are cell arrays of the same size, each element of old is replaced by the corresponding element of new. If E and old are scalars and new is an array or cell array, the scalars are expanded to produce an array result.

If you want to tell MATLAB that *f* is a function of the variable *t*, type syms f(t). Thereafter, f behaves like a function of t, and you can manipulate it with the toolbox commands. For example, to create a new function $g(t) = f(t + 2) - f(t)$, the session is

```
>>syms f(t)
>>g = subs(f,t,t+2)-f
g =
  f(t+2)-f(t)
```

Once a specific function is defined for *f(t)*, the function *g(t)* will be available. We will use this technique with the *Laplace transform* in Section 11.5.

**LAPLACE
TRANSFORM**

To perform multiple substitutions, enclose the new and old elements in braces. For example, to substitute *a* = *x* and *b* = *2* into the expression *E* = *a* sin *b*, the session is

```
>>syms a b x
>>E = a*sin(b);
>>F = subs(E,{a, b}, {x, 2})
F =
    x*sin(2)
```

Evaluating Expressions

In most applications we eventually want to obtain numerical values or a plot from the symbolic expression. Use the subs and double functions to evaluate an expression numerically. First use subs(E,old,new) to replace old with a numeric value new. Then use the double function to convert the expression E to a numeric form. For example,

```
>>syms x
>>E = x^2+6*x+7;
>>G = subs(E,x,2) % G is a symbolic constant.
G =
    23
>>class(G)
ans =
    sym
>>H = double(G) % H is a numeric quantity.
H =
    23
>>class(H)
ans =
    double
```

Sometimes MATLAB will display all zeros as the result of evaluating an expression, whereas in fact the value can be nonzero but so small that you need to evaluate the expression with more accuracy to see that it is nonzero. You can use the digits and the vpa functions to change the number of digits used by MATLAB for calculating and evaluating expressions. The accuracy of individual arithmetic operations in MATLAB is about 16 digits, whereas symbolic operations can be carried out to an arbitrary number of digits. The default is 32 digits. Type digits(d) to change the number of digits used to d. Be aware that larger values of d will require more time and computer memory to perform operations. Type vpa(E) to compute the expression E to the number of digits specified by the default value of 32 or the current setting of digits. Type vpa(E,d) to compute the expression E using d digits. (The abbreviation vpa stands for "variable precision arithmetic.")

Plotting Expressions

The MATLAB function fplot(E) generates a plot of a symbolic expression E, which is a function of one variable. The default range of the independent variable is the interval [−5, 5], unless this interval contains a singularity. The optional form fplot(E,[xmin xmax]) generates a plot over the range from xmin to xmax. Of course, you can enhance the plot generated by fplot by using the plot format commands discussed in Chapter 5; for example, the axis, xlabel, and ylabel commands.

For example,

```
>>syms x
>>E = x^2-6*x+7;
>>fplot(E,[-2 6])
```

Sometimes the automatic selection of the ordinate scale is not satisfactory. To obtain an ordinate scale from −5 to 25, and to place a label on the the ordinate, you would type

```
>>fplot(E),axis([-2 6 -5 25],ylabel('E')
```

Table 11.1–1 Functions for Creating and Evaluating Symbolic Expressions

Command	Description
class(E)	Returns the class of the expression E.
digits(d)	Sets the number of decimal digits used to do variable precision arithmetic. The default is 32 digits.
double(E)	Converts the expression E to numeric form.
findsym(E)	Finds the symbolic variables in a symbolic expression or matrix, where E is a scalar or matrix symbolic expression, and returns a string containing all of the symbolic variables appearing in E. The variables are returned in alphabetical order and are separated by commas. If no symbolic variables are found, findsym returns the empty string.
findsym(E,n)	Returns the n symbolic variables in E closest to x, with the tie breaker going to the variable closer to z.
fplot(E)	Generates a plot of a symbolic expression E, which is a function of one variable. The default range of the independent variable is the interval [−5, 5], unless this interval contains a singularity. The optional form fplot(E,[xmin xmax]) generates a plot over the range from xmin to xmax.
[num den] = numden(E)	Returns two symbolic expressions that represent the numerator expression num and denominator expression den for the rational representation of the expression E.
x = sym('x')	Creates the symbolic variable with name x. Typing x = sym('x', 'real') tells MATLAB to assume that x is real.
syms x y u v	Creates the symbolic variables x, y, u, and v. When used without arguments, syms lists the symbolic objects in the workspace.
vpa(E,d)	Sets the number of digits used to evaluate the expression E to d. Typing vpa(E) causes E to be evaluated to the number of digits specified by the default value of 32 or by the current setting of digits.

Order of Precedence

MATLAB does not always arrange expressions in a form that we normally would use. For example, MATLAB might provide an answer in the form: $-c+b$ whereas we would normally write $b-c$. The order of precedence used by MATLAB must be constantly kept in mind to avoid misinterpreting the output. MATLAB frequently expresses results in the form $1/a*b$ whereas we would normally write b/a. MATLAB sometimes fails to group terms like $x^(1/2)*y^(1/2)$ instead of writing them as $(x*y)^(1/2)$, and often fails to cancel negative signs where possible, as in $-a/(-b*c-d)$ instead of $a/(b*c+d)$.

Tables 11.1–1 and 11.1–2 summarize the functions for creating, evaluating, and manipulating symbolic expressions.

Test Your Understanding

T11.1–1 Given the expressions: $E_1 = x^3 - 15x^2 + 75x - 125$ and $E_2 = (x + 5)^2 - 20x$, use MATLAB to

(a) find the product E_1E_2 and express it in its simplest form.

(b) find the quotient E_1/E_2 and express it in its simplest form.

(c) evaluate the sum $E_1 + E_2$ at $x = 7.1$ in symbolic form and in numeric form.

(Answers: (a) $(x - 5)^5$ (b) $x - 5$ (c) 13671/1000 in symbolic form, 13.6710 in numeric form)

Table 11.1–2 Functions for Manipulating Symbolic Expressions

Command	Description
collect(E)	Collects coefficients of like powers in the expression **E**.
expand(E)	Expands the expression **E** by carrying out powers.
factor(E)	Factors the expression **E**.
poly2sym(p)	Converts a polynomial coefficient vector **p** to a symbolic polynomial. The form poly2sym(p, 'v') generates the polynomial in terms of the variable **v**.
simplify(E)	Attempts to simplify the expression **E**.
subs(E,old,new)	Substitutes new for old in the expression **E**, where old can be a symbolic variable or expression, new can be a symbolic variable, expression, or matrix, or a numeric value or matrix.
sym2poly(E)	Converts the expression **E** to a polynomial coefficient vector.

11.2 Algebraic and Transcendental Equations

The Symbolic Math Toolbox can solve algebraic and transcendental equations, as well as systems of such equations. A *transcendental* equation is one that contains one or more transcendental functions, such as $\sin x$, e^x, or $\log x$. The appropriate function to solve such equations is the `solve` function.

The function `solve(E)` solves a symbolic expression or equation represented by the expression E. If E represents an *equation*, the equation's expression must be enclosed in single quotes. If E represents an expression, then the solution obtained will be the roots of the expression E; that is, the solution of the equation $E = 0$. Multiple expressions or equations can be solved by separating them with a comma, as `solve(E1,E2, ..., En)`. Note that you need not declare the symbolic variable with the `sym` or `syms` function before using `solve`.

To solve the equation $x + 5 = 0$, one way is

```
>>syms x
>>solve(x+5==0)
ans =
    -5
```

Another way is

```
>>syms x
>>eqn = x+5==0;
>>solve(eqn)
ans =
    -5
```

You can store the result in a named variable as follows.

```
>>syms x
>>x = solve(x+5==0)
x =
   -5
```

To solve the equation $e^{2x} + 3e^x = 54$, the session is

```
>>syms x
>>solve(exp(2*x)+3*exp(x)==54)
ans =
    log(9) + pi*1i
    log(6)
```

Note that the first answer is $\ln(9) + \pi i$, which is equivalent to $\ln(-9)$. To see this, in MATLAB type `log(-9)` to obtain $2.1972 + 3.1416i$. So we obtained two solutions, instead of one, and now we must decide if both are meaningful. This depends on the application that produced the original equation. If the application requires a real number for a solution, then we should choose `log(6)` as the answer.

The following sessions provide some more examples of the use of these functions.

```
>>syms y
>>eqn1 = y^2+3*y+2==0;
>>solve(eqn1)
ans =
    -2
    -1
>>syms x
>>eqn2 = x^2+9*y^4==0;
>>solve(eqn2)  % x is presumed to be the unknown variable.
ans =
    -y^2*3*i
    y^2*3*i
```

When there is more than one variable in the expression, MATLAB assumes that the variable closest to x in the alphabet is the variable to be found. You can specify the solution variable using the syntax solve(E, 'v'), where v is the solution variable. For example,

```
>>syms b c
>>solve(b^2+8*c+2*b==0)   % Solves for c.
ans =
-1*b^2/8-b/4
>> solve(b^2+8*c+2*b==0,b)   % Solves for b.
ans =
-(1-8*c)^(1/2)-1
(1-8*c)^(1/2)-1
```

Thus the solution of $b^2 + 8c + 2b = 0$ for c is $c = -(b^2 + 2b)/8$. The solution for b is $b = -1 \pm \sqrt{1 - 8c}$.

You can save the solutions as vectors by using the form [x, y] = solve(eqn1, eqn2). Note the difference in the output formats in the following example. The first format gives the solution as a structure.

```
>>syms x y
>>eqn3 = 6*x+2*y==14;
>>eqn4 = 3*x+7*y==31;
>>solve(eqn3,eqn4)
ans =
    x: [1x1 sym]
    y: [1x1 sym]
>>x = ans.x
x =
    1
>>y = ans.y
```

```
y =
   4
>>[x, y] = solve(eqn3,eqn4)
x =
   1
y =
   4
```

Solution structure You can save the solution in a structure with named fields (see Chapter 3, Section 3.7 for a discussion of structures and fields). The individual solutions are saved in the fields. For example, continue the above session as follows.

```
>>S = solve(eqn3,eqn4)
S =
     x: [1x1 sym]
     y: [1x1 sym]
>>S.x
ans =
     1
>>S.y
ans =
     4
```

Test Your Understanding

T11.2–1 Use MATLAB to solve the equation $\sqrt{1-x^2} = x$. (Answer: $x = \sqrt{2}/2$)

T11.2–2 Use MATLAB to solve the equation set: $x + 6y = a$, $2x - 3y = 9$ in terms of the parameter a. (Answer: $x = (a + 18)/5$, $y = (2a - 9)/15$)

EXAMPLE 11.2–1	Intersection of Two Circles

We want to find the intersection points of two circles. The first circle has a radius of 2 and is centered at $x = 3$, $y = 5$. The second circle has a radius b and is centered at $x = 5$, $y = 3$. See Figure 11.2–1.

(a) Find the (x, y) coordinates of the intersection points in terms of the parameter b.

(b) Evaluate the solution for the case where $b = \sqrt{3}$.

■ **Solution**

(a) The intersection points are found from the solutions of the two equations for the circles. These equations are

$$(x - 3)^2 + (y - 5)^2 = 4$$

for the first circle, and

$$(x - 5)^2 + (y - 3)^2 = b^2$$

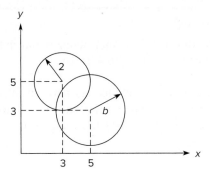

Figure 11.2–1 Intersection points of two circles.

The session to solve these equations is the following. Note that the result `x: [2x1 sym]` indicates that there are two solutions for x. Similarly, there are two solutions for y.

```
>>syms x y b
>>S = solve((x-3)^2+(y-5)^2-4,(x-5)^2+(y-3)^2-b^2)
ans =
S
    x: [2x1 sym]
    y: [2x1 sym]
>>simplify(S.x)
ans =
    9/2-b^2/8+(-16+24*b^2-b^4)^(1/2)/8
    -(-16+24*b^2- b^4)^(1/2)/8-b^2/8+9/2
```

The solution for the x coordinates of the intersection points is

$$x = \frac{9}{2} - \frac{1}{8}b^2 \pm \frac{1}{8}\sqrt{-16 + 24b^2 - b^4}$$

The solution for the y coordinates can be found in a similar way by typing `S.y`.

(b) Continue the above session by substituting $b = \sqrt{3}$ into the expression for x.

```
>>subs(S.x,b,sqrt(3));
>>simplify(ans)
ans =
    33/8-47^(1/2)/8
    47^(1/2)/8 +33/8
>>double(ans)
ans =
    3.2680
    4.9820
```

Thus the x coordinates of the two intersection points are $x = 4.982$ and $x = 3.268$. The y coordinates can be found in a similar way.

Test Your Understanding

T11.2–3 Find the y coordinates of the intersection points in Example 11.2–1. Use $b = \sqrt{3}$. (Answer: $y = 4.7320, 3.0180$)

Equations containing periodic functions can have an infinite number of solutions. In such cases the `solve` function restricts the solution search to solutions near zero. For example, to solve the equation $\sin 2x - \cos x = 0$, the session is

```
>>solve(sin(2*x)-cos(x)==0)
ans =
    pi/2
    pi/6
```

Note that $x = -\pi/2$ and $x = 5\pi/6$ are also solutions.

EXAMPLE 11.2–2   Positioning a Robot Arm

Figure 11.2–2 shows a robot arm having two joints and two links. The angles of rotation of the motors at the joints are θ_1 and θ_2. From trigonometry we can derive the following expressions for the (x, y) coordinates of the hand.

$$y = L_1 \sin\theta_1 + L_2 \sin(\theta_1 + \theta_2)$$

$$x = L_1 \cos\theta_1 + L_2 \cos(\theta_1 + \theta_2)$$

Suppose that the link lengths are $L_1 = 4$ ft and $L_2 = 3$ ft.

(a) Compute the motor angles required to position the hand at $x = 6$ ft, $y = 2$ ft.

(b) It is desired to move the hand along the straight line where x is constant at 6 ft, and y varies from $y = 0.1$ to $y = 3.6$ ft. Obtain a plot of the required motor angles as a function of y.

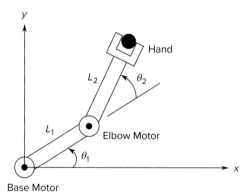

Figure 11.2–2 A robot arm having two joints and two links.

■ Solution

(a) Substituting the given values of L_1, L_2, x, and y into the above equations gives

$$6 = 4 \cos \theta_1 + 3 \cos(\theta_1 + \theta_2)$$
$$2 = 4 \sin \theta_1 + 3 \sin(\theta_1 + \theta_2)$$

The following session solves these equations. The variables `th1` and `th2` represent θ_1 and θ_2.

```
>> syms th1 th2
>>S = solve(4*cos(th1)+3*cos(th1+th2)==6,...
   4*sin(th1)+3*sin(th1+th2)==2)
S =
   th1:[2x1 sym]
   th2:[2x1 sym]
>>double(S.th1)*(180/pi) % convert to degrees.
ans =
   40.1680
   -3.2981
>>double(S.th2)*(180/pi) % convert to degrees.
ans =
   -51.3178
   51.3178
```

Thus there are two solutions. The first solution is $\theta_1 = 40.168°$, $\theta_2 = -51.3178°$. This is called the "elbow up" solution. The second is $\theta_1 = -3.2981°$, $\theta_2 = 51.3178°$. This is called the "elbow down" solution, which is illustrated in Figure 11.2–2. When a problem can be solved numerically, as in this case, the `solve` function will not perform a symbolic solution. In part (b), however, the symbolic solution capabilities of the `solve` function are put to use.

(b) First we find the solutions for the motor angles in terms of the variable y. Then we evaluate the solution for numerical values of y, and plot the results. The script file is shown below. Note that because there are three symbolic variables in the problem, we must tell the `solve` function that we want to solve for θ_1 and θ_2.

```
syms y
S = solve(4*cos(th1)+3*cos(th1+th2)==6,...
   4*sin(th1)+3*sin(th1+th2)==y, th1,th2)
yr = 1:0.1:3.6;
th1r = subs(S.th1,y,yr);
th2r = subs(S.th2,y,yr);
th1r = (180/pi)*double(th1r);
th2r = (180/pi)*double(th2r);
subplot(2,1,1)
plot(yr,th1r,2, -3.2981,x,2,40.168,'o'),...
      xlabel('y (feet)'),ylabel('Theta1 (degrees)')
subplot(2,1,2)
```

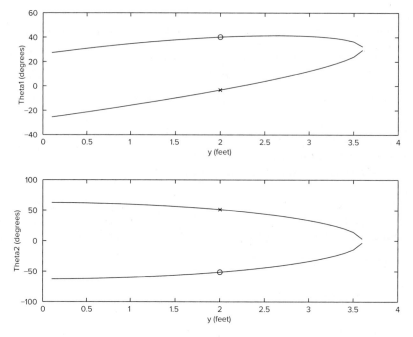

Figure 11.2–3 Plot of the motor angles for the robot hand moving along a vertical line.

```
plot(yr,th2r,2, -51.3178,'o',2,51.3178,'x'),...
    xlabel('y (feet)'),ylabel('Theta2 (degrees)')
```

The results are shown in Figure 11.2–3, where we have marked the solutions from part (a) to check the validity of the symbolic solutions. The elbow-up solutions are marked with an "o", and the elbow-down solutions are marked with an "x". We could have printed the expression for the solutions for θ_1 and θ_2 as functions of y, but the expressions are cumbersome and unnecessary if all we want is the plot.

MATLAB is powerful enough to solve the robot arm equations for arbitrary values of the hand coordinates (x, y). However, the resulting expressions for θ_1 and θ_2 are complicated.

Table 11.2–1 summarizes the `solve` function.

Table 11.2–1 Functions for Solving Algebraic and Transcendental Equations

Command	Description
`solve(E)`	Solves a symbolic expression or equation represented by the expression **E**. If **E** represents an *equation*, the equation's expression must contain the equality symbol (==). If **E** represents an *expression*, then the solution obtained will be the roots of the expression **E**; that is, the solution of the equation $E = 0$.
`solve(E1,...,En)`	Solves multiple expressions or equations.
`S = solve(E)`	Saves the solution in the structure S.

11.3 Calculus

In Chapter 9 we discussed techniques for performing numerical differentiation and numerical integration; this section covers differentiation and integration of symbolic expressions to obtain closed form results for the derivatives and integrals.

Differentiation

The `diff` function is used to obtain the symbolic derivative. Although this function has the same name as the function used to compute numerical differences (see Chapter 9), MATLAB detects whether or not a symbolic expression is used in the argument, and directs the calculation accordingly. The basic syntax is `diff(E)`, which returns the derivative of the expression E with respect to the default independent variable.

For example, the derivatives

$$\frac{dx^n}{dx} = nx^{n-1}$$

$$\frac{d \ln x}{dx} = \frac{1}{x}$$

$$\frac{d \sin^2 x}{dx} = 2 \sin x \cos x$$

$$\frac{d \sin y}{dy} = \cos y$$

$$\frac{d[\sin(xy)]}{dx} = y \cos(xy)$$

are obtained with the following session.

```
>>syms n x y
>>diff(x^n)
ans =
    n*x^(n-1)
>>diff(log(x))
ans =
    1/x
>>diff((sin(x))^2)
ans =
    2*cos(x)*sin(x)
>>diff(sin(y))
ans =
    cos(y)
```

If the expression contains more than one variable, the `diff` function operates on the variable `x` or the variable closest to x, unless told to do otherwise. When there is more than one variable, the `diff` function computes the *partial* derivative. For example, if

$$f(x, y) = \sin(xy)$$

then

$$\frac{\partial f}{\partial x} = y\, \cos(xy)$$

The corresponding session is

```
>>syms x y
>>diff(sin(x*y))
ans =
    y*cos(x*y)
```

There are three other forms of the `diff` function. The function `diff(E,v)` returns the derivative of the expression E with respect to the variable v. For example,

$$\frac{\partial[x\, \sin(xy)]}{\partial y} = x^2 \cos(xy)$$

is given by

```
>>syms x y
>>diff(x*sin(x*y),y)
ans =
    x^2*cos(x*y)
```

The function `diff(E,n)` returns the *n*th derivative of the expression E with respect to the default independent variable. For example,

$$\frac{d^2(x^3)}{dx^2} = 6x$$

is given by

```
>>syms x
>>diff(x^3,2)
ans =
    6*x
```

The function `diff(E,v,n)` returns the *n*th derivative of the expression E with respect to the variable v. For example,

$$\frac{\partial^2[x\, \sin(xy)]}{\partial y^2} = -x^3\, \sin(xy)$$

is given by

```
>>syms x y
>>diff(x*sin(x*y),y,2)
ans =
    -x^3*sin(x*y)
```

Table 11.3–1 Symbolic Calculus Functions

Command	Description
diff(E)	Returns the derivative of the expression **E** with respect to the default independent variable.
diff(E,v)	Returns the derivative of the expression **E** with respect to the variable **v**.
diff(E,n)	Returns the *n*th derivative of the expression **E** with respect to the default independent variable.
diff(E,v,n)	Returns the *n*th derivative of the expression **E** with respect to the variable **v**.
int(E)	Returns the integral of the expression **E** with respect to the default independent variable.
int(E,v)	Returns the integral of the expression **E** with respect to the variable **v**.
int(E,a,b)	Returns the integral of the expression **E** with respect to the default independent variable over the interval [*a*, *b*], where *a* and *b* are numeric quantities.
int(E,v,a,b)	Returns the integral of the expression **E** with respect to the variable **v** over the interval [*a*, *b*], where *a* and *b* are numeric quantities.
int(E,m,n)	Returns the integral of the expression **E** with respect to the default independent variable over the interval [*m*, *n*], where *m* and *n* are symbolic expressions.
limit(E)	Returns the limit of the expression **E** as the default independent variable goes to 0.
limit(E,a)	Returns the limit of the expression **E** as the default independent variable goes to **a**.
limit(E,v,a)	Returns the limit of the expression **E** as the variable **v** goes to **a**.
limit(E,v,a,'d')	Returns the limit of the expression **E** as the variable **v** goes to **a** from the direction specified by **d**, which may be `right` or `left`.
symsum(E)	Returns the symbolic summation of the expression **E**.
taylor(f,x,a)	Gives the fifth-order Taylor series for the function defined in the expression **f**, evaluated at the point $x = a$. If the parameter **a** is omitted the function returns the series evaluated at $x = 0$.

Table 11.3–1 summarizes the differentiation functions.

Max-Min Problems

The derivative can be used to find the maximum or minimum of a continuous function, say $f(x)$, over an interval $a \leq x \leq b$. A *local* maximum or local minimum (one that does not occur at one of the boundaries $x = a$ or $x = b$) can occur only at a *critical point*, which is a point where either $df/dx = 0$ or df/dx does not exist. If $d^2f/dx^2 > 0$, the point is a relative minimum; if $d^2f/dx^2 < 0$, the point is a relative maximum. If $d^2f/dx^2 = 0$, the point is a neither a minimum nor a maximum, but is an *inflection point*. If multiple candidates exist, you must evaluate the function at each point to determine the *global* maximum and global minimum.

EXAMPLE 11.3–1

Topping the Green Monster

The "Green Monster" is a wall 37 ft high in left field at Fenway Park in Boston. The wall is 310 ft from home plate down the left field line. Assuming that the batter hits the ball 4 ft above the ground, and neglecting air resistance, determine the *minimum* speed the batter must give to the ball in order to hit it over the Green Monster. In addition, find the angle at which the ball must be hit. (See Figure 11.3–1.)

■ Solution

The equations of motion for a projectile launched with a speed v_0 at an angle θ relative to the horizontal are:

$$x(t) = (v_0\cos\theta)t \qquad y(t) = -\frac{gt^2}{2} + (v_0\sin\theta)t$$

where $x = 0$, $y = 0$ is the location of the ball when it is hit. Because we are not concerned with the time of flight in this problem, we can eliminate t and obtain an equation for y in terms of x. To do this, easily solve the x equation for t and substitute this into the y equation to obtain

$$y(t) = -\frac{g}{2}\frac{x^2(t)}{v_0^2\cos^2\theta} + x(t)\tan\theta$$

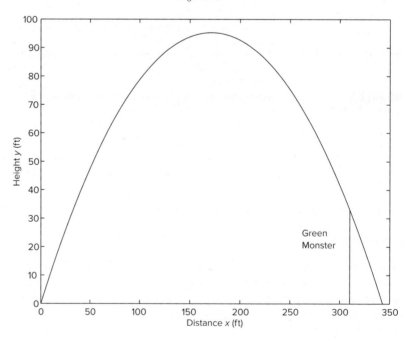

Figure 11.3–1 A baseball trajectory to clear the Green Monster.

(You could use MATLAB to do this simple algebra if you wish. We will use MATLAB to do the more difficult task to follow.)

Because the ball is hit 4 feet above the ground, the ball must rise $37 - 4 = 33$ ft to clear the wall. Let h represent the relative height of the wall (33 ft). Let d represent the distance to the wall (310 ft). Use $g = 32.2$ ft/sec^2. When $x = d$, $y = h$. Thus the previous equation gives

$$h = -\frac{g}{2}\frac{d^2}{v_0^2 \cos^2\theta} + d \tan\theta$$

which can easily be solved for v_0^2 as follows.

$$v_0^2 = \frac{g}{2}\frac{d^2}{\cos^2\theta(d\tan\theta - h)}$$

Because $v_0 > 0$, minimizing v_0^2 is equivalent to minimizing v_0. Note also that $gd^2/2$ is a multiplicative factor in the expression for v_0^2. Thus the minimizing value of θ is independent of g, and can be found by minimizing the function

$$f = \frac{1}{\cos^2\theta(d\tan\theta - h)}$$

The session to do this is as follows. The variable th represents the angle θ of the ball's velocity vector relative to the horizontal. The first step is to calculate the derivative $df/d\theta$, and solve the equation $df/d\theta = 0$ for θ.

```
>>syms d g h th
>>f = (1/(((cos(th))^2)*(d*tan(th)-h)));
>>dfdth = diff(f,th);
>>thmin = solve(dfdth,th);
>>thmin = double(subs(thmin,{d,h},{310,33}))
thmin =
    -0.7324
     2.4092
    -2.3032
     0.8384
```

Obviously, the solution must lie between 0 and $\pi/2$ radians, so the only solution candidate is $\theta = 0.8384$ radians, or about 48°. To verify that this is a minimum solution, and not a maximum or an inflection point, we can check the second derivative $d^2f/d\theta^2$. If this derivative is positive, the solution represents a minimum. To check this and to find the speed required, continue the session as follows.

```
>>second = diff(f,2,th); % Second derivative.
>>second = double(subs(second,{th,d,h},...
    {thmin(4),310,33}))
second =
    0.0321
>>v2 = (g*d^2/2)*f;
```

```
>>v2min = subs(v2,{d,h,g},{310,33,32.2});
>>vmin =sqrt(v2min);
>>vmin = double(subs(vmin(1),{th,d,h,g},...
    {thmin(4),310,33,32.2}))
vmin =
    105.3613
```

Because the second derivative (second) is positive, the solution is a minimum. Thus the minimum speed (vmin) required is 105.3613 ft/sec, or about 72 mph. A ball hit with this speed will clear the wall only if it is hit at an angle of approximately 48°.

Test Your Understanding

T11.3–1   Given that $y = \sinh(3x) \cosh(5x)$, use MATLAB to find dy/dx at $x = 0.2$.

(Answer: 9.2288)

T11.3–2   Given that $z = 5 \cos(2x) \ln(4y)$, use MATLAB to find dz/dy.

(Answer: $5 \cos(2x)/y$)

Integration

The int(E) function is used to integrate a symbolic expression E. It attempts to find the symbolic expression I such that diff(E) = I. It is possible that the integral does not exist in closed form, or that MATLAB cannot find the integral even if it exists. In such cases, the function will return the expression unevaluated.

The function int(E) returns the integral of the expression E with respect to the default independent variable. For example, you can obtain the following integrals with the session shown below.

$$\int x^n dx = \frac{x^{n+1}}{n+1} \quad \text{if } n \neq -1$$
$$\int \frac{1}{x} dx = \ln x$$
$$\int \cos x \, dx = \sin x$$
$$\int \sin y \, dy = -\cos y$$

```
>>syms n x y
int(x^n)
ans =
    piecewise(n == -1, log(x), n ~= -1, x^(n+1)/(n+1))
>>int(1/x)
```

```
ans =
    log(x)
>>int(cos(x))
ans =
    sin(x)
>>int(sin(y))
ans =
    -cos(y)
```

The form `int(E,v)` returns the integral of the expression E with respect to the variable v. For example, the result

$$\int x^n dn = \frac{x^n}{\ln x}$$

can be obtained with the session

```
>>syms n x
>>int(x^n,n)
ans =
    x^n/log(x)
```

The form `int(E,a,b)` returns the integral of the expression E with respect to the default independent variable evaluated over the interval [a, b], where a and b are numeric expressions. For example, the result

$$\int_2^5 x^2 dx = \frac{x^3}{3}\bigg|_2^5 = 39$$

is obtained as follows.

```
>>syms x
>>int(x^2,2,5)
ans =
    39
```

The form `int(E,v,a,b)` returns the integral of the expression E with respect to the variable v evaluated over the interval [a, b], where a and b are numeric quantities. For example, the result

$$\int_0^5 xy^2 dy = x\frac{y^3}{3}\bigg|_0^5 = \frac{125}{3}x$$

is obtained from

```
>>syms x y
>>int(xy^2,y,0,5)
ans =
    (125*x)/3
```

The result

$$\int_a^b x^2 \, dx = \frac{b^3}{3} - \frac{a^3}{3}$$

is obtained from

```
>>syms a b x
>>int(x^2,a,b)
ans =
    b^3/3-a^3/3
```

The form `int(E,m,n)` returns the integral of the expression `E` with respect to the default independent variable evaluated over the interval $[m, n]$, where `m` and `n` are symbolic expressions. For example,

$$\int_1^t x \, dx = \left. \frac{x^2}{2} \right|_1^t = \frac{1}{2}t^2 - \frac{1}{2}$$

$$\int_t^{e^t} \sin x \, dx = -\cos x \Big|_t^{e^t} = -\cos(e^t) + \cos t$$

are given by the session:

```
>>syms t x
>>int(x,1,t)
ans =
    t^2/2-1/2
int(sin(x),t,exp(t))
ans =
    cos(t)-cos(exp(t))
```

The following session gives an example for which no integral can be found. The indefinite integral exists, but the definite integral does not exist if the limits of integration include the singularity at $x = 1$. The integral is

$$\int \frac{1}{x-1} \, dx = \ln|x - 1|$$

The session is

```
>>syms x
>>int(1/(x-1))
ans =
    log(x-1)
>>int(1/(x-1),0,2)
    NaN
```

The `NaN` result ("not a number") indicates that a solution could not be found (because it would involve the undefined function $\ln(-1)$).

Table 11.3–1 summarizes the integration functions.

Test Your Understanding

T11.3–3   Given that $y = x \sin(3x)$, use MATLAB to find $\int y \, dx$.

(Answer: $(\sin(3x) - 3x \cos(3x))/9$)

T11.3–4   Given that $z = 6y^2 \tan(8x)$, use MATLAB to find $\int z \, dy$.

(Answer: $2y^3 \tan(8x)$)

T11.3–5   Use MATLAB to evaluate

$$\int_{-2}^{5} x \sin(3x) \, dx$$

(Answer: 0.6672)

Taylor Series

Taylor's theorem states that a function $f(x)$ can be represented in the vicinity of $x = a$ by the expansion

$$f(x) = f(a) + \left(\frac{df}{dx} \right)\Big|_{x=a} (x - a) + \frac{1}{2} \left(\frac{d^2 f}{dx^2} \right)\Big|_{x=a} (x - a)^2 + \cdots \qquad (11.3\text{–}1)$$

$$+ \frac{1}{k!} \left(\frac{d^k f}{dx^k} \right)\Big|_{x=a} (x - a)^k + \cdots + R_n$$

The term R_n is the remainder and is given by

$$R_n = \frac{1}{n!} \left(\frac{d^n f}{dx^n} \right)\Big|_{x=b} (x - a)^n \qquad (11.3\text{–}2)$$

where b lies between a and x.

These results hold if $f(x)$ has continuous derivatives through order n. If R_n approaches zero for large n, the expansion is called the *Taylor* series for $f(x)$ about $x = a$. If $a = 0$, the series is sometimes called the *Maclaurin* series.

Some common examples of the Taylor series are

$$\sin x = x - \frac{x^3}{3!} + \frac{x^5}{5!} - \frac{x^7}{7!} + \cdots, \quad -\infty < x < \infty$$

$$\cos x = 1 - \frac{x^2}{2!} + \frac{x^4}{4!} - \frac{x^6}{6!} + \cdots, \quad -\infty < x < \infty$$

$$e^x = 1 + x + \frac{x^2}{2!} + \frac{x^3}{3!} + \frac{x^4}{4!} + \cdots, \quad -\infty < x < \infty$$

where $a = 0$ in all three examples.

The function `taylor(f,x)` gives the fifth-order Taylor series approximation of `f` about $x = 0$, whereas `taylor(f,x,a)` gives the fifth-order Taylor series approximation of f about $x = a$. To compute the Taylor approximation of order $n - 1$ about $x = 0$, use the function `taylor(f,x,'order',n)`. Here are some examples.

```
>>syms x
>>f = exp(x);
>>taylor(f,x);
ans =
      x^5/120 + x^4/24 + x^3/6 +x^2/2 + x + 1
```

The answer is

$$1 + x + \frac{x^2}{2} + \frac{x^3}{6} + \frac{x^4}{24} + \frac{x^5}{120}$$

Continuing this session we have

```
>>simplify(taylor(f,x,2))
ans =
(exp(2)*(x^5 - 5*x^4 + 20*x^3 - 20*x^2 + 40*x + 8))/120
```

   This expression corresponds to

$$\frac{e^2}{120}\left(x^5 - 5x^4 + 20x^3 - 20x^2 + 40x + 8\right) \qquad (11.3-3)$$

The Live Editor is useful for obtaining the standard mathematical expressions from the code. For example, the Live Editor operations producing Equation (11.3–3) are shown in Figure 11.3–2.

Figure 11.3–2 The Live Editor displaying the operations used to obtain Equation (11.3–3).
Source: MATLAB

Sums

The `symsum(E)` function returns the symbolic summation of the expression `E`; that is

$$\sum_{x=0}^{x-1} E(x) = E(0) + E(1) + E(2) + \cdots + E(x-1)$$

The `symsum(E,a,b)` function returns the sum of the expression `E` as the default symbolic variable varies from `a` to `b`. That is, if the symbolic variable is `x`, then `S = symsum(E,a,b)` returns

$$\sum_{x=a}^{b} E(x) = E(a) + E(a+1) + E(a+2) + \cdots + E(b)$$

Here are some examples. The summations

$$\sum_{k=0}^{10} k = 0 + 1 + 2 + 3 + \cdots + 9 + 10 = 55$$

$$\sum_{k=0}^{n-1} k = 0 + 1 + 2 + 3 + \cdots + n - 1 = \frac{1}{2}n^2 - \frac{1}{2}n$$

$$\sum_{k=1}^{4} k^2 = 1 + 4 + 9 + 16 = 30$$

are given by

```
>>syms k n
>>symsum(k,0,10)
ans =
    55
>>symsum(k^2, 1, 4)
ans =
    30
>>symsum(k,0,n-1)
ans =
    n*(n-1)/2
```

The latter expression is the standard form of the result.

Limits

The function `limit(E,a)` returns the limit

$$\lim_{x \to a} E(x)$$

if `x` is the symbolic variable. There are several variations of this syntax. The basic form, `limit(E)`, finds the limit as $x \to 0$. For example

$$\lim_{x \to 0} \frac{\sin(ax)}{x} = a$$

is given by

```
>>syms a x
>>limit(sin(a*x)/x)
ans =
    a
```

The form `limit(E,v,a)` finds the limit as $v \to a$. For example,

$$\lim_{x \to 3} \frac{x-3}{x^2-9} = \frac{1}{6}$$

$$\lim_{x \to 0} \frac{\sin(x+h) - \sin(x)}{h}$$

are given by

```
>>syms h x
>>limit((x-3)/(x^2-9),3)
ans =
    1/6
>>limit((sin(x+h)-sin(x))/h,h,0)
ans =
    cos(x)
```

The forms `limit(E,v,a,'right')` and `limit(E,v,a,'left')` specify the direction of the limit. For example,

$$\lim_{x \to 0-} \frac{1}{x} = -\infty$$

$$\lim_{x \to 0+} \frac{1}{x} = \infty$$

are given by

```
>>syms x
>>limit(1/x,x,0,'left')
ans =
    -inf
>>limit(1/x,x,0,'right')
ans =
    inf
```

Table 11.3–1 summarizes the series and limit functions.

Test Your Understanding

T11.3–6   Use MATLAB to find the first three nonzero terms in the Taylor series for cos x.

(Answer: $1 - x^2/2 + x^4/24$)

T11.3–7   Use MATLAB to find a formula for the sum

$$\sum_{m=0}^{m-1} m^3$$

(Answer: $m^4/4 - m^3/2 + m^2/4$)

T11.3–8   Use MATLAB to evaluate

$$\sum_{n=0}^{7} \cos(\pi n)$$

(Answer: 0)

T11.3–9   Use MATLAB to evaluate

$$\lim_{x \to 5} \frac{2x - 10}{x^3 - 125}$$

(Answer: 2/75)

11.4 Differential Equations

A first-order ordinary differential equation (ode) can be written in the following form:

$$\frac{dy}{dt} = f(t, y)$$

where t is the independent variable and y is a function of t. A solution to such an equation is a function $y = g(t)$ such that $dg/dt = f(t,g)$, and the solution will contain one arbitrary constant. This constant becomes determined when we apply an additional condition of the solution by requiring that the solution have a specified value $y(t_1)$ when $t = t_1$. The chosen value t_1 is often the smallest, or starting value, of t, and if so, the condition is called the *initial condition* (quite often $t_1 = 0$). The general term for such a requirement is a *boundary condition*, and MATLAB lets us specify conditions other than initial conditions. For example, we can specify the value of the dependent variable at $t = t_2$, where $t_2 > t_1$.

INITIAL CONDITION

BOUNDARY CONDITION

Methods for obtaining a numerical solution to differential equations were covered in Chapter 9. However, we prefer to obtain an analytical solution whenever possible, because it is more general, and thus more useful for designing engineering devices or processes.

A second-order ode has the form

$$\frac{d^2 y}{dt^2} = f\left(t, y, \frac{dy}{dt}\right)$$

Its solution will have two arbitrary constants that can be determined once two additional conditions are specified. These conditions are often the specified values of y and dy/dt at $t = 0$. The generalization to third order and higher equations is straightforward.

We will occasionally use the following abbreviations for the first and second-order derivatives.

$$\dot{y} = \frac{dy}{dt} \qquad \ddot{y} = \frac{d^2 y}{dt^2}$$

MATLAB provides the dsolve function for solving ordinary differential equations. Its various forms differ according to whether they are used to solve single equations or sets of equations, whether or not boundary conditions are specified, and whether or not the default independent variable t is acceptable. Note that t is the default independent variable, and not x as with the other symbolic functions. This is because many ode models of engineering applications have time t as the independent variable.

Solving a Single Differential Equation

The dsolve function's syntax for solving a single equation is dsolve('eqn'). The function returns a symbolic solution of the ode specified by the symbolic expression eqn. Use the uppercase letter D to represent the first derivative; use D2 to represent the second derivative, and so on. Any character immediately following the differentiation operator is taken to be the dependent variable. Thus Dw represents dw/dt. Because of this syntax, you cannot use uppercase D as a symbolic variable when using the dsolve function.

The arbitrary constants in the solution are denoted by C1, C2, and so on. The number of such constants is the same as the order of the ode. For example, the equation

$$\frac{dy}{dt} + 2y = 12$$

has the solution

$$y(t) = 6 + C_1 e^{-2t}$$

This solution can be found with the following session.

```
>>syms y(t)
>>dsolve(diff(y,t)+2*y==12)
ans =
    C1*exp(-2*t)+6
```

There can be symbolic constants in the equation. For example,

$$\frac{dy}{dt} = \sin(at)$$

has the solution

$$y(t) = -\frac{\cos(at)}{a} + C_1$$

It can be found as follows.

```
>>syms y(t) a
>>dsolve(diff(y,t)==sin(a*t))
ans =
    C1-cos(a*t)/a
```

Here is a second-order example.

$$\frac{d^2y}{dt^2} = c^2y$$

The solution $y(t) = C_1e^{ct} + C_2e^{-ct}$ can be found with the session:

```
>>syms y(t) c
>>dsolve(diff(y,t,2)==c^2*y)
ans =
    C1*exp(-c*t)+C2*exp(c*t)
```

Solving Sets of Equations

Sets of equations can be solved with dsolve. The appropriate syntax is dsolve(eqn1,eqn2,...). The function returns a symbolic solution of the set of equations specified by the symbolic expressions eqn1 and eqn2.

For example, the set

$$\frac{dx}{dt} = 3x + 4y$$

$$\frac{dy}{dt} = -4x + 3y$$

has the solution

$$x(t) = C_1e^{3t}\cos 4t + C_2e^{3t}\sin 4t, \; y(t) = -C_1e^{3t}\sin 4t + C_2e^{3t}\cos 4t.$$

The session is

```
>>syms x(t) y(t)
>>eqn1 = diff(x,t)==3*x+4*y;
>>eqn2 = diff(y,t)==-4*x+3*y;
>>[x, y] = dsolve(eqn1,eqn2)
    x = C1*exp(3*t)*cos(4*t)+C2*exp(3*t)*sin(4*t)
    y = -C1*exp(3*t)*sin(4*t)+C2*exp(3*t)*cos(4*t)
```

Specifying Initial and Boundary Conditions

Conditions on the solutions at specified values of the independent variable are specified as the second argument in dsolve. The form dsolve(eqn,cond1, cond2,...) returns a symbolic solution of the ode specified by the symbolic expression eqn, subject to the conditions specified in the expressions cond1, cond2, and so on. If `y` is the dependent variable, let `Dy = diff(y,t)`, `D2y=diff(y,t,2)`, and so on. These conditions are specified as follows: `cond = [y(a) = b, Dy(a) = c, D2y(a) = d]`, and so on. These correspond to $y(a)$, $\dot{y}(a)$, $\ddot{y}(a)$, and so on. If the number of conditions is less than the order of the equation, the returned solution will contain arbitrary constants `C1`, `C2`, and so on.

For example, the problem

$$\frac{dy}{dt} = \sin bt, \ \ y(0) = 0$$

has the solution $y(t) = (1 - \cos bt)/b$. It can be found as follows.

```
>>syms y(t) b
>>cond = y(0)==0;
>>eqn = diff(y,t)==sin(b*t);
>>dsolve(eqn, cond)
ans =
     1/b-cos(b*t)/b
```

The problem

$$\frac{d^2 y}{dt^2} = c^2 y, \ \ y(0) = 1, \ \dot{y}(0) = 0$$

has the solution $y(t) = (e^{ct} + e^{-ct})/2$. The session is

```
>>syms y(t) c
>>eqn = diff(y,t,2)==c^2*y;
>>Dy = diff(y,t);
>>cond = [y(0)==1, Dy(0)==0];
>>dsolve(eqn,cond)
ans =
     1/2*exp(c*t)+1/2*exp(-c*t)
```

Arbitrary boundary conditions, such as $y(0) = c$, can be used. For example, the solution of the problem

$$\frac{dy}{dt} + ay = b, \ \ y(0) = c$$

is

$$y(t) = \frac{b}{a} + \left(c - \frac{b}{a}\right)e^{-at}$$

The session is

```
>>syms y(t) a b c
>>eqn = diff(y,t)+a*y==b;
>>cond = y(0)==c;
>>dsolve(eqn, cond)
ans =
     (b-exp(-a*t)*(b-a*c))/a
```

Plotting the Solution

The `fplot` function can be used to plot the solution, just as with any other symbolic expression, provided no undetermined constants such as `C1` are present. For example, the problem

$$\frac{dy}{dt} + 10y = 10 + 4\ \sin 4t, \quad y(0) = 0$$

has the solution

$$y(t) = 1 - \frac{4}{29}\cos 4t + \frac{10}{29}\sin 4t - \frac{25}{29}e^{-10t}$$

The session is

```
>>syms y(t)
>>Dy = diff(y,t);
>>eqn = Dy+10*y==10+4*sin(4*t);
>>cond = y(0)==0;
>>y = dsolve(eqn, cond);
>>fplot(y),axis([0 5 0 2]),xlabel('t')
```

You can type this code either in the Command window or in the Live Editor to produce the plot shown in Figure 11.4–1.

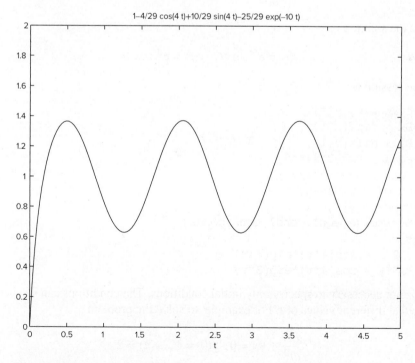

Figure 11.4–1 Plot of the solution of $\dot{y} + 10y = 10 + 4\sin 4t$, $y(0) = 0$.

Sometimes the fplot function uses too few values of the independent variable and thus does not produce a smooth plot. Instead, you can use the subs function to substitute an array of values for the independent variable, and then use the plot function to evaluate the result numerically. For example, you can continue the previous session as follows.

```
>>syms t
>>x = [0:0.05:5];
>>P = subs(y,t,x);
>>plot(x,P),axis([0 5 0 2]),xlabel('t')
```

Equation Sets with Boundary Conditions

Sets of equations with specified boundary conditions can be solved as follows. The function dsolve(eqn1,eqn2, ... ,cond1,cond2, ...) returns a symbolic solution of a set of equations specified by the symbolic expressions eqn1, eqn2, and so on, subject to the initial conditions specified in the expressions cond1, cond2, and so on.

For example, the problem

$$\frac{dx}{dt} = 3x + 4y, \quad x(0) = 0$$

$$\frac{dy}{dt} = -4x + 3y, \quad y(0) = 1$$

has the solution

$$x(t) = e^{3t}\sin 4t, \quad y(t) = e^{3t}\cos 4t$$

The session is

```
>>syms x(t) y(t)
>>Dx = diff(x,t);
>>Dy = diff(y,t);
>>eqn1 = Dx==3*x+4*y;
>>eqn2 = Dy==-4*x+3*y;
>>cond1 = x(0)==0;
>>cond2 = y(0)==1;
>>S = solve(eqn1,cond1,eqn2,cond2)
ans =
     S.x = sin(4*t)*exp(3*t)
     S.y = cos(4*t)*exp(3*t)
```

It is not necessary to specify only initial conditions. The conditions can be specified at different values of t. For example, to solve the problem

$$\frac{d^2y}{dt^2} + 9y = 0, \quad y(0) = 1, \quad \dot{y}(\pi) = 2$$

the session is

```
>>syms y(t)
>>Dy = diff(y,t);
>>D2y = diff(Dy,t);
>>cond1 = y(0)==1;
>>cond2 = Dy(pi)==2;
>>dsolve(eqn,cond1,cond2)
ans =
    cos(3*t)-2*sin(3*t)/3
```

So the solution is

$$y = \cos 3t - \frac{2}{3}\sin 3t$$

Using Other Independent Variables

Although the default independent variable is t, you can use a different independent variable. For example, the solution of the equation

$$\frac{dv}{dx} + 2v = 12$$

is $v(x) = 6 + C_1 e^{-2x}$. The session is

```
>>syms v(x) x
>>Dv = diff(v,x);
>>eqn = Dv+2*v==12;
>>dsolve(eqn)
ans =
    C1*exp(-2*x)+6
```

Test Your Understanding

T11.4–1   Use MATLAB to solve the equation

$$\frac{d^2y}{dt^2} + b^2y = 0$$

Check the answer by hand or with MATLAB.

(Answer: $y(t) = C_1 \sin bt + C_2 \cos bt$)

T11.4–2   Use MATLAB to solve the problem

$$\frac{d^2y}{dt^2} + b^2y = 0, \quad y(0) = 1, \quad \dot{y}(0) = 0$$

Check the answer by hand or with MATLAB.

(Answer: $y(t) = \cos bt$)

Solving Nonlinear Equations

MATLAB can solve many nonlinear first-order differential equations. For example, the problem

$$\frac{dy}{dt} = 4 + y^2 \quad y(0) = 1 \tag{11.4–1}$$

can be solved with the following session.

```
>>syms y(t)
>>Dy = diff(y,t);
>>eqn = Dy==4 + y^2;
>>cond = y(0)==1;
>>dsolve(eqn,cond)
ans =
     2*(tan(2*t+atan(1/2))
```

which is equivalent to

$$y(t) = 2 \, \tan(2t + \phi), \quad \phi = \tan^{-1}(1/2)$$

Not all nonlinear equations can be solved in closed form. For example, the following equation is the equation of motion of a specific pendulum.

$$\frac{d^2 y}{dt^2} + 9 \, \sin y = 0, \;\; y(0) = 1, \;\; \dot{y}(0) = 0$$

If you attempt to solve this with MATLAB you will get a message indicating that a solution could not be found. In fact, no such solution exists in terms of elementary functions. A tabulated (numerical) solution has been found and it is called the *elliptic integral*.

Table 11.4–1 summarizes the functions for solving differential equations.

Table 11.4–1 The dsolve Function

Command	Description
dsolve(eqn)	Returns a symbolic solution of the ode specified by the symbolic expression eqn. You can use the abbreviations syms y(t), Dy=diff(y,t), D2y=diff(y,t,2), and so on to represent the first and second derivatives, and so on.
dsolve(eqn1,eqn2, ...)	Returns a symbolic solution of the set of differential equations specified by the symbolic expressions eqn1 and eqn2.
dsolve(eqn,cond1, cond2, ...)	Returns a symbolic solution of the ode specified by the symbolic expression eqn, subject to the conditions specified in the expressions cond1, cond2, and so on. If y is the dependent variable, these conditions are specified as follows: y(a) = b, Dy(a) = c, D2y(a) = d, and so on.
dsolve(eqn1,eqn2, ..., cond1,cond2, ...)	Returns a symbolic solution of set of equations specified by the symbolic expressions eqn1, eqn2, and so on, subject to the initial conditions specified in the expressions cond1, cond2, and so on.

11.5 Laplace Transforms

This section shows how to use the *Laplace transform* with MATLAB. The Laplace transform can be used to solve some types of differential equations that cannot be solved with `dsolve`. Application of the Laplace transform converts a linear differential equation problem into an algebraic problem. With proper algebraic manipulation of the resulting quantities, the solution of the differential equation can be recovered in an orderly fashion by inverting the transformation process to obtain a function of time. We assume that you are familiar with the fundamentals of differential equations outlined in Chapter 9, Sections 9.3 and 9.4.

Laplace transform   The Laplace transform $\mathcal{L}[y(t)]$ of a function $y(t)$ is defined to be

$$\mathcal{L}[y(t)] = \int_0^\infty y(t)\,e^{-st}\,dt \qquad (11.5\text{--}1)$$

The integration removes t as a variable, and the transform is thus a function of only the Laplace variable s, which may be a complex number. The integral exists for most of the commonly encountered functions if suitable restrictions are placed on s. An alternative notation is the use of the uppercase symbol to represent the transform of the corresponding lowercase symbol; that is,

$$Y(s) = \mathcal{L}[y(t)]$$

Step function   We will use the *one-sided* transform, which assumes that the variable $y(t)$ is zero for $t < 0$. For example, the *step function* is such a function. Its name comes from the fact that its graph looks like a stair step (see Figure 11.5–1).

The height of the step is M, and is called the *magnitude*. The *unit-step function*, denoted $u_s(t)$, has a height of $M = 1$, and is defined as follows:

$$u_s(t) = \begin{cases} 0 & t < 0 \\ 1 & t > 0 \\ \text{indeterminate} & t = 0 \end{cases}$$

The engineering literature generally uses the term *step* function, whereas in the mathematical literature the name *Heaviside* function is used. The Symbolic

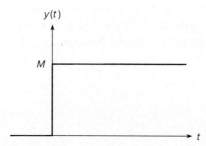

Figure 11.5–1 A step function of magnitude M.

Math toolbox includes the `heaviside(t)` function, which produces a unit-step function.

A *step function* of height M can be written as $y(t) = Mu_s(t)$. Its transform is

$$\mathscr{L}[y(t)] = \int_0^\infty M u_s(t) e^{-st} dt = M \int_o^\infty e^{-st} dt = M \left. \frac{e^{-st}}{s} \right|_0^\infty = \frac{M}{s}$$

where we have assumed that the real part of s is greater than zero, so that the limit of e^{-st} exists as $t \to \infty$. Similar considerations of the region of convergence of the integral apply for other functions of time. However, we need not concern ourselves with this here, because the transforms of all the common functions have been calculated and tabulated. They can be obtained in MATLAB with the Symbolic Math Toolbox by typing `laplace(function)`, where `function` is a symbolic expression representing the function $y(t)$ in Equation (11.5–1). The default independent variable is `t` and the default return is a function of `s`. The optional form is `syms x y, laplace(function,x,y)`, where `function` is a function of `x`, and `y` is the Laplace variable.

Here is a session with some examples. The functions are t^3, e^{-at}, and sin bt.

```
>>syms b t
>>laplace(t^3)
ans =
    6/s^4
>>laplace(exp(-b*t))
ans =
    1/(s+b)
>>laplace(sin(b*t))
ans =
    b/(s^2+b^2)
```

Because the transform is an integral, its has the properties of integrals. In particular, it has the *linearity property*, which states that if a and b are not functions of t, then

$$\mathscr{L}[af_1(t) + bf_2(t)] = a\mathscr{L}[f_1(t)] + b\mathscr{L}[f_2(t)] \tag{11.5–2}$$

The *inverse Laplace transform* $\mathscr{L}^{-1}[Y(s)]$ is that time function $y(t)$ whose transform is $Y(s)$; that is, $y(t) = \mathscr{L}^{-1}[Y(s)]$. The inverse operation is also linear. For example, the inverse transform of $10/s + 4/(s + 3)$ is $10 + 4e^{-3t}$. Inverse transforms can be found using the `ilaplace` function. For example,

```
>>syms b s
>>ilaplace(1/s^4)
ans =
    t^3/6
>>ilaplace(1/(s+b))
ans =
    exp(-b*t)
```

```
>>ilaplace(b/(s^2+b^2)
ans =
   sin(b*t)
```

The transforms of derivatives are useful for solving differential equations. Applying integration by parts to the definition of the transform, we obtain

$$\mathcal{L}\left(\frac{dy}{dt}\right) = \int_0^\infty \frac{dy}{dt}e^{-st}\,dt = y(t)\,e^{-st}\big|_0^\infty + s\int_0^\infty y(t)\,e^{-st}\,dt$$

$$= s\mathcal{L}[y(t)] - y(0) = sY(s) - y(0) \tag{11.5-3}$$

This procedure can be extended to higher derivatives. For example, the result for the second derivative is

$$\mathcal{L}\left(\frac{d^2y}{dt^2}\right) = s^2\,Y(s) - sy(0) - \dot{y}(0) \tag{11.5-4}$$

The general result for any order derivative is

$$\mathcal{L}\left(\frac{d^n y}{dt^n}\right) = s^n\,Y(s) - \sum_{k=1}^n s^{n-k}g_{k-1} \tag{11.5-5}$$

where

$$g_{k-1} = \frac{d^{k-1}y}{dt^{k-1}}\bigg|_{t=0} \tag{11.5-6}$$

Application to Differential Equations

The derivative and linearity properties can be used to solve the differential equation

$$a\dot{y} + y = bv(t) \tag{11.5-7}$$

If we multiply both sides of the equation by e^{-st} and then integrate over time from $t = 0$ to $t = \infty$, we obtain

$$\int_0^\infty (a\dot{y} + y)\,e^{-st}\,dt = \int_0^\infty bv(t)\,e^{-st}\,dt$$

or

$$\mathcal{L}(a\dot{y} + y) = \mathcal{L}[bv(t)]$$

or, using the linearity property,

$$a\mathcal{L}(\dot{y}) + \mathcal{L}(y) = b\mathcal{L}[v(t)]$$

Using the derivative property and the alternative transform notation, the above equation can be written as

$$a[sY(s) - y(0)] + Y(s) = bV(s)$$

where $V(s)$ is the transform of v. This equation is an algebraic equation for $Y(s)$ in terms of $V(s)$ and $y(0)$. Its solution is

$$Y(s) = \frac{ay(0)}{as + 1} + \frac{b}{as + 1} V(s) \tag{11.5–8}$$

Applying the inverse transform to Equation (11.5–8) gives

$$y(t) = \mathcal{L}^{-1}\left[\frac{ay(0)}{as + 1}\right] + \mathcal{L}^{-1}\left[\frac{b}{as + 1} V(s)\right] \tag{11.5–9}$$

From the transform given earlier, it can be seen that

$$\mathcal{L}^{-1}\left[\frac{ay(0)}{as + 1}\right] = \mathcal{L}^{-1}\left[\frac{y(0)}{s + 1/a}\right] = y(0)\, e^{-t/a}$$

which is the free response. The forced response is given by

$$\mathcal{L}^{-1}\left[\frac{b}{as + 1} V(s)\right] \tag{11.5–10}$$

This cannot be evaluated until $V(s)$ is specified. Suppose $v(t)$ is a unit-step function. Then $V(s) = 1/s$ and Equation (11.5–10) becomes

$$\mathcal{L}^{-1}\left[\frac{b}{s(as + 1)}\right]$$

To find the inverse transform, enter

```
>>syms a b s
>>ilaplace(b/(s*(a*s+1)))
ans =
    b*(1-exp(-t/a))
```

Thus the forced response of Equation (11.5–7) to a unit-step input is $b(1 - e^{-t/a})$.

You can use the `heaviside` function with the `dsolve` function to find the step response, but the resulting expressions are more complicated than those obtained with the Laplace transform method.

Consider the second-order model

$$\ddot{x} + 1.4\dot{x} + x = f(t) \tag{11.5–11}$$

Transforming this equation gives

$$[s^2 X(s) - sx(0) - \dot{x}(0)] + 1.4[sX(s) - x(0)] + X(s) = F(s)$$

Solve for $X(s)$.

$$X(s) = \frac{x(0)s + \dot{x}(0) + 1.4x(0)}{s^2 + 1.4s + 1} + \frac{F(s)}{s^2 + 1.4s + 1}$$

The free response is obtained from

$$x(t) = \mathcal{L}^{-1}\left[\frac{x(0)s + \dot{x}(0) + 1.4x(0)}{s^2 + 1.4s + 1}\right]$$

Suppose the initial conditions are $x(0) = 2$ and $\dot{x}(0) = -3$. Then the free response is obtained from

$$x(t) = \mathscr{L}^{-1}\left(\frac{2s - 0.2}{s^2 + 1.4s + 1}\right) \tag{11.5–12}$$

It can be found by typing

```
>>ilaplace((2*s-0.2)/(s^2+1.4*s+1))
```

The free response is

$$x(t) = e^{-0.7t}\left[2\cos\left(\frac{\sqrt{51}}{10}t\right) - \frac{16\sqrt{51}}{51}\sin\left(\frac{\sqrt{51}}{10}t\right)\right]$$

The forced response is obtained from

$$x(t) = \mathscr{L}^{-1}\left[\frac{F(s)}{s^2 + 1.4s + 1}\right]$$

If $f(t)$ is a unit-step function, $F(s) = 1/s$, and the forced response is

$$x(t) = \mathscr{L}^{-1}\left[\frac{1}{s(s^2 + 1.4s + 1)}\right]$$

To find the forced response, enter

```
>>ilaplace(1/(s*(s^2+1.4*s+1)))
```

The answer obtained is

$$x(t) = 1 - e^{-0.7t}\left[\cos\left(\frac{\sqrt{51}}{10}t\right) + \frac{7\sqrt{51}}{51}\sin\left(\frac{\sqrt{51}}{10}t\right)\right] \tag{11.5–13}$$

Input Derivatives

Two similar mechanical systems are shown in Figure 11.5–2. In both cases the input is a displacement $y(t)$. Their equations of motion are

$$m\ddot{x} + c\dot{x} + kx = ky \tag{11.5–14}$$

$$m\ddot{x} + c\dot{x} + kx = ky + c\dot{y} \tag{11.5–15}$$

The only difference between these systems is that the system in Figure 11.5–2a has an equation of motion containing the derivative of the input function $y(t)$. Both systems are examples of the more general differential equation

$$m\ddot{x} + c\dot{x} + kx = dy + g\dot{y} \tag{11.5–16}$$

As noted earlier, you can use the heaviside function with the dsolve function to find the step response of equations containing derivatives of the input, but the resulting expressions are more complicated than those obtained with the Laplace transform method.

Figure 11.5–2 Two mechanical systems. The model for (a) contains the derivative of the input $y(t)$; the model for (b) does not.

We now demonstrate how to use the Laplace transform to find the step response of equations containing derivatives of the input. Suppose the initial conditions are zero. Then transforming Equation (11.5–16) gives

$$X(s) = \frac{d + gs}{ms^2 + cs + k} Y(s) \tag{11.5–17}$$

Let us compare the unit-step response of Equation (11.5–16) for two cases using the values $m = 1$, $c = 1.4$, and $k = 1$, with zero initial conditions. The two cases are $g = 0$ and $g = 5$.

With $Y(s) = 1/s$, Equation (11.5–17) gives

$$X(s) = \frac{1 + gs}{s(s^2 + 1.4s + 1)} \tag{11.5–18}$$

The response for the case $g = 0$ was found earlier. It is given by Equation (11.5–13). The response for $g = 5$ is found by typing

```
>>syms s
>>ilaplace((1+5*s)/(s*(s^2+1.4*s+1)))
```

The response obtained is

$$x(t) = 1 - e^{0.7t}\left[\cos\left(\frac{\sqrt{51}}{10}t\right) + \frac{43\sqrt{51}}{51}\sin\left(\frac{\sqrt{51}}{10}t\right)\right] \tag{11.5–19}$$

Figure 11.5–3 shows the responses given by Equations (11.5–13) and (11.5–19). The effect of differentiating the input is an increase in the response's peak value.

Impulse Response

The area A under the curve of the *pulse function* shown in Figure 11.5–4a is called the *strength* of the pulse. If we let the pulse duration T approach zero, while keeping the area A constant, we obtain the *impulse* function of strength A, represented by Figure 11.5–4b. If the strength is 1, we have a *unit* impulse.

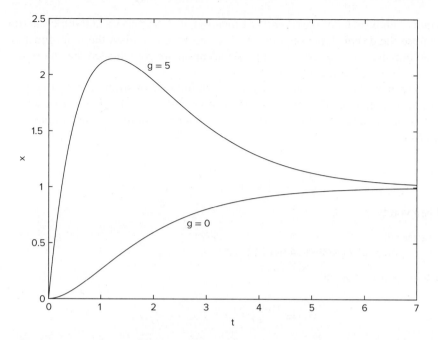

Figure 11.5–3 The step response of the model $\ddot{x} + 1.4\dot{x} + x = u + g\dot{u}$ for $g = 0$ and $g = 5$.

The impulse can be thought of as the derivative of the step function, and is a mathematical abstraction for convenience in analyzing the response of systems subjected to an input that is applied and removed suddenly, such as the force from a hammer blow.

The engineering literature generally uses the term *impulse* function, whereas in the mathematical literature the name *Dirac delta* function is used. The Symbolic

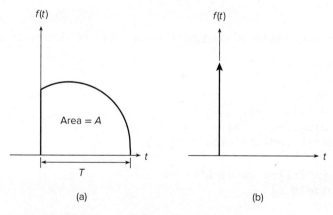

Figure 11.5–4 Pulse and impulse functions.

Math toolbox includes the `dirac(t)` function, which returns a unit impulse. You can use the `dirac` function with the `dsolve` function when the input function is an impulse, but the resulting expressions are more complicated than those obtained with the Laplace transform.

It can be shown that the transform of an impulse of strength A is simply A. So, for example, to find the impulse response of $\ddot{x} + 1.4\,\dot{x} + x = f(t)$, where $f(t)$ is an impulse of strength A, for zero inital conditions, first obtain the transform.

$$X(s) = \frac{1}{s^2 + 1.4s + 1} \quad F(s) = \frac{A}{s^2 + 1.4s + 1}$$

Then you type

```
>>syms A s
>>ilaplace(A/(s^2+1.4*s+1))
```

The response obtained is

$$x(t) = \frac{10A\sqrt{51}}{51} e^{-0.7t} \sin\left(\frac{\sqrt{51}}{10} t\right)$$

Direct Method

Instead of performing by hand the algebra required to find the response transform, we could use MATLAB to do the algebra for us. We now demonstrate the most direct way of using MATLAB to solve an equation with the Laplace transform. One advantage of this method is that we are not required to use the transforms for the derivatives. Let us solve the equation

$$a\frac{dy}{dt} + y = f(t) \tag{11.5–20}$$

with $f(t) = \sin t$, in terms of an unspecified value for $y(0)$. Here is the session.

```
>>syms a L s y(t)
>>Dy = diff(y,t);
>>F = sin(t);
>>eqn = a*Dy+y-F==0;
>>E = laplace(eqn,t,s);
>>E = subs(E,laplace(y(t),t,s),L)
E =
  L - 1/(s^2+1)-a*(y(0)-L*s)==0
>>L = solve(E,L)
L =
  (a*y(0)+1/(s^2+1))/(a*s+1)
```

```
>>I = ilaplace(L)
I =
(sin(t)-a*cos(t))/(a^2+1)+(exp(-t/a)*(y(0)*a^2+a+y(0)))/(a^2+1)
```

The answer is

$$y(t) = \frac{1}{a^2 + 1}\left\{ \sin t - a \cos t + e^{-t/a}\left[a^2 y(0) + a + y(0)\right]\right\}$$

Note that this session consists of the following steps:

1. Define the symbolic variables, including the derivatives that appear in the equation. Note that $y(t)$ is explicitly expressed as a function of t in these definitions.
2. Move all terms to the left side of the equation, and define the left side as a symbolic expression.
3. Apply the Laplace transformation to the differential equation to obtain an algebraic equation.
4. Substitute a symbolic variable, here L, for the expression `laplace(y(t),t,s)` in the algebraic equation. Then solve the equation for the variable L, which is the transform of the solution.
5. Invert L to find the solution as a function of t.

Note that this procedure can be used to solve sets of equations.

Test Your Understanding

T11.5–1 Find the Laplace transform of the following functions: $1 - e^{-at}$ and $\cos bt$. Use the `ilaplace` function to check your answers.

T11.5–2 Use the Laplace transform to solve the problem $5\ddot{y} + 20\dot{y} + 15y = 30u - 4\dot{u}$, where $u(t)$ is a unit-step function, and $y(0) = 5$, $\dot{y}(0) = 1$.

(Answer: $y(t) = -1.6e^{-3t} + 4.6e^{-t} + 2$)

Table 11.5–1 summarizes the Laplace transform functions.

Table 11.5–1 Laplace Transform Functions

Command	Description
`ilaplace(function)`	Returns the inverse Laplace transform of `function`
`laplace(function)`	Returns the Laplace transform of `function`.
`laplace(function,x,y)`	Returns the Laplace transform of `function`, which is a function of `x`, in terms of the Laplace variable `y`.

11.6 Symbolic Linear Algebra

You can perform operations with symbolic matrices in much the same way as with numeric matrices. Here we will give examples of finding matrix products, the matrix inverse, eigenvalues, and the characteristic polynomial of a matrix.

Remember that using symbolic matrices avoids numerical imprecision in subsequent operations. You can create a symbolic matrix from a numeric matrix in several ways, as shown in the following session.

```
>>A = sym([3, 5; 2, 7]);
>>syms a b c d
>>B = [a,b; c, d];
>>C = [3, 5; 2, 7];
>>D = sym(C);
```

The matrix A represents the most direct method. The matrix B can be used for further symbolic manipulation in terms of the variable a, b, c, and d. The matrix D can be used to preserve matrix C in symbolic form. The matrices A, B, and D are all symbolic. The matrix C looks like the A and D, but is numeric of class double.

You can create a symbolic matrix consisting of functions. For example, the relationship between the coordinates (x_2, y_2) of a coordinate system rotated counterclockwise through an angle a relative to the (x_1, y_1) coodinate system is

$$x_2 = x_1 \cos a + y_1 \sin a$$
$$y_2 = y_1 \cos a - x_1 \sin a$$

These equations can be expressed in matrix form as

$$\begin{bmatrix} x_2 \\ y_2 \end{bmatrix} = \begin{bmatrix} \cos a & \sin a \\ -\sin a & \cos a \end{bmatrix} \begin{bmatrix} x_1 \\ y_1 \end{bmatrix} = \mathbf{R} \begin{bmatrix} x_1 \\ y_1 \end{bmatrix}$$

where the rotation matrix $\mathbf{R}(a)$ is defined as

$$\mathbf{R}(a) = \begin{bmatrix} \cos a & \sin a \\ -\sin a & \cos a \end{bmatrix} \tag{11.6-1}$$

The symbolic matrix $\mathbf{R}$ can be defined in MATLAB as follows:

```
>>syms a
>>R = [cos(a), sin(a); -sin(a), cos(a)]
R =
   [ cos(a), sin(a) ]
   [ -sin(a), cos(a) ]
```

If we rotate the coordinate system twice by the same angle to produce a third coordinate system (x_3, y_3), the result is the same as a single rotation with twice the angle. Let us see if MATLAB gives that result. The vector matrix equation is

$$\begin{bmatrix} x_3 \\ y_3 \end{bmatrix} = \mathbf{R} \begin{bmatrix} x_2 \\ y_2 \end{bmatrix} = \mathbf{RR} \begin{bmatrix} x_1 \\ y_1 \end{bmatrix}$$

Thus, $\mathbf{R}(a)\,\mathbf{R}(a)$ should be the same as $\mathbf{R}(2a)$. Continue the previous session as follows.

```
>>Q = R*R
Q =
   [ cos(a)^2-sin(a)^2,  2*cos(a)*sin(a) ]
   [ -2*cos(a)*sin(a),  cos(a)^2-sin(a)^2 ]
>>Q = simplify(Q)
Q =
   [ cos(2*a),  sin(2*a) ]
   [ -sin(2*a), cos(2*a) ]
```

The matrix $\mathbf{Q}$ is the same as $\mathbf{R}(2a)$, as we suspected.

To evaluate a matrix numerically, use the `subs` and `double` functions. For example, for a rotation of $a = \pi/4$ radians (45°),

```
>>R = double(subs(R,a,pi/4))
R =
   0.7071 0.0701
   -0.7071 0.0701
```

Characteristic Polynomial and Roots

Sets of first order differential equations can be expressed in vector-matrix notation as

$$\dot{\mathbf{x}} = \mathbf{Ax} + \mathbf{Bf}(t)$$

where $\mathbf{x}$ is the vector of dependent variables and $\mathbf{f}(t)$ is a vector containing the forcing function. For example, the equation set

$$\dot{x}_1 = x_2$$
$$\dot{x}_2 = -kx_1 - 2x_2 + f(t)$$

comes from the equation of motion of a mass connected to a spring and sliding on a surface having viscous friction. The term $f(t)$ is the applied force acting on the mass. For this set, the vector $\mathbf{x}$ and the matrices $\mathbf{A}$ and $\mathbf{B}$ are

$$\mathbf{x} = \begin{bmatrix} x_1 \\ x_2 \end{bmatrix}$$

$$\mathbf{A} = \begin{bmatrix} 0 & 1 \\ -k & -2 \end{bmatrix} \quad \mathbf{B} = \begin{bmatrix} 0 \\ 1 \end{bmatrix}$$

The equation $|s\mathbf{I} - \mathbf{A}| = 0$ is the characteristic equation of the model, where s represents the characteristic roots of the model. The function `charpoly(A)` gives the coefficients if the characteristic polynomial in decreasing powers of the variable. For example,

```
>>syms k
>>A = [0 ,1;-k, -2];
```

```
>>charpoly(A)
ans =
   [ 1, 2, k]
```

which corresponds to the polynomial $x^2 + 2x + k$.

The statements syms s, charpoly(A,s) find the polynomial in terms of the symbolic variable s. For example, to find the characteristic equation and solve for the roots in terms of the spring constant k, use the following session.

```
>>syms k s
>>A = [0 ,1;-k, -2];
>>charpoly(A, s))
ans =
   s^2+2*s+k
>>solve(ans)
ans =
   -(1-k)^(1/2)-1
   (1-k)^(1/2)-1
```

Thus the polynomial is $s^2 + 2s + k$, and the roots are $s = -1 \pm \sqrt{1-k}$.

Use the eig(A) function to find the roots directly without finding the characteristic equation ("eig" stands for "eigenvalue," which is another term for "characteristic root"). For example,

```
>>syms k
>>A = [0 ,1;-k, -2];
>>eig(A)
ans =
   -(1-k)^(1/2)-1
   (1-k)^(1/2)-1
```

You can use the inv(A) and det(A) functions to invert and find the determinant of a matrix symbolically. For example, using the same matrix A from the previous session,

```
>>inv(A)
ans =
   [ -2/k, -1/k ]
   [ 1, 0 ]
>>A*ans    % Verify that the inverse is correct.
ans =
   [ 1, 0 ]
   [ 0, 1 ]
>>det(A)
ans =
   k
```

Solving Linear Algebraic Equations

You can use matrix methods in MATLAB to solve linear algebraic equations symbolically. You can use the matrix inverse method, if the inverse exists, or the left-division method (see Chapter 8 for a discussion of these methods). For example, to solve the set

$$2x - 3y = 3$$
$$5x + 4y = 19$$

using both methods, the session is

```
>>A = sym([2, -3; 5, 4]);
>>b = sym([3; 19]);
>>x = inv(A)*b    % The matrix inverse method.
x =
   3
   1
>>x = A\b    % The left-division method.
x =
   3
   1
```

Table 11.6–1 summarizes the functions used in this section. Note that their syntax is identical to the numeric versions used in earlier chapters.

Test Your Understanding

T11.6–1 Consider three successive coordinate rotations using the same angle a. Show that the product **RRR** of the rotation matrix **R**(a) given by Equation (11.6–1) equals **R**(3a).

T11.6–2 Find the characteristic polynomial and roots of the following matrix.

$$\mathbf{A} = \begin{bmatrix} -2 & 1 \\ -3k & -5 \end{bmatrix}$$

(Answers: $s^2 + 7s + 10 + 3k$ and $s = (-7 \pm \sqrt{9 - 12k})/2$)

Table 11.6–1 Linear Algebra Functions

Command	Description
det(A)	Returns the determinant of the matrix A in symbolic form.
eig(A)	Returns the eigenvalues (characteristic roots) of the matrix A in symbolic form.
inv(A)	Returns the inverse of the matrix A in symbolic form.
charpoly(A,s)	Returns the characteristic polynomial of the matrix A in symbolic form in terms of the variable s.

T11.6–3 Use the matrix inverse and the left-division method to solve the following set.

$$-4x + 6y = -2$$
$$7x - 4y = 23$$

(Answer: $x = 5, y = 3$)

11.7 Summary

This chapter covers a subset of the capabilities of the Symbolic Math toolbox, specifically

- symbolic algebra,
- symbolic methods for solving algebraic and transcendental equations,
- symbolic methods for solving ordinary differential equations,
- symbolic calculus, including integration, differentiation, limits, and series,
- Laplace transforms, and
- selected topics in linear algebra, including symbolic methods for obtaining determinants, matrix inverses, and eigenvalues.

Now that you have finished this chapter, you should be able to use MATLAB to

- create symbolic expressions and manipulate them algebraically,
- obtain symbolic solutions to algebraic and transcendental equations,
- perform symbolic differentiation and integration,
- evaluate limits and series symbolically,
- obtain symbolic solutions to ordinary differential equations,
- obtain Laplace transforms, and
- perform symbolic linear algebra operations, including obtaining expressions for determinants, matrix inverses, and eigenvalues.

Table 11.7–1 is a guide by category to the functions introduced in this chapter.

Table 11.7–1 Guide to MATLAB Commands Introduced in This Chapter

Creating and Evaluating Expressions	See Table 11.1–1
Manipulating Expressions	See Table 11.1–2
Solving Algebraic and Transcendental Equations	See Table 11.2–1
Symbolic Calculus Functions	See Table 11.3–1
Solving Differential Equations	See Table 11.4–1
Laplace Transforms	See Table 11.5–1
Linear Algebra	See Table 11.6–1

Miscellaneous Functions

IMPULSE FUNCTION	`dirac(t)`	Dirac delta function (unit *impulse function* at $t = 0$).
	`heaviside(t)`	Heaviside function (unit-step function making a transition from 0 to 1 at $t = 0$.)

Key Terms

Problems

Section 11.1

1. Use MATLAB to prove the following identities.
   (a) $\sin^2 x + \cos^2 x = 1$
   (b) $\sin(x + y) = \sin x \cos y + \cos x \sin y$
   (c) $\sin 2x = 2 \sin x \cos x$
   (d) $\cosh^2 x - \sinh^2 x = 1$

2. Use MATLAB to express $\cos 5\theta$ as a polynomial in x, where $x = \cos\theta$.

3.* Two polynomials in the variable x are represented by the coefficient vectors p1 = [6, 2, 7, -3] and p2 = [10, -5, 8].
   (a) Use MATLAB to find the product of these two polynomials, and express it in its simplest form.
   (b) Use MATLAB to find the numeric value of the product if $x = 2$.

4.* The equation of circle of radius r centered at $x = 0$, $y = 0$ is
$$x^2 + y^2 = r^2$$
Use the subs and other MATLAB functions to find the equation of a circle of radius r centered at the point $x = a$, $y = b$. Rearrange the equation into the form $Ax^2 + Bx + Cxy + Dy + Ey^2 = F$, and find the expressions for the coefficients in terms of a, b, and r.

5. The equation for a curve called the "lemniscate" in polar coordinates (r, θ) is
$$r^2 = a^2 \cos(2\theta)$$
Use MATLAB to find the equation for the curve in terms of Cartesian coordinates (x, y), where $x = r \cos\theta$ and $y = r \sin\theta$.

Section 11.2

6.* The law of cosines for a triangle states that $a^2 = b^2 + c^2 - 2bc \cos A$, where a is the length of the side opposite the angle A, and b and c are the lengths of the other sides.
   (a) Use MATLAB to solve for b.
   (b) Suppose that $A = 60°$, $a = 5$ m, and $c = 2$ m. Determine b.

7. (a) Use MATLAB to solve the polynomial equation $x^3 + 8x^2 + ax + 10 =$ for x in terms of the parameter a.

    (b) Evaluate your solution for the case $a = 17$. Use MATLAB to check the answer.

8.* The equation for an ellipse centered at the origin of the Cartesian coordinates (x, y) is

$$\frac{x^2}{a^2} + \frac{y^2}{b^2} = 1$$

where a and b are constants that determine the shape of the ellipse.

   (a) In terms of the parameter b, use MATLAB to find the points of intersection of the two ellipses described by

$$x^2 + \frac{y^2}{b^2} = 1$$

and

$$\frac{x^2}{100} + 4y^2 = 1$$

   (b) Evaluate the solution obtained in part (a) for the case: $b = 2$.

9. The equation

$$r = \frac{p}{1 - \epsilon \cos\theta}$$

describes the polar coordinates of an orbit with the coordinate origin at the sun. If $\epsilon = 0$, the orbit is circular; if $0 < \epsilon < 1$, the orbit is elliptical. The planets have orbits that are nearly circular; comets have orbits that are highly elongated with ϵ nearer to 1. It is of obvious interest to see whether or not a comet's or asteroid's orbit will intersect that of a planet. For each of the two cases below, use MATLAB to determine whether or not orbits A and B intersect, and if they do, determine the polar coordinates of the intersection point. The units of distance are AU, where 1 AU is the mean distance of the earth from the sun.

   (a) Orbit A: $p = 1$, $\epsilon = 0.01$. Orbit B: $p = 0.1$, $\epsilon = 0.9$.

   (b) Orbit A: $p = 1$, $\epsilon = 0.01$. Orbit B: $p = 1.5$, $\epsilon = 0.5$.

10. Figure 11.2–2 in Section 11.2 shows a robot arm having two joints and two links. The angles of rotation of the motors at the joints are θ_1 and θ_2. From trigonometry we can derive the following expressions for the (x, y) coordinates of the hand.

$$x = L_1 \cos\theta_1 + L_2 \cos(\theta_1 + \theta_2)$$
$$y = L_1 \sin\theta_1 + L_2 \sin(\theta_1 + \theta_2)$$

Suppose that the link lengths are $L_1 = 3$ ft and $L_2 = 2$ ft.

(*a*) Compute the motor angles required to position the hand at $x = 3$ ft, $y = 1$ ft. Identify the elbow-up and elbow-down solutions.

(*b*) Suppose you want to move the hand along a straight, horizontal line at $y = 1$, for $2 \leq x \leq 4$. Plot the required motor angles versus x. Label the elbow-up and elbow-down solutions.

Section 11.3

11. Use MATLAB to find all the values of x where the graph of $y = 3x - 2x$ has a horizontal tangent line.

12. Use MATLAB to determine all the local minima and local maxima, and all the inflection points where $dy/dx = 0$, of the following function.

$$y = x^4 - \frac{16}{3}x^3 + 8x^2 - 4$$

13. The surface area of a sphere of radius r is $S = 4\pi r^2$. Its volume is $V = 4\pi r^3/3$.

(*a*) Use MATLAB to find the expression for dS/dV.

(*b*) A spherical balloon expands as air is pumped into it. What is the rate of increase in the balloon's surface area with volume when its volume is 30 cubic inches?

14. Use MATLAB to find the point on the line $y = 2 - x/3$ that is closest to the point $x = -3$, $y = 1$.

15. A particular circle is centered at the origin and has a radius of 5. Use MATLAB to find the equation of the line that is tangent to the circle at the point $x = 3$, $y = 4$.

16. Ship A is traveling north at 6 mph, and ship B is traveling west at 12 mph. When ship A was dead ahead of ship B, it was 6 miles away. Use MATLAB to determine how close the ships come to each other.

17. Suppose you have a wire of length L. You cut a length x to make a square, and use the remaining length $L - x$ to make a circle. Use MATLAB to find the length x that maximizes the sum of the areas enclosed by the square and the circle.

18.* A certain spherical street lamp emits light in all directions. It is mounted on a pole of height h (see Figure P18). The brightness B at point P on the sidewalk is directly proportional to sin θ, and inversely proportional to the square of the distance d from the light to the point. Thus

$$B = \frac{c}{d^2} \sin \theta$$

Figure P18

where c is a constant. Use MATLAB to determine how high h should be to maximize the brightness at point P, which is 30 ft from the base of the pole.

19.* A certain object has a mass $m = 100$ kg and is acted on by a force $f(t) = 500[2 - e^{-t} \sin(5\pi t)]$ N. The mass is at rest at $t = 0$. Use MATLAB to compute the object's velocity v at $t = 5$ s. The equation of motion is $m\dot{v} = f(t)$.

20. A rocket's mass decreases as it burns fuel. The equation of motion for a rocket in vertical flight can be obtained from Newton's law, and is

$$m(t)\frac{dv}{dt} = T - m(t)g$$

where T is the rocket's thrust and its mass as a function of time is given by $m(t) = m_0(1 - rt/b)$. The rocket's initial mass is m_0, the burn time is b, and r is the fraction of the total mass accounted for by the fuel.

Use the values $T = 48{,}000$ N, $m_0 = 2200$ kg, $r = 0.8$, $g = 9.81$ m/s^2, and $b = 40$ s.

(a) Use MATLAB to compute the rocket's velocity as a function of time, for $t \le b$.

(b) Use MATLAB to compute the rocket's velocity at burnout.

21. The equation for the voltage $v(t)$ across a capacitor as a function of time is

$$v(t) = \frac{1}{C}\left[\int_0^t i(t)dt + Q_0\right]$$

where $i(t)$ is the applied current and Q_0 is the initial charge. Suppose that $C = 10^{-7}$ F and that $Q_0 = 0$. If the applied current is $i(t) = 0.3 + 0.1e^{-5t} \sin(25\pi t)$, use MATLAB to compute and plot the voltage $v(t)$ for $0 \le t \le 7$ seconds.

22. The power P dissipated as heat in a resistor R as a function of the current $i(t)$ passing through it is $P = i^2R$. The energy $E(t)$ lost as a function of time is the time integral of the power. Thus

$$E(t) = \int_0^t P(t)dt = R\int_0^t i^2(t)dt$$

If the current is measured in amperes, the power is in watts and the energy is in joules (1 watt = 1 joule/second). Suppose that a current $i(t) = 0.2[1 + \sin(0.2t)]$ amperes is applied to the resistor.

(a) Determine the energy $E(t)$ dissipated as a function of time.

(b) Determine the energy dissipated in one minute if $R = 1000\ \Omega$.

23. The RLC circuit shown in Figure P23 can be used as a *narrow-band filter*. If the input voltage $v_i(t)$ consists of a sum of sinusoidally varying voltages with different frequencies, the narrow-band filter will allow to pass only those voltages whose frequencies lie within a narrow range. These filters are used in tuning circuits, such as those used in AM radios, to allow reception only of the carrier signal of the desired radio station. The magnification ratio M of a circuit is the ratio of the amplitude of the output voltage $v_o(t)$ to the amplitude of the input voltage $v_i(t)$. It is a function of the radian frequency ω of the input voltage. Formulas for M are derived in elementary electrical circuits courses. For this particular circuit, M is given by

$$M = \frac{RC\omega}{\sqrt{(1 - LC\omega^2)^2 + (RC\omega)^2}}$$

Figure P23

The frequency at which M is a maximum is the frequency of the desired carrier signal.

(a) Determine this frequency as a function of R, C, and L.

(b) Plot M versus ω for two cases where $C = 10 \times 10^{-5}$ F and $L = 5 \times 10^{-3}$ H. For the first case, $R = 1000\ \Omega$. For the second case, $R = 10\ \Omega$. Comment on the filtering capability of each case.

24. The shape of a cable hanging with no other load other than its own weight is a *catenary* curve. A particular bridge cable is described by the catenary $y(x) = 10 \cosh((x - 20)/10)$ for $0 \le x \le 50$, where x and y are the horizontal and vertical coordinates measured in feet. (See Figure P24.) It is desired to hang plastic sheeting from the cable to protect passersby while the bridge is being repainted. Use MATLAB to determine how many

square feet of sheeting are required. Assume that the bottom edge of the sheeting is located along the x axis at $y = 0$.

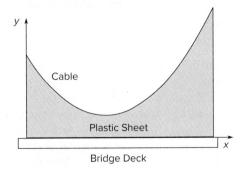

Figure P24

25. The shape of a cable hanging with no other load other than its own weight is a *catenary* curve. A particular bridge cable is described by the catenary $y(x) = 10 \cosh((x - 20)/10)$ for $0 \le x \le 50$, where x and y are the horizontal and vertical coordinates measured in feet.

    The length L of a curve described by $y(x)$ for $a \le x \le b$ can be found from the integral

    $$L = \int_a^b \sqrt{1 + \left(\frac{dy}{dx}\right)^2}\, dx$$

    Determine the length of the cable.

26. Use the first five nonzero terms in the Taylor series for e^{ix}, $\sin x$, and $\cos x$ about $x = 0$ to demonstrate the validity of Euler's formula $e^{ix} = \cos x + \sin x$.

27. Find the Taylor series for $e^x \sin x$ about $x = 0$ in two ways: (a) by multiplying the Taylor series for e^x and that for $\sin x$, and (b) by using the `taylor` function directly on $e^x \sin x$.

28. Integrals that cannot be evaluated in closed form sometimes can be evaluated approximately by using a series representation for the integrand. For example, the following integral is used for some probability calculations (see Chapter 7, Section 7.2).

    $$I = \int_0^1 e^{-x^2} dx$$

    (a) Obtain the Taylor series for e^{-x^2} about $x = 0$, and integrate the first six nonzero terms in the series to find I. Use the seventh term to estimate the error.

(b) Compare your answer with that obtained with the MATLAB `erf(t)` function, defined as

$$\text{erf}(t) = \frac{2}{\sqrt{\pi}} \int_0^t e^{-x^2} dx$$

29.* Use MATLAB to compute the following limits.

(a) $\lim\limits_{x \to 1} \dfrac{x^2 - 1}{x^2 - x}$

(b) $\lim\limits_{x \to -2} \dfrac{x^2 - 4}{x^2 + 4}$

(c) $\lim\limits_{x \to 0} \dfrac{x^4 + 2x^2}{x^3 + x}$

30. Use MATLAB to compute the following limits.

(a) $\lim\limits_{x \to 0+} x^x$

(b) $\lim\limits_{x \to 0+} (\cos x)^{1/\tan x}$

(c) $\lim\limits_{x \to 0+} \left(\dfrac{1}{1 - x} \right)^{-1/x^2}$

(d) $\lim\limits_{x \to 0-} \dfrac{\sin x^2}{x^3}$

(e) $\lim\limits_{x \to 5-} \dfrac{x^2 - 25}{x^2 - 10x + 25}$

(f) $\lim\limits_{x \to 1+} \dfrac{x^2 - 1}{\sin (x - 1)^2}$

31. Use MATLAB to compute the following limits.

(a) $\lim\limits_{x \to \infty} \dfrac{x + 1}{x}$

(b) $\lim\limits_{x \to -\infty} \dfrac{3x^3 - 2x}{2x^3 + 3}$

32. Find the expression for the sum of the geometric series:

$$\sum_{k=0}^{n-1} r^k$$

for $r \neq 1$.

33. A particular rubber ball rebounds to one-half its original height when dropped on a floor.

(a) If the ball is initially dropped from a height h and is allowed to continue to bounce, find the expression for the total distance traveled by the ball, after the ball hits the floor for the nth time.

(b) How far will the ball have traveled after it hits the floor for the eighth time, if it is initially dropped from a height of 10 feet?

Section 11.4

34. The equation for the voltage y across the capacitor of an RC circuit is

$$RC\frac{dy}{dt} + y = v(t)$$

where $v(t)$ is the applied voltage. Suppose that $RC = 0.2$ s and that the capacitor voltage is initially 2 V. If the applied voltage goes from 0 to 10 V at $t = 0$, use MATLAB to determine and plot the voltage $y(t)$ for $0 \le t \le 1$ second.

35. The following equation describes the temperature $T(t)$ of a certain object immersed in a liquid bath of temperature $T_b(t)$.

$$10\frac{dT}{dt} + T = T_b$$

Suppose the object's temperature is initially $T(0) = 70°F$ and the bath temperature is $170°F$. Use MATLAB to answer the following questions.

(a) Determine $T(t)$.

(b) How long will it take for the object's temperature T to reach $168°F$?

(c) Plot the object's temperature $T(t)$ as a function of time.

36.* The following equation describes the motion of a mass connected to a spring, with viscous friction on the surface.

$$m\ddot{y} + c\dot{y} + ky = f(t)$$

where $f(t)$ is an applied force. The position and velocity of the mass at $t = 0$ are denoted by x_0 and v_0. Use MATLAB to answer the following questions.

(a) What is the free response in terms of x_0 and v_0 if $m = 3$, $c = 18$, $k = 102$?

(b) What is the free response in terms of x_0 and v_0 if $m = 3$, $c = 39$, and $k = 120$?

37. The equation for the voltage y across the capacitor of an RC circuit is

$$RC\frac{dy}{dt} + y = v(t)$$

where $v(t)$ is the applied voltage. Suppose that $RC = 0.2$ s and that the capacitor voltage is initially 2 s. If the applied voltage is $v(t) = 10[2 - e^{-t} \sin(5\pi t)]$, use MATLAB to compute and plot the voltage $y(t)$ for $0 \le t \le 5$ s.

38. The following equation describes a certain dilution process, where $y(t)$ is the concentration of salt in a tank of fresh water to which salt brine is being added.

$$\frac{dy}{dt} + \frac{2}{10 + 2t}y = 4$$

Suppose that $y(0) = 0$. Use MATLAB to compute and plot $y(t)$ for $0 \le t \le 10$.

39. The following equation describes the motion of a certain mass connected to a spring, with viscous friction on the surface.

$$3\ddot{y} + 18\dot{y} + 102y = f(t)$$

where $f(t)$ is an applied force. Suppose that $f(t) = 0$ for $t < 0$ and $f(t) = 10$ for $t \ge 0$.

(a) Use MATLAB to compute and plot $y(t)$ when $y(0) = \dot{y}(0) = 0$.
(b) Use MATLAB to compute and plot $y(t)$ when $y(0) = 0$ and $\dot{y}(0) = 10$.

40. The following equation describes the motion of a certain mass connected to a spring, with viscous friction on the surface.

$$3\ddot{y} + 39\dot{y} + 120y = f(t)$$

where $f(t)$ is an applied force. Suppose that $f(t) = 0$ for $t < 0$ and $f(t) = 10$ for $t \ge 0$.

(a) Use MATLAB to compute and plot $y(t)$ when $y(0) = \dot{y}(0) = 0$.
(b) Use MATLAB to compute and plot $y(t)$ when $y(0) = 0$ and $\dot{y}(0) = 10$.

41. The equations for an armature-controlled dc motor are the following. The motor's current is i and its rotational velocity is ω.

$$L\frac{di}{dt} = -Ri - K_e\omega + v(t)$$

$$I\frac{d\omega}{dt} = K_T i - c\omega$$

where L, R, and I are the motor's inductance, resistance, and inertia, K_T and K_e are the torque constant and back emf constant, c is a viscous damping constant, and $v(t)$ is the applied voltage.

Use the values $R = 0.8\ \Omega$, $L = 0.003\ \text{H}$, $K_T = 0.05\ \text{N} \cdot \text{m/A}$, $K_e = 0.05\ \text{V} \cdot \text{s/rad}$, $c = 0$, and $I = 8 \times 10^{-5}\ \text{N} \cdot \text{m} \cdot \text{s}^2$.

Suppose the applied voltage is 20 V. Use MATLAB to compute and plot the motor's speed and current versus time. Choose a final time large enough to show the motor's speed becoming constant.

Section 11.5

42. The RLC circuit described in Problem 23 and shown in Figure P23 has the following differential equation model.

$$LC\ddot{v}_o + RC\dot{v}_o + v_o = RC\dot{v}_i(t)$$

Use the Laplace transform method to solve for the unit-step response of $v_o(t)$ for zero initial conditions, where $C = 10 \times 10^{-5}\ \text{F}$ and $L = 5 \times 10^{-3}\ \text{H}$. For the first case (a broadband filter), $R = 1000\ \Omega$. For the second case (a narrowband filter), $R = 10\ \Omega$. Compare the step responses of the two cases.

43. The differential equation model for a certain speed control system for a vehicle is

$$\ddot{v} + (1 + K_p)\dot{v} + K_I v = K_p \dot{v}_d + K_I v_d$$

where the actual speed is v, the desired speed is $v_d(t)$, and K_p and K_I are constants called the "control gains." Use the Laplace transform method to find the unit-step response (that is, $v_d(t)$ is a unit-step function). Use zero initial conditions. Compare the response for three cases:

(a) $K_p = 9$, $K_I = 50$
(b) $K_p = 9$, $K_I = 25$
(c) $K_p = 54$, $K_I = 250$

44. The differential equation model for a certain position control system for a metal cutting tool is

$$\frac{d^3 x}{dt^3} + (6 + K_D)\frac{d^2 x}{dt^2} + (11 + K_p)\frac{dx}{dt} + (6 + K_I)x$$
$$= K_D \frac{d^2 x_d}{dt^2} + K_p \frac{dx_d}{dt} + K_I x_d$$

where the actual tool position is x, the desired position is $x_d(t)$, and K_p, K_I, and K_D are constants called the "control gains." Use the Laplace transform method to find the unit-step response (that is, $x_d(t)$ is a unit-step function). Use zero initial conditions. Compare the response for three cases:

(a) $K_p = 30$, $K_I = K_D = 0$
(b) $K_p = 27$, $K_I = 17.18$, $K_D = 0$
(c) $K_p = 36$, $K_I = 38.1$, $K_D = 8.52$

45. The differential equation model for the motor torque $m(t)$ required for a certain speed control system is

$$4\ddot{m} + 4K\dot{m} + K^2 m = K^2 \dot{v}_d$$

where the desired speed is $v_d(t)$, and K is a constant called the "control gain."

(a) Use the Laplace transform method to find the unit-step response (that is, $v_d(t)$ is a unit-step function). Use zero initial conditions.
(b) Use symbolic manipulation in MATLAB to find the value of the gain K that minimizes the peak torque that must be supplied by the motor. In addition, compute the value of the peak torque.

Section 11.6

46. Show that $\mathbf{R}^{-1}(a)\mathbf{R}(a) = \mathbf{I}$, where $\mathbf{I}$ is the identity matrix, and $\mathbf{R}(a)$ is the rotation matrix given by Equation (11.6–1). This shows that the inverse coordinate transformation returns you to the original coordinate system.

47. Show that $\mathbf{R}^{-1}(a) = \mathbf{R}(-a)$. This shows that a rotation through a negative angle is equivalent to an inverse transformation.

48.* Find the characteristic polynomial and roots of the following matrix.

$$\mathbf{A} = \begin{bmatrix} -6 & 2 \\ 3k & -7 \end{bmatrix}$$

49.* Use the matrix inverse and the left-division method to solve the following set.

$$4cx + 5y = 43$$
$$3x - 4y = -22$$

50. The currents i_1, i_2, and i_3 in the circuit shown in Figure P50 are described by the following equation set if all the resistances are equal to R.

$$\begin{bmatrix} 2R & -R & 0 \\ -R & 3R & -R \\ 0 & R & -2R \end{bmatrix} \begin{bmatrix} i_1 \\ i_2 \\ i_3 \end{bmatrix} = \begin{bmatrix} v_1 \\ 0 \\ v_2 \end{bmatrix}$$

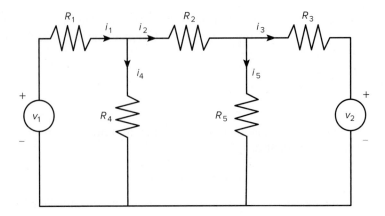

Figure P50

where v_1 and v_2 are applied voltages. The other two currents can be found from $i_4 = i_1 - i_2$ and $i_5 = i_2 - i_3$.

(*a*) Use both the matrix inverse method and the left-division method to solve for the currents in terms of the resistance R and the voltages v_1 and v_2.

(*b*) Find the numerical values for the currents if $R = 1000\ \Omega$, $v_1 = 100$ V, and $v_2 = 25$ V.

51.   The equations for the armature-controlled dc motor shown in Figure P51 are the following. The motor's current is i and its rotational velocity is ω.

$$L\frac{di}{dt} = -Ri - K_e\omega + v(t)$$

$$I\frac{d\omega}{dt} = K_T i - c\omega$$

Figure P51

where L, R, and I are the motor's inductance, resistance, and inertia, K_T and K_e are the torque constant and back emf constant, c is a viscous damping constant, and $v(t)$ is the applied voltage.

(*a*) Find the characteristic polynomial and the characteristic roots.

(b) Use the values $R = 0.8\ \Omega$, $L = 0.003$ H, $K_T = 0.05$ N · m/A, $K_e = 0.05$ V · s/rad, and $I = 8 \times 10^{-5}$ N · m · s². The damping constant c is often difficult to determine with accuracy. For these values find the expressions for the two characteristic roots in terms of c.

(c) Using the parameter values in part (b), determine the roots for the following values of c (in newton meter second): $c = 0$, $c = 0.01$, $c = 0.1$, and $c = 0.2$. For each case, use the roots to estimate how long the motor's speed will take to become constant, and discuss whether or not the speed will oscillate before it becomes constant.

APPENDIX

Guide to Commands and Functions in This Text

Operators and special characters

Item	Description	Pages
+	Plus; addition operator.	8
−	Minus; subtraction operator.	8
*	Scalar and matrix multiplication operator.	8
.*	Array multiplication operator.	63
^	Scalar and matrix exponentiation operator.	8
.^	Array exponentiation operator.	63
\	Left-division operator.	8, 63
/	Right-division operator.	8, 63
.\	Array left-division operator.	63
./	Array right-division operator.	63
:	Colon; generates regularly spaced elements and represents an entire row or column.	12, 53
( )	Parentheses; enclose function arguments and array indices; override precedence.	9, 128
[ ]	Brackets; enclose array elements.	19, 52
{ }	Braces; enclose cell elements.	90
. . .	Ellipsis; line-continuation operator.	12
,	Comma; separates statements, and elements in a row of an array.	12
;	Semicolon; separates columns in an array, and suppresses display.	12, 53
%	Percent sign; designates a comment, and specifies formatting.	27, 583
`	Quote sign and transpose operator.	53, 55
.'	Nonconjugated transpose operator.	55
=	Assignment (replacement) operator.	10
@	Creates a function handle.	128

560

Logical and relational operators

Item	Description	Pages
==	Relational operator: equal to.	161
~=	Relational operator: not equal to.	161
<	Relational operator: less than.	161
<=	Relational operator: less than or equal to.	161
>	Relational operator: greater than.	161
>=	Relational operator: greater than or equal to.	161
&	Logical operator: AND.	164
&&	Short-circuit AND.	164
|	Logical operator: OR.	164
||	Short-circuit OR.	164
~	Logical operator: NOT.	164

Special variables and constants

Item	Description	Pages
ans	Most recent answer.	15
eps	Accuracy of floating-point precision.	15
i,j	The imaginary unit $\sqrt{-1}$.	15
Inf	Infinity.	15
NaN	Undefined numerical result (not a number).	15
pi	The number π.	15

Commands for managing a session

Item	Description	Pages
clc	Clears Command window.	12
clear	Removes variables from memory.	12
doc	Displays documentation.	34
exist	Checks for existence of file or variable.	12
global	Declares variables to be global.	127
help	Displays Help text in the Command window.	33
lookfor	Searches Help entries for a keyword.	33
namelengthmax	Returns the maximum number of characters allowed in a name.	12
quit	Stops MATLAB.	12
who	Lists current variables.	12
whos	Lists current variables (long display).	12

System and file commands

Item	Description	Pages
cd	Changes current directory.	26
date	Displays current date.	124
dir	Lists all files in current directory.	26
importdata	Imports several different file types.	144
load	Loads workspace variables from a file.	26
path	Displays search path.	26
pwd	Displays current directory.	26
readtable	Creates a table from a file.	144
save	Saves workspace variables in a file.	26
what	Lists all MATLAB files.	26
which	Displays the path name.	26
xlsread	Imports an Excel workbook file.	143
xlswrite	Writes an array to an Excel file.	143

Input/output commands

Item	Description	Pages
disp	Displays contents of an array or string.	30
format	Controls screen display format.	16, 30
fprintf	Performs formatted writes to screen or file.	584
input	Displays prompts and waits for input.	30, 177
;	Suppresses screen printing.	12

Numeric display formats

Item	Description	Pages
format short	Four decimal digits (default).	16, 26
format long	16 decimal digits.	16, 26
format short e	Five digits plus exponent.	16, 26
format long e	16 digits plus exponent.	16, 26
format bank	Two decimal digits.	16, 26
format +	Positive, negative, or zero.	16, 26
format rat	Rational approximation.	16, 26
format compact	Suppresses some line feeds.	16, 26
format loose	Resets to less compact display mode.	16, 26

Array functions

Item	Description	Pages
cat	Concatenates arrays.	62
find	Finds indices of nonzero elements.	58, 60, 161
length	Computes number of elements.	20, 58
linspace	Creates regularly spaced vector.	58
logspace	Creates logarithmically spaced vector.	58
max	Returns largest element.	58
min	Returns smallest element.	58
ndims(A)	Returns the number of dimensions of A.	58
numel(A)	Returns the total number of elements in the array A.	58
norm(x)	Computes the geometric length of x.	58
openvar	Opens the Variable Editor.	61
size	Computes array size.	58
sort	Sorts each column.	58, 60, 296
sum	Sums each column.	58

Special matrices

Item	Description	Pages
eye	Creates an identity matrix.	81
ones	Creates an array of 1s.	81
zeros	Creates an array of 0s.	81

Matrix functions for solving linear equations

Item	Description	Pages
det	Computes determinant of an array.	351
inv	Computes inverse of a matrix.	351
pinv	Computes pseudoinverse of a matrix.	361
rank	Computes rank of a matrix.	353
rref	Computes reduced row echelon form.	364

Exponential and logarithmic functions

Item	Description	Pages
exp(x)	Exponential; e_x.	19, 114
log(x)	Natural logarithm; $\ln x$.	114
log10(x)	Common (base 10) logarithm; $\log x = \log_{10} x$.	114
sqrt(x)	Square root; $\sqrt{x}$	114

Complex functions

Item	Description	Pages
abs(x)	Absolute value.	114
angle(x)	Angle of a complex number x.	114
conj(x)	Complex conjugate of x.	114
imag(x)	Imaginary part of a complex number x.	114
real(x)	Real part of a complex number x.	114

Numeric functions

Item	Description	Pages
ceil	Rounds to the nearest integer toward ∞.	114
fix	Rounds to the nearest integer toward zero.	114
floor	Rounds to the nearest integer toward $-\infty$.	114
round	Rounds toward the nearest integer.	114
sign	Signum function.	114

Trigonometric functions using radian measure (functions using degree measure have a d appended, as sind(x) and asind(x)).

Item	Description	Pages
acos(x)	Inverse cosine; $\arccos x = \cos^{-1}x$.	19, 118
acot(x)	Inverse cotangent; $\text{arccot } x = \cot^{-1}x$.	118
acsc(x)	Inverse cosecant; $\text{arccsc } x = \csc^{-1}x$.	118
asec(x)	Inverse secant; $\text{arcsec } x = \sec^{-1}x$.	118
asin(x)	Inverse sine; $\arcsin x = \sin^{-1}x$.	19, 118
atan(x)	Inverse tangent; $\arctan x = \tan^{-1}x$.	19, 118
atan2(y,x)	Four-quadrant inverse tangent.	118
cos(x)	Cosine; $\cos x$.	19, 118
cot(x)	Cotangent; $\cot x$.	118
csc(x)	Cosecant; $\csc x$.	118
sec(x)	Secant; $\sec x$.	118
sin(x)	Sine; $\sin x$.	118
tan(x)	Tangent; $\tan x$.	118

Hyperbolic functions

Item	Description	Pages
acosh(x)	Inverse hyperbolic cosine; $\cosh^{-1}x$.	119
acoth(x)	Inverse hyperbolic cotangent; $\coth^{-1}x$.	119
acsch(x)	Inverse hyperbolic cosecant; $\text{csch}^{-1}x$.	119
asech(x)	Inverse hyperbolic secant; $\text{sech}^{-1}x$.	119
asinh(x)	Inverse hyperbolic sine; $\sinh^{-1}x$.	119
atanh(x)	Inverse hyperbolic tangent; $\tanh^{-1}x$.	119
cosh(x)	Hyperbolic cosine; $\cosh x$.	119
coth(x)	Hyperbolic cotangent; $\cosh x/\sinh x$.	119
csch(x)	Hyperbolic cosecant; $1/\sinh x$.	119
sech(x)	Hyperbolic secant; $1/\cosh x$.	119
sinh(x)	Hyperbolic sine; $\sinh x$.	119
tanh(x)	Hyperbolic tangent; $\sinh x/\cosh x$.	119

Polynomial functions

Item	Description	Pages
conv	Computes product of two polynomials.	85
deconv	Computes ratio of polynomials.	85
eig	Computes the eigenvalues of a matrix.	416
poly	Computes polynomial from roots.	85
polyfit	Fits a polynomial to data.	276, 285
polyval	Evaluates polynomial.	285
roots	Computes polynomial roots.	85

Logical functions

Item	Description	Pages
any	True if any elements are nonzero.	167
all	True if all elements are nonzero.	167
find	Finds indices of nonzero elements.	167
finite	True if elements are finite.	167
ischar	True if elements are a character array.	167
isempty	True if matrix is empty.	167
isinf	True if elements are infinite.	167
isnan	True if elements are undefined.	167
isnumeric	True if elements have numeric values.	167
isreal	True if all elements are real.	167
logical	Converts a numeric array to a logical array.	167
xor	Exclusive OR.	167

Miscellaneous mathematical functions

Item	Description	Pages
cross	Computes cross products.	83
dot	Computes dot products.	84
function	Creates a user-defined function.	120
nargin	Number of function input arguments.	176
nargout	Number of function output arguments.	177

Cell and structure functions

Item	Description	Pages
cell	Creates cell array.	88
fieldnames	Returns field names in a structure array.	93
isfield	Identifies a structure array field.	93
isstruct	Identifies a structure array.	93
rmfield	Removes a field from a structure array.	93
struct	Creates a structure array.	93

Basic *xy* plotting commands

Item	Description	Pages
axis	Sets axis limits and other axis properties.	228, 231
fplot	Intelligent plotting of functions.	229, 231
ginput	Reads coordinates of the cursor position.	23
grid	Displays grid lines.	231
plot	Generates *xy* plot.	23, 231
print	Prints plot or saves plot to a file.	231
title	Puts text at top of plot.	231
xlabel	Adds text label to *x* axis.	226, 231
ylabel	Adds text label to *y* axis.	226, 231

Plot enhancement commands

Item	Description	Pages
colormap	Sets the color map of the current figure.	573
gtext	Enables label placement by mouse.	238, 240
hold	Freezes current plot.	239, 240
legend	Places legend by mouse.	237, 240
subplot	Creates plots in subwindows.	234, 240
text	Places string in figure.	238, 240

Specialized plot functions

Item	Description	Pages
bar	Creates bar chart.	244, 312
errorbar	Plots error bars.	246
fimplicit	Plots an implicit function.	234
loglog	Creates log-log plot.	234
polarplot	Creates polar plot.	234
publish	Creates reports in a variety of formats.	247
semilogx	Creates semilog plot (logarithmic abscissa).	234
semilogy	Creates semilog plot (logarithmic ordinate).	234
stairs	Creates stairs plot.	234
stem	Creates stem plot.	234
yyaxis	Enables plotting on left and right axes.	234

Three-dimensional plotting functions using function inputs

Function	Description	Pages
fcontour((f)	Creates a contour plot.	261
fimplicit3(f)	Plots an implicit 3-D function.	261
fmesh(f)	Creates a 3-D surface plot.	261
fplot3(fx,fy,fz)	Creates a 3-D line plot.	261
fsurf(f)	Creates a shaded 3-D surface plot.	261

Three-dimensional plotting functions using array inputs

Item	Description	Pages
contour	Creates contour plot.	260
mesh	Creates three-dimensional mesh surface plot.	260
meshc	Same as mesh with contour plot underneath.	260
meshgrid	Creates rectangular grid.	260
meshz	Same as mesh with vertical lines underneath.	260
plot3	Creates three-dimensional plots from lines and points.	260
shading	Specifies type of shading.	573
surf	Creates shaded three-dimensional mesh surface plot.	260
surfc	Same as surf with contour plot underneath.	260
surfl	Same as surf but with lighting.	573
view	Sets the angle of the view.	572
waterfall	Same as mesh with mesh lines in one direction.	260
zlabel	Adds text label to z axis.	255

Program flow control

Item	Description	Pages
break	Terminates execution of a loop.	183
case	Provides alternate execution paths within switch structure.	195
continue	Passes control to the next iteration of a for or while loop.	183
else	Delineates alternate block of statements.	172
elseif	Conditionally executes statements.	174
end	Terminates for, while, and if statements.	171
for	Repeats statements a specific number of times.	178
if	Executes statements conditionally.	171
otherwise	Provides optional control within a switch structure.	195
switch	Directs program execution by comparing input with case expressions.	195
while	Repeats statements an indefinite number of times.	190

Debugging commands

Item	Description	Pages
dbclear	Remove a breakpoint.	199
dbquit	Quit debug mode.	200
dbstep	Executes one or more lines.	200
dbstop	Sets a breakpoint.	200
echo	Traces program execution.	200

Optimization and root-finding functions

Item	Description	Pages
fminbnd	Finds the minimum of a function of one variable.	130
fminsearch	Finds the minimum of a multivariable function.	131
fzero	Finds the zero of a function.	128

Histogram functions

Item	Description	Pages
bar	Creates a bar chart.	244, 315
histogram	Creates a histogram plot.	312, 315

Statistical functions

Item	Description	Pages
cumsum	Computes the cumulative sum across a row.	318
erf	Computes the error function erf(x).	320
mean	Calculates the mean.	312, 319
median	Calculates the median.	312, 319
mode	Calculates the mode.	312
std	Calculates the standard deviation.	319

Random number functions

Item	Description	Pages
rand	Generates uniformly distributed random numbers between 0 and 1.	325
randi	Generates non-unique random integers.	325
randn	Generates normally distributed random numbers.	325
randperm	Generates random permutation of unique integers.	325
rng	Initializes the random number generator.	325

Polynomial functions

Item	Description	Pages
poly	Computes the coefficients of a polynomial from its roots.	85
polyfit	Fits a polynomial to data.	276, 285
polyval	Evaluates a polynomial and generates error estimates.	85, 285
roots	Computes the roots of a polynomial from its coefficients.	85

Interpolation functions

Item	Description	Pages
interp1	Linear and cubic spline interpolation of a function of one variable.	336, 339
interp2	Linear interpolation of a function of two variables.	336
pchip	Interpolation using Hermite Polynomials.	339
spline	Cubic spline interpolation.	339
unmkpp	Computes the coefficients of cubic spline polynomials.	339

Numerical integration functions

Item	Description	Pages
integral	Numerical integration of a single integral.	391
integral2	Numerical integration of a double integral.	391
integral3	Numerical integration of a triple integral.	391
polyint	Integration of a polynomial.	391
trapz	Numerical integration with the trapezoidal rule.	391

Numerical differentiation functions

Item	Description	Pages
del2	Computes the Laplacian from data.	402
diff(x)	Computes the differences between adjacent elements in the vector x.	399, 402
gradient	Computes the gradient from data.	400, 402
polyder	Differentiates a polynomial, a polynomial product, or a polynomial quotient.	399, 402

ODE solvers

Item	Description	Pages
ode45	Nonstiff, medium-order solver.	405, 413
ode15s	Stiff, variable-order solver.	405
odeset	Creates integrator options structure for ODE solvers.	413

LTI object functions

Item	Description	Pages
ss	Creates an LTI object in state-space form.	417
ssdata	Extracts state-space matrices from an LTI object.	417
tf	Creates an LTI object in transfer-function form.	417
tfdata	Extracts equation coefficients from an LTI object.	417

LTI ODE solvers

Item	Description	Pages
impulse	Computes and plots the impulse response of an LTI object.	419, 420
initial	Computes and plots the free response of an LTI object.	419
linearSystemAnalyzer	Invokes an interactive user interface for analyzing LTI systems.	426
lsim	Computes and plots the response of an LTI object to a general input.	419, 423
step	Computes and plots the step response of an LTI object.	419, 421

Predefined input functions

Item	Description	Pages
gensig	Generates a periodic sine, square, or pulse input.	426

Functions for creating and evaluating symbolic expressions

Item	Description	Pages
class	Returns the class of an expression.	498, 504
digits	Sets the number of decimal digits used to do variable precision arithmetic.	504
double	Converts an expression to numeric form.	502, 504
findsym	Finds the symbolic variables in a symbolic expression.	499, 504
fplot	Generates a plot of a symbolic expression.	504, 529
numden	Returns the numerator and denominator of an expression.	501, 504
sym	Creates a symbolic variable.	497, 504
syms	Creates one or more symbolic variables.	497, 504
vpa	Sets the number of digits used to evaluate expressions.	503, 504

Functions for manipulating symbolic expressions

Item	Description	Pages
collect	Collects coefficients of like powers in an expression.	500, 505
expand	Expands an expression by carrying out powers.	500, 505
factor	Factors an expression.	500, 505
poly2sym	Converts a polynomial coefficient vector to a symbolic polynomial.	502, 505
simplify	Simplifies an expression.	501, 505
subs	Substitutes variables or expressions.	502, 505
sym2poly	Converts an expression to a polynomial coefficient vector.	502, 505

Symbolic solution of algebraic and transcendental equations

Item	Description	Pages
solve	Solves symbolic equations.	506, 512

Symbolic calculus functions

Item	Description	Pages
diff	Returns the derivative of an expression.	513, 516
dirac	Dirac delta function (unit impulse).	540, 546
heaviside	Heaviside function (unit step).	534, 546
int	Returns the integral of an expression.	515, 518
limit	Returns the limit of an expression.	515, 523
symsum	Returns the symbolic summation of an expression.	515
taylor	Returns the Taylor series of a function.	515, 521

Symbolic solution of differential equations

Item	Description	Pages
dsolve	Returns a symbolic solution of a differential equation or set of equations.	526, 532

Laplace transform functions

Item	Description	Pages
ilaplace	Returns the inverse Laplace transform.	535, 541
laplace	Returns the Laplace transform.	534, 541

Symbolic linear algebra functions

Item	Description	Pages
charpoly	Returns the characteristic polynomial of a matrix.	633, 635, 543
det	Returns the determinant of a matrix.	351, 544
eig	Returns the eigenvalues (characteristic roots) of a matrix.	416, 544
inv	Returns the inverse of a matrix.	351, 544

Animation functions

Item	Description	Pages
addpoints	Adds points to an animated line.	574
animatedline	Creates and adds an animated line to the current axes.	574
clearpoints	Removes points from an animated line.	574
drawnow	Initiates immediate plotting.	574
gca	Returns the current axes properties.	576
get	Returns a complete list of an object's properties.	575
getframe	Captures current figure in a frame.	572
getpoints	Retrieves points from an animated line.	574
movie	Plays recorded movie frames.	572
pause	Pauses the display.	574
set	Used with a handle to set an object's properties.	575
view	Sets the angle of the view.	572

Sound functions

Item	Description	Pages
audioplayer	Creates a handle for a WAVE file.	581
audioread	Reads a WAVW file.	576
audiorecorder	Records sounds.	581
audiowrite	Creates a WAVE file.	576
play	Plays a WAVE file using its handle.	581
recordblocking	Holds control until recording completes.	582
sound	Plays a vector as sound.	580
soundsc	Scales data and plays as sound.	581

B

Animation and Sound in MATLAB

B.1 Animation

Animation can be used to display the behavior of an object over time. Some of the MATLAB demos are M-files that perform animation. After completing this section, which has simple examples, you may study the demo files, which are more advanced. Two methods can be used to create animations in MAT-LAB. The first method uses the `movie` function. The second method uses the `drawnow` command.

Creating Movies in MATLAB

The `getframe` command captures, or takes a snapshot of, the current figure to create a single frame for the movie. The `getframe` function is usually used in a `for` loop to assemble an array of movie frames. The `movie` function plays back the frames after they have been captured.

To create a movie, use a script file of the following form.

```
for k = 1:n
    plotting expressions
    M(k) = getframe; % Saves current figure in array M
end
movie(M)
```

For example, the following script file creates 20 frames of the function $te^{-t/b}$ for $0 \le t \le 100$ for each of 20 values of the parameter b from $b = 1$ to $b = 20$.

```
% Program movie1.m
% Animates the function t*exp(-t/b).
```

```
t = 0:0.05:100;
for b = 1:20
  plot(t,t.*exp(-t/b)),axis([0 100 0 10]),xlabel('t');
  M(:,b) = getframe;
end
```

The line M(:,b) = getframe; acquires and saves the current figure as a column of the matrix M. Once this file is run, the frames can be replayed as a movie
by typing movie(M). The animation shows how the location and height of the
function peak changes as the parameter *b* is increased.

Rotating a 3D Surface

The following example rotates a three-dimensional surface by changing the
viewpoint. The data are created using the built-in function peaks.

```
% Program movie2.m
% Rotates a 3D surface.
[X,Y,Z] = peaks(50);      % Create data.
surfl(X,Y,Z)       % Plot the surface.
axis([-3 3 -3 3 -5 5])% Retain same scaling for each frame.
axis vis3d off % Set the axes to 3D and turn off tick marks,
               % and so forth.
shading interp      % Use interpolated shading.
colormap(winter)      % Specify a color map.
for k = 1:60 % Rotate the viewpoint and capture each frame.
       view(-37.5+0.5*(k-1),30)
        M(k) = getframe;
end
cla      % Clear the axes.
movie(M)     % Play the movie.
```

The colormap(map) function sets the current figure's color map to map.
Type help graph3d to see a number of color maps to choose for map. The choice
winter provides blue and green shading. The view function specifies the 3D
graph viewpoint. The syntax view(az,el) sets the angle of the view from
which an observer sees the current 3D plot, where az is the azimuth or horizontal rotation and el is the vertical elevation (both in degrees). Azimuth revolves
about the *z* axis, with positive values indicating counterclockwise rotation of the
viewpoint. Positive values of elevation correspond to moving above the object;
negative values move below. The choice az = -37.5, el = 30 is the default
3D view.

Extended Syntax of the movie Function

The function movie(M) plays the movie in array M once, where M must be an array
of movie frames (usually acquired with getframe). The function movie(M,n)

plays the movie n times. If n is negative, each "play" is once forward and once backward. If n is a vector, the first element is the number of times to play the movie, and the remaining elements are a list of frames to play in the movie. For example, if M has four frames, then n = [10 4 4 2 1] plays the movie 10 times, and the movie consists of frame 4 followed by frame 4 again, followed by frame 2 and finally frame 1.

The function movie (M,n,fps) plays the movie at fps frames per second. If fps is omitted, the default is 12 frames per second. Computers that cannot achieve the specified fps will play the movie as fast as they can. The function movie(h,...) plays the movie in object h, where h is a handle to a figure or an axis. Handles are discussed in Section 2.2.

The function movie(h,M,n,fps,loc) specifies the location to play the movie, relative to the lower left corner of object h and in pixels, regardless of the value of the object's Units property, where loc = [x y unused unused] is a four-element position vector, of which only the x and y coordinates are used, but all four elements are required. The movie plays back using the width and height in which it was recorded.

The disadvantage of the movie function is that it might require too much memory if many frames or complex images are stored.

The drawnow Command

The drawnow command causes the previous graphics command to be executed immediately. If the drawnow command were not used, MATLAB would complete all other operations before performing any graphics operations and would display only the last frame of the animation.

Animation speed depends of the intrinsic speed of the computer and on what and how much is being plotted. Symbols such as o, *, or + will be plotted slower than a line. The number of points being plotted also affects the animation speed. The animation can be slowed by using the pause(n) function, which pauses the program execution for n seconds.

The command animatedline creates an animated line that has no data, and adds it to the current axes. Add points to the line in a loop to create a line animation. Use addpoints, getpoints, and clearpoints to add more points, retrieve the points, and clear the points from the animated line, respectively. The following program illustrates the process.

```
% animated_line_1.m
h = animatedline;axis([0,10,0,2]),xlabel('t'),ylabel('y')
t = linspace(0,10,500);
y = 1 + exp(-t/2).*sin(2*t);
for k = 1:length(t)
addpoints(h,t(k),y(k));
drawnow
end
```

Handle Graphics

MATLAB treats graphs as composed of layers. Consider what you do when drawing a graph by hand. First you select a piece of paper, then you draw a set of axes with scales on the paper, and then you draw the plot, for example a line or curve. In MATLAB the first layer is the Figure window, which is like the piece of paper. MATLAB draws the axes on the second layer, and draws the plot on the third layer. This is what occurs when you use the plot function.

An expression of the form

```
p = plot(...)
```

assigns the results of the plot function to the variable p, which is a figure identifier called a *figure handle*. This stores the figure and makes it available for future use. Any valid variable name may be assigned to a handle. A figure handle is a specific type of *object handle*. Handles may be assigned to other types of objects. For example, later we will create a handle for the axes.

The set function can be used with the handle to change the object properties. This function has the general format

```
set(object handle, 'PropertyName', 'PropertyValue', ...)
```

If the object is an entire figure, its handle also contains the specifications for line color and type and marker size. Two of the properties of the figure specify the data to be plotted. Their property names are XData and YData. The following example shows how to use these properties.

Graphs in MATLAB can be modified using handle graphics. A *handle* is simply a name attached to an object such as a graph, so that we may reference it. We can assign a handle to a graph as shown in the following program and resulting output.

```
>>x = 1:10;
>>y = 5*x;
>>h = plot(x,y)
h =
    Line with properties:
                Color: [0 0.4470 0.7410]
            LineStyle: '-'
            LineWidth: 0.5000
               Marker: 'none'
           MarkerSize: 6
      MarkerFaceColor: 'none'
                XData: [1 2 3 4 5 6 7 8 9 10]
                YData: [5 10 15 20 25 30 35 40 45 50]
                ZData: [1×0 double]
```

The plot handle is h. This handle refers to the plotted line. Since we did not put a semicolon after the plot function, MATLAB displays some of the properties of the graph. If you type get(h) you will see a very long list of properties.

The line color is indicated by the RGB triplet (red, green, blue). The triplet [0 0 0] indicates black, [1 1 1] indicates white, [0 0 1] indicates blue, and so on. Prior to R2016b the first line drawn was blue; now it is a blue-green [0 0.4470 0.7410]. Note that the data used to make the plot are given in the arrays XData and YData.

This handle refers to the plotted line. To get the handle for the figure window, use the figure function. If this is the first plot, type fig_handle = figure(1). You will see

```
 Number: 1
   Name: ''
  Color: [0.9400 0.9400 0.9400]
Position: [1 1 1184 347]
  Units: 'pixels'
```

The figure background contains equal amounts of red, green, and blue, which give an almost white background.

The command gca returns the current axes for the current figure. For example,

```
>>axes_handle = gca
axes_handle  =
    Axes with properties:
                  XLim: [1 10]
                  YLim: [5 50]
                XScale: 'linear'
                YScale: 'linear'
         GridLineStyle: '-'
              Position: [0.1300 0.1100 0.7750 0.8150]
                 Units: 'normalized'
```

Typing get(axes_handle) returns a very extensive list of axes properties.

Animating a Function

Consider the function $te^{-t/b}$, which was used in the first movie example. This function can be animated as the parameter b changes with the following program.

```
% Program animate1.m
% Animates the function t*exp(-t/b).
t = 0:0.05:100;
b = 1;
p = plot(t,t.*exp(-t/b));axis([0 100 0 10]),xlabel('t');
for b = 2:20
    set(p,'XData',t,'YData',t.*exp(-t/b),...
    axis([0 100 0 10]),xlabel('t');
    drawnow
    pause(0.1)
end
```

In this program the function $te^{-t/b}$ is first evaluated and plotted over the range $0 \le t \le 100$ for $b = 1$, and the figure handle is assigned to the variable p. This establishes the plot format for all following operations, for example, line type and color, labeling, and axis scaling. The function $te^{-t/b}$ is then evaluated and plotted over the range $0 \le t \le 100$ for $b = 2, 3, 4, \ldots$ in the for loop, and the previous plot is erased. Each call to set in the for loop causes the next set of points to be plotted.

Animating Projectile Motion

This following program illustrates how user-defined functions and subplots can be used in animations. The following are the equations of motion for a projectile launched with a speed s_0 at an angle θ above the horizontal, where x and y are the horizontal and vertical coordinates, g is the acceleration due to gravity, and t is time.

$$x(t) = (s_0 \cos \theta)t \qquad y(t) = -\frac{gt^2}{2} + (s_0 \sin \theta)t$$

By setting $y = 0$ in the second expression, we can solve for t and obtain the following expression for the maximum time the projectile is in flight t_{max}.

$$t_{max} = \frac{2s_0}{g} \sin \theta$$

The expression for $y(t)$ may be differentiated to obtain the expression for the vertical velocity:

$$v_{vert} = \frac{dy}{dt} = -gt + s_0 \sin \theta$$

The maximum distance x_{max} may be computed from $x(t_{max})$, the maximum height y_{max} may be computed from $y(t_{max}/2)$, and the maximum vertical velocity occurs at $t = 0$.

The following functions are based on these expressions, where s0 is the launch speed s_0 and th is the launch angle θ.

```
function x = xcoord(t,s0,th);
% Computes projectile horizontal coordinate.
x = s0*cos(th)*t;

function y = ycoord(t,s0,th,g);
% Computes projectile vertical coordinate.
y = -g*t.^2/2+s0*sin(th)*t;

function v = vertvel(t,s0,th,g);
% Computes projectile vertical velocity.
v = -g*t+s0*sin(th);
```

The following program uses these functions to animate the projectile motion in the first subplot, while simultaneously displaying the vertical velocity in the second subplot, for the values $\theta = 45°$, $s_0 = 105$ ft/sec, and $g = 32.2$ ft/sec^2. Note that the values of xmax, ymax, and vmax are computed and used to set the axes scales. The figure handles are h1 and h2.

```
% Program animate2.m
% Animates projectile motion.
% Uses functions xcoord, ycoord, and vertvel.
th = 45*(pi/180);
g = 32.2; s0 = 105;
%
tmax = 2*s0*sin(th)/g;
xmax = xcoord(tmax,s0,th);
ymax = ycoord(tmax/2,s0,th,g);
vmax = vertvel(0,s0,th,g);
w = linspace(0,tmax,500);
%
subplot(2,1,1)
plot(xcoord(w,s0,th),ycoord(w,s0,th,g)),hold,
h1 = plot(xcoord(w,s0,th),ycoord(w,s0,th,g),'o'),...
    axis([0 xmax 0 1.1*ymax]),xlabel('x'), ylabel('y')
subplot(2,1,2)
plot(xcoord(w,s0,th),vertvel(w,s0,th,g)),hold,
h2 = plot(xcoord(w,s0,th),vertvel(w,s0,th,g),'s');...
    axis([0 xmax -1.1*vmax 1.1*vmax]),xlabel('x'),...
    ylabel('Vertical Velocity')
for t = 0:0.01:tmax
    set(h1,'XData',xcoord(t,s0,th),'YData',ycoord(t,s0,th,g))
    set(h2,'XData',xcoord(t,s0,th),'YData',vertvel(t,s0,th,g))
    drawnow
    pause(0.005)
end
hold
```

You should experiment with different values of the pause function argument.

Animation with Arrays

Thus far we have seen how the function to be animated may be evaluated in the set function with an expression or with a function. A third method is to compute the points to be plotted ahead of time and store them in arrays. The following program shows how this is done, using the projectile application. The plotted points are stored in the arrays x and y.

```
% Program animate3.m
% Animation of a projectile using arrays.
```

```
th = 70*(pi/180);
g = 32.2; s0=100;
tmax = 2*s0*sin(th)/g;
xmax = xcoord(tmax,s0,th);
ymax = ycoord(tmax/2,s0,th,g);
%
w = linspace(0,tmax,500);
x = xcoord(w,s0,th);y = ycoord(w,s0,th,g);
plot(x,y),hold,
h1 = plot(x,y,'o');...
    axis([0 xmax 0 1.1*ymax]),xlabel('x'),ylabel( 'y')
%
kmax = length(w);
for k =1:kmax
    set(h1,'XData',x(k),'YData',y(k))
    drawnow
    pause(0.001)
end
hold
```

B.2 Sound

MATLAB provides a number of functions for creating, recording, and playing sound on the computer. This section gives a brief introduction to these functions.

A Model of Sound

Sound is the fluctuation of air pressure as a function of time t. If the sound is a *pure tone*, the pressure $p(t)$ oscillates sinusoidally at a single frequency, that is,

$$p(t) = A \sin(2\pi ft + \phi)$$

where A is the pressure amplitude (the "loudness"), f is the sound frequency in cycles per second (Hz), and ϕ is the phase shift in radians. The *period* of the sound wave is $P = 1/f$.

Because sound is an analog variable (one having an infinite number of values), it must be converted to a finite set of numbers before it can be stored and used in a digital computer. This conversion process involves *sampling* the sound signal into discrete values and *quantizing* the numbers so that they can be represented in binary form. Quantization is an issue when you are using a microphone and analog-to-digital converter to capture real sound, but we will not discuss it here because we will produce only simulated sounds in software.

You use a process similar to sampling whenever you plot a function in MATLAB. To plot the function, you should evaluate it at enough points to produce a smooth plot. So, to plot a sine wave, we should "sample" or evaluate it many times over the period. The frequency at which we evaluate it is the *sampling frequency*. So, if we use a time step of 0.1 s, our sampling frequency is 10 Hz. If the sine wave has a period of 1 s, then we are "sampling" the function 10 times every period. So we see that the higher the sampling frequency, the better is our representation of the function.

Creating Sound in MATLAB

The MATLAB function sound(sound_vector,Fs) plays the signal in the vector sound_vector, created with the sampling frequency Fs, on the computer's speaker. MATLAB includes some sound files. For example, load the MAT-file chirp.mat and play the sound as follows:

```
>>load chirp
>>sound(y,Fs)
```

Note that the sound vector has been stored in the MAT-file as the row vector y and the sampling frequency has been stored as the scalar Fs. You can also try the files gong.mat and handel.mat, which contain a small sample of Handel's *Messiah*.

Use of the sound function is demonstrated with the following user-defined function, which plays a simple tone.

```
function playtone(freq,Fs,amplitude,duration)
% Plays a simple tone.
% freq = frequency of the tone (in Hz).
% Fs = sampling frequency (in Hz).
% amplitude = sound amplitude (dimensionless).
% duration = sound duration (in seconds).
t = 0:1/Fs:duration;
sound_vector = amplitude*sin(2*pi*freq*t);
sound(sound_vector,sf)
```

Try this function with the following values: freq = 1000, Fs = 10000, amplitude = 1, and duration = 10. The sound function truncates or "clips" any values in sound_vector that lie outside the range −1 to +1. Try using amplitude = 0.1 and amplitude = 5 to see the effect on the loudness of the sound.

Of course, real sound contains more than one tone. You can create a sound having two tones by adding two vectors created from sine functions having different frequencies and amplitudes. Just make sure that they are sampled with the same frequency, have the same number of samples, and that their sum lies in the range −1 to +1. You can play two different sounds in sequence by concatenating them in a row vector, as sound([sound_vector_1, sound_vector_2], Fs) if they have the same sampling frequency. You can play two different sounds simultaneously in stereo by concatenating them in a column vector, as sound([sound_vector_1', sound_vector_2'], Fs). For example, to play the *Messiah* segment followed by a chirp, use the following script. Note that the y returned by the load command is a *row* vector.

```
% Program sounds.m
load handel
S = load('chirp.mat')
y1 = S.y
Fs1 = S.Fs
sound([y',y1'],Fs) % Note that Fs = Fs1 here.
```

A related function is soundsc(sound_vector,Fs). This function scales the signal in sound_vector to the range −1 to +1 so that the sound is played as loudly as possible without clipping.

Reading and Playing WAVE Files

The MATLAB functions audiowrite and audioread create and read a Microsoft WAVE file having the extension .wav. For example, to create a wav file from the file handel.mat, you type

```
>>load handel.mat
>> audiowrite('handel.wav',y,Fs);
>>[y1, Fs1] = audioread('handel.wav');
>>sound(y1,Fs1)
```

Many computers have WAVE files to play bells, beeps, chimes, etc., to signal you when certain actions occur. For example, to load and play the WAVE file chimes.wav located in C:\windows\media on some Windows systems, you type

```
>>[y, Fs] = audioread('c:\windows\media\chimes.wav');
>>sound(y, Fs)
```

You can also use the audioplayer and play functions instead of sound, as follows.

```
>>p = audioplayer(y, Fs);
>>play(p)
```

The sound function lets you play the sound only at a given sampling rate, but the audioplayer function enables you to do more than that, such as pause the playback and resume it, and set properties of the object.

Recording and Writing Sound Files

You can use MATLAB to record sound and write sound data to a WAVE file. The audiorecorder function records sound from a PC-based audio input device. The audiorecorder function holds control until recording completes. By default, the audiorecorder function creates an 8000 Hz, 8-bit, 1-channel object. The following program shows how to record your voice for 5 seconds. The recordblocking function records audio from an input device for a specified number of seconds, holding control until recording completes.

```
% Record your voice for 5 seconds.
my_voice = audiorecorder;
disp('Start speaking.')
recordblocking(my_voice, 5);
disp('End of Recording.');
% Play back the recording.
play(my_voice);
```

The extended syntax audiorecorder(Fs, nBits, nChannels) sets the sample rate Fs (in Hz), the sample size nBits, and the number of channels nChannels. Typical values supported by most sound cards are 8000, 11,025, 22,050, 44,100, 48,000, and 96,000 Hz. For example, to record your voice for 5 seconds at 11,025 Hz on channel 1, replace the second line in the previous program with the following two lines.

```
Fs = 11025;
my_voice = audiorecorder(Fs,5*Fs, 1);
```

Formatted Output in MATLAB

The disp and format commands provide simple ways to control the screen output. However, some users might require greater control over the screen display. In addition, some users might want to write formatted output to a data file. The fprintf function provides this capability. Its syntax is count = fprintf(fid, format,A,...), which formats the data in the real part of matrix A (and in any additional matrix arguments) under control of the specified string format, and writes the data to the file associated with file identifier fid. A count of the number of bytes written is returned in the variable count. The argument fid is an integer file identifier obtained from fopen. (It may also be 1 for standard output—the screen—or 2 for standard error. See fopen for more information.)

Omitting fid from the argument list causes output to appear on the screen, and is the same as writing to standard output (fid = 1). The string format specifies notation, alignment, significant digits, field width, and other aspects of output format. It can contain ordinary alphanumeric characters, along with escape characters, conversion specifiers, and other characters, organized as shown in the following examples. Table C.1 summarizes the basic syntax of fprintf. Consult MATLAB Help for more details.

Suppose the variable Speed has the value 63.2. To display its value using three digits with one digit to the right of the decimal point, along with a message, the session is

```
>>fprintf('The speed is: %3.1f\n',Speed)
The speed is: 63.2
```

Here the "field width" is 3, because there are three digits in 63.2. You may want to specify a wide enough field to provide blank spaces or to accommodate an

Table C.1 Display formats with the `fprintf` function

Syntax	Description
`fprintf('format',A, ...)`	Displays the elements of the array `A`, and any additional array arguments, according to the format specified in the string `'format'`.
`'format'` arrangement	`%[-][number1.number2]C`, where `number1` specifies the minimum field width, `number2` specifies the number of digits to the right of the decimal point, and `C` contains control codes and format codes. Items in brackets are optional. `[-]` specifies left-justified.

Control codes		Format codes	
Code	Description	Code	Description
`\n`	Start new line.	`%e`	Scientific format with lowercase e.
`\r`	Beginning of new line.	`%E`	Scientific format with uppercase E.
`\b`	Backspace.	`%f`	Fixed-point notation.
`\t`	Tab.	`%g`	`%e` or `%f`, whichever is shorter.
`' '`	Apostrophe.	`%s`	String of characters
`\\`	Backslash.		

unexpectedly large numerical value. The % sign tells MATLAB to interpret the following text as codes. The code \n tells MATLAB to start a new line after displaying the number.

The output can have more than one column, and each column can have its own format. For example,

```
>>r = 2.25:20:42.25;
>>circum = 2*pi*r;
>>y = [r;circum];
>>fprintf('%5.2f %11.5g\n',y)
   2.25      14.137
  22.25      139.8
  42.25      265.46
```

Note that the `fprintf` function displays the *transpose* of the matrix `y`.

Format code can be placed within text. For example, note how the period after the code `%6.3f` appears in the output at the end of the displayed text.

```
>>fprintf('The first circumference is %6.3f.\n',circum(1))
The first circumference is 14.137.
```

An apostrophe in displayed text requires two single quotes. For example:

```
>>fprintf('The second circle''s radius %15.3e is large.\n',r(2))
The second circle's radius       2.225e+001 is large.
```

A percent sign in displayed text requires two percent signs. Otherwise the single percent sign will be interpreted as a placeholder for data. For example, typing

```
fprintf('The inflation rate was %3.2f %%. \n', 3.15)
```

gives the output:

```
The inflation rate was 3.15 %.
```

A minus sign in the format code causes the output to be left-justified within its field. Compare the following output with the preceding example:

```
>>fprintf('The second circle''s radius %-15.3e is large.\n',r(2))
The second circle's radius 2.225e+001 is large.
```

Control codes can be placed within the format string. The following example uses the tab code (\t).

```
>>fprintf('The radii are:%4.2f \t %4.2f \t %4.2f\n',r)
The radii are:   2.25   22.25   42.25
```

The `disp` function sometimes displays more digits than necessary. We can improve the display by using the `fprintf` function instead of `disp`. Consider the program:

```
p = 8.85; A = 20/100^2;
d = 4/1000; n = [2:5];
C = ((n - 1).*p*A/d);
table (:,1) = n';
table (:,2) = C';
disp (table)
```

The `disp` function displays the number of decimal places specified by the `format` command (4 is the default value).

If we replace the line `disp(table)` with the following three lines

```
E='';
fprintf('No.Plates Capacitance (F) X e12 %s\n',E)
fprintf('%2.0f \t \t \t %4.2f\n',table')
```

we obtain the following display:

```
2          4.42
3          8.85
4          13.27
5          17.70
```

The empty matrix E is used because the syntax of the `fprintf` statement requires that a variable be specified. Because the first `fprintf` is needed to display the table title only, we need to fool MATLAB by supplying it with a variable whose value will not display.

Note that the `fprintf` command truncates the results, instead of rounding them. Note also that we must use the transpose operation to interchange the rows and columns of the `table` matrix in order to display it properly.

Only the real part of complex numbers will be displayed with the `fprintf` command. For example,

```
>>z = -4+9i;
>>fprintf('Complex number: %2.2f \n',z)
Complex number:    -4.00
```

Instead you can display a complex number as a row vector. For example, if $w = -4+9i$,

```
>>w = [-4,9];
>>fprintf('Real part is %2.0f. Imaginary part is %2.0f. \n',w)
Real part is -4. Imaginary part is 9.
```

MATLAB also has the `sprintf` function, which assigns a name to the formatted string, instead of sending it to the Command window. Its syntax is similar to that of `fprintf`. This can be used with the `text` function to place a label on a plot without knowing the exact wording ahead of time. For example, a script file would be

```
x = 1:10;y = (x + 2.3).^2;
mean_y = mean(y);
label = sprintf('Mean of y is:%4.0f \n',mean_y);
plot(x,y),text(2,100,label)
```

References

[Brown, 1994] Brown, T. L.; H. E. LeMay, Jr.; and B. E. Bursten. *Chemistry: The Central Science*. 6th ed. Upper Saddle River, NJ: Prentice-Hall, 1994.

[Eide, 2008] Eide, A. R.; R. D. Jenison; L. L. Northup; and S. Mickelson. *Introduction to Engineering Problem Solving*. 5th ed. New York: McGraw-Hill, 2008.

[Felder, 1986] Felder, R. M., and R. W. Rousseau. *Elementary Principles of Chemical Processes*. New York: John Wiley & Sons, 1986.

[Garber, 1999] Garber, N. J., and L. A. Hoel. *Traffic and Highway Engineering*. 2nd ed. Pacific Grove, CA: PWS Publishing, 1999.

[Jayaraman, 1991] Jayaraman, S. *Computer-Aided Problem Solving for Scientists and Engineers*. New York: McGraw-Hill, 1991.

[Kreyzig, 2009] Kreyzig, E. *Advanced Engineering Mathematics*. 9th ed. New York: John Wiley & Sons, 1999.

[Kutz, 1999] Kutz, M., editor. *Mechanical Engineers' Handbook*. 2nd ed. New York: John Wiley & Sons, 1999.

[Palm, 2014] Palm, W. *System Dynamics*. 3rd ed. New York: McGraw-Hill, 2014.

[Rizzoni, 2007] Rizzoni, G. *Principles and Applications of Electrical Engineering*. 5th ed. New York: McGraw-Hill, 2007.

[Starfield, 1990] Starfield, A. M.; K. A. Smith; and A. L. Bleloch. *How to Model It: Problem Solving for the Computer Age*. New York: McGraw-Hill, 1990.

Answers to Selected Problems

Chapter 1

2. (a) −13.3333; (b) 0.6; (c) 15; (d) 1.0323.

12. (a) $x + y = -3.0000 - 2.0000i$; (b) $xy = -13.0000 - 41.0000i$; (c) $x/y = -1.7200 + 0.0400i$.

25. $x = -15.685$ and $x = 0.8425 \pm 3.4008i$.

Chapter 2

3.
$$A = \begin{bmatrix} 0 & 6 & 12 & 18 & 24 & 30 \\ -20 & -10 & 0 & 10 & 20 & 30 \end{bmatrix}$$

7. (a) Length is 3. Absolute values = [2 4 7];
(b) Same as (a); (c) Length is 3. Absolute values = [5.8310 5.0000 7.2801].

11. (b) The largest elements in the first, second, and third layers are 10, 9, and 10, respectively. The largest element in the entire array is 10.

15. (a)
$$A + B + C = \begin{bmatrix} -6 & -3 \\ 23 & 15 \end{bmatrix}$$
(b) $A + B + C = \begin{bmatrix} -14 & -7 \\ 1 & 19 \end{bmatrix}$

17. (a) A.*B = [784, -128; 144, 32];
(b) A/B = [76, -168; -12, 32];
(c) B.^3 = [2744, -64; 216, -8].

23. (a) F.*D = [1200, 275, 525, 750, 3000] J; (b) sum(F.*D) = 5750 J.

36. (a) A*B = [-47, -78; 39, 64];
(b) B*A = [-5, -3, 48, 22].

39. 60 tons of copper, 67 tons of magnesium, 6 tons of manganese, 76 tons of silicon, and 101 tons of zinc.

42. $M = 675$ N·m if **F** is in newtons and **r** is in meters.

49. [q,r] = deconv ([14,-6,3,9], [5,7,-4]), q = [2.8, -5.12], r = [0, 0, 50.04, -11.48]. The quotient is $2.8x - 5.12$ with a remainder of $50.04x - 11.48$.

50. 2.0458.

Chapter 3

1. (a) 3, 3.1623, 3.6056;
(b) $1.7321i$, $0.2848 + 1.7553i$, $0.5503 + 1.8174i$;
(c) $15 + 21i$, $22 + 16i$, $29 + 11i$;
(d) $-0.4 - 0.2i$, $-0.4667 - 0.0667i$, $-0.5333 + 0.0667i$.

2. (a) $|xy| = 105$, $\angle xy = -2.6$ rad.
(b) $|x/y| = 0.84$, $\angle x/y = -1.67$ rad.

3. (a) 1.01 rad (58°); (b) 2.13 rad (122°)
(c) −1.01 rad (−58°); (d) −2.13 rad (−122°).

7. $F_1 = 197.5217$ N.

9. 2.7324 sec while ascending; 7.4612 sec while descending.

Chapter 4

4. (*a*) z = 1; (*b*) z = 0; (*c*) z = 1; (*d*) z = 1.

5. (*a*) z = 0; (*b*) z = 1; (*c*) z = 0; (*d*) z = 4; (*e*) z = 1; (*f*) z = 5; (*g*) z = 1; (*h*) z = 0.

6. (*a*) z = [0, 1, 0, 1, 1];
(*b*) z = [0, 0, 0, 1, 1];
(*c*) z = [0, 0, 0, 1, 0];
(*d*) z = [1, 1, 1, 0, 1].

11. (*a*) z = [1, 1, 1, 0, 0, 0];
(*b*) z = [1, 0, 0, 1, 1, 1];
(*c*) z = [1, 1, 0, 1, 1, 1];
(*d*) z = [0, 1, 0, 0, 0, 0].

13. (*a*) $7300; (*b*) $5600; (*c*) 1200 shares; (*d*) $15,800.

29. Best location: $x = 9$, $y = 16$. Minimum cost: $294.51. There is only one solution.

35. After 33 years, the amount will be $1,041,800.

37. $W = 300$ and $T = [428.5714, 471.4286, 266.6667, 233.3333, 200, 100]$

49. Weekly inventory for cases (*a*) and (*b*):

Week	1	2	3	4	5
Inventory (*a*)	50	50	45	40	30
Inventory (*b*)	30	25	20	20	10
Week	6	7	8	9	10
Inventory (*a*)	30	30	25	20	10
Inventory (*b*)	10	5	0	0	(<0)

Chapter 5

1. Production is profitable for $Q \geq 10^8$ gal/yr. The profit increases linearly with Q, so there is no upper limit on the profit.

3. $x = -0.4795$, 1.1346, and 3.8318.

5. 37.622 m above the left-hand point, and 100.6766 m above the right-hand point.

10. 0.54 rad (31°).

14. The steady-state value of y is $y = 1$. $y = 0.98$ at $t = 4/b$.

17. (*a*) The ball will rise 1.68 m and will travel 9.58 m horizontally before striking the ground after 1.17 s.

Chapter 6

2. (*a*) $y = 53.5x - 1354.5$;
(*b*) $y = 3582.1x^{-0.9764}$;
(*c*) $y = 2.0622 \times 10^5 (10)^{-0.0067x}$

4. (*a*) $b = 1.2603 \times 10^{-4}$; (*b*) 836 years.
(*c*) Between 760 and 928 years ago.

8. If unconstrained to pass through the origin, $f = 0.3998x - 0.0294$. If constrained to pass through the origin, $f = 0.3953x$.

10. $d = 0.0509v^2 + 1.1054v + 2.3571$; $J = 10.1786$; $S = 57,550$; $r^2 = 0.9998$.

11. $y = 40 + 9.6x_1 - 6.75x_2$. Maximum percent error is 7.125 percent.

Chapter 7

7. (*a*) 96%; (*b*) 68%.

11. (*a*) Mean pallet weight is 3000 lb. Standard deviation is 10.95 lb; (*b*) 8.55 percent.

18. Mean yearly profit is $64,609. Minimum expected profit is $51,340. Maximum expected profit is $79,440. Standard deviation of yearly profit is $5967.

24. The estimated temperatures at 5 p.m. and 9 p.m. are 22.5° and 16.5°.

Chapter 8

2. (*a*) $\mathbf{C} = \mathbf{B}^{-1}(\mathbf{A}^{-1}\mathbf{B} - \mathbf{A})$.
(*b*) C = [-0.8536, -1.6058; 1.5357, 1.3372].

5. (*a*) $x = 3c$, $y = -2c$, $z = c$
(*b*) The plot consists of three straight lines that intersect at (0,0).

8. $T_1 = 19.7596°C$, $T_2 = -7.0214°C$, $T_3 = -9.7462°C$. Heat loss in watts is 66.785.

13. Infinite number of solutions: $x = -1.3846z + 4.9231$, $y = 0.0769z - 1.3846$.

18. Unique solution: $x = 8$ and $y = 2$.

20. Least-squares solution; $x = 6.0928$ and $y = 2.2577$.

Chapter 9

1. 23,690 m.

7. 13.65 ft.

10. 1363 m/s.

27. 150 m/s.

Chapter 11

3. (a) $60x^5 - 10x^4 + 108x^3 - 49x^2 + 71x - 24$; (b) 2546.

4. $A = 1$, $B = -2a$, $C = 0$, $D = -2b$, $E = 1$, and $F = r^2 - a^2 - b^2$.

6. (a) $b = c \cos A \pm \sqrt{a^2 - c^2 \sin^2 A}$; (b) $b = 5.6904$.

8. (a) $x = \pm 10 \sqrt{(4b^2 - 1)/(400b^2 - 1)}$, $y = \pm \sqrt{99/(400b^2 - 1)}$; (b) $x = \pm 0.9685$, $y = \pm 0.4976$.

18. $h = 21.2$ ft $\theta = 0.6155$ rad $(35.26°)$.

19. 49.6808 m/s.

29. (a) 2; (b) 0; (c) 0.

36. (a) $(3x_0/5 + v_0/5)e^{-3t}\sin 5t + x_0 e^{-3t}\cos 5t$; (b) $e^{-5t}(8x_0/3 + v_0/3) + (-5x_0/3 - v_0/3)e^{-8t}$.

48. $s^2 + 13s + 42 - 6k$, $s = (-13 \pm \sqrt{1 + 24k})/2$.

49.
$$x = \frac{62}{16c + 15} \quad y = \frac{129 + 88c}{16c + 15}$$

INDEX

MATLAB Symbols

Simulink Blocks

Topics